Multilevel and Longitudinal Modeling Using Stata

Second Edition

Multilevel and Longitudinal Modeling Using Stata

Second Edition

SOPHIA RABE-HESKETH
University of California, Berkeley
Institute of Education, University of London

ANDERS SKRONDAL
London School of Economics
Norwegian Institute of Public Health

A Stata Press Publication
StataCorp LP
College Station, Texas

 First edition 2005
Second edition 2008

Published by Stata Press, 4905 Lakeway Drive, College Station, Texas 77845
Typeset in LaTeX 2ε
Printed in the United States of America
10 9 8 7 6 5 4 3 2

ISBN-10: 1-59718-040-8
ISBN-13: 978-1-59718-040-5

Acknowledgments

We thank The Research Council of Norway for awarding us a grant, which made it possible to spend time together writing this book. We are also grateful to Nina Bregnegaard, Leonardo Grilli, Bobby Gutierrez, Joe Hilbe, Katrin Hohl, and Carla Rampichini for their extensive and extremely useful comments on drafts of the book and to Lisa Gilmore and others for efficient publishing. We thank several cohorts of students at the University of California, Berkeley and the London School of Economics for feedback helping to improve the second edition, and Charlie Hallahan, Nick Horton, Thomas Loughlin, and Rory Wolfe, among others, for comments. We also like to thank Andrew Pickles with whom we have collaborated for many years. Readers are encouraged to provide feedback on the current edition to improve future editions.

We are grateful to the many people who have contributed datasets to this book, either by making them publicly available themselves or by allowing us to do so. G. Dunn, J. Neuhaus, A. Pebley, G. Rodriguez, D. Stott, T. Touloupulou, and M. Yang kindly contributed their datasets to this book. Some of the datasets we use accompany software packages and are freely downloadable. For instance, BUGS (Spiegelhalter et al. 1996a,b), HLM (Raudenbush et al. 2004), Latent GOLD (Vermunt and Magidson 2005), MLwiN (Rasbash et al. 2005), and the MIX programs (e.g., Hedeker and Gibbons 1996) all provide exciting, real datasets.

Some journals, such as *Biometrics*, *Journal of Applied Econometrics*, *Journal of the Royal Statistical Society (Series A)*, and *Statistical Modelling* encourage authors to make their data available on the journal web site. We are grateful to J. Abrevaya, N. Goldman, E. Lesaffre, D. Moore, E. Ross, G. Rodriguez, B. Spiessens, F. Vella, M. Verbeek, and R. Winkelmann for making their data available this way.

We also used datasets from textbooks with accompanying web pages, some of which are made available (together with worked examples using various software packages, including Stata) by UCLA Technology Services
(see http://www.ats.ucla.edu/stat/examples/default.htm). We found the books by Allison (1995, 2005), Baltagi (2005), Cameron and Trivedi (2005), De Boeck and Wilson (2004), Davis (2002), DeMaris (2004), Dohoo, Martin, and Stryhn (2003), Fitzmaurice, Laird, and Ware (2004), Fox (1997), Frees (2004), Gelman and Hill (2007), Johnson and Albert (1999), Hand et al. (1994), Littell et al. (2006), Singer and Willett (2003), West, Welch, and Galecki (2007), and Wooldridge (2002) particularly helpful in this regard.

Sometimes data are printed in papers or books, allowing patient people like us to type them. Data were provided in this form by D. Altman, M. Bland, G. Koch, S. MacNab, R. Mare, and P. Sham.

We thank Chapman & Hall/CRC for permission to use figures and tables from our book *Generalized Latent Variable Modeling: Multilevel, Longitudinal and Structural Equation Models.*

Any omissions are purely accidental.

Contents

Tables

Figures

Preface

This is a book about applied multilevel and longitudinal modeling. Other terms for multilevel models include hierarchical models, random-effects or random-coefficient models, mixed-effects models, or simply mixed models. Longitudinal data are also referred to as panel data, repeated measures, or cross-sectional time series. A popular type of multilevel model for longitudinal data is the growth-curve model.

The common theme of the book is regression modeling when data are clustered in some way. In cross-sectional settings, students may be nested in schools, people in neighborhoods, employees in firms, or twins in twin-pairs. Longitudinal data are by definition clustered since multiple observations over time are nested within units, typically subjects.

Such clustered designs often provide rich information on processes operating at different levels, for instance, people's characteristics interacting with institutional characteristics. Importantly, the standard assumption of independent observations is likely to be violated due to dependence among observations within the same cluster. The multilevel and longitudinal methods discussed in this book extend conventional regression to handle such dependence and exploit the richness of the data.

Our emphasis is on explaining the models and their assumptions, applying the methods to real data, and interpreting results. Many of the issues are conceptually demanding but do not require that you understand complex mathematics. Wherever possible, we therefore introduce ideas through examples and graphical illustrations, keeping the technical descriptions as simple as possible, often confining formulas to subsections that can be skipped. Some sections that go beyond an introductory course on multilevel and longitudinal modeling are tagged with the symbol ❖. For an advanced and comprehensive treatment, we refer to Skrondal and Rabe-Hesketh (2004), which uses the same notation as this book.

This book shows how all the analyses described can be performed using Stata. There are many advantages of using a general-purpose statistical package such as Stata. First, for those already familiar with Stata, it is convenient not having to learn a new stand-alone package. Second, conducting multilevel-analysis within a powerful package has the advantage that it allows complex data manipulation to be performed, alternative estimation methods to be used, and publication-quality graphics to be produced, all without having to switch packages. Finally, Stata is a natural choice for multilevel and longitudinal modeling since it has gradually become perhaps the most powerful general-purpose statistics package for such models.

Each chapter is based on one or more research problems and real datasets. We walk through the analysis using Stata, pausing when statistical issues arise that need further explanation. Stata can be used either via a graphical user interface (GUI) or through commands. We recommend using commands interactively—or preferably in do-files—for serious analysis in Stata. For this reason, and because the GUI is fairly self-explanatory, this book exclusively uses commands. However, the GUI can be useful for learning the Stata syntax. Generally, we use the typewriter font `command` to refer to Stata commands, syntax, and variables. A "dot" prompt followed by a command indicates that you can type verbatim what is displayed after the dot (in context) to replicate the results in the book. Some readers may find it useful to intersperse reading with running these commands.

The commands used for data manipulation and graphics are explained to some extent, but the purpose of this book is not to teach Stata from scratch. For basic introductions to Stata, we refer to Kohler and Kreuter (2005) or Rabe-Hesketh and Everitt (2007). Other books and further resources for learning Stata are listed at

http://www.stata.com

We have included applications from a wide range of disciplines, including medicine, economics, education, sociology, and psychology. The interdisciplinary nature of the book is also reflected in the choice of models and topics covered. If a chapter is primarily based on an application from one discipline, we try to balance this by including exercises with real data from other disciplines. We encourage users to write do-files for solving the data analysis exercises since this is standard practice for professional data analysis.

All datasets used in this book are freely available for download from

http://www.gllamm.org/books

These datasets can be downloaded into a local directory on your computer. Alternatively, individual datasets can be loaded directly into 'net-aware' Stata by specifying the complete URL. For example,

```
. use http://www.stata-press.com/data/mlmus2/pefr
```

If you have stored the datasets in a local directory, omit the path, and type

```
. use pefr
```

We will generally describe all Stata commands that can be used for a given problem, discussing their advantages and disadvantages. We make extensive use of the Stata commands `xtreg`, `xtmixed`, `xtlogit`, `xtpoisson`, `xtmelogit`, `xtmepoisson`, and our own command `gllamm`. Using `xtmixed` requires Stata release 9 or later, `xtmelogit` and `xtmepoisson` require release 10 or later, and `gllamm` requires that the program be installed. If you are connected to the Internet, the easiest way of installing `gllamm` is by issuing the Stata command:

```
. ssc install gllamm
```

Since `gllamm` is not part of official Stata and therefore only mentioned but not described in the Stata manuals, we include detailed descriptions of the syntax for `gllamm` and its postestimation commands `gllapred` and `gllasim` in the appendices. For quick and easy reference, we have placed the bare essentials in appendix A. We refer to the *Stata Longitudinal/Panel-Data Reference Manual* (StataCorp 2007) for detailed information on all the official Stata commands for multilevel and longitudinal modeling.

We assume that readers have a good knowledge of linear regression modeling, in particular the use and interpretation of dummy variables and interactions. However, we have included a new first chapter in this edition which reviews linear regression and can serve as a refresher.

The book consists of four parts: I Preliminaries, II Two-level linear models, III Two-level generalized linear models, and IV Models with nested and crossed random effects. Part I is a review of linear regression modeling and prepares the reader for the rest of the book. For readers who are new to multilevel and longitudinal modeling, the chapters in part II should be read sequentially and can form the basis of an introductory course on this topic. The remaining chapters can then be read in any order, except that chapter 6 should be read before chapter 7 and chapter 10 before chapter 11. A course on linear models could cover part II as well as the first half of chapter 10, and perhaps chapter 11, in part IV.

Errata for different editions and printings of the book can be downloaded from http://www.stata-press.com/books/errata/mlmus2.html, and answers to exercises are available for instructors; see http://www.stata-press.com/books/mlmus2-answers.html.

The second edition of the book includes 3 new chapters, comprehensive updates for Stata 10, 38 new exercises and 27 new datasets. All chapters of the previous edition have been substantially revised.

Berkeley and London
December 2007

Sophia Rabe-Hesketh
Anders Skrondal

Part I

Preliminaries

1 Review of linear regression

1.1 Introduction

In this chapter, we review the statistical models underlying independent-samples t tests, analysis of variance (ANOVA), analysis of covariance (ANCOVA), simple regression, and multiple regression, and formulate all these models as linear regression models.

We deliberately take a model-based approach to statistical inference in this chapter because this is the approach used in multilevel modeling. We focus on model assumptions and interpretation of regression coefficients and discuss in detail the use of dummy variables and interactions.

The regression models considered here are essential building blocks for multilevel models. Although linear multilevel or mixed models for continuous responses are sometimes viewed from an ANOVA perspective, the regression perspective is beneficial because it is easily generalizable to other response types. Furthermore, the Stata commands for multilevel modeling follow a regression syntax.

This chapter is not intended as a first introduction to linear regression, but rather as a refresher for readers already familiar with most of the ideas. Even experienced regression modelers should benefit from reading this chapter because it introduces our notation and terminology as well as Stata commands used in later chapters. If you are able to correctly answer the self-assessment exercise 1.6, you should be well-prepared for the rest of the book.

1.2 Is there gender discrimination in faculty salaries?

DeMaris (2004) analyzed data on the salaries of faculty (academic staff) at Bowling Green State University (BGSU) in Ohio, U.S.A. in the academic year 1993–1994. The data are provided with his book *Regression with Social Data: Modeling Continuous and Limited Variables* and have previously been analyzed by Balzer et al. (1996) and Boudreau et al. (1997). The primary purpose of these studies was to investigate whether there was any evidence of gender inequity in faculty salaries at BGSU.

The data considered here are a subset comprising $n = 514$ of the faculty members provided by DeMaris (2004), excluding faculty from the Fireland campus, nonprofessors (instructors/lecturers), those who are not on graduate faculty, and three professors hired as "Ohio Board of Regents Eminent Scholars". We will use the following variables:

- `salary`: academic year (9 month) salary in U.S. dollars
- `male`: gender (1: male; 0: female)
- `market`: marketability of discipline, defined as the ratio of the national average salary paid in the discipline to the national average across all disciplines
- `yearsdg`: time since degree (in years)
- `rank`: academic rank (1: assistant professor; 2: associate professor; 3: full professor)

1.3 Independent-samples t test

An obvious first thing to do is to compare mean salaries between male and female professors at BGSU. We start by reading the data into Stata

```
. use http://www.stata-press.com/data/mlmus2/faculty
```

and use the `tabstat` command to produce a table of means, standard deviations, and sample sizes by gender:

```
. tabstat salary, by(male) statistics(mean sd n)
Summary for variables: salary
     by categories of: male
```

male	mean	sd	N
Women	42916.6	9161.61	128
Men	53499.24	12583.48	386
Total	50863.87	12672.77	514

We see that the male faculty at BGSU earn, on average, over $10,000 more than the female faculty. The standard deviation is also considerably larger for the men than for the women.

Due to chance or sampling variation, the large difference between the mean salary $\overline{y}_1$ of the n_1 men and the mean salary $\overline{y}_0$ of the n_0 women in the sample does not imply that the corresponding population means or *expectations* μ_1 and μ_0 for male and female faculty differ (the Greek letter μ is pronounced mu). Here *population* refers either to an imagined infinite population from which the data can be viewed as sampled, or to the statistical model that is viewed as the data-generating mechanism for the observed data.

To define a statistical model, let y_i and x_i denote the salary and gender of the ith professor, respectively, where $x_i = 1$ for men and $x_i = 0$ for women. A standard model for the current problem can then be specified as

$$y_i|x_i \sim N(\mu_{x_i}, \sigma^2_{x_i})$$

Here "$y_i|x_i \sim$" means "y_i, for a given value of x_i, is distributed as", and $N(\mu_{x_i}, \sigma^2_{x_i})$ stands for a normal distribution with mean parameter μ_{x_i} and variance parameter

$\sigma^2_{x_i}$ (the Greek letter σ is pronounced sigma). The term *conditional* simply means that we are considering only the subset of the population for which some condition is satisfied; here that x_i takes on the same value for all individuals. In other words, we are considering the distribution of salaries for men ($x_i = 1$) separately from the distribution for women ($x_i = 0$).

When conditioning on a categorical variable like gender, the expression "conditional on gender" can be replaced by "within gender" or "separately for each gender". Since x_i takes on only two values, we can be more explicit and write the statistical model as

$$\begin{aligned} y_i|x_i = 0 &\sim N(\mu_0, \sigma_0^2) \\ y_i|x_i = 1 &\sim N(\mu_1, \sigma_1^2) \end{aligned}$$

Each conditional distribution has its own conditional expectation, or conditional population mean

$$\mu_{x_i} \equiv E(y_i|x_i)$$

μ_1 for men and μ_0 for women ($\equiv$ stands for "defined as"). Each conditional distribution also has its own conditional variance

$$\sigma^2_{x_i} \equiv \text{Var}(y_i|x_i)$$

denoted σ_1^2 for men and σ_0^2 for women. A final assumption is that y_i is conditionally independent from $y_{i'}$, given the values of x_i and $x_{i'}$, for different professors i and i'. This means that knowing one professor's salary does not help us predict another professor's salary if we already know that other professor's gender and the corresponding gender-specific mean salary.

By modeling y_i conditional on x_i, we are treating y_i as a *response variable* (sometimes also called dependent variable or criterion variable) and x_i as an *explanatory variable* or *covariate* (sometimes referred to as independent variable, predictor, or regressor).

The normality assumptions stated above are usually assessed by inspecting the conditional sample distributions for the men and women, using boxplots

```
. graph box salary, over(male) ytitle(Academic salary) asyvars
```

(see left panel of figure 1.1) or histograms

```
. histogram salary, by(male, rows(2)) xtitle(Academic salary)
```

(see left panel of figure 1.2). We see that both distributions are somewhat positively skewed.

(*Continued on next page*)

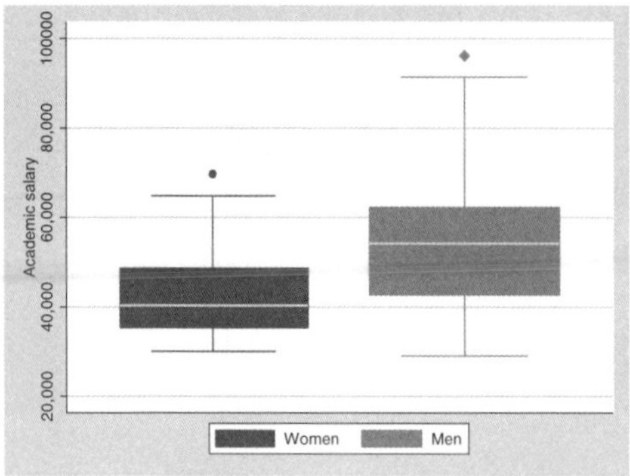

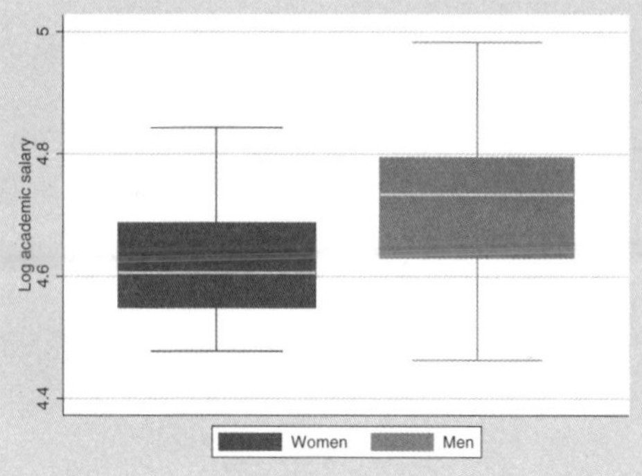

Figure 1.1: Boxplots of salary and log salary by gender

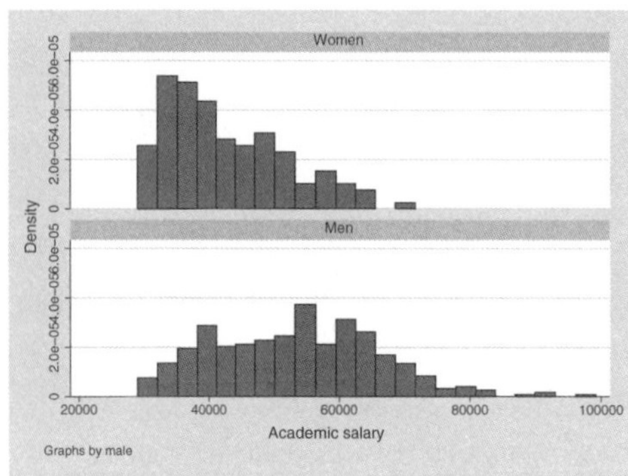

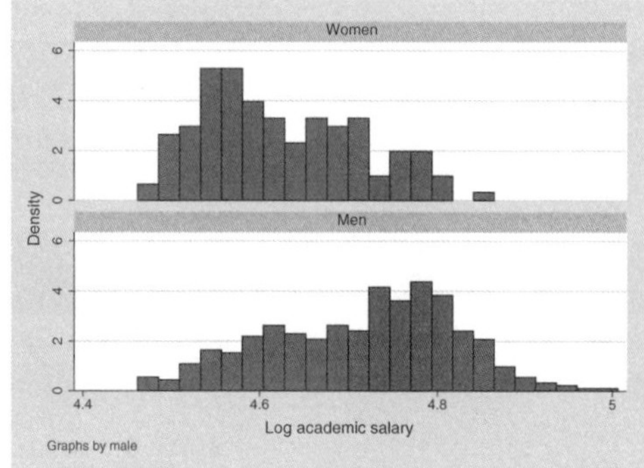

Figure 1.2: Histograms of salary and log salary by gender

A logarithmic transformation of salary makes the distributions more symmetric (like normal distributions):

```
. generate lsalary = log10(salary)
. graph box lsalary, over(male) ytitle(Log academic salary) asyvars
. histogram lsalary, by(male, rows(2)) xtitle(Log academic salary)
```

(see right panels of figures 1.1 and 1.2).

Assuming that the statistical model is true (or relying on the sample mean becoming normally distributed in large samples), we can use the independent-samples t test command `ttest` in Stata. This command estimates the population means, produces confidence intervals, and tests the null hypothesis that the two population means are the same

$$H_0\colon \mu_0 = \mu_1 \qquad \text{or} \qquad H_0\colon \mu_0 - \mu_1 = 0$$

against the two-sided alternative that they are different,

$$H_a\colon \mu_0 \neq \mu_1 \qquad \text{or} \qquad H_a\colon \mu_0 - \mu_1 \neq 0$$

The most popular version of the t test makes the additional assumption that the conditional variances are equal, $\sigma_0^2 = \sigma_1^2$, which we denote by dropping the subscript of $\sigma_{x_i}^2$, i.e.,

$$\text{Var}(y_i|x_i) \;=\; \sigma^2$$

so that the model becomes

$$y_i|x_i \;\sim\; N(\mu_{x_i}, \sigma^2)$$

The equal-variance assumption seems more reasonable for the log-transformed salaries than for the salaries on their original scale. However, salary in dollars is more interpretable than its log transformation, so for simplicity we will work with the untransformed variable in this chapter.

We can perform t tests with and without the assumption of equal variances, using

```
. ttest salary, by(male)
Two-sample t test with equal variances
```

Group	Obs	Mean	Std. Err.	Std. Dev.	[95% Conf.	Interval]
Women	128	42916.6	809.7795	9161.61	41314.2	44519.01
Men	386	53499.24	640.4822	12583.48	52239.96	54758.52
combined	514	50863.87	558.972	12672.77	49765.72	51962.03
diff		-10582.63	1206.345		-12952.63	-8212.636

```
    diff = mean(Women) - mean(Men)                                t =  -8.7725
Ho: diff = 0                                     degrees of freedom =      512

    Ha: diff < 0                 Ha: diff != 0                 Ha: diff > 0
 Pr(T < t) = 0.0000         Pr(|T| > |t|) = 0.0000          Pr(T > t) = 1.0000
```

and

```
. ttest salary, by(male) unequal
Two-sample t test with unequal variances
```

Group	Obs	Mean	Std. Err.	Std. Dev.	[95% Conf.	Interval]
Women	128	42916.6	809.7795	9161.61	41314.2	44519.01
Men	386	53499.24	640.4822	12583.48	52239.96	54758.52
combined	514	50863.87	558.972	12672.77	49765.72	51962.03
diff		-10582.63	1032.454		-12614.48	-8550.787

```
    diff = mean(Women) - mean(Men)                                t = -10.2500
Ho: diff = 0                      Satterthwaite's degrees of freedom =  297.227

    Ha: diff < 0                 Ha: diff != 0                 Ha: diff > 0
 Pr(T < t) = 0.0000         Pr(|T| > |t|) = 0.0000          Pr(T > t) = 1.0000
```

For both versions of the t test, the population means are estimated by the sample means,

$$\widehat{\mu}_0 = \overline{y}_0, \quad \widehat{\mu}_1 = \overline{y}_1$$

and the t statistic is given by

$$t = \frac{\widehat{\mu}_0 - \widehat{\mu}_1}{\widehat{\mathrm{SE}}(\widehat{\mu}_0 - \widehat{\mu}_1)}$$

the estimated difference in population means (the difference in sample means) divided by the estimated standard error of the estimated difference in population means. Under the null hypothesis, the statistic has a t distribution with df degrees of freedom, where $\mathrm{df} = n-2$ (sample size n minus 2 for two estimated means) for the test assuming equal variances.

The 95% confidence interval for the difference in population means $\mu_0 - \mu_1$ is

$$\widehat{\mu}_0 - \widehat{\mu}_1 \ \pm \ t_{.975,\mathrm{df}}\, \widehat{\mathrm{SE}}(\widehat{\mu}_0 - \widehat{\mu}_1)$$

where $t_{.975,\mathrm{df}}$ is the 97.5th percentile of the t distribution with df degrees of freedom.

The standard error is estimated as \$1,206.345 under the equal-variance assumption $\sigma_0^2 = \sigma_1^2$ and as \$1,032.454 under the unequal-variance assumption and the degrees of freedom also differ. In both cases, the two-tailed p-value (given under `Ha: diff != 0`) is less than 0.0005 (typically reported as $p < 0.001$) leading to clear rejection of the null hypothesis at say the 5% level. The 95% confidence intervals for the difference in population mean salary for men and women are from −\$12,953 to −\$8,213 under the equal-variance assumption and from −\$12,615 to −\$8,551 under the unequal-variance assumption. Repeating the analysis for log-salary (not shown) also gives $p < 0.001$ and a smaller relative difference between the estimated standard errors for the two versions of the t test.

1.4 One-way analysis of variance

The model underlying the t test with equal variances is also called a one-way analysis-of-variance (ANOVA) model.

Analysis of variance involves partitioning the *total sum of squares* (TSS), the sum of squared deviations of the y_i from their overall mean,

$$\mathrm{TSS} = \sum_{i=1}^{n} (y_i - \overline{y})^2$$

into the *model sum of squares* (MSS) and the *sum of squared errors* (SSE).

The group-specific sample means can be viewed as predictions, $\widehat{y}_i = \widehat{\mu}_{x_i} = \overline{y}_{x_i}$, representing the best guess of the salary when all that is known about the professor is the gender. The MSS, also known as regression sum of squares, becomes

$$\mathrm{MSS} = \sum_{i=1}^{n} (\widehat{y}_i - \overline{y})^2 = \sum_{i,x_i=0} (\overline{y}_0 - \overline{y})^2 + \sum_{i,x_i=1} (\overline{y}_1 - \overline{y})^2 = n_0(\overline{y}_0 - \overline{y})^2 + n_1(\overline{y}_1 - \overline{y})^2$$

the sum of squared deviations of the sample means from the overall mean, interpretable as the between-group sum of squares (here "i, x_i=0" and "i, x_i=1" means that the sums are taken over females and males, respectively).

The SSE, also known as residual sum of squares, is

$$\text{SSE} = \sum_{i=1}^{n}(y_i - \widehat{y}_i)^2 = \sum_{i,x_i=0}(y_i - \overline{y}_0)^2 + \sum_{i,x_i=1}(y_i - \overline{y}_1)^2 = (n_0 - 1)s_0^2 + (n_1 - 1)s_1^2$$

the sum of squared deviations of responses from their respective sample means, interpretable as the within-group sum of squares (s_0 and s_1 are the within-group sample standard deviations). The sample means minimize the SSE and are therefore referred to as *ordinary least-squares* (OLS) estimates. When evaluating the quality of the predictions $\widehat{y}_i$, the SSE becomes the sum of squared prediction errors.

The total sum of squares equals the model sum of squares plus the sum of squared errors:

$$\text{TSS} = \text{MSS} + \text{SSE}$$

The deviations contributing to each of these sums of squares are shown in figure 1.3 for an observation y_i (shown as •) in a hypothetical dataset. These deviations add up in the same way as the corresponding sums of squares.

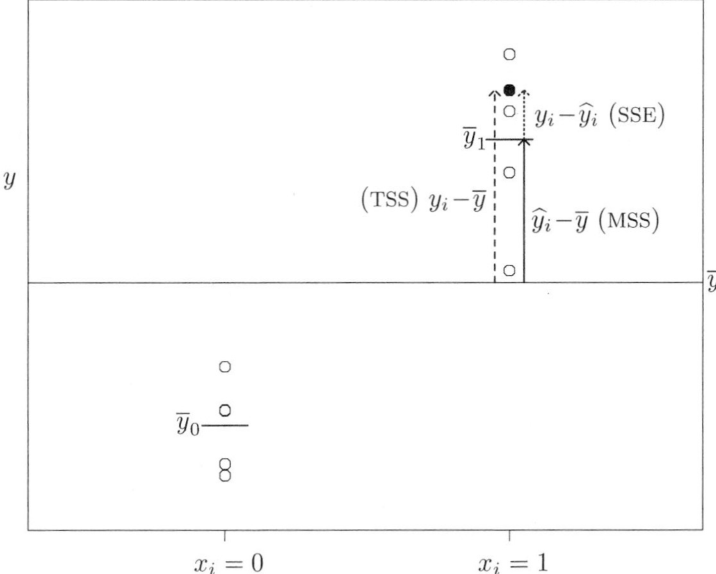

Figure 1.3: Illustration of deviations contributing to total sum of squares (TSS), model sum of squares (MSS), and sum of squared errors (SSE)

The model mean square (MMS) and mean squared error (MSE) can be obtained from the corresponding sums of squares by dividing by the appropriate degrees of freedom as shown in table 1.1 for the general case of g groups.

Table 1.1: Sums of squares and mean squares for one-way ANOVA

Source	SS	df	MS= $\frac{SS}{\text{df}}$	
Model	MSS	$g-1$	MMS	Between-group
Error	SSE	$n-g$	MSE	Within-group
Total	TSS	$n-1$		

The MSE is the pooled within-group sample variance, which is an estimator for the population variance parameter σ^2:

$$\widehat{\sigma^2} = \text{MSE}$$

The F statistic for the null hypothesis that the population means are the same (against the two-sided alternative) then is

$$F(g-1, n-g) = \frac{\text{MMS}}{\text{MSE}}$$

Under the null hypothesis, this statistic has an F distribution with $g-1$ and $n-g$ degrees of freedom. When $g=2$ as in our example, the F statistic is the square of the t statistic from the independent-samples t test under the equal-variance assumption and the p-values from both tests are identical.

We can perform a one-way ANOVA in Stata using

```
. anova salary male

                        Number of obs =      514     R-squared     =  0.1307
                        Root MSE      = 11827.4     Adj R-squared =  0.1290

               Source |  Partial SS    df       MS           F     Prob > F
           -----------+----------------------------------------------------
                Model |  1.0765e+10     1  1.0765e+10     76.96     0.0000
                      |
                 male |  1.0765e+10     1  1.0765e+10     76.96     0.0000
                      |
             Residual |  7.1622e+10   512   139887048
           -----------+----------------------------------------------------
                Total |  8.2387e+10   513   160599133
```

which gives a statistically significant difference in population mean salaries for men and women as before ($F(1, 512) = 76.96$, $p < 0.001$). The estimate $\sqrt{\widehat{\sigma^2}} = 11827$ is given under `Root MSE` in the output. Estimates of μ_0 and μ_1 are not shown but can be obtained using the postestimation command `adjust`:

```
. adjust, by(male)

Dependent variable: salary     Command: anova

     male          xb

    Women     42916.6
      Men     53499.2

     Key:  xb  =  Linear Prediction
```

1.5 Simple linear regression

Salaries can vary considerably between academic departments. Some disciplines are more marketable than others, perhaps because there are highly paid jobs in those disciplines outside academia or because there is a low supply of qualified graduates. The dataset contains a variable, `market`, for the marketability of the discipline, defined as the mean U.S. faculty salary in that discipline divided by the mean salary across all disciplines.

Let us now investigate the relationship between the salaries at BGSU and marketability of the discipline. Marketability ranges from 0.71 to 1.33 in this sample, taking on 46 different values. A one-way ANOVA model, allowing for a different mean salary, μ_{x_i}, for each value of marketability, x_i, would have a large number of parameters and many groups containing only one individual and would hence be *overparameterized*. There are two popular ways of dealing with this problem: (1) categorizing the continuous explanatory variable into intervals, thus producing fewer and larger groups, and (2) assuming a parametric, typically linear, relationship between y_i and x_i.

Taking the latter approach, the *simple linear regression model* can be written as

$$y_i|x_i \;\sim\; N(\mu_{x_i}, \sigma^2) \tag{1.1}$$

where

$$\mu_{x_i} \equiv E(y_i|x_i) \;=\; \beta_1 + \beta_2 x_i \tag{1.2}$$

or alternatively, as

$$y_i \;=\; \beta_1 + \beta_2 x_i + \epsilon_i, \qquad \epsilon_i|x_i \;\sim\; N(0, \sigma^2)$$

where ϵ_i is the residual or error term for the ith professor, assumed to be independent of the residuals for other professors. Here β_1 (the Greek letter β is pronounced beta) is called the *intercept* (often denoted β_0 or α) and represents the conditional expectation of y_i when $x_i = 0$

$$E(y_i|x_i = 0) \;=\; \beta_1$$

β_2 is called the *slope*, or *regression coefficient*, of x_i and represents the difference in conditional expectations when x_i increases one unit, for instance from some value a to $a+1$:

$$E(y_i|x_i=a+1) - E(y_i|x_i=a) = \beta_1 + \beta_2(a+1) - (\beta_1 + \beta_2 a) = \beta_2$$

β_2 is also often called the "effect" of x_i, although a causal interpretation assumes that the model is the data-generating mechanism. In this book we use the term effect casually (not necessarily causally) as a synonym of coefficient. We refer to $\beta_1 + \beta_2 x_i$ as the *fixed part* and ϵ_i as the *random part* of the model.

In addition to assuming that the conditional expectations fall on a straight line, the model assumes that the conditional variances, or residual variances, of the y_i are equal for all x_i,

$$\text{Var}(y_i|x_i) = \text{Var}(\epsilon_i|x_i) = \sigma^2$$

known as the *homoskedasticity* assumption (in contrast to *heteroskedasticity*).

Note that $\epsilon_i|x_i \sim N(0, \sigma^2)$ follows from (1.1) and (1.2), and in particular that $E(y_i|x_i) = \beta_1 + \beta_2 x_i$ implies that $E(\epsilon_i|x_i) = 0$. The covariate x_i is called *exogenous* if the assumption $E(\epsilon_i|x_i) = 0$ is satisfied and said to be *endogenous* if this assumption is violated. The exogeneity assumption implies that the correlation between the residual and the covariate is zero, $\text{Cor}(\epsilon_i, x_i) = 0$, and this weaker requirement is sometimes used to define exogeneity.

A graphical illustration of the simple linear regression model is given in figure 1.4.

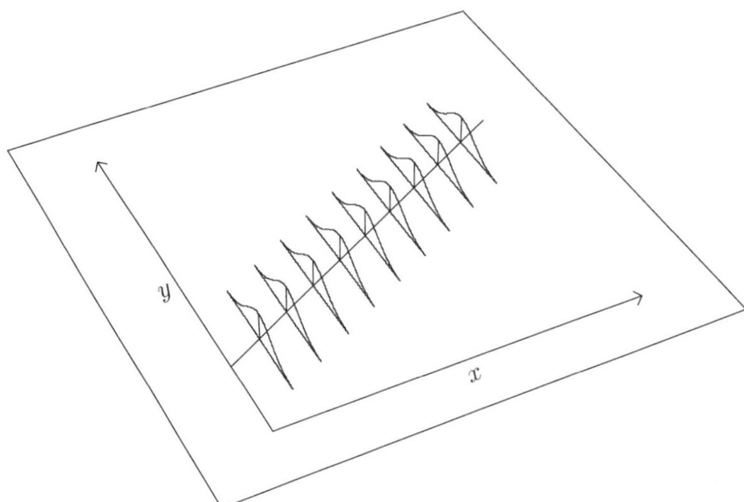

Figure 1.4: Illustration of simple linear regression model

Here the line represents the conditional expectation $E(y_i|x_i)$ as a function of x_i and the density curves represent the conditional distributions of $y_i|x_i$ shown only for some values of x_i.

For the OLS estimates of the regression coefficients, we can partition the TSS into MSS and SSE exactly as in section 1.4, the only difference being the form of the predicted value of y_i

$$\widehat{y}_i \;=\; \widehat{\beta}_1 + \widehat{\beta}_2 x_i \tag{1.3}$$

The contribution to each sum of squares is shown in figure 1.5 for an observation (shown as •) in a hypothetical dataset. The OLS estimates of β_1 and β_2 are obtained by minimizing the SSE and the estimate of σ^2 is again the MSE, where the error degrees of freedom are $n-2$ (number of observations minus 2 estimated regression coefficients). Maximum likelihood estimation, used for more complex models in later chapters, gives the same estimates of the regression coefficients as OLS but a smaller estimate of the residual variance because the latter is given by the SSE divided by n instead of divided by $n-2$ as for OLS.

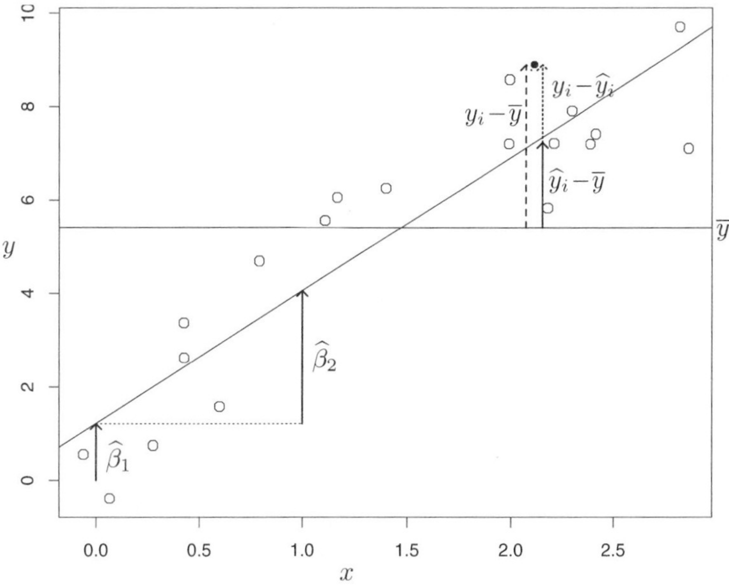

Figure 1.5: Illustration of sums of squares for simple linear regression

The *coefficient of determination* R^2 is defined as

$$R^2 \equiv \frac{\text{MSS}}{\text{TSS}} \;=\; \frac{\text{TSS} - \text{SSE}}{\text{TSS}}$$

and can be motivated as the proportion of the total variability (TSS) that is "explained" by the model that includes the covariate x_i (MSS). Alternatively, if TSS is viewed as the sum of squared prediction errors when the predictions are simply $\widehat{y}_i = \overline{y}$ (not using x_i to make predictions) whereas SSE is the sum of squared prediction errors taking into account x_i as in (1.3), then R^2 can be interpreted as the proportional reduction in prediction error variability due to knowing x_i.

The Stata command and output for the simple linear regression model are

```
. regress salary market

      Source |       SS       df       MS              Number of obs =     514
-------------+------------------------------           F(  1,   512) =  101.77
       Model |  1.3661e+10     1  1.3661e+10           Prob > F      =  0.0000
    Residual |  6.8726e+10   512   134231433           R-squared     =  0.1658
-------------+------------------------------           Adj R-squared =  0.1642
       Total |  8.2387e+10   513   160599133           Root MSE      =   11586

------------------------------------------------------------------------------
      salary |      Coef.   Std. Err.      t    P>|t|     [95% Conf. Interval]
-------------+----------------------------------------------------------------
      market |   34545.22   3424.333    10.09   0.000     27817.75    41272.69
       _cons |   18096.99   3288.009     5.50   0.000     11637.35    24556.64
------------------------------------------------------------------------------
```

The estimated coefficient of marketability is given next to `market` in the regression table and the estimated intercept is given next to `_cons`. The reason for the label `_cons` is that the intercept can be thought of as the coefficient of a variable that is equal to 1 for each observation, referred to as a constant (because it does not vary). The estimated regression coefficients, their estimated standard errors, and the estimated residual standard deviations are also given in the first two columns under "Section 1.5" in table 1.2.

Table 1.2: Ordinary least-squares (OLS) estimates for salary data (in U.S. dollars)

	Section 1.5				Section 1.6		Section 1.7				Section 1.8	
	Est	(SE)	Est	(SE)	Est	(SE)	Est	(SE)	Est	(SE)	Est	(SE)
[_cons]	18,097	(3,288)	50,864	(511)	42,917	(1,045)	44,324	(983)	34,834	(734)	36,774	(1,072)
[market]	34,545	(3,424)										
[marketc]			34,545	(3,424)			29,972	(3,302)	38,402	(2,172)	38,437	(2,161)
[male]					10,583	(1,206)	8,708	(1,139)	2,040	(783)	−593	(1,320)
[yearsdg]									949	(36)	763	(83)
[male×years]											227	(92)
[associate]												
[full]												
[male×associate]												
[male×full]												
[market×years]												
[yearsdg2]												
σ	11,586		11,586		11,827		10,986		7,148		7,112	

	Section 1.9				Section 1.10				Section 1.11	
	Est	(SE)	Est	(SE)	Est	(SE)	Est	(SE)	Est	(SE)
[_cons]	39,866	(746)	37,493	(989)	37,371	(1,029)	38,040	(984)	38,838	(1,027)
[market]										
[marketc]			36,987	(1,975)	36,899	(1,981)	48,557	(3,477)	46,578	(3,544)
[male]			−1,043	(1,215)	−759	(1,298)	−1,970	(1,219)	−1,182	(1,252)
[yearsdg]			405	(87)	370	(111)	334	(87)	39	(145)
[m_years]			184	(85)	231	(125)	249	(85)	178	(89)
[associate]	7,285	(1,026)	3,349	(872)	4,305	(1,480)	3,671	(863)	4,812	(968)
[full]	21,267	(966)	11,168	(1,168)	11,943	(2,437)	11,691	(1,158)	12,791	(1,230)
[male_ass]					−1,467	(1,835)				
[male_full]					−1,211	(2,790)				
[market_yrs]							−869	(216)	−726	(222)
[yearsdg2]									10	(4)
σ	8,917		6,482		6,491		6,387		6,353	

We obtain the estimates $\widehat{\beta}_1$ = \$18,096.99, $\widehat{\beta}_2$ = \$34,545.22, and $\sqrt{\widehat{\sigma^2}}$ = \$11,586. The estimated intercept is the estimated population mean salary when marketability is zero, a value that does not occur in this sample and is meaningless. Before interpreting the estimates, we therefore refit the model after mean-centering `market`:

```
. egen mn_market = mean(market)

. generate marketc = market - mn_market

. regress salary marketc

      Source |       SS       df       MS              Number of obs =     514
-------------+------------------------------           F(  1,   512) =  101.77
       Model |  1.3661e+10     1  1.3661e+10           Prob > F      =  0.0000
    Residual |  6.8726e+10   512   134231433           R-squared     =  0.1658
-------------+------------------------------           Adj R-squared =  0.1642
       Total |  8.2387e+10   513   160599133           Root MSE      =   11586

------------------------------------------------------------------------------
      salary |      Coef.   Std. Err.      t    P>|t|     [95% Conf. Interval]
-------------+----------------------------------------------------------------
     marketc |   34545.22   3424.333    10.09   0.000     27817.75    41272.69
       _cons |   50863.87    511.029    99.53   0.000      49859.9    51867.85
------------------------------------------------------------------------------
```

Here `mn_market`, the sample mean of `market`, was subtracted from `market` so that `marketc` equals zero when `market` is equal to its sample mean. The estimated regression coefficients with standard errors are given in the last two columns under "Section 1.5" in table 1.2. The only estimate that is affected by the centering is the intercept that now represents the estimated population mean salary when marketability is equal to its sample mean (when `marketc` is zero). The estimated standard error of the intercept is considerably smaller after mean-centering because we are no longer extrapolating outside the range of the data.

For each unit increase in marketability, the population mean salary is estimated to increase by \$34,545 (with 95% confidence interval from \$27,818 to \$41,273). Since marketability ranges from 0.71 to 1.33, with no two people differing by as much as 1 unit, we could consider the effect of a 0.1 point increase in marketability, which is associated with an estimated increase in population mean salary of \$3,454 ($=\widehat{\beta}_2/10$). Alternatively, we could standardize marketability (giving it a standard deviation of one in addition to a mean of zero) in which case the estimated regression parameter would be interpreted as the estimated increase in population mean salary when marketability increases by one standard deviation.

If we standardize both the response variable `salary` and the covariate `market`, the estimated regression coefficient becomes a *standardized regression coefficient*, interpreted as the estimated change in standard deviations of salary when marketability is changed by one standard deviation. Standardized regression coefficients can also be obtained by using the `beta` option in the `regress` command. However, they should be used with caution because they depend on sample-specific standard deviations, which invalidates comparisons across samples. A similar issue arises for mean-centering based on sample means. It might have been preferable to subtract 1 (where the mean salary for the discipline equals the mean salary across disciplines) from `marketability` instead of the sample mean 0.9485214. Also, a standard deviation change in a categorical covariate such as gender does not make any sense. See Greenland, Schlesselman, and Criqui (1986) for further discussion.

To test the null hypothesis that the regression coefficient of marketability is zero, H_0: $\beta_2 = 0$, against the two-sided alternative H_a: $\beta_2 \neq 0$, we again use a t statistic, now given by

$$t = \frac{\widehat{\beta}_2}{\widehat{\mathrm{SE}}(\widehat{\beta}_2)}$$

If the null hypothesis is true, this statistic has a t distribution with degrees of freedom given by the error degrees of freedom, denoted `Residual df` in the regression output. Here $t = 99.53$, df $= 512$, and $p < 0.001$ so we can clearly reject the null hypothesis at the 5% level of significance. The null hypothesis for the intercept H_0: $\beta_1 = 0$ is usually irrelevant and p-values for intercepts are thus ignored.

To visualize the fitted model, we can calculate predicted values $\widehat{y}_i$ using the postestimation command `predict` with the `xb` option (were `xb` is short for "xes times betas" and stands for $\widehat{\beta}_1 + \widehat{\beta}_2 x_i$ here)

```
. predict yhat, xb
```

We can then produce a scatterplot of the data points (y_i, x_i) together with a line connecting the points ($\widehat{y}_i$, x_i) with the following command:

```
. twoway (scatter salary market) (line yhat market, sort),
> ytitle(Academic salary) xtitle(Marketability)
```

The graph in figure 1.6 shows that a straight line appears to fit reasonably well and that the constant-variance assumption does not appear to be violated.

(*Continued on next page*)

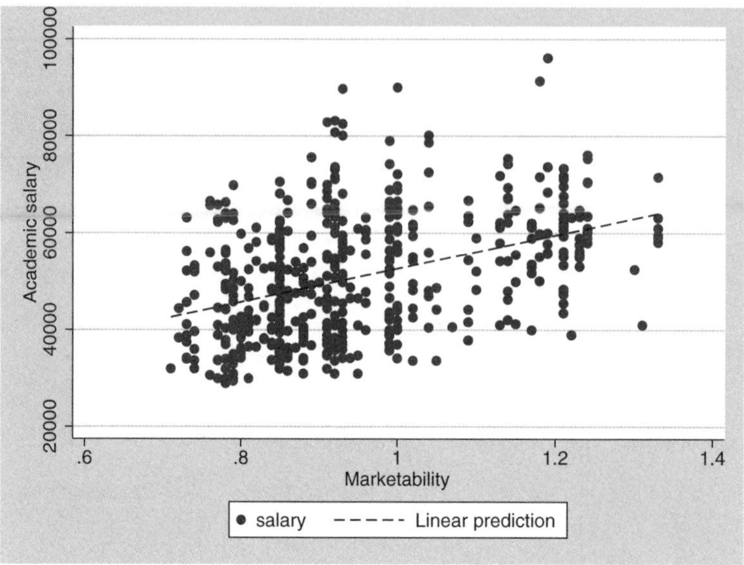

Figure 1.6: Scatterplot with predicted line from simple regression

There is considerable scatter around the regression line with only 16.6% of the variability in salaries being explained by marketability.

1.6 Dummy variables

Instead of using a t test to compare the population mean salaries between men and women, we can use simple linear regression. This becomes obvious by considering the diagram in figure 1.7 where we have simply used the variable $x_i = 1$ for men and $x_i = 0$ for women and connected the corresponding conditional expectations by a straight line. We are not making any assumption regarding the relationship between the conditional means and x here because any two means can be connected by a straight line. (In contrast, assuming in the previous section that the conditional means for the 46 values of marketability lay on a straight line was a strong assumption.) We are, however, assuming equal conditional variances for the two populations due to the homoskedasticity assumption discussed earlier.

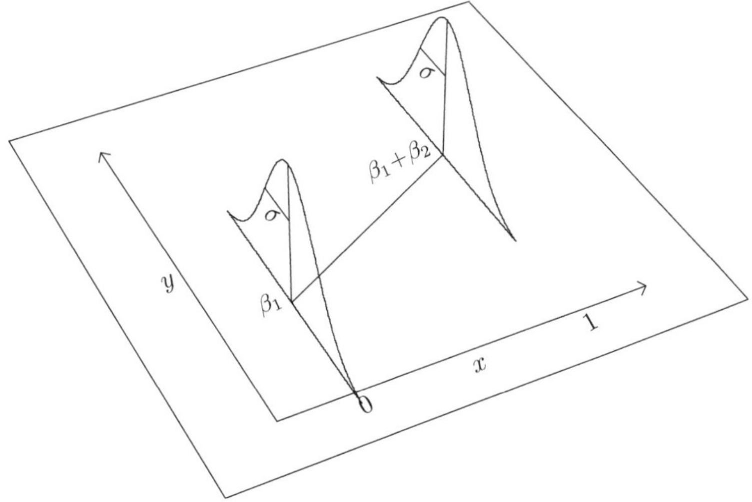

Figure 1.7: Illustration of simple linear regression with a dummy variable

The model can be written as

$$\mu_{x_i} \equiv E(y_i|x_i) \;=\; \beta_1 + \beta_2 x_i, \qquad y_i|x_i \;\sim\; N(\mu_{x_i}, \sigma^2)$$

so that

$$\begin{aligned} \mu_0 &= \beta_1 + \beta_2 \times 0 = \beta_1 \\ \mu_1 &= \beta_1 + \beta_2 \times 1 = \beta_1 + \beta_2 \end{aligned}$$

The intercept β_1 can now be interpreted as the expectation for the group coded 0, the *reference group* (here women), whereas the slope β_2 represents the difference in expectations $\beta_2 = \mu_1 - \mu_0$ between the group coded 1 (here men) and the reference group.

When a dichotomous variable, coded 0 and 1, is used in a regression model like this, it is referred to as a *dummy variable*. A useful convention is to give the dummy variable the name of the group for which it is 1 and to describe it as a dummy variable for being in that group, here a dummy variable for being male.

The null hypothesis that $\beta_2 = 0$ is equivalent to the null hypothesis that the population means are the same $\mu_1 - \mu_0 = 0$. We can therefore use simple regression to obtain the same result as previously produced for the two-sided independent-samples t test with the equal-variances assumption:

```
. regress salary male

      Source |       SS       df       MS              Number of obs =     514
-------------+------------------------------           F(  1,   512) =   76.96
       Model |  1.0765e+10     1  1.0765e+10           Prob > F      =  0.0000
    Residual |  7.1622e+10   512   139887048           R-squared     =  0.1307
-------------+------------------------------           Adj R-squared =  0.1290
       Total |  8.2387e+10   513   160599133           Root MSE      =   11827

------------------------------------------------------------------------------
      salary |      Coef.   Std. Err.      t    P>|t|     [95% Conf. Interval]
-------------+----------------------------------------------------------------
        male |   10582.63   1206.345     8.77   0.000     8212.636    12952.63
       _cons |    42916.6   1045.403    41.05   0.000      40862.8    44970.41
------------------------------------------------------------------------------
```

The t statistic has the same value as that previously shown for the independent-sample t test (with the equal-variance assumption), apart from the sign that depends on whether $\mu_0 - \mu_1$ or $\mu_1 - \mu_0$ is estimated and is thus arbitrary. The estimated regression coefficients with standard errors are also given under "Section 1.6" in table 1.2.

We can also relax the homoskedasticity (and normality) assumption in linear regression by replacing the conventional *model-based estimator* for the standard errors by the so-called *sandwich estimator*. Simply add the `vce(robust)` option (where `vce` stands for "variance covariance matrix of estimates") to the `regress` command:

```
. regress salary male, vce(robust)
Linear regression                                      Number of obs =     514
                                                       F(  1,   512) =  105.26
                                                       Prob > F      =  0.0000
                                                       R-squared     =  0.1307
                                                       Root MSE      =   11827

------------------------------------------------------------------------------
             |               Robust
      salary |      Coef.   Std. Err.      t    P>|t|     [95% Conf. Interval]
-------------+----------------------------------------------------------------
        male |   10582.63   1031.462    10.26   0.000     8556.213    12609.05
       _cons |    42916.6    808.184    53.10   0.000     41328.84    44504.37
------------------------------------------------------------------------------
```

The resulting t statistic is similar to that using the `ttest` command with the `unequal` option in section 1.3 (10.26 compared with 10.25). In the rest of the chapter, we will stick to the conventional model-based estimator for the standard errors although the more robust sandwich estimator may often be preferable.

The utility of using a regression model with a dummy variable instead of the t test to compare two groups becomes evident in the next section where we want to "control" for other variables, which is straightforward in a multiple regression model.

1.7 Multiple linear regression

An important question when investigating gender discrimination is whether the men and women being compared are similar in the variables that justifiably affect salaries.

As we have seen, there is some variability in marketability, and marketability has an effect on salary. Could the lower mean salary for women be due to women tending to work in disciplines with lower marketability? This is a possible explanation only if women do have lower mean marketability than men, which is indeed the case:

```
. tabstat marketc, by(male) statistics(mean sd)
Summary for variables: marketc
     by categories of: male

    male |      mean        sd
---------+--------------------
   Women | -.0469589  .1314393
     Men |  .0155718  .1518486
---------+--------------------
   Total | -2.96e-08    .14938
------------------------------
```

We could render the comparison of salaries more fair by matching each woman to a man with the same value of marketability, but this would be cumbersome and we may not find matches for everyone. Instead, we can assume that marketability has the same, linear, effect on salary for both genders and check whether gender has any additional effect after allowing for the effect of marketability.

This can be accomplished by specifying a multiple linear regression model

$$y_i \;=\; \beta_1 + \beta_2 x_{2i} + \beta_3 x_{3i} + \epsilon_i, \qquad \epsilon_i | x_{2i}, x_{3i} \;\sim\; N(0, \sigma^2)$$

where we have multiple covariates, a dummy variable x_{2i} for being a man and mean-centered marketability x_{3i}. (We number explanatory variables beginning with 2 to correspond to the regression coefficients.)

For disciplines with mean marketability ($x_{3i} = 0$), the model specifies that the expected salary is β_1 for women ($x_{2i} = 0$) and $\beta_1 + \beta_2$ for men ($x_{2i} = 1$). Therefore, β_2 can be interpreted as the difference in population mean salary between men and women in disciplines with mean marketability. Fortunately, β_2 has an even more general interpretation as the difference in population mean salary between men and women in disciplines with *any* level of marketability, as long as both genders have the same value for marketability, e.g., $x_{3i} = a$:

$$E(y_i|x_{2i} = 1, x_{3i}{=}a) - E(y_i|x_{2i} = 0, x_{3i}{=}a) \;=\; (\beta_1 + \beta_2 + \beta_3 a) - (\beta_1 + \beta_3 a) \;=\; \beta_2$$

When comparing genders, we are now *controlling* for, *adjusting* for, or *keeping constant* marketability. Marketability is said to be a *confounder* for the association between gender and salary if $\beta_3 \neq 0$ and marketability is associated with gender. Ignoring marketability then produces a biased estimate of β_2 when the above model is true, sometimes referred to as *omitted-variable bias.*

The Stata command and output for the model are

```
. regress salary male marketc

      Source |       SS       df       MS              Number of obs =     514
-------------+------------------------------           F(  2,   511) =   85.80
       Model |  2.0711e+10     2  1.0356e+10           Prob > F      =  0.0000
    Residual |  6.1676e+10   511   120696838           R-squared     =  0.2514
-------------+------------------------------           Adj R-squared =  0.2485
       Total |  8.2387e+10   513   160599133           Root MSE      =   10986

------------------------------------------------------------------------------
      salary |      Coef.   Std. Err.      t    P>|t|     [95% Conf. Interval]
-------------+----------------------------------------------------------------
        male |   8708.423   1139.411     7.64   0.000     6469.917    10946.93
     marketc |    29972.6   3301.766     9.08   0.000     23485.89     36459.3
       _cons |   44324.09   983.3533    45.07   0.000     42392.17       46256
------------------------------------------------------------------------------
```

The estimated regression coefficients with standard errors are given in the first two columns under "Section 1.7" in table 1.2. The difference in population mean salaries between men and women, controlling for marketability, is estimated as $8,708, considerably smaller than the unadjusted difference of $10,583 that we estimated in the previous section. The estimated coefficient of `marketc` has also gone down and is now interpretable as the effect of marketability for a given gender or the within-gender effect of marketability. The coefficient of determination is now $R^2 = 0.251$ compared with $R^2 = 0.166$ for the model containing only `marketc` (R^2 cannot decrease when more covariates are added).

We can obtain predicted salaries $\widehat{y}_i$ and plot them with the observed salaries y_i against mean-centered marketability x_{3i} separately for each gender x_{2i} using

```
. predict yhat2, xb
. twoway (scatter salary market if male==1, msymbol(o))
>        (line yhat2 market if male==1, sort lpatt(solid))
>        (scatter salary market if male==0, msymbol(oh))
>        (line yhat2 market if male==0, sort lpatt(dash)),
> ytitle(Academic salary) xtitle(Marketability)
> legend(order(1 " " 2 "Men" 3 " " 4 "Women"))
```

which produces the graph in figure 1.8.

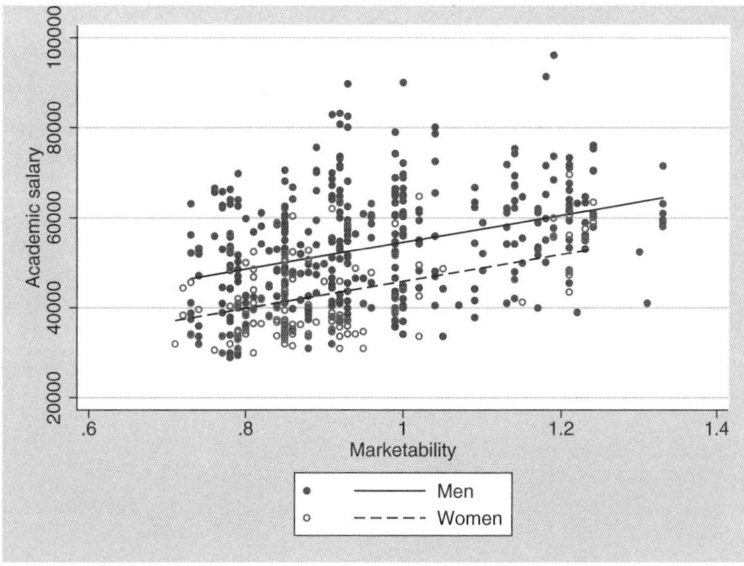

Figure 1.8: Scatterplot with predicted lines from multiple regression

The vertical distance between the regression lines is $\widehat{\beta}_2$ and the slope of both regression lines is $\widehat{\beta}_3$.

Figure 1.9 illustrates the idea of confounding using hypothetical data. There are two groups whose responses are represented by hollow and filled circles. In the top panel, the vertical distance between the lines represents the unadjusted difference in population means for the groups. It is clear from the figure that the continuous covariate x is a confounder for the association between the response variable and group, since x is associated with group (the mean of x is larger for the group represented by filled circles) as well as with the response (the mean of y is larger for larger values of x). Thus the estimated adjusted difference in population means, given by the vertical distance between the population regression lines in the lower panel, is different from the unadjusted counterpart in the top panel. If the regression lines coincide (with no vertical distance between them), the association between the response variable and group is said to be "spurious".

(*Continued on next page*)

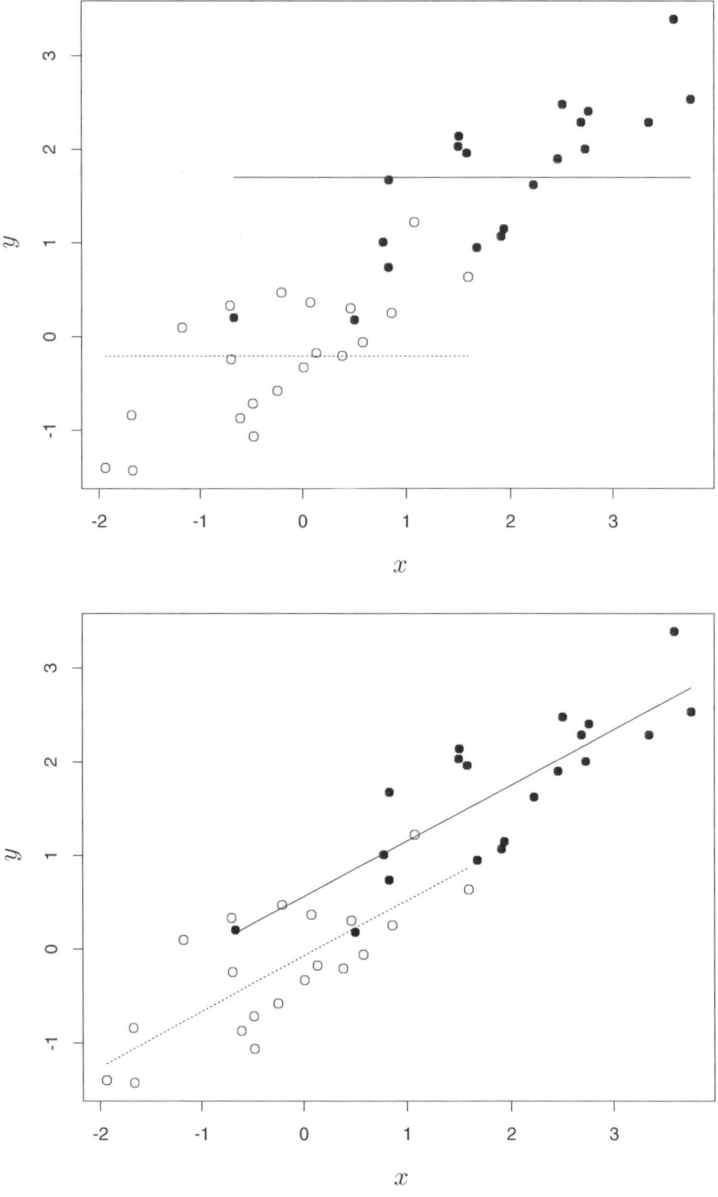

Figure 1.9: Illustration of confounding. Top panel shows unadjusted differences in population means between groups. Bottom panel shows differences in population means between groups adjusted for covariate x.

The model used in this section is sometimes called an *analysis of covariance* (ANCOVA) model because in addition to the usual categorical explanatory variable or *factor* gender, there is a continuous covariate marketability. Departing from the ANOVA/ANCOVA terminology, we use the word "covariate" for any observed explanatory variable, also including dummy variables, throughout this book.

Another covariate we should perhaps control for is time since the degree (in years), `yearsdg`:

```
. regress salary male marketc yearsdg
      Source |       SS       df       MS              Number of obs =     514
-------------+------------------------------           F(  3,   510) =  367.56
       Model |  5.6333e+10     3  1.8778e+10           Prob > F      =  0.0000
    Residual |  2.6054e+10   510  51087083.4           R-squared     =  0.6838
-------------+------------------------------           Adj R-squared =  0.6819
       Total |  8.2387e+10   513   160599133           Root MSE      =  7147.5

------------------------------------------------------------------------------
      salary |      Coef.   Std. Err.      t    P>|t|     [95% Conf. Interval]
-------------+----------------------------------------------------------------
        male |   2040.211    783.122     2.61   0.009     501.6684    3578.753
     marketc |   38402.39   2171.689    17.68   0.000     34135.83    42668.95
     yearsdg |   949.2583   35.94867    26.41   0.000     878.6326    1019.884
       _cons |    34834.3   733.7898    47.47   0.000     33392.68    36275.93
------------------------------------------------------------------------------
```

The estimated regression coefficients with standard errors are given in the last two columns under "Section 1.7" in table 1.2. For a given marketability and time since degree, the estimated population mean salary for men is \$2,040 greater than for women, a substantial reduction in the estimated gender gap due to controlling for time since degree `yearsdg`. For a given gender and time since degree, the estimated effect of marketability is \$38,402 extra mean salary per unit increase in marketability. Comparing professors of the same gender from disciplines with the same marketability, the estimated effect of time since degree is \$949 extra mean salary per year since degree. It is tempting to interpret this as an effect of experience, but it is important to note that those with, for instance, `yearsdg` = 40 differ from those with `yearsdg` = 0 not only in terms of experience, but also also because they were recruited in a different epoch. As we will discuss in section 5.4, we cannot separate such cohort effects from age or experience effects using cross-sectional data.

We can obtain adjusted mean estimates for the two genders for certain values of the other covariates in the model, for instance, `marketc` = 0 and `yearsdg` = 10, using the `adjust` command:

(*Continued on next page*)

```
. adjust marketc=0 yearsdg=10, by(male)

      Dependent variable: salary     Command: regress
Covariates set to value: marketc = 0, yearsdg = 10


      male            xb

     Women       44326.9
       Men       46367.1

      Key:  xb  =  Linear Prediction
```

1.8 Interactions

The models considered in the previous section assumed that the effects of different covariates were additive. For instance, if the dummy variable x_{2i} changes from 0 (women) to 1 (men), the mean salary increases by an amount β_2 regardless of the values of the other variables. This assumption translates to parallel regression lines with vertical distance given by β_2.

However, this assumption can be violated. For instance, starting salaries could be similar for men and women but men might receive larger increases or increases at shorter time intervals. We can investigate this by including an *interaction* between gender and time since degree. An interaction between two variables implies that the effect of each variable depends on the value of the other variable: the effect of gender depends on time since degree and the effect of time since degree depends on gender.

We can incorporate an interaction in a regression model (with the usual assumptions) by simply including the product of `male` (x_{2i}) and `yearsdg` (x_{4i}) as a further covariate with regression coefficient β_5:

$$\begin{aligned}
y_i &= \beta_1 + \beta_2 \texttt{male}_i + \beta_3 x_{3i} + \beta_4 \texttt{yearsdg}_i + \beta_5 \texttt{male}_i \times \texttt{yearsdg}_i + \epsilon_i \\
&= \beta_1 + (\beta_2 + \beta_5 \texttt{yearsdg}_i)\texttt{male}_i + \beta_3 x_{3i} + \beta_4 \texttt{yearsdg}_i + \epsilon_i \qquad (1.4) \\
&= \beta_1 + \beta_2 \texttt{male}_i + \beta_3 x_{3i} + (\beta_4 + \beta_5 \texttt{male}_i)\texttt{yearsdg}_i + \epsilon_i \qquad (1.5)
\end{aligned}$$

From (1.4) we see that the effect of `male` is given by $\beta_2 + \beta_5$`yearsdg` and hence depends on time since degree if $\beta_5 \neq 0$. From (1.5) we see that the effect of `yearsdg` is given by $\beta_4 + \beta_5$`male` and hence depends on gender if $\beta_5 \neq 0$. We can describe time since degree as a *moderator* or *effect modifier* of the effect of gender or vice versa.

When including an interaction between two variables, it is essential to keep both variables in the model. For instance, dropping `male` or setting $\beta_2 = 0$ would force the gender gap to be exactly 0 when time since degree is 0. This is a completely arbitrary constraint unless it corresponds to a specific research question.

An illustration of the model is given in figure 1.10. If β_5 were 0, we would obtain two parallel regression lines with vertical distance β_2. We see that β_5 represents the additional slope for men compared with women, or the additional gender gap when **yearsdg** increases by one unit.

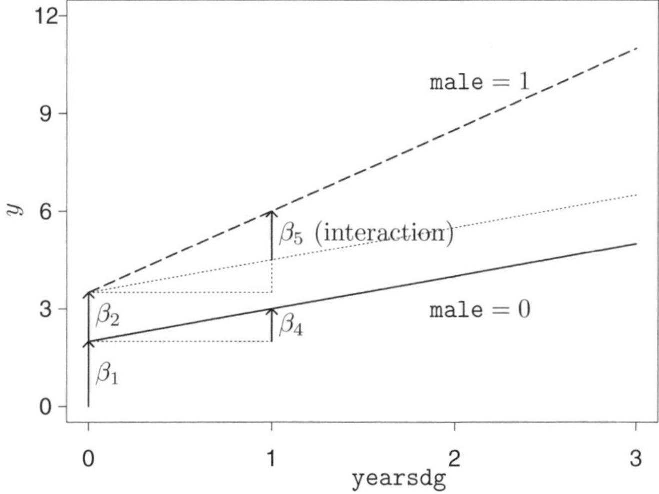

Figure 1.10: Illustration of interaction between **male** (x_{2i}) and **yearsdg** (x_{4i}) for **marketc** (x_{3i}) equal to 0

To fit this model in Stata, we generate the interaction as the product of gender and years since graduation

```
. generate m_years = male*yearsdg
```

and include it in the regression

```
. regress salary male marketc yearsdg m_years
```

Source	SS	df	MS
Model	5.6641e+10	4	1.4160e+10
Residual	2.5746e+10	509	50581607.4
Total	8.2387e+10	513	160599133

Number of obs	=	514
F(4, 509)	=	279.95
Prob > F	=	0.0000
R-squared	=	0.6875
Adj R-squared	=	0.6850
Root MSE	=	7112.1

salary	Coef.	Std. Err.	t	P>|t|	[95% Conf.	Interval]
male	-593.3088	1320.911	-0.45	0.654	-3188.418	2001.8
marketc	38436.65	2160.963	17.79	0.000	34191.14	42682.15
yearsdg	763.1896	83.4169	9.15	0.000	599.3057	927.0734
m_years	227.1532	91.99749	2.47	0.014	46.41164	407.8947
_cons	36773.64	1072.395	34.29	0.000	34666.78	38880.51

The estimated regression coefficients and their estimated standard errors are given under "Section 1.8" in table 1.2. Here it is natural to interpret the interaction in terms of the effect of gender. When time since degree is 0 years, the population mean salary for men minus the population mean salary for women (after adjusting for the other covariates) is estimated as $-\$593$. For every additional year since completing the degree, we add about \$227 to the difference, giving a zero difference after a little over 2 years, a difference of about $-\$593 + \$227 \times 10 = \$1,677$ after 10 years, a difference of $-\$593 + \$227 \times 20 = \$3,947$ after 20 years, and a difference of $-\$593 + \$227 \times 30 = \$6{,}217$ after 30 years. Although the estimated gender gap at 0 years is not statistically significant at the 5% level ($t = -0.45$, df $= 509$, $p = 0.65$), the change in gender gap with years since degree (or interaction) is significant ($t = -2.47$, df $= 509$, $p = 0.01$).

We might wonder if the gender gap is statistically significant for faculty with 10 years of experience, hence testing the null hypothesis H_0: $\beta_2 + \beta_4 \times 10 = 0$ against the two-sided alternative. The null hypothesis involves a linear combination of coefficients, and we can use the `lincom` command (which stands for "linear combination") to perform a test:

```
. lincom male + m_years*10
 ( 1)  male + 10 m_years = 0
```

salary	Coef.	Std. Err.	t	P>|t|	[95% Conf.	Interval]
(1)	1678.223	792.9094	2.12	0.035	120.4449	3236.001

giving $t = 2.12$, df $= 509$, $p = 0.04$. We also obtain a 95% confidence interval for the difference in population mean salaries 10 years since the degree that ranges from \$120 to \$3,236.

The regression model now includes three variables, and it is difficult to represent it using a two-dimensional graph. However, we can hold certain variables constant, for example, `marketc`, and display the estimated population mean salary as a function of the other variables when `marketc` is zero. We could do this by setting `marketc` to zero and then using `predict`, or we can use Stata's `twoway function` command

```
. twoway (function Women = 36773 + 763.19*x, range(0 41) lpatt(dash))
>        (function Men = 36773 + -593.31 + (763.19 + 227.15)*x,
>         range(0 41) lpatt(solid)),
>  xtitle(Time since degree (years)) ytitle(Mean salary)
```

to produce figure 1.11. Here we have typed in the predicted regression lines for women and men as a function of `yearsdg` (here referred to as `x`), using (1.5) with `male`$=0$ for women and `male`$=1$ for men and with $x_{3i}=0$.

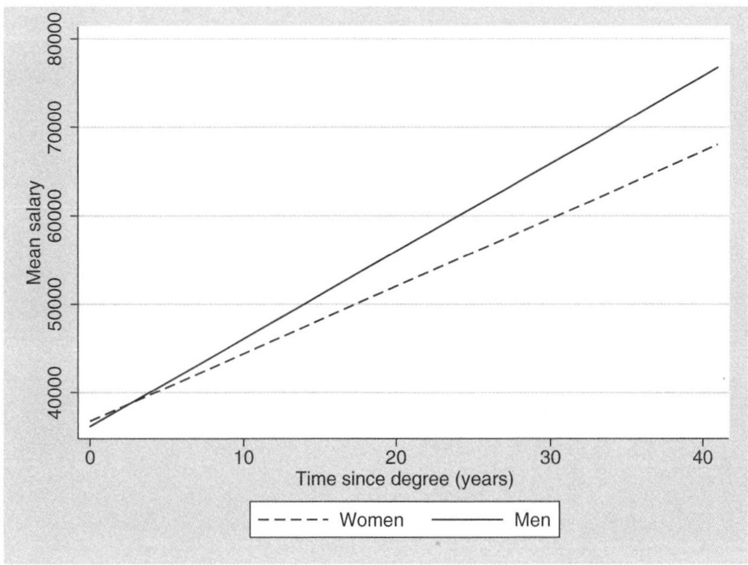

Figure 1.11: Estimated effect of gender and time since degree on mean salary for disciplines with mean marketability

Since we have just fitted the model in Stata, we can also refer to the regression coefficients as `_b[_cons]` for the intercept, `_b[yearsdg]` for the coefficient of `yearsdg`, etc., saving us from having to type in all the coefficients, which is error prone. The Stata command then looks like this:

```
twoway (function Women =_b[_cons] + _b[yearsdg]*x, range(0 41) lpatt(dash))
  (function Men =_b[_cons] + _b[male] + (_b[yearsdg] + _b[m_years])*x,
  range(0 41) lpatt(solid)), xtitle(Time since degree (years)) ytitle(Mean salary)
```

We could of course also include an interaction between `marketc` and `male` and, as will be discussed in section 1.10.1, an interaction between the two continuous covariates `marketc` and `yearsdg`. In addition to these three *two-way interactions*, we could consider the *three-way interaction* represented by the product of all three covariates. However, such higher-order interactions are rarely used because they are difficult to interpret.

1.9 Dummies for more than two groups

Another important explanatory variable for salary is academic rank, coded in the variable `rank` as 1: assistant professor, 2: associate professor, and 3: full professor.

Using the variable `rank` as a continuous covariate in a simple regression model would force a constraint on the population mean salaries for the three groups, namely, that the mean salary of associate professors is halfway between the mean salaries of assistant and full professors:

$$E(y_i|x_i) = \beta_1 + \beta_2 x_i = \begin{cases} \beta_1 + \beta_2 \times 1 & \text{for assistant professors} \\ \beta_1 + \beta_2 \times 2 & \text{for associate professors} \\ \beta_1 + \beta_2 \times 3 & \text{for full professors} \end{cases}$$

Such linearity is a strong assumption for an ordinal variable such as rank and a meaningless assumption for unordered categorical covariates such as ethnicity, where the values assigned to categories are arbitrary.

It thus makes sense to estimate the mean of each rank freely by treating one of the ranks, for instance, assistant professor, as reference category and creating dummy variables for the other two ranks:

```
. generate associate = rank==2 if rank < .
. generate full = rank==3 if rank < .
```

Here the logical expression `rank==2` evaluates to 1 if it is true and zero otherwise. This yields a 0 when `rank` is 1, 3, or missing but we do not want to interpret a missing value, so the `if` condition is necessary to ensure that missing values in `rank` translate to missing values in the dummy variables. Denoting these dummy variables x_{2i} and x_{3i}, respectively, we specify the model

$$E(y_i|x_{2i}, x_{3i}) = \beta_1 + \beta_2 x_{2i} + \beta_3 x_{3i} = \begin{cases} \beta_1 & \text{for assistant professors} \\ \beta_1 + \beta_2 & \text{for associate professors} \\ \beta_1 + \beta_3 & \text{for full professors} \end{cases}$$

showing that the intercept β_1 represents the population mean salary for the reference category (assistant professors), β_2 the difference in mean salaries between associate and assistant professors, and β_3 the difference in mean salaries between full and assistant professors. The coefficient of each dummy variable represents the population mean of the corresponding group minus the population mean of the reference group.

Estimates for the regression model with dummy variables for academic rank are obtained by using

```
. regress salary associate full
      Source |       SS       df       MS              Number of obs =     514
-------------+------------------------------           F(  2,   511) =  262.54
       Model |  4.1753e+10     2  2.0877e+10           Prob > F      =  0.0000
    Residual |  4.0634e+10   511  79518710.1           R-squared     =  0.5068
-------------+------------------------------           Adj R-squared =  0.5049
       Total |  8.2387e+10   513   160599133           Root MSE      =  8917.3

------------------------------------------------------------------------------
      salary |      Coef.   Std. Err.      t    P>|t|     [95% Conf. Interval]
-------------+----------------------------------------------------------------
   associate |   7285.121    1026.19     7.10   0.000     5269.049    9301.192
        full |   21267.11   965.8886    22.02   0.000     19369.51    23164.71
       _cons |   39865.86   745.7043    53.46   0.000     38400.84    41330.88
------------------------------------------------------------------------------
```

and given in the first two columns under "Section 1.9" in table 1.2. We see that the estimated difference in population means between associate professors and assistant professors is \$7,285 and that the estimated difference in population means between full professors and assistant professors is \$21,267. We can obtain the estimated population mean salary for associate professors, $\widehat{\beta}_1 + \widehat{\beta}_2$, by using

```
. lincom _cons + associate
 ( 1)  associate + _cons = 0

------------------------------------------------------------------------------
      salary |      Coef.   Std. Err.      t    P>|t|     [95% Conf. Interval]
-------------+----------------------------------------------------------------
         (1) |   47150.98   704.9766    66.88   0.000     45765.97    48535.99
------------------------------------------------------------------------------
```

and the estimated difference in population mean salary between full and associate professors, $\widehat{\beta}_3 - \widehat{\beta}_2$, by using

```
. lincom full - associate
 ( 1) - associate + full = 0

------------------------------------------------------------------------------
      salary |      Coef.   Std. Err.      t    P>|t|     [95% Conf. Interval]
-------------+----------------------------------------------------------------
         (1) |   13981.99   934.8036    14.96   0.000     12145.45    15818.52
------------------------------------------------------------------------------
```

The estimated difference in population mean salaries between full and associate professors is considerably larger than the difference between associate and assistant professors, suggesting that the constraint imposed if rank were treated as a continuous covariate is unreasonable.

We can fit the same regression model without forming dummy variables ourselves by using the `xi` prefix command and then preceding the categorical covariate(s) with `i.`:

```
. xi: regress salary i.rank
i.rank            _Irank_1-3          (naturally coded; _Irank_1 omitted)
      Source |       SS       df       MS              Number of obs =     514
-------------+------------------------------           F(  2,    511) =  262.54
       Model |  4.1753e+10     2  2.0877e+10           Prob > F      =  0.0000
    Residual |  4.0634e+10   511  79518710.1           R-squared     =  0.5068
-------------+------------------------------           Adj R-squared =  0.5049
       Total |  8.2387e+10   513   160599133           Root MSE      =  8917.3

------------------------------------------------------------------------------
      salary |      Coef.   Std. Err.      t    P>|t|     [95% Conf. Interval]
-------------+----------------------------------------------------------------
    _Irank_2 |   7285.121    1026.19     7.10   0.000     5269.049    9301.192
    _Irank_3 |   21267.11   965.8886    22.02   0.000     19369.51    23164.71
       _cons |   39865.86   745.7043    53.46   0.000     38400.84    41330.88
------------------------------------------------------------------------------
```

Here `_Irank_2` and `_Irank_3` are dummy variables for `rank` being 2 and 3, respectively. Although using the `xi` prefix is convenient, the advantage of constructing your own dummy variables with meaningful names should be obvious.

Our preferred method for producing dummy variables is using the `tabulate` command with the `generate()` option as follows:

```
tabulate rank, generate(r)
rename r2 associate
rename r3 full
```

The `generate(r)` option produces dummy variables for each unique value of `rank`, here `r1`, `r2`, and `r3` for the values 1, 2, and 3. (The naming of the dummy variables would have been the same if the unique values had been 0, 1, and 4.) An advantage of using `tabulate` is that it places missing values into dummy variables whenever the original variable is missing—this requires extra caution when using the `generate` command.

Instead of using regression with dummy variables, we could use one-way ANOVA with $g=3$ groups (see table 1.1):

```
. anova salary rank

                           Number of obs =      514     R-squared     =  0.5068
                           Root MSE      = 8917.33     Adj R-squared =  0.5049

                  Source   Partial SS    df       MS            F     Prob > F

                   Model   4.1753e+10     2  2.0877e+10      262.54     0.0000

                    rank   4.1753e+10     2  2.0877e+10      262.54     0.0000

                Residual   4.0634e+10   511  79518710.1

                   Total   8.2387e+10   513   160599133
```

This command produces the same sums of squares, mean squares, and F statistic as given at the top of the regression output, but no estimates of population means or their differences. The F test is a test of the null hypothesis that all three population means are the same, or in other words that the coefficients β_2 and β_3 of the dummy variables are both zero. The alternative hypothesis is that at least one of the coefficients differs from zero. Such simultaneous hypotheses can also be tested after fitting the regression model using `testparm` (see below).

The one-way ANOVA model with g groups is sometimes written as

$$y_{ij} \;=\; \beta + \alpha_j + \epsilon_i, \qquad \sum_{i=1}^{g} \alpha_j = 0, \quad \epsilon_i \sim N(0, \sigma^2)$$

which corresponds to the model considered in section 1.3 with $\mu_j = \beta + \alpha_j$.

Adding the dummy variables for academic rank constructed above to the model from the previous section, we obtain

```
. regress salary male marketc yearsdg m_years associate full

      Source |       SS       df       MS              Number of obs =     514
-------------+------------------------------           F(  6,   507) =  242.32
       Model |  6.1086e+10     6  1.0181e+10           Prob > F      =  0.0000
    Residual |  2.1301e+10   507  42014709.8           R-squared     =  0.7414
-------------+------------------------------           Adj R-squared =  0.7384
       Total |  8.2387e+10   513   160599133           Root MSE      =  6481.9

------------------------------------------------------------------------------
      salary |      Coef.   Std. Err.      t    P>|t|     [95% Conf. Interval]
-------------+----------------------------------------------------------------
        male |  -1043.394   1215.034    -0.86   0.391    -3430.516    1343.727
     marketc |   36987.08   1974.888    18.73   0.000     33107.11    40867.06
     yearsdg |   405.2749   86.72844     4.67   0.000     234.8835    575.6663
     m_years |   184.3764   85.06732     2.17   0.031     17.24853    351.5042
   associate |   3349.005   871.6155     3.84   0.000     1636.582    5061.428
        full |   11168.26   1167.809     9.56   0.000     8873.923     13462.6
       _cons |   37493.09    988.658    37.92   0.000     35550.72    39435.46
------------------------------------------------------------------------------
```

The estimated regression coefficients with standard errors are given in the last two columns under "Section 1.9" in table 1.2. The estimated coefficients of `associate` and `full` are considerably lower than before because they now represent the estimated *adjusted* or *partial* differences in population means, holding the other variables in the model constant.

We can test the null hypothesis that the coefficients of these dummy variables are both zero using

```
. testparm associate full
 ( 1)  associate = 0
 ( 2)  full = 0

       F(  2,   507) =   52.89
            Prob > F =    0.0000
```

The F statistic is equal to the difference in MSS between the models that do and do not contain the two dummy variables for academic rank (but contain all the other terms of the model), divided by the product of the difference in model degrees of freedom and the MSE of the larger model.

After controlling for academic rank (and the other covariates), the difference in mean salary between men and women is estimated as $-1043 + 184 \times$ `yearsdg`, which is lower for every year since degree than the estimate of $-593 + 227 \times$ `yearsdg` before adjusting for academic rank. For example, at 10 years since degree, the difference in mean salary is now estimated as about \$800,

```
. lincom male + m_years*10
 ( 1)  male + 10 m_years = 0

------------------------------------------------------------------------------
      salary |      Coef.   Std. Err.      t    P>|t|     [95% Conf. Interval]
-------------+----------------------------------------------------------------
         (1) |   800.3697   727.9737     1.10   0.272    -629.8468    2230.586
------------------------------------------------------------------------------
```

whereas the corresponding estimate not adjusting for academic rank was about \$1,678.

We see that the estimated gender effect is smaller when it is adjusted for academic rank. However, adjusting for academic rank could be problematic if it is a mediating or intervening variable on the causal pathway from gender to salary, i.e., if gender affects promotions, which in turn affect salary. Here we must decide whether we are interested in the direct effect of gender on salary or the total effect of gender on salary (the sum of the direct effect and the indirect effect mediated by rank). If we are interested in the direct effect, we should control for rank whereas we should not control for rank if we are interested in the total effect. Boudreau et al. (1997) discuss a study showing that gender does not appear to affect academic rank at BGSU, implying that the direct effect should be the same as the total effect.

1.10 Other types of interactions

1.10.1 Interaction between dummy variables

Could the differences in population mean salaries between the different academic ranks differ between men and women? This question can be answered by including the two interaction terms `male`×`associate` ($x_{2i}x_{6i}$) and `male`×`full` ($x_{2i}x_{7i}$) in the model

$$\begin{aligned} y_i &= \beta_1 + \beta_2 \texttt{male}_i + \cdots + \beta_6 \texttt{associate}_i + \beta_7 \texttt{full}_i + \beta_8 \texttt{male}_i \times \texttt{associate}_i \\ &\quad + \beta_9 \texttt{male}_i \times \texttt{full}_i + \epsilon_i \\ &= \beta_1 + \beta_2 \texttt{male}_i + \cdots + (\beta_6 + \beta_8 \texttt{male}_i) \texttt{associate}_i + (\beta_7 + \beta_9 \texttt{male}_i) \texttt{full}_i + \epsilon_i \qquad (1.6) \\ &= \beta_1 + (\beta_2 + \beta_8 \texttt{associate}_i + \beta_9 \texttt{full}_i) \texttt{male}_i + \cdots + \beta_6 \texttt{associate}_i + \beta_7 \texttt{full}_i + \epsilon_i \qquad (1.7) \end{aligned}$$

If the other terms denoted "$\cdots$" above are omitted, this model becomes a two-way ANOVA model with main effects and an interaction between academic rank and gender.

An interaction between dummy variables can be interpreted as a difference of a difference. For instance, we see from (1.6) that β_8 represents the difference between men and women of the difference between the population mean salaries of associate and assistant professors.

We now construct the interactions

```
. generate male_ass = male*associate
. generate male_full = male*full
```

and fit the regression model including these interactions:

```
. regress salary male marketc yearsdg m_years associate full male_ass male_full

      Source |       SS       df       MS              Number of obs =     514
-------------+------------------------------           F(  8,   505) =  181.33
       Model |  6.1113e+10     8  7.6391e+09           Prob > F      =  0.0000
    Residual |  2.1274e+10   505  42127285.4           R-squared     =  0.7418
-------------+------------------------------           Adj R-squared =  0.7377
       Total |  8.2387e+10   513   160599133           Root MSE      =  6490.6

------------------------------------------------------------------------------
      salary |      Coef.   Std. Err.      t    P>|t|     [95% Conf. Interval]
-------------+----------------------------------------------------------------
        male |  -758.7161   1297.934    -0.58   0.559    -3308.732      1791.3
     marketc |   36898.71   1980.616    18.63   0.000     33007.45    40789.97
     yearsdg |   370.3911   111.2443     3.33   0.001     151.8325    588.9498
     m_years |   231.2697    125.128     1.85   0.065    -14.56586    477.1052
   associate |   4304.504   1480.349     2.91   0.004     1396.104    7212.905
        full |   11942.88   2436.995     4.90   0.000     7154.983    16730.78
    male_ass |  -1467.288    1834.53    -0.80   0.424    -5071.539    2136.964
   male_full |  -1211.408   2790.304    -0.43   0.664    -6693.441    4270.625
       _cons |   37371.46   1029.014    36.32   0.000     35349.79    39393.14
------------------------------------------------------------------------------
```

The estimated regression coefficients with standard errors are given in the first two columns under "Section 1.10" in table 1.2. From (1.7), we see that when time since degree **yearsdg** is zero, the estimated difference in population mean salaries between men and women reduces to $\widehat{\beta}_2 + \widehat{\beta}_8$**associate** + $\widehat{\beta}_9$**full**. In other words, the estimated coefficient $\widehat{\beta}_9$ of **male_ass** can be interpreted as the difference in estimated gender gap between associate and assistant professors, and similarly for **male_full**. Neither interaction coefficient is significant at the 5% level. However, it is not considered good practice to include only some of the interaction terms for a group of dummy variables representing one categorical variable, and we hence should test both coefficients simultaneously:

```
. testparm male_ass male_full
 ( 1)  male_ass = 0
 ( 2)  male_full = 0

       F(  2,   505) =    0.32
            Prob > F =    0.7244
```

There is little evidence for an interaction between gender and academic rank ($F(2, 505) = 0.32$, $p = 0.72$).

1.10.2 Interaction between continuous covariates

The effect of marketability, **marketc** (x_{3i}), could increase or decrease with time since degree, **yearsdg** (x_{4i}). We can include an interaction between these two continuous covariates in a regression model (with the usual assumptions)

$$\begin{aligned} y_i &= \beta_1 + \beta_2 x_{2i} + \beta_3 \texttt{marketc}_i + \beta_4 \texttt{yearsdg}_i + \cdots + \beta_8 \texttt{marketc}_i \times \texttt{yearsdg}_i + \epsilon_i \\ &= \beta_1 + \beta_2 x_{2i} + (\beta_3 + \beta_8 \texttt{yearsdg}_i)\texttt{marketc}_i + \beta_4 \texttt{yearsdg}_i + \cdots + \epsilon_i && (1.8) \\ &= \beta_1 + \beta_2 x_{2i} + \beta_3 \texttt{marketc}_i + (\beta_4 + \beta_8 \texttt{marketc}_i)\texttt{yearsdg}_i + \cdots + \epsilon_i && (1.9) \end{aligned}$$

and fit this model in Stata using the commands

```
. generate market_yrs = marketc*yearsdg
. regress salary male marketc yearsdg m_years associate full market_yrs
      Source |       SS       df       MS              Number of obs =     514
-------------+------------------------------           F(  7,   506) =  216.20
       Model |  6.1744e+10     7  8.8205e+09           Prob > F      =  0.0000
    Residual |  2.0644e+10   506  40797696.2           R-squared     =  0.7494
-------------+------------------------------           Adj R-squared =  0.7460
       Total |  8.2387e+10   513   160599133           Root MSE      =  6387.3

------------------------------------------------------------------------------
      salary |      Coef.   Std. Err.      t    P>|t|     [95% Conf. Interval]
-------------+----------------------------------------------------------------
        male |  -1969.756    1219.33    -1.62   0.107    -4365.329    425.8174
     marketc |   48556.75   3476.913    13.97   0.000     41725.79    55387.72
     yearsdg |   334.3462   87.26943     3.83   0.000     162.8911    505.8012
     m_years |   248.8044   85.34796     2.92   0.004     81.12438    416.4844
   associate |   3670.866    862.631     4.26   0.000     1976.087    5365.646
        full |   11691.02   1158.111    10.09   0.000     9415.717    13966.31
  market_yrs |  -869.0199   216.4177    -4.02   0.000    -1294.208   -443.8319
       _cons |   38039.86   983.7037    38.67   0.000     36107.22    39972.51
------------------------------------------------------------------------------
```

The estimated regression coefficients with standard errors are given in the last two columns under "Section 1.10" in table 1.2. Using (1.8), we see that the estimated effect of marketability, $\widehat{\beta}_3 + \widehat{\beta}_4$**yearsdg**, decreases from \$48,557 for faculty who have just completed their degree to \$48,557 − \$869.02 × 30 = \$22,486 for faculty who completed their degree 30 years ago.

1.11 Nonlinear effects

We have assumed that the relationship between population mean salary and each of the continuous covariates `marketc` and `yearsdg` is linear after controlling for the other variables. However, the difference in population mean salary for each extra year is likely to increase with time since degree (for instance, if percentage increases are constant).

Such a nonlinear relationship can be modeled by including the square of `yearsdg` in the model in addition to `yearsdg` itself:

```
. generate yearsdg2 = yearsdg^2

. regress salary male marketc yearsdg m_years associate full
>  market_yrs yearsdg2

      Source |       SS       df       MS              Number of obs =     514
-------------+------------------------------           F(  8,   505) =  192.04
       Model |  6.2005e+10     8  7.7507e+09           Prob > F      =  0.0000
    Residual |  2.0382e+10   505    40360475           R-squared     =  0.7526
-------------+------------------------------           Adj R-squared =  0.7487
       Total |  8.2387e+10   513   160599133           Root MSE      =    6353

------------------------------------------------------------------------------
      salary |      Coef.   Std. Err.      t    P>|t|     [95% Conf. Interval]
-------------+----------------------------------------------------------------
        male |  -1181.825   1251.647    -0.94   0.346    -3640.901    1277.251
     marketc |    46578.2   3544.482    13.14   0.000     39614.45    53541.95
     yearsdg |   39.04014   144.8758     0.27   0.788    -245.5933    323.6736
     m_years |   177.7129   89.36428     1.99   0.047     2.141361    353.2845
   associate |   4811.533   967.9374     4.97   0.000     2909.853    6713.213
        full |      12791   1230.256    10.40   0.000     10373.95    15208.05
  market_yrs |  -726.3863   222.4265    -3.27   0.001    -1163.382   -289.3911
    yearsdg2 |    10.1092   3.970824     2.55   0.011     2.307829    17.91057
       _cons |   38837.69   1027.381    37.80   0.000     36819.23    40856.16
------------------------------------------------------------------------------
```

The estimated regression coefficients with standard errors are given under "Section 1.11" in table 1.2. The estimated coefficient of `yearsdg2` is significantly different from 0 at, say, the 5% level ($t = 2.55$, df $= 505$, $p = 0.01$), whereas the coefficient of `yearsdg` is no longer statistically significant. It should nevertheless be retained to form a flexible quadratic curve since the maximum or minimum of the curve is otherwise arbitrarily forced to occur when `yearsdg`$= 0$.

Setting the covariates `marketc` (and hence `market_yrs`), `associate`, and `full` to zero, we can visualize the relationship between salary and time since degree for male and female assistant professors in disciplines with mean marketability using the `twoway function` command

```
. twoway (function Women = _b[_cons] + _b[yearsdg]*x + _b[yearsdg2]*x^2,
>          range(0 41) lpatt(dash))
>        (function Men = _b[_cons] + _b[male] + (_b[yearsdg] + _b[m_years])*x
>          + _b[yearsdg2]*x^2, range(0 41) lpatt(solid)),
>  xtitle(Time since degree (years)) ytitle(Mean salary)
```

producing figure 1.12.

(Continued on next page)

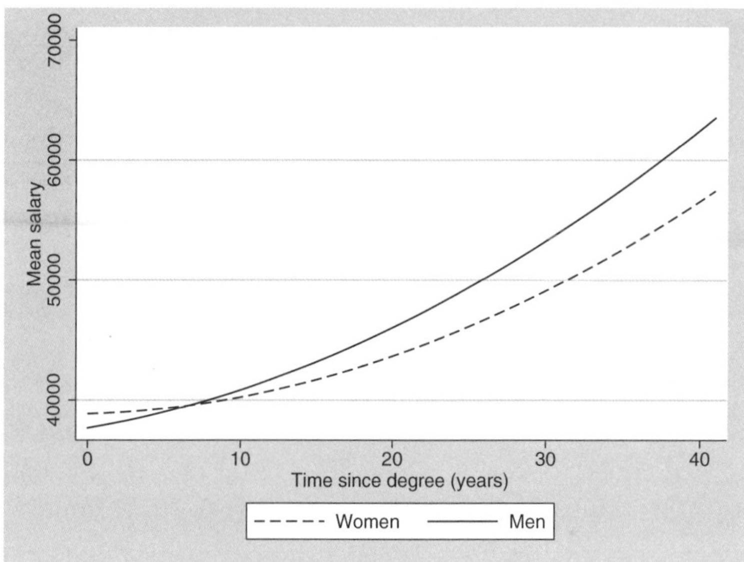

Figure 1.12: Estimated effects of gender and time since degree on mean salary for assistant professors in disciplines with mean marketability

A more flexible curve can be produced by using *higher-order polynomials*, also including `yearsdg` cubed (`yearsdg`^{3}) and possibly higher powers.

Finally, we note that the effect of gender disappears after adding `yearsrank`, the number of years spent at the current academic rank, to the final regression model presented here.

1.12 Residual diagnostics

Observed or estimated residuals are defined as the differences between the observed and predicted responses

$$\widehat{\epsilon}_i = y_i - (\widehat{\beta}_1 + \widehat{\beta}_2 x_{2i} + \cdots + \widehat{\beta}_p x_{pi})$$

and estimated standardized residuals are obtained as

$$r_i = \frac{\widehat{\epsilon}_i}{\sqrt{\widehat{\sigma^2}}}$$

Estimated residuals or standardized residuals can be used to investigate if model assumptions such as homoskedasticity and normally distributed errors are violated. Estimated standardized residuals have the advantage that they have an approximate standard normal distribution if the model assumptions are true. For instance, a value greater than 3 should occur only about 0.1% of the time and may therefore be an outlier.

The postestimation command `predict` with the `residual` option provides estimated residuals for the last regression model that was fitted:

```
. predict res, residual
```

A histogram of the estimated residuals with an overlayed normal distribution is produced by the `histogram` command with the `normal` option

```
. histogram res, normal
```

and presented in figure 1.13.

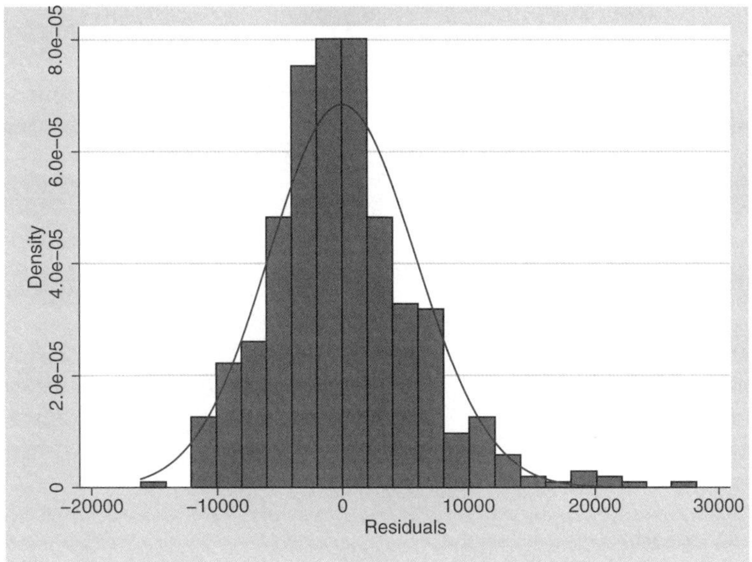

Figure 1.13: Estimated residuals with overlayed normal distribution

The distribution is somewhat skewed, suggesting that salary should perhaps be log-transformed as discussed in section 1.3. Again it may be advisable to use `vce(robust)`, which provides standard errors that are not just robust to heteroskedasticity but also to other violations of the distributional assumptions.

Standardized residuals can be obtained by using the `rstandard` option of `predict`.

1.13 Summary and further reading

In this chapter, we have shown how linear regression can be used to model the relationship between a continuous response variable and explanatory variables of different types, including continuous, dichotomous, and ordered or unordered polytomous (multicategory) variables. Special cases of such regression models are one-way ANOVA and the model underlying an independent-samples t test.

We have also shown how dummy variables can be used to represent categories of categorical explanatory variables and how products of variables can be used to model interactions, where the effect of each variable is moderated by the other variable. Finally, we have shown how nonlinear relationships between the response and a continuous covariate can be modeled using polynomials.

A good elementary introduction to linear models is provided by Agresti and Finlay (2007). More advanced, but accessible, introductions that also include regression for other response types such as logistic regression include DeMaris (2004) for social science, Vittinghoff et al. (2006) for medicine, and Cameron and Trivedi (2005) for economics.

Brief introductions to logistic regression for dichotomous and ordinal responses, discrete-time hazard models for survival or duration data, and Poisson regression for counts, are given in the beginning of chapters 6–9 of the book, before we embark on multilevel versions of these models.

1.14 Exercises

1.1 High-school-and-beyond data

The data considered here are from the High School and Beyond Survey. They are discussed and analyzed in Raudenbush et al. (2004) and accompany the HLM program (Raudenbush et al. 2004).

The variables in the dataset `hsb.dta` that we will use are

- `schoolid`: school identifier
- `mathach`: a measure of mathematics achievement
- `ses`: socioeconomic status based on parental education, occupation, and income
- `minority`: dummy variable for student being nonwhite

1. Keep only data on the five schools with the lowest values of `schoolid` (1224, 1288, 1296, 1308, and 1317). Also drop the variables not listed above.
2. Obtain the means and standard deviations for the continuous variables and frequency tables for the categorical variables. Also obtain the mean and standard deviation of the continuous variables for each of the five schools (using the `table` or `tabstat` command).
3. Produce a histogram and a boxplot of `mathach`.
4. Produce a scatterplot of `mathach` versus `ses`. Also produce a scatterplot for each school (using the `by()` option).
5. Treating `mathach` as the response variable y_i and `ses` as an explanatory variable x_i, consider the linear regression model of y_i on x_i:

$$y_i = \beta_1 + \beta_2 x_i + \epsilon_i, \quad \epsilon_i | x_i \sim N(0, \sigma^2)$$

a. Fit the model.
b. Report and interpret the estimates of the three parameters of this model.
c. Interpret the confidence interval and p-value associated with β_2.

6. Using the `predict` command, create a new variable `yhat` equal to the predicted values $\widehat{y}_i$ of `mathach`:

$$\widehat{y}_i = \widehat{\beta}_1 + \widehat{\beta}_2 x_i$$

7. Produce a scatterplot of `mathach` versus `ses` with the regression line (`yhat` versus `ses`) superimposed. Produce the same scatterplot by school. Does it appear as if schools differ in their mean math achievement after controlling for `ses`?
8. Extend the regression model from step 5 by including dummy variables for four of the five schools.
   a. Fit the model.
   b. Describe what the coefficients of the school dummies represent.
   c. Test the null hypothesis that the true coefficients of all four dummy variables are zero (use `testparm`).

See also exercises 3.8.

1.2 Anorexia data

Hand et al. (1994) provide data on young girls receiving three different treatments for anorexia: cognitive behavioral therapy, family therapy, and control (treatment as usual). The response variable is the girls' weight in kilograms after treatment, and we also have their weights before treatment.

The variables in `anorexia.dta` are

- `treat`: treatment group (1: cognitive behavioral therapy; 2: control; 3: family therapy)
- `weight1`: weight before treatment (in kilograms)
- `weight2`: weight after treatment (in kilograms)

1. Produce a table of the means and standard deviations of `weight2`, as well as sample sizes by treatment group.
2. Plot boxplots and histograms of `weight2` by group as shown in section 1.3.
3. Create dummy variables `cbt` and `ft` for cognitive behavioral therapy and family therapy, respectively.
4. Fit a regression model with `weight2` as the response variable and `weight1`, `cbt`, and `ft` as covariates.
5. Interpret the estimated regression coefficients.
6. For which of the three pairs of treatment groups is there any evidence, at the 5% level, that one treatment is better than the other?

7. Fit the model again, this time relaxing the homoskedasticity assumption. Does this alter your answer for step 6?
8. Plot a histogram of the estimated residuals for this model with a normal density curve superimposed.

1.3 Smoking and birthweight data

Here we consider a subset of data on birth outcomes provided by Abrevaya (2006), which are analyzed in chapter 3. The data were derived from birth certificates by the U.S. National Center for Health Statistics.

The variables in `smoking.dta` that we will consider here are

- `momid`: mother identifier
- `idx`: chronological numbering of multiple children to the same mother in the database (1: first child; 2: second child; 3: third child)
- `birwt`: birthweight (in grams)
- `mage`: mother's age at the birth of the child (in years)
- `smoke`: dummy variable for mother smoking during pregnancy (1: smoking; 0: not smoking)
- `male`: dummy variable for baby being male (1: male; 0: female)
- `hsgrad`: dummy variable for mother having graduated from high school
- `somecoll`: dummy variable for mother having some college education (but no degree)
- `collgrad`: dummy variable for mother having graduated from college
- `black`: dummy variable for mother being black (1: black; 0: white)

1. Keep only the data on each mother's first birth in the dataset, i.e., where `idx` is 1.
2. Create a variable `education` taking the value 1 if `hsgrad` is 1, 2 if `somecoll` is 1, 3 if `collgrad` is 1, and 0 otherwise.
3. Produce a table of the means and standard deviations of `birwt` for all the subgroups defined by `smoke`, `education`, `male`, and `black`. Hint: use the `table` command with `smoke` as *rowvar*, `education` as *colvar*, and `male` and `black` as *superrowvars*; see `help table`.
4. Produce boxplots for the same groups. Hint: use the `over()` option for each grouping variable except the last (starting with `over(education)`) and use the `by()` option for the last grouping variable. Use the `nooutsides` option to suppress the display of outliers making the graph easier to interpret. What do you observe?
5. Regress `birwt` on `smoke` and interpret the estimated regression coefficients.
6. Add `mage`, `male`, `black`, `hsgrad`, `somecoll`, and `collgrad` to the model in step 5.
7. Interpret each of the estimated regression coefficients from step 6.

8. Discuss the difference in the estimated coefficient of `smoke` from steps 5 and 6.
9. Use the `adjust` command to produce a table of estimated population means for girls born to white mothers of average age by smoking status and education. Hint: use the variable `education` in the `by()` option although this variable is not used in the model.
10. Extend the model from step 6 to investigate whether the adjusted difference in mean birthweight between boys and girls differs between black and white mothers. Is there any evidence, at the 5% level, that it does?

1.4 Interaction I

The table below gives estimates for a multiple regression model fitted to data from the 2000 Program for International Student Assessment (PISA). Specifically, the reading score for students from three of the countries (the United States, the United Kingdom, and Germany) was regressed on a dummy variable `test_lan` for the test language (English or German depending on the country) being spoken at home, dummy variables `usa` and `uk` for the United States and the United Kingdom, respectively, and interactions `lan_usa` = `test_lan` × `usa` and `lan_uk` = `test_lan` × `uk`.

	Est	(SE)
β_1	423.40	11.06
β_2 [test_lan]	98.53	11.31
β_3 [usa]	42.37	14.84
β_4 [uk]	114.79	17.89
β_5 [lan_usa]	−35.66	15.33
β_6 [lan_uk]	−98.54	18.17

1. Interpret each estimated coefficient.
2. Plot the estimated relationship between mean reading score and `test_lan` for each of the countries. You may do this using the `twoway function` command in Stata.

1.5 Interaction II

The following estimates were obtained by regressing a continuous measure y_i of fear of crime on the variables $\texttt{age}_i$, $\texttt{fem}_i$ and their interaction (using data from the British Crime Survey, 2001–2002):

$$\widehat{y}_i = -0.19 + 0.66\texttt{fem}_i - 0.02\texttt{age}_i - 0.07\texttt{fem}_i \times \texttt{age}_i$$

Here $\texttt{fem}_i$ is a dummy variable for being female and $\texttt{age}_i$ is the number of 10-year intervals since age 16, i.e., $(\texttt{age} - 16)/10$.

1. What is the predicted fear of crime for males and females at age 16?
2. What is the predicted fear of crime for males and females at age 80?
3. Interpret the estimated coefficient of $\texttt{fem}_i \times \texttt{age}_i$.
4. Plot the predicted relationship between fear of crime and age (for the age-range 16–90 years) using two separate lines for the two genders. You may use Stata's `twoway function` command.
5. Adding $\texttt{age}_i^2$ to the model gave these estimates:

$$\widehat{y}_i = -0.24 + 0.66\texttt{fem}_i + 0.02\texttt{age}_i - 0.006\texttt{age}_i^2 - 0.06\texttt{fem}_i \times \texttt{age}_i$$

   Use Stata to plot the predicted relationship between fear of crime and age (for the age-range 16–90 years) using two separate curves for the two genders.

1.6 Self-assessment exercise

1. In the simple linear regression model shown below, salary (y_i) is regressed on years of experience (x_i):

$$y_i = \beta_1 + \beta_2 x_i + \epsilon_i, \quad \epsilon_i | x_i \sim N(0, \sigma^2)$$

   a. What are the usual terms used to describe β_1, β_2, ϵ_i, and σ^2?
   b. State the assumptions of this model in words.
   c. If the variables are called `salary` (y_i) and `yearsexp` (x_i) in Stata, write down the command for fitting the model.

2. In a simple linear regression model for salary, the coefficient of a dummy variable for being male is estimated as $3,623.

   a. Interpret the estimated coefficient.
   b. If a dummy variable for being female had been used instead of a dummy variable for being male, what would have been the value of the estimated regression coefficient?
   c. The estimated standard error for the estimated coefficient of the male dummy variable was $1,912. Can you reject the null hypothesis that the population mean salary is the same for men and women using a two-sided test at the 5% level?

3. What is the relationship between one-way ANOVA and multiple linear regression. Is one model a special case of the other? Explain.
4. In the output for a multiple linear regression model, an F test is given with $F(4, 268) = 12.63$. State the null and alternative hypothesis being tested.
5. In a regression of salary (in U.S. dollars) on age (in years), the intercept is estimated as −$2,000. Explain how this is possible given that salary must be positive.
6. Using the United States sample of the 2000 Program for International Student Assessment (PISA), the difference in population mean (English) reading score between those who do and do not speak English at home is estimated as 63

with a standard error of 10. When controlling for socioeconomic status, the adjusted difference in population mean reading score is estimated as 49 with a standard error of 10.

a. What does it mean to "control" or "adjust for" socioeconomic status?
b. Under what circumstances does controlling for a variable x_1 alter the estimated regression coefficient of another variable x_2?
c. In the context of this example, explain why the adjusted estimate differs from the unadjusted estimate, paying attention to the direction of the difference.

7. The salaries of a company's employees were regressed on a dummy variable for being male, `male`, and years of experience, `yearsexp`, and their interaction, giving the following results:

$$\widehat{y}_i = \$30{,}000 + \$2{,}000\ \texttt{male}_i + \$600\ \texttt{yearsexp}_i - \$100\ \texttt{male}_i \times \texttt{yearsexp}_i$$

a. Interpret each estimated coefficient.
b. What is the estimated difference in population mean salary between men and women who have 10 years of experience?

8. Regression output for data from the 2000 PISA study is given below. The sample analyzed here included children from the United States, the United Kingdom, and Germany. `wleread` is the reading score, `usa` is a dummy variable for the child being from the United States, `uk` is a dummy variable for the child being from the United Kingdom, and `female` is a dummy variable for the child being female.

```
. regress wleread usa uk female

      Source |       SS       df       MS              Number of obs =    4528
-------------+------------------------------           F(  3,  4524) =   44.21
       Model |  1158538.51     3  386179.502           Prob > F      =  0.0000
    Residual |  39518824.7  4524  8735.37239           R-squared     =  0.0285
-------------+------------------------------           Adj R-squared =  0.0278
       Total |  40677363.2  4527  8985.50104           Root MSE      =  93.463

------------------------------------------------------------------------------
     wleread |      Coef.   Std. Err.      t    P>|t|     [95% Conf. Interval]
-------------+----------------------------------------------------------------
         usa |    4.00653   3.715402     1.08   0.281    -3.277473    11.29053
          uk |   20.30115   3.159347     6.43   0.000     14.10728    26.49501
      female |   26.13154   2.781327     9.40   0.000     20.67878     31.5843
       _cons |   504.7174   2.672635   188.85   0.000     499.4778    509.9571
------------------------------------------------------------------------------
```

a. Write down the linear regression model being fitted.
b. Interpret the estimated coefficient of `uk`.
c. Estimate the difference in population mean reading scores between the United States and the United Kingdom.
d. What percentage of the variability in reading scores is explained by nationality and gender?
e. What is the magnitude of the estimated residual variance?

f. Write down the necessary Stata commands for investigating whether the difference in population mean reading scores between girls and boys differs between countries.

Solutions for self-assessment exercise

1. a. β_1 is the intercept, β_2 is the slope or coefficient of x_i, ϵ_i is the residual or error term for subject i and σ^2 is the residual variance.

   b. For given x_i, the expectation of y_i is linearly related to x_i, the residuals ϵ_i are normally distributed with zero mean and variance σ^2 and independent for different units i. The residual variance is constant for all values of x_i, an assumption called homoskedasticity.

   c. `regress salary yearsexp`

2. a. The estimated difference in population mean salaries between men and women is $3,623.

   b. −$3,623

   c. No, the t statistic is $t = 3,623/1,912 = 1.89$. The smallest t statistics that yields a two-sided p-value of $p \leq 0.05$ is 1.96, when the degrees of freedom are very large (larger t statistics are required for smaller degrees of freedom), so the test is not significant at the 5% level.

3. One-way ANOVA is a special case of multiple linear regression where the only explanatory variables are the dummy variables for each category of a categorical variable, except the reference category.

4. The null hypothesis is that all true regression coefficients, except the intercept, are zero. The alternative hypothesis is that at least one of the true regression coefficients (except the intercept) is nonzero.

5. The intercept is the estimated population mean salary, or predicted salary, at age 0. This is an extrapolation well outside the range of the data. Even if it were possible to have a salary at age 0, there would be no reason to believe that the relationship between salary and age is linear beginning with age 0.

6. a. Controlling for socioeconomic status means attempting to estimate the difference in mean reading score for two (sub)populations (native and nonnative speakers) having the same socioeconomic status.

   b. When x_1 is associated with both x_2 and y.

   c. Native speakers may have higher mean socioeconomic status, and socioeconomic status may be positively correlated with reading scores, so some of the apparent advantage of native speakers is actually due to their higher mean socioeconomic status.

7. a. The estimated population mean salary is $30,000 for females with no experience and $2,000 greater for males with no experience. For each extra year of experience, the estimated population mean salary increases $600 for females and $100 less (i.e., $500) for males.

   b. The difference in population means for males and females after 10 years is estimated as $2,000 − $100 × 10 = $1,000.

8. a. $y_i = \beta_1 + \beta_2 \texttt{usa}_i + \beta_3 \texttt{uk}_i + \beta_4 \texttt{female}_i + \epsilon_i$ where, for given covariates, ϵ_i has a normal distribution with mean 0 and constant variance σ^2, and is independent of $\epsilon_{i'}$ for another student i'.

   b. The estimated coefficient of `uk` represents the estimated difference in population mean reading scores between children in the United Kingdom and children in Germany.

   c. The estimated difference in population mean reading scores between the United States and the United Kingdom is the difference between the estimated coefficients of `usa` and `uk`, giving $4.01 - 20.30 = -16.29$.

d. 2.85%

e. $\widehat{\sigma^2} = 8735.37$ (under MS for Residual).

f.
```
generate fem_usa = female*usa
generate fem_uk = female*uk
regress wleread usa uk female fem_usa fem_uk
testparm fem_usa fem_uk
```

Part II

Two-level linear models

2 Variance-components models

2.1 Introduction

Units of observation often fall into groups or clusters. For example, individuals could be nested in families, hospitals, schools, neighborhoods, or firms. Longitudinal data also consist of clusters of observations made at different occasions for the same subject. For two examples of clustered data, the nesting structure is depicted in figure 2.1.

In clustered data, it is usually important to allow for dependence or correlations among the responses observed for units belonging to the same cluster. For example, the adult heights of siblings are likely to be correlated because siblings are genetically related to each other and have usually been raised within the same family. Variance-components models are designed to model and estimate such within-cluster correlations.

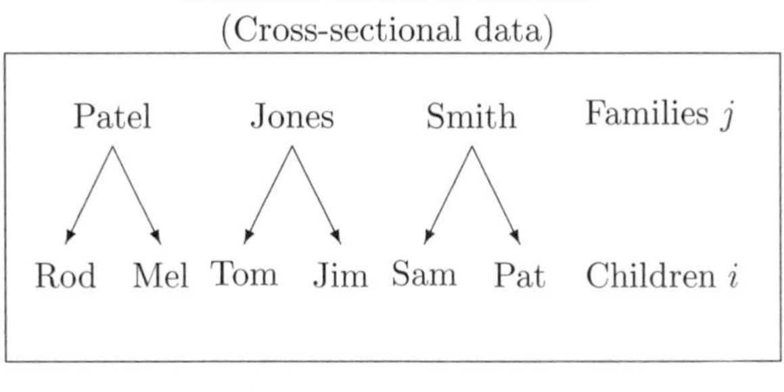

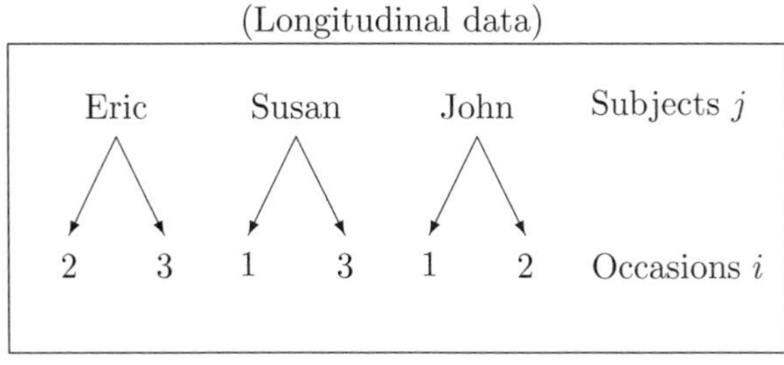

Figure 2.1: Examples of clustered data

In this chapter, we consider the simple situation of clustered data without explanatory variables. This situation is important in its own right and is also useful for introducing and motivating the notions of random effects and variance components. We also describe basic principles of estimation and prediction in this simple setting. However, this means that some parts of the chapter may be a bit demanding, and you may like to skip sections 2.7 and 2.9 on first reading.

We also introduce the Stata commands `xtreg`, `xtmixed`, and `gllamm`, which will be used in subsequent chapters.

2.2 How reliable are peak-expiratory-flow measurements?

The data come from a reliability study conducted by Martin Bland using 17 of his family and colleagues as subjects. The purpose was to illustrate a way of assessing the quality of two instruments for measuring people's peak-expiratory-flow rate (PEFR). The PEFR, which is roughly speaking how strongly subjects can breathe out, is a central clinical measure in respiratory medicine.

The subjects had their PEFR measured twice (in liters per minute) using the standard Wright peak flow meter and twice using the new Mini Wright peak flow meter. The methods were used in random order to avoid confounding practice (prior experience) effects with method effects. If the new method agrees sufficiently well with the old, the old may be replaced with the more convenient Mini meter. Somewhat remarkably, the paper reporting this study (Bland and Altman 1986) is the most cited paper in the *Lancet*, one of the most prestigious medical journals.

In this chapter, we analyze the two sets of measurements using the Mini Wright peak flow meter. Analyses comparing the standard Wright and Mini Wright peak flow meters are discussed in chapter 10.

The data are presented in table 2.1 and are in `pefr.dta` in the same form as in the table, with the following variable names:

- `id`: subject identifier
- `wp1`: Wright peak flow meter, occasion 1
- `wp2`: Wright peak flow meter, occasion 2
- `wm1`: Mini Wright flow meter, occasion 1
- `wm2`: Mini Wright flow meter, occasion 2

Table 2.1: Peak-expiratory-flow rate measured on two occasions using both the Wright and the Mini Wright peak flow meters

	Wright peak flow meter		Mini Wright peak flow meter	
Subject	First	Second	First	Second
1	494	490	512	525
2	395	397	430	415
3	516	512	520	508
4	434	401	428	444
5	476	470	500	500
6	557	611	600	625
7	413	415	364	460
8	442	431	380	390
9	650	638	658	642
10	433	429	445	432
11	417	420	432	420
12	656	633	626	605
13	267	275	260	227
14	478	492	477	467
15	178	165	259	268
16	423	372	350	370
17	427	421	451	443

First, we load the data into Stata using the command

```
. use http://www.stata-press.com/data/mlmus2/pefr
```

The first and second recordings on the Mini Wright peak flow meter can be plotted against the subject identifier with a horizontal line representing the overall mean by using

```
. generate mean_wm = (wm1+wm2)/2

. summarize mean_wm

    Variable |       Obs        Mean    Std. Dev.       Min        Max
-------------+--------------------------------------------------------
     mean_wm |        17    453.9118    111.2912      243.5        650

. twoway (scatter wm1 id, msymbol(circle))
>        (scatter wm2 id, msymbol(circle_hollow)),
> xtitle(Subject id)  xlabel(1/17) ytitle(Mini Wright measurements)
> legend( order(1 "Occasion 1" 2 "Occasion 2")) yline(453.9118)
```

The resulting graph is shown in figure 2.2.

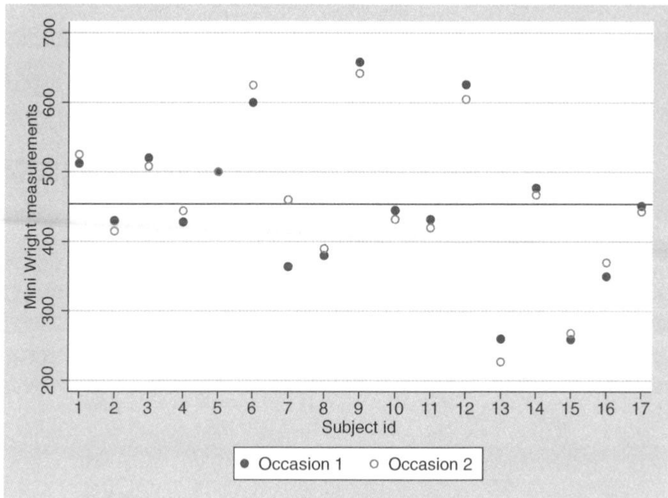

Figure 2.2: First and second recordings of peak expiratory flow using Mini Wright meter versus subject number (the horizontal line represents the overall mean)

It is clear from the figure that repeated measurements on the same subject tend to be closer to each other than to the measurements on a different subject. Indeed, if this were not the case, the Mini Wright peak flow meter would be useless as a tool for discriminating between the individuals in this particular sample. Because there are large differences between subjects (for example, compare subjects 9 and 15) and only small differences within subjects, the responses for occasions 1 and 2 on the same subject tend to lie on the same side of the overall mean, shown as a horizontal line in the figure, and are therefore correlated (see also section 2.3.3). We can also see that there is within-subject dependence by considering prediction of a subject's response at occasion 2 if all we know are the subjects' responses at occasion 1. If the response for a given subject at occasion 2 were independent of his or her response at occasion 1, a good prediction would be the mean response at occasion 1 across all subjects. However, it is clear that a much better prediction here is the subject's own response at occasion 1 because the responses are dependent within subjects. We see that the within-subject dependence is due to between-subject heterogeneity. If all subjects were more or less alike (for example, pick subjects 2, 4, 10, 11, 14, and 17), there would be much less within-subject dependence.

2.3 The variance-components model

2.3.1 Model specification and path diagram

It may be tempting to model the responses y_{ij} of subject j on occasion i using a standard regression model without covariates

$$y_{ij} = \beta + \xi_{ij} \tag{2.1}$$

where ξ_{ij} are residuals or error terms that are independent over both subjects and occasions (the Greek letter ξ is pronounced xi). However, this specification is unreasonable since, as we have seen in figure 2.2, measurements are expected to be more similar within than between subjects or in other words be dependent within subjects.

We can model this dependence by splitting the residual ξ_{ij} into two components: a component ζ_j (ζ is pronounced zeta), which is specific to each subject j and constant across occasions i, and a component ϵ_{ij}, which is specific to each subject j at each occasion i

$$y_{ij} = \beta + \zeta_j + \epsilon_{ij} \tag{2.2}$$

as shown for a single subject in figure 2.3. Here ζ_j is the random deviation of subject j's mean measurement (over a hypothetical population of measurement occasions) from the overall mean β. The component ζ_j, often called a random effect of subject or a *random intercept*, has zero population mean and variance ψ (pronounced psi) over subjects and is assumed to be independent over subjects. The component ϵ_{ij}, often called the residual or within-subject residual, is the random deviation of y_{ij} from subject j's mean. This residual has zero population mean and variance θ (pronounced theta) over occasions and subjects, and is assumed to be independent over both subjects and occasions. We can interpret ψ is the between-subject variance and θ as the within-subject variance.

In classical psychometric test theory, (2.2) represents a measurement model where $\beta + \zeta_j$ is the *true score* for subject j, defined as the long-term mean measurement (e.g., Streiner and Norman 2003).

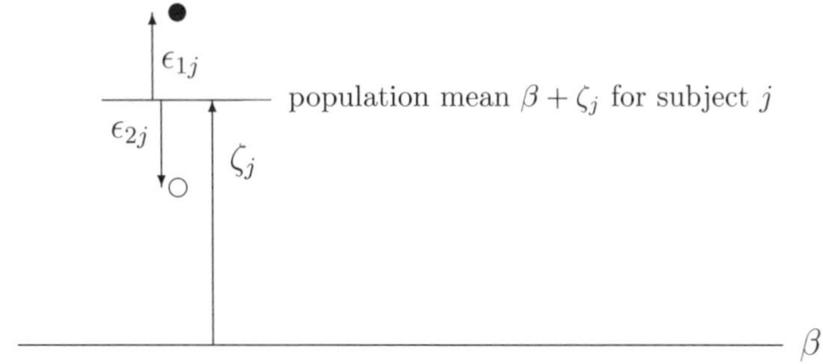

Figure 2.3: Illustration of random-intercept model for a subject j

We assume that the random intercepts ζ_j are normally distributed,

$$\zeta_j \sim N(0, \psi)$$

and that the ϵ_{ij} are normally distributed

$$\epsilon_{ij} \sim N(0, \theta)$$

These distributional assumptions are illustrated for a subject j in figure 2.4. The random intercept ζ_j has a normal distribution with mean zero and variance ψ (see the top distribution in the figure). Drawing a realization from this distribution for subject j determines the mean $\beta+\zeta_j$ of the distributions from which responses y_{ij} for this subject are subsequently drawn. At a given measurement occasion i, a response y_{ij} is therefore sampled from a normal distribution with mean $\beta + \zeta_j$ and variance θ (see the bottom distribution in the figure). Equivalently, a residual (or measurement error) ϵ_{ij} is drawn from a normal distribution with mean zero and variance θ.

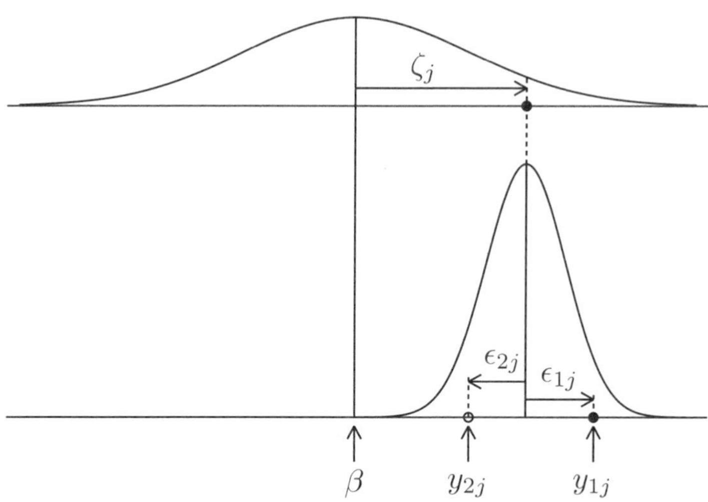

Figure 2.4: Illustration of distributions of error components for a subject j

This hierarchical model is a simple example of a two-level model, where occasions are level-1 units and subjects are level-2 clusters. The random intercept ζ_j is then referred to as the level-2 residual with level-2 (between-subject) variance ψ and ϵ_{ij} as the level-1 residual with level-1 (between-occasion, within-subject) variance θ.

The model is often motivated in terms of two-stage survey sampling where the randomness (between hypothetical repeated samples) is due to two-stage random sampling of clusters with fixed ζ_j from the population of clusters and then units with fixed ϵ_{ij} from the population of units within the clusters. This way of thinking is not useful when the term "population" is interpreted too literally as a *finite* population, which would leave no randomness at level 2 when the clusters are for instance all U.S. states and no randomness at level 1 when the clusters are people's heads and the units are both eyes on each head. In this book we assume that readers want to make inferences regarding data-generating mechanisms or wider populations.

We can display the random part of the model (every term except β) using a path diagram as shown in figure 2.5. Here the rectangles represent the observed responses y_{1j} and y_{2j} for each subject j, where the j subscript is implied by the label "subject j" inside the frame surrounding the diagram. The long arrows from ζ_j to the responses represent regressions with slopes equal to 1, and the short arrows pointing at the responses from below represent the additive level-1 residuals ϵ_{1j} and ϵ_{2j}.

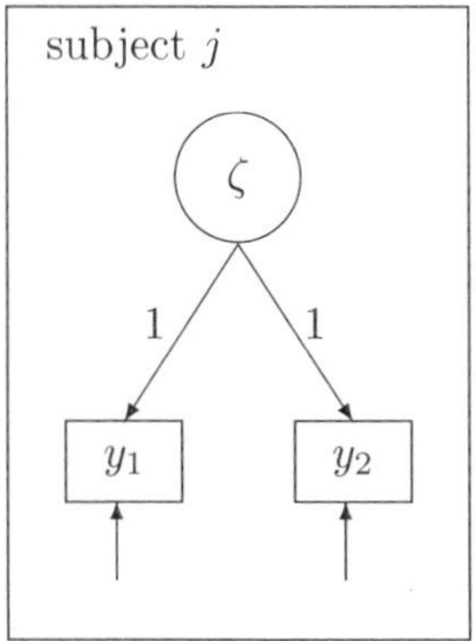

Figure 2.5: Path diagram of random part of random-intercept model

The path diagram makes it clear that the two responses on each subject are *conditionally independent* given ζ_j since there is no arrow directly connecting them. It follows that the responses are conditionally uncorrelated when controlling for ζ_j (i.e., holding it constant across subjects)

$$\text{Cor}(y_{ij}, y_{i'j}|\zeta_j) = 0$$

This can also be seen by imagining that the data in figure 2.2 were generated by the model depicted in figure 2.3, where the dependence is solely due to the measurements being shifted up or down by the shared random intercept ζ_j for each cluster j.

The (marginal) within-subject correlation is induced by ζ_j because this is shared by all responses for the same subject. As we will see in later chapters, path diagrams are useful for conveying the structure of complex models involving several random effects.

2.3.2 Error components, variance components, and reliability

Each response differs from the overall mean β by a total residual or error ξ_{ij}, the sum of two error terms or *error components* ζ_j and ϵ_{ij}

$$\xi_{ij} \equiv \zeta_j + \epsilon_{ij}$$

The random intercept ζ_j is shared between measurement occasions on the same subject j, whereas ϵ_{ij} is unique for each occasion i (and subject).

Since the error components are independent, the total variance is the sum of the *variance components*

$$\text{Var}(y_{ij}) = \text{Var}(\beta + \zeta_j + \epsilon_{ij}) = \underbrace{\text{Var}(\beta)}_{0} + \text{Var}(\underbrace{\zeta_j + \epsilon_{ij}}_{\xi_{ij}}) = \psi + \theta$$

the between-subject and within-subject variances. The proportion of the total variance that is between subjects, or due to subjects, is

$$\rho = \frac{\text{Var}(\zeta_j)}{\text{Var}(y_{ij})} = \frac{\psi}{\psi + \theta} \tag{2.3}$$

In the measurement context, ψ is the variance between subjects' *true scores* $\beta + \zeta_j$, θ is the *measurement error variance* (the squared *standard error of measurement*), and ρ is a *reliability*, here a test–retest reliability (see, for example, Streiner and Norman 2003). The reliability can be thought of as the proportion of the total variance that is "explained" by subjects, analogously to the coefficient of determination R^2 in linear regression discussed in section 1.5.

2.3.3 Intraclass correlation

Consider first the marginal (not conditional on ζ_j) covariance between the measurements on two occasions i and i' for the same subject, defined as

$$\text{Cov}(y_{ij}, y_{i'j}) = E[\{y_{ij} - E(y_{ij})\}\{y_{i'j} - E(y_{i'j})\}]$$

The corresponding marginal correlation is the above covariance divided by the product of the standard deviations

$$\text{Cor}(y_{ij}, y_{i'j}) = \frac{\text{Cov}(y_{ij}, y_{i'j})}{\sqrt{\text{Var}(y_{ij})}\sqrt{\text{Var}(y_{i'j})}} \tag{2.4}$$

It follows from the variance-components model that the population means are constrained to be equal to β and the standard deviations to be equal to $\sqrt{\psi + \theta}$ for responses y_{ij} and $y_{i'j}$ at two occasions i and i'. Under the variance-components model, the marginal (not conditional on ζ_j) covariance between the measurements therefore becomes

$$
\begin{aligned}
\mathrm{Cov}(y_{ij}, y_{i'j}) &= E\{(y_{ij} - \underbrace{\beta}_{E(y_{ij})})(y_{i'j} - \underbrace{\beta}_{E(y_{i'j})})\} = E\{(\zeta_j + \epsilon_{ij})(\zeta_j + \epsilon_{i'j})\} \\
&= E(\zeta_j^2) + \underbrace{E(\zeta_j \epsilon_{i'j})}_{0} + \underbrace{E(\epsilon_{ij}\zeta_j)}_{0} + \underbrace{E(\epsilon_{ij}\epsilon_{i'j})}_{0} = E(\zeta_j^2) = \psi
\end{aligned}
$$

The corresponding correlation, called the *intraclass correlation*, becomes

$$
\mathrm{Cor}(y_{ij}, y_{i'j}) = \frac{\mathrm{Cov}(y_{ij}, y_{i'j})}{\sqrt{\mathrm{Var}(y_{ij})}\sqrt{\mathrm{Var}(y_{i'j})}} = \frac{\psi}{\sqrt{\psi+\theta}\sqrt{\psi+\theta}} = \frac{\psi}{\psi+\theta} = \rho
$$

Thus ρ previously given in (2.3) also represents the within-cluster correlation, which cannot be negative because $\psi \geq 0$. We see that between-cluster heterogeneity and within-cluster correlations are different ways of describing the same phenomenon; both are zero when there is no between-cluster variance $\psi = 0$ and both increase when the between-cluster variance increases relative to the within-cluster variance.

The intraclass correlation is estimated by simply plugging in estimates for the unknown parameters

$$
\widehat{\rho} = \frac{\widehat{\psi}}{\widehat{\psi}+\widehat{\theta}}
$$

Figure 2.6 shows data with an estimated intraclass correlation $\widehat{\rho} = 0.58$ and data with an estimated intraclass correlation $\widehat{\rho} = 0.87$.

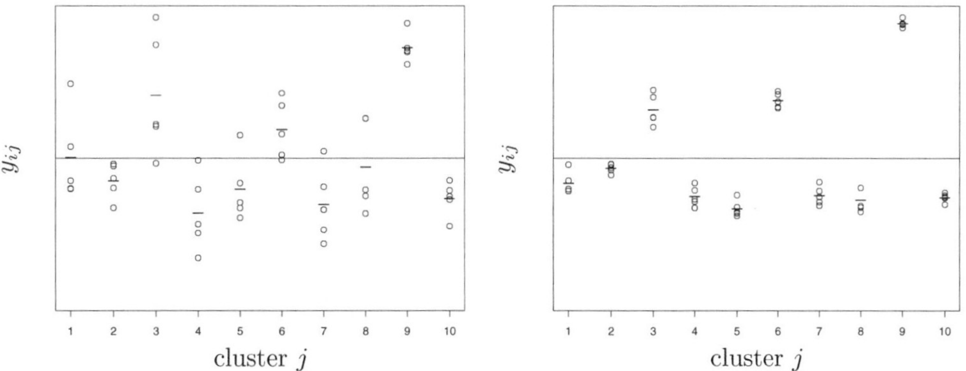

Figure 2.6: Illustration of lower intraclass correlation (left) and higher intraclass correlation (right)

In contrast to the estimated intraclass correlation, the Pearson correlation r is obtained by replacing the expectations in (2.4) by sample means and plugging in *separate* sample means $\overline{y}_{i\cdot}$ and $\overline{y}_{i'\cdot}$ and sample standard deviations s_{y_i} and $s_{y_{i'}}$ for the two occasions,

$$r = \frac{\frac{1}{J-1}\sum_{j=1}^{J}(y_{ij}-\overline{y}_{i\cdot})(y_{i'j}-\overline{y}_{i'\cdot})}{s_{y_i}s_{y_{i'}}}$$

where J is the number of clusters.

To give more insight into the interpretation of the estimated intraclass correlation and Pearson correlation, consider what happens if we alter the second Mini Wright peak flow measurements by adding 100 to them, as shown in figure 2.7. (Such a systematic increase could, for instance, be due to a practice effect.) For the variance-components model it is obvious that the within-cluster variance has increased giving a much smaller intraclass correlation than for the original data (estimated as 0.63 instead of 0.97). In contrast, the Pearson correlation r is 0.97 in both cases (figures 2.2 and 2.7) since it is based on deviations of the first and second measurements from their *respective* means. In contrast, the intraclass correlation is based on deviations from the *overall* or pooled mean.

The Pearson correlation can be thought of as a measure of *relative agreement*, which refers to how well rankings of subjects based on each measure agree and is therefore not affected by linear transformations of the measurements. In contrast, the intraclass correlation is a measure of *absolute agreement*.

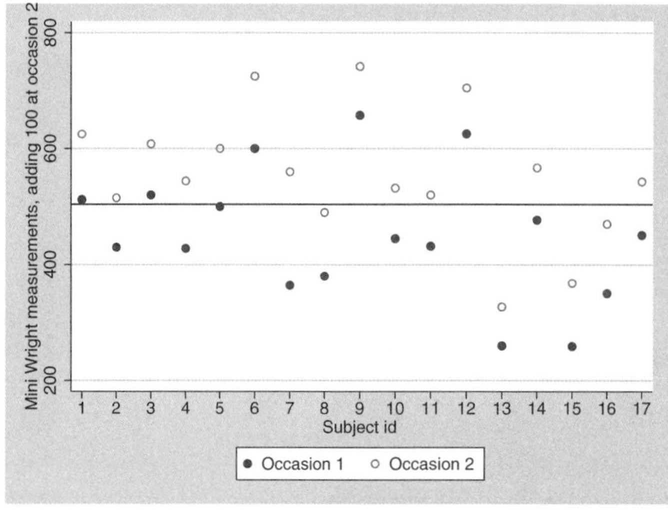

Figure 2.7: First recording of Mini Wright meter and second recording plus 100 versus subject number (the horizontal line represents the overall mean)

The intraclass correlation is useful when the units i are *exchangeable* with identical means and standard deviations. For instance, for twin data, there may not even be such a thing as the first and second twin (presuming that birth order is either irrelevant or unknown), and whereas the Pearson correlation can only be obtained by making an arbitrary assignment to y_{1j} and y_{2j} for each twin-pair, the intraclass correlation does not require this. Twins are an example of exchangeable dyads, where the intraclass correlation is more appropriate, whereas married couples are an example of nonexchangeable or distinguishable dyads where the Pearson correlation between husbands y_{1j} and wives y_{2j} is more appropriate because it is usually difficult to justify that husbands and wives have the same population mean β. Another difference between the intraclass correlation and the Pearson correlation is that the latter is only defined for pairs of variables whereas the former summarizes dependence for clusters of size larger than 2 and clusters of variable sizes; see for example exercise 2.4.

2.4 Fixed versus random effects

Each subject has a different effect ζ_j on the measured peak-expiratory-flow rates. In analysis of variance (ANOVA) terminology (see sec. 1.4 and 1.9), the subjects can therefore be thought of as the levels of a factor or categorical explanatory variable. Since the effects of subjects are random, the variance-components model is therefore sometimes referred to as a one-way random-effects ANOVA model.

The one-way *random*-effects ANOVA model can be written as

$$y_{ij} = \beta + \zeta_j + \epsilon_{ij}, \qquad \epsilon_{ij}|\zeta_j \sim N(0,\theta) \quad \zeta_j \sim N(0,\psi) \tag{2.5}$$

where ζ_j is a random intercept. In contrast, the one-way *fixed*-effects ANOVA model becomes

$$y_{ij} = \beta + \alpha_j + \epsilon_{ij}, \qquad \epsilon_{ij} \sim N(0,\theta) \quad \sum_{j=1}^{J} \alpha_j = 0 \tag{2.6}$$

where α_j are unknown cluster-specific parameters. In the random-effects model, the random intercepts are independent across clusters and independent of the level-1 residuals, and in both models the level-1 residuals are independent across units. Both random-effects models and fixed-effects models include cluster-specific intercepts, ζ_j and α_j respectively, to account for unobserved heterogeneity. Thus a natural question is whether to use a random- or fixed-effects approach.

One way of answering this question is by being explicit about the target of inference, namely, whether interest concerns the *population* of clusters or the particular clusters in the *dataset*. If we are interested in the variance ψ for the population of clusters or inference regarding the population mean β when both clusters and units are viewed as sampled from respective populations, a random-effects approach must be used. In contrast, if we are interested in the "effects" α_j of the clusters in a particular dataset and inferences regarding β when only units (and not clusters) are taken as sampled, a fixed-effects approach must be used.

Whether the clusters are viewed as sampled from a population mostly affects the estimated standard error of $\widehat{\beta}$ as we will see in section 2.7.2 but can also affect $\widehat{\beta}$ itself.

We can also make inferences regarding the particular clusters in the dataset when using a random-effects approach, as will be discussed in section 2.9.

It is often said that the random-effects approach should only be used if there is a sufficient number of clusters in the sample, typically more than 10 or 20. This is true if the variance components are of interest since ψ will be poorly estimated with a small number of clusters. However, if a random-effects approach is used merely to make appropriate inferences regarding β, a smaller number of clusters may suffice.

Regarding cluster sizes, these should be large in the fixed-effects approach if the α_j are of interest. However, in random-effects models, it is only required that there are a good number of clusters of size 2 or more. It does not matter if there are also "clusters" of size 1. Such singleton clusters do not provide information on the within-cluster correlation or on how the total variance is partitioned into ψ and θ, but they do contribute to the estimation of β and $\psi + \theta$.

In the peak-expiratory-flow application, the one-way fixed-effects ANOVA model has 19 parameters (β, $\alpha_1, \ldots, \alpha_{17}$,$\theta$) and one constraint ($\sum_j \alpha_j = 0$). The one-way random-effects ANOVA model is thus more parsimonious, having only 3 parameters (β, ψ, θ).

2.5 Estimation using Stata

In Stata, maximum likelihood estimates for variance-components models can be obtained using `xtreg` with the `mle` option, `xtmixed` with the `mle` option, or `gllamm`. Restricted maximum likelihood (REML) estimates can be obtained using `xtmixed, reml` (the default method), and generalized least-squares (GLS) estimates can be obtained using `xtreg, re` (the default method). See section 2.7.1 for information on these estimation methods.

`xtreg` is undoubtedly the most computationally efficient procedure for variance-components models. The postestimation command `predict` for `xtmixed` and the postestimation command `gllapred` for `gllamm` are useful for predicting random intercepts. However, at the time of writing this book only `gllapred` produces different kinds of standard errors for the predictions.

We do not generally recommend using `gllamm` for linear variance-components models because `xtreg` and `xtmixed` are more computationally efficient and sometimes more accurate than `gllamm` for such models.

2.5.1 Data preparation

We now set up the data for estimation in Stata. Currently, the responses for occasions 1 and 2 are in *wide format* as two separate variables, `wp1` and `wp2` for the Wright peak flow meter and `wm1` and `wm2` for the Mini Wright peak flow meter:

```
. list if id < 6, clean noobs

    id   wp1   wp2   wm1   wm2   mean_wm
     1   494   490   512   525     518.5
     2   395   397   430   415     422.5
     3   516   512   520   508       514
     4   434   401   428   444       436
     5   476   470   500   500       500
```

However, we need to stack the occasion 1 and 2 measurements using a given meter into one variable. We can use the `reshape` command to obtain such a *long format* with one variable `wp` for both Wright peak flow meter measurements, one variable `wm` for both Mini Wright peak flow meter measurements, and a variable `occasion` (equal to 1 and 2) for the measurement occasion:

```
. reshape long wp wm, i(id) j(occasion)
(note: j = 1 2)
Data                               wide   ->   long
-----------------------------------------------------------------------------
Number of obs.                       17   ->      34
Number of variables                   6   ->       5
j variable (2 values)                     ->   occasion
xij variables:
                                wp1 wp2   ->   wp
                                wm1 wm2   ->   wm
-----------------------------------------------------------------------------
```

The data for the first five subjects now look like this:

```
. list if id < 6, clean noobs

    id   occasion    wp    wm   mean_wm
     1          1   494   512     518.5
     1          2   490   525     518.5
     2          1   395   430     422.5
     2          2   397   415     422.5
     3          1   516   520       514
     3          2   512   508       514
     4          1   434   428       436
     4          2   401   444       436
     5          1   476   500       500
     5          2   470   500       500
```

In the above `reshape` command, `i()` is used to specify clusters, denoted j in this book, and `j()` is used to specify units within clusters, denoted i in this book.

2.5.2 Using xtreg

We can now estimate the parameters of the variance-components model (2.2) using the `xtreg` command with the `mle` option, which stands for maximum likelihood estimation (see sec. 2.7.1).

As in the `regress` command, the response variable `wm` and explanatory variables are listed after the command name. In the variance-components models, the fixed part is just the intercept β, which is included by default, so we do not specify any explanatory variables. The random part includes a random intercept ζ_j whose cluster identifier j can be specified in the `i()` option, here `i(id)`. The level-1 residual ϵ_{ij} need not be specified because it is always included. Therefore the command is

```
. xtreg wm, i(id) mle
Random-effects ML regression                    Number of obs      =        34
Group variable: id                              Number of groups   =        17
Random effects u_i ~ Gaussian                   Obs per group: min =         2
                                                               avg =       2.0
                                                               max =         2
                                                Wald chi2(0)       =      0.00
Log likelihood  = -184.57839                    Prob > chi2        =         .

------------------------------------------------------------------------------
          wm |      Coef.   Std. Err.      z    P>|z|     [95% Conf. Interval]
-------------+----------------------------------------------------------------
       _cons |   453.9118   26.18616    17.33   0.000     402.5878    505.2357
-------------+----------------------------------------------------------------
    /sigma_u |   107.0464   18.67858                       76.0406    150.6949
    /sigma_e |   19.91083   3.414659                       14.2269     27.8656
         rho |   .9665602   .0159494                      .9210943    .9878545
------------------------------------------------------------------------------
Likelihood-ratio test of sigma_u=0: chibar2(01)=   46.27 Prob>=chibar2 = 0.000
```

Instead of specifying the cluster identifier in the `i()` option, we could have declared the data to be "panel data" and `id` to be the "group variable" using the `xtset` command before the `xtreg` command.

We see in the output that there are 34 observations belonging to 17 groups (the clusters, here subjects) and that the log likelihood is −184.58. In the Stata output and in the Stata documentation for `xtreg`, the i subscript is used for clusters (instead of j used in this book), u_i for the random intercept (instead of ζ_j) and the t subscript for occasions (instead of i used in this book). Clusters are called "panels" in the documentation.

The estimate of the overall population mean β, given next to `_cons` in the output, is 453.91. The estimate of the between-subject standard deviation $\sqrt{\psi}$ of the random intercepts of subjects, referred to as `/sigma_u`, is 107.05, and the estimate of the within-subject standard deviation, $\sqrt{\theta}$, referred to as `/sigma_e`, is 19.91. It follows that the intraclass correlation is estimated as

$$\widehat{\rho} = \frac{\widehat{\psi}}{\widehat{\psi}+\widehat{\theta}} = \frac{107.05^2}{19.91^2 + 107.05^2} = 0.97$$

which is referred to as `rho` in the output. This estimate is close to one, indicating that the Mini Wright peak flow meter is very reliable. However, the reliability is not just a characteristic of the instrument; it also depends on the between-subject variance, ψ, and this can differ between populations. The parameter estimates are also given in table 2.2 under "`xtreg`, `xtmixed`".

2.5.3 Using xtmixed

The variance-components model considered here is a simple special case of a linear mixed-effects models that can be fitted using the `xtmixed` command (available as of Stata 9).

The fixed part of the model, here β, is specified as in any estimation command in Stata (response variable followed by list of explanatory variables). The random part, except the residual ϵ_{ij}, is specified after two vertical bars `||`. To include a random intercept ζ_j, which varies between subjects whose identifier is in the variable `id`, the syntax is simply `id:` because a random intercept ζ_j is included by default (it can be excluded using the `noconstant` option). Finally, we can request maximum likelihood estimation using the `mle` option:

```
. xtmixed wm || id:, mle

Mixed-effects ML regression                     Number of obs      =        34
Group variable: id                              Number of groups   =        17

                                                Obs per group: min =         2
                                                               avg =       2.0
                                                               max =         2

                                                Wald chi2(0)       =         .
Log likelihood = -184.57839                     Prob > chi2        =         .

------------------------------------------------------------------------------
          wm |      Coef.   Std. Err.      z    P>|z|     [95% Conf. Interval]
-------------+----------------------------------------------------------------
       _cons |   453.9118   26.18616    17.33   0.000     402.5878    505.2357
------------------------------------------------------------------------------

------------------------------------------------------------------------------
  Random-effects Parameters  |   Estimate   Std. Err.     [95% Conf. Interval]
-----------------------------+------------------------------------------------
id: Identity                 |
                   sd(_cons) |   107.0464   18.67857       76.0406    150.6949
-----------------------------+------------------------------------------------
                sd(Residual) |   19.91083   3.414679      14.22688    27.86565
------------------------------------------------------------------------------
LR test vs. linear regression: chibar2(01) =    46.27 Prob >= chibar2 = 0.0000
```

The table of estimates for the fixed part has the same form as that for `xtreg` and all Stata estimation commands. The random part is given under `Random-effects Parameters`. Here `sd(_cons)` is the estimate of the random-intercept standard deviation $\sqrt{\psi}$, and `sd(Residual)` is the estimate of the standard deviation $\sqrt{\theta}$ of the level-1 residuals. All these estimates are identical to the estimates using `xtreg` and are given under "`xtreg`, `xtmixed`" in table 2.2. We could also obtain estimated variances (instead of standard deviations) with their standard errors using the `variance` option.

There are some differences in the terminology and notation used in this book and the Stata documentation for `xtmixed`. Using the usual multilevel or hierarchical modeling conventions, we use the indices i for occasions and j for subjects and call the corresponding levels 1 and 2. In contrast, the `xtmixed` documentation uses i for subjects and j for occasions and calls subjects level 1. Contrary to common terminology where multilevel models for such data are called two-level models, the `xtmixed` documentation calls these models single-level models because the lowest level, here occasions, is not considered a level.

2.5.4 Using gllamm

We now introduce the `gllamm` command, which will be used extensively for models with categorical or discrete responses in later chapters.

To check if `gllamm` is installed on your computer, type

```
. which gllamm
```

If the following message appears,

```
command gllamm not found as either built-in or ado-file
```

install `gllamm` (assuming that you have a net-aware Stata) by using the `ssc` command:

```
. ssc install gllamm
```

Occasionally, you should update `gllamm` using `ssc` with the `replace` option:

```
. ssc install gllamm, replace
```

The `gllamm` command for fitting the variance-components model (2.2) resembles the `xtreg` command. We add two options to ensure accurate estimates: the `nip(12)` and `adapt` options (see sec. 6.11.1 for more details).

```
. gllamm wm, i(id) nip(12) adapt
number of level 1 units = 34
number of level 2 units = 17

Condition Number = 152.64774

gllamm model

log likelihood = -184.57839
```

wm	Coef.	Std. Err.	z	P>|z|	[95% Conf.	Interval]
_cons	453.9116	26.18394	17.34	0.000	402.592	505.2312

```
Variance at level 1
------------------------------------------------------------------------------

  396.70879 (136.11609)

Variances and covariances of random effects
------------------------------------------------------------------------------


***level 2 (id)

    var(1): 11456.828 (3997.7689)
------------------------------------------------------------------------------
```

The output from gllamm first shows the number of units at each level, here 34 units at level 1 (the total number of measurements) and 17 units at level 2 (the subjects). If the Condition Number is very large, the model may not be well identified, but here it is not alarming.

Next the maximized log likelihood is given as -184.58 followed by a regression table giving the estimated fixed regression coefficient $\widehat{\beta}$ next to _cons.

Estimates and standard errors for the random part of the model are given under the headings "Variance at level 1" for the variance θ of the level-1 residuals ϵ_{ij} and "Variances and covariances of random effects" and "***level 2 (id)" for the variance ψ of the random intercept ζ_j. Variance estimates from gllamm are presented under "gllamm" in table 2.2.

xtreg and xtmixed display the estimated standard deviations instead of variances. We can convert these standard deviations to variances $\widehat{\theta} = 19.91083^2 = 396.44115$ and $\widehat{\psi} = 107.0464^2 = 11,458.932$, which differ slightly from the estimates using gllamm. The reason for the discrepancy is that gllamm uses numerical integration, whereas xtreg and xtmixed exploit the closed form of the likelihood for random-effects models with normally distributed continuous responses. The accuracy of the gllamm estimates can be improved by increasing the number of integration points (see section 6.11.1) using the nip() option.

Table 2.2: Maximum likelihood estimates for Mini Wright peak flow meter

	Est	(SE)
Fixed part		
β	453.91	(26.18)
Random part		
xtreg, xtmixed		
$\sqrt{\psi}$	107.05	
$\sqrt{\theta}$	19.91	
gllamm		
ψ	11,456.83	
θ	396.71	
Log likelihood	−184.58	

Slight discrepancies between estimates due to numerical integration in gllamm

Before doing further analyses, we save the gllamm estimates using estimates store:

```
. estimates store RI
```

We can `restore` these estimates later without having to refit the model. This will be useful in section 2.9.3, where we use `gllamm`'s prediction command `gllapred`. Storing estimates means that they remain available during a Stata session; if we require the estimates again in a later Stata session, we can save them in a file using `estimates save` *filename* (a command introduced in Stata release 10).

2.6 Hypothesis tests and confidence intervals

2.6.1 Hypothesis test and confidence interval for the population mean

In the regression tables produced by `xtreg`, `xtmixed`, and `gllamm`, z statistics are reported for β instead of the t statistics given by the `regress` command.

As the t statistic in ordinary linear regression, the z statistic is given by

$$z = \frac{\widehat{\beta}}{\widehat{\mathrm{SE}}(\widehat{\beta})}$$

(where the standard error takes a different form than in linear regression as discussed in section 1.5).

The reason this statistic is called z instead of t is that a standard normal sampling distribution is assumed under the null hypothesis that $\beta = 0$ instead of a t distribution. The t distribution is a finite sample distribution whose shape depends on the degrees of freedom. For the variance-components model, the finite sample distribution does not have a simple form, so Stata's commands use the asymptotic (large-sample) sampling distribution. (Some other software packages approximate the finite-sample distribution by a t distribution where the degrees of freedom are some function of the data.) The null hypothesis that the population mean β is zero is not of interest in the peak-expiratory-flow example.

An asymptotic 95% confidence interval is given by

$$\widehat{\beta} \pm z_{.975}\widehat{\mathrm{SE}}(\widehat{\beta})$$

where $z_{.975}$ is the 97.5th percentile of the standard normal distribution, i.e., $z_{.975} = 1.96$. This kind of confidence interval based on assuming a normal sampling distribution is often called a Wald confidence interval. In the Mini Wright application, the 95% Wald confidence interval for the population mean β is from 402.59 to 505.24, as shown for instance in the output from `xtreg` on page 64.

As for linear regression there are two versions of estimated standard errors; a model-based version and a "robust" version based on the so-called sandwich estimator. The latter can be obtained for the GLS estimator in `xtreg` using the `vce(robust)` option. Robust standard errors for maximum likelihood estimates are produced by `gllamm` when the `robust` option is specified.

2.6.2 Hypothesis test and confidence interval for the between-cluster variance

We now consider testing hypotheses regarding the between-cluster variance ψ. In particular, we are often interested in the hypotheses

$$H_0\colon \psi = 0 \quad \text{against} \quad H_a\colon \psi > 0$$

This null hypothesis is equivalent to the hypothesis that $\zeta_j = 0$ or that there is no random intercept in the model. If the null hypothesis is true, we can use ordinary regression instead of a variance-components model.

Likelihood-ratio tests are typically used with the test statistic

$$L = 2(l_1 - l_0)$$

where l_1 is the maximized log likelihood for the variance-components model (which includes ζ_j) and l_0 is the maximized log likelihood for a model without ζ_j. Importantly, the distribution of L under H_0 is not χ^2 with 1 degree of freedom as usual. This is because the null hypothesis is on the boundary of the parameter space since $\psi \geq 0$, which renders standard statistical test theory invalid.

If the variance-components model is used for replicated datasets generated under the null hypothesis, we would expect positive correlations among the responses about half of the time and negative correlations the other half of the time. Thus ψ would be estimated as positive half of the time and as zero (since negative correlations cannot be produced by nonnegative ψ in the variance-components model) the other half of the time. The correct sampling distribution under the null hypothesis hence takes a simple form, being a 50:50 mixture of a spike at 0 and a χ^2 with 1 df. The correct p-value can be obtained by simply dividing the "naive" p-value, based on the χ^2 with 1 df, by 2.

This p-value is given at the bottom of the `xtreg` and `xtmixed` output, where the correct sampling distribution is referred to as `chibar2(01)` (click on `chibar2(01)`, which is shown in blue in the Stata output window to find an explanation). We can also perform the likelihood-ratio test ourselves by fitting the variance-components model, storing the estimates, then fitting the model without the random intercept, and comparing the models using the `lrtest` command:

```
. quietly xtmixed wm || id:, mle
. estimates store m1
. quietly xtmixed wm, mle
. lrtest m1 .
Likelihood-ratio test                          LR chibar2(01)  =     46.27
(Assumption: . nested in m1)                   Prob > chibar2  =    0.0000
```

Here the `quietly` prefix command is used to suppress output from `xtmixed`. In the `lrtest` command, `m1` refers to the estimates stored under that name and "`.`" refers to the current (or last) estimates. For the peak-expiratory-flow application, we see that the test of the null hypothesis $\psi = 0$ has a very small p-value and the null hypothesis is rejected at standard significance levels.

An alternative tests for ψ is the *score test* (sometimes called the Lagrange multiplier test) which is based on a quadratic approximation of the likelihood at $\psi = 0$, and can also be used to construct confidence intervals. Score tests and confidence intervals for variance components are not provided by official Stata at the time of writing but Bottai and Orsini (2004) have provided the postestimation command `xtvc` for `xtreg` with the `mle` option. We first install the command by typing

```
. ssc install xtvc
```

and then run it by typing

```
. quietly xtreg wm, mle
. xtvc
```

```
         wm    ML Estimate      [95% Conf. Interval]

   /sigma_u       107.0464     84.36725     200.3493

Score test of sigma_u=0: chi2(1)= 30242.81 Prob>=chi2 = 0.000
```

The null hypothesis is clearly rejected, and the estimated 95% confidence interval for the standard deviation $\sqrt{\psi}$ is from 84.37 to 200.35. This confidence interval is different from that produced by `xtreg`. The latter confidence interval is obtained by exponentiating the limits of the Wald confidence interval for the log standard deviation.

We can also base the test for unexplained between-cluster heterogeneity on the fixed-effects model, testing the null hypothesis that the fixed effects α_j of the clusters are all zero. This test, described by for instance Wooldridge (2002), can be obtained using `xtreg` with the `fe` option:

```
. xtreg wm, i(id) fe
Fixed-effects (within) regression               Number of obs      =        34
Group variable: id                              Number of groups   =        17
R-sq:  within  = 0.0000                         Obs per group: min =         2
       between = 0.0000                                        avg =       2.0
       overall = 0.0000                                        max =         2
                                                F(0,17)            =      0.00
corr(u_i, Xb)  = 0.0000                         Prob > F           =         .

          wm       Coef.   Std. Err.       t    P>|t|     [95% Conf. Interval]

       _cons    453.9118   3.414679   132.93   0.000     446.7074    461.1161

     sigma_u   111.29118
     sigma_e   19.910831
         rho   .96898482   (fraction of variance due to u_i)

F test that all u_i=0:     F(16, 17) =    62.48              Prob > F = 0.0000
```

We see from the bottom of the output that the null hypothesis is clearly rejected using the F test. (Use of the `xtreg` command with the `fe` option, which stands for "fixed effects", is discussed in more detail in section 3.7.2.)

If the uncertainty in the estimated level-1 residual variances is ignored and $\widehat{\theta}$ is treated as the true θ, the F test can be replaced by a χ^2 test. A similar χ^2 test is described in Raudenbush and Bryk (2002) and implemented in the HLM software of Raudenbush et al. (2004). Instead of basing the χ^2 statistic on the estimated mean $\widehat{\beta}$ from the fixed-effects model, they base it on the estimated mean from the variance-components model.

It does not make sense to test the null hypothesis that $\theta = 0$, or in other words that all $\epsilon_i = 0$, because this would force all responses y_{ij} for the same cluster j to be identical. The estimated standard errors reported for the estimated between-cluster and within-cluster standard deviations $\sqrt{\psi}$ and $\sqrt{\theta}$ by `xtreg` and `xtmixed` and for the corresponding variances ψ and θ by `gllamm` should not be used to construct test statistics or confidence intervals. In particular, when the estimates are small and/or there are few clusters, the estimators do not have Gaussian or normal sampling distributions.

2.7 More on statistical inference

2.7.1 ❖ Different estimation methods

A classical method for estimating the parameters of statistical models is maximum likelihood (ML). The likelihood is just the joint probability density of all the observed responses y_{ij}, $(i=1,\ldots,n_j)$, $(j=1,\ldots,N)$, as a function of the model parameters β, ψ, and θ. The idea is to find parameter estimates $\widehat{\beta}$, $\widehat{\psi}$, and $\widehat{\theta}$ that maximize this likelihood function, thus making the responses appear as "likely" as possible. It has been shown that ML estimators have good properties, such as consistency (the estimates approach the true values as the sample size increases) and efficiency (estimates have the smallest possible sampling variance in large samples).

When the data are balanced with the same number of units $n_j=n$ in each of the J clusters ($n=2$ occasions in the current application), the ML estimators for the two-level variance-components model have relatively simple expressions. The expressions are in terms of the model sum of squares (MSS) and sum of squared errors (SSE) from a one-way ANOVA, treating subjects as a fixed factor (see sec. 1.4). Here the MSS is the sum of squared deviations of cluster means from the overall mean

$$\text{MSS} \;=\; \sum_{j=1}^{J}\sum_{i=1}^{n}(\overline{y}_{\cdot j}-\overline{y}_{\cdot\cdot})^2, \qquad \overline{y}_{\cdot\cdot} = \frac{1}{Jn}\sum_{j=1}^{J}\sum_{i=1}^{n} y_{ij}$$

and the SSE is the sum of squared deviations of responses from their cluster means

$$\text{SSE} \;=\; \sum_{j=1}^{J}\sum_{i=1}^{n}(y_{ij}-\overline{y}_{\cdot j})^2, \qquad \overline{y}_{\cdot j} = \frac{1}{n}\sum_{i=1}^{n} y_{ij}$$

The population mean β is estimated by the sample mean

$$\widehat{\beta} \;=\; \overline{y}_{\cdot\cdot}$$

and the ML estimator of the within-cluster variance θ is

$$\widehat{\theta} = \frac{1}{J(n-1)}\text{SSE} = \text{MSE}$$

where MSE is the mean squared error from the one-way ANOVA.

The ML estimator of the between-cluster variance ψ is given by

$$\widehat{\psi} = \frac{\text{MSS}}{Jn} - \frac{\widehat{\theta}}{n}$$

The ML estimators for β and θ are unbiased if the model is true, whereas the estimator for ψ has downward bias.

The unbiased moment estimator or analysis of variance (ANOVA) estimator of ψ is given by

$$\widehat{\psi}^M = \frac{\text{MSS}}{(J-1)n} - \frac{\widehat{\theta}}{n} = \frac{1}{n}(\text{MMS} - \text{MSE})$$

where MMS is the model mean square from the one-way ANOVA. The between-cluster sum of squares is now divided by the model degrees of freedom $(J-1)n$ instead of Jn. The difference between the biased maximum likelihood estimator $\widehat{\psi}$ and the unbiased moment estimator $\widehat{\psi}^M$ becomes small when the number of clusters J is large. In the example considered in this chapter, there are only $J = 17$ clusters, so the difference between ML and ANOVA estimates will not be negligible (see exercise 2.9).

For balanced data, the ANOVA estimator is also the restricted maximum likelihood (REML) estimator. For unbalanced data, the ANOVA, REML, and ML estimators are all different, and the latter two estimators are preferable because they are more efficient. The difference between REML and ML is that REML estimates the random-intercept variance taking into account the loss of 1 degree of freedom resulting from the estimation of the overall mean β. In models considered in the next chapters that include explanatory variables, further degrees of freedom are lost due to estimation of additional regression coefficients. Contrary to common belief, REML is not unbiased for ψ in unbalanced designs.

Another estimation method, popular in econometrics, is generalized least squares (GLS). Here the fixed parameter β (usually there are more fixed parameters; see chapter 3) is estimated by a weighted version of ordinary least squares using weights that depend on the variance-component estimates. For the simple model and the balanced case considered here, the variance-component estimates are identical to the ANOVA and REML estimates.

2.7.2 Inference for β

Estimate and standard error: Balanced case

We first consider the balanced case where $n_j = n$. As mentioned in the previous section, the maximum likelihood estimator $\widehat{\beta}$ of β under the variance-components model is simply the overall sample mean

$$\widehat{\beta} = \frac{1}{Jn}\sum_{j=1}^{J}\sum_{i=1}^{n} y_{ij} = \frac{1}{J}\sum_{j=1}^{J}\overline{y}_{\cdot j}$$

an unweighted mean of the cluster means. The estimated standard error is given by

$$\widehat{\mathrm{SE}}(\widehat{\beta}) = \sqrt{\frac{n\widehat{\psi}+\widehat{\theta}}{Jn}}$$

The corresponding estimator for the model without the random intercept, which we refer to as the ordinary least-squares (OLS) estimator $\widehat{\beta}^{\mathrm{OLS}}$, is the same as the maximum likelihood (ML) estimator but the estimated standard error is now approximately

$$\widehat{\mathrm{SE}}(\widehat{\beta}^{\mathrm{OLS}}) \approx \sqrt{\frac{\widehat{\psi}+\widehat{\theta}}{Jn}}$$

where the OLS estimate of the residual variance $\widehat{\sigma^2}$ is approximated by the sum of the estimated variance components $\widehat{\psi}+\widehat{\theta}$ (the approximation is better for larger n). We see that $\widehat{\mathrm{SE}}(\widehat{\beta}^{\mathrm{OLS}}) \leq \widehat{\mathrm{SE}}(\widehat{\beta})$, the difference becoming larger as the between-cluster variance $\widehat{\psi}$ and cluster size n get larger. This makes sense since the estimated standard error for the variance-components model acknowledges that the estimate $\widehat{\beta}$ can change considerably from sample to sample if the clusters included in the sample change, and if the clusters differ from one another considerably (large ψ), affecting large portions of the overall sample (large n).

In the one-way fixed-effects model (with the random ζ_j replaced by fixed α_j, see section 2.4), the estimator of β is again the same but now the estimated standard error is

$$\widehat{\mathrm{SE}}(\widehat{\beta}^{\mathrm{F}}) = \sqrt{\frac{\widehat{\theta}}{Jn}}$$

which is even smaller than the standard error of the OLS estimator if $\widehat{\psi} > 0$. This again makes intuitive sense because fixed-effects inference is for repeated sampling of units from within the *same* clusters, so the between-cluster variance does not contribute to the standard error.

For the peak-expiratory-flow application, we see from the output of `xtreg` with the `mle` option on page 64 that $\widehat{\mathrm{SE}}(\widehat{\beta}) = 26.19$ and from the output of `xtreg` with the `fe` option on page 70 that $\widehat{\mathrm{SE}}(\widehat{\beta}^{\mathrm{F}}) = 3.41$. We obtain the estimated model-based standard error for the OLS estimator by using the `regress` command

```
. regress wm
      Source |       SS       df       MS              Number of obs =      34
-------------+------------------------------           F(  0,    33) =    0.00
       Model |           0     0           .           Prob > F      =       .
    Residual |  403082.735    33  12214.6283           R-squared     =  0.0000
-------------+------------------------------           Adj R-squared =  0.0000
       Total |  403082.735    33  12214.6283           Root MSE      =  110.52

------------------------------------------------------------------------------
          wm |      Coef.   Std. Err.      t    P>|t|     [95% Conf. Interval]
-------------+----------------------------------------------------------------
       _cons |   453.9118   18.95399    23.95   0.000     415.3496    492.4739
------------------------------------------------------------------------------
```

which gives $\widehat{\text{SE}}(\widehat{\beta}^{\text{OLS}}) = 18.95$. For the application, we see that $\widehat{\text{SE}}(\widehat{\beta}) > \widehat{\text{SE}}(\widehat{\beta}^{\text{OLS}}) > \widehat{\text{SE}}(\widehat{\beta}^{\text{F}})$ as expected.

Although the clustered nature of the data is not taken into account in the OLS estimator $\widehat{\beta}^{\text{OLS}}$ of the population mean, a *sandwich estimator* can be used, which takes the clustering into account to produce "robust" standard errors for $\widehat{\beta}^{\text{OLS}}$. Using the `regress` command with the `vce(cluster id)` option produces the following:

```
. regress wm, vce(cluster id)
Linear regression                                      Number of obs =      34
                                                       F(  0,    16) =    0.00
                                                       Prob > F      =       .
                                                       R-squared     =  0.0000
                                                       Root MSE      =  110.52

                                  (Std. Err. adjusted for 17 clusters in id)
------------------------------------------------------------------------------
             |               Robust
          wm |      Coef.   Std. Err.      t    P>|t|     [95% Conf. Interval]
-------------+----------------------------------------------------------------
       _cons |   453.9118   26.99208    16.82   0.000     396.6911    511.1324
------------------------------------------------------------------------------
```

The estimated "robust" standard error is 26.99, which is close to the estimated model-based standard error of 26.19 from maximum likelihood estimation of the variance-components model. (We have previously used a sandwich estimator to obtain "robust" standard errors for nonclustered data in section 1.6.) By fitting an ordinary regression model with robust standard errors for clustered data instead of fitting variance-components models, we are treating the within-cluster dependence as a "nuisance", not as a phenomenon we are interested in. We learn nothing about the between and within-cluster variances or intraclass correlation.

Estimate: Unbalanced case

In the unbalanced case, the maximum likelihood estimator of β under the variance-components model becomes a weighted mean of the cluster means

$$\widehat{\beta} = \frac{\sum_{j=1}^{J} w_j \overline{y}_{\cdot j}}{\sum_{j=1}^{J} w_j} \qquad \text{where} \quad w_j = \frac{1}{\widehat{\psi} + \widehat{\theta}/n_j}$$

Small clusters have a similar weight as large clusters if $\widehat{\theta}$ is small compared with $\widehat{\psi}$. In contrast, the conventional OLS estimator, which disregards clustering and treats the data as single-level, is

$$\widehat{\beta}^{\text{OLS}} = \frac{\sum_{j=1}^{J} n_j \overline{y}_{\cdot j}}{\sum_{j=1}^{J} n_j}$$

with cluster means given weights equal to the cluster sizes n_j. Thus a variance-components model tends to give more weight to smaller clusters than an ordinary regression model.

2.8 Crossed versus nested effects

So far, we have considered the random (or fixed) effects of a single factor "subjects". Another potential factor in the dataset is the measurement occasion with 2 levels, occasions 1 and 2. In the variance-components model, occasion was allowed to have an effect on the response variable only via the residual term ϵ_{ij}, which takes on a different value for each combination of subject and occasion. We have therefore implicitly treated occasions as *nested* within subjects meaning that occasion (2 versus 1) does not have the same effect for all subjects.

If all subjects had been measured and remeasured in the same sessions and if there were anything specific to the session (e.g., time of day, temperature, or calibration of the measurement instrument) that could influence measurements on all subjects in a similar way, then subjects and occasions would be *crossed*. We would then include an occasion-specific term in the model that takes on the same value for all subjects.

The distinction between nested and crossed factors is illustrated in figure 2.8.

(*Continued on next page*)

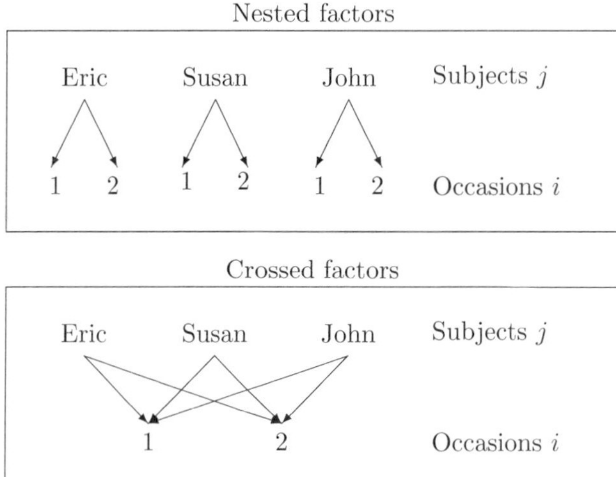

Figure 2.8: Illustration of nested and crossed factors

In the nested case, the effect of occasion 1 (or 2) is different for every subject, whereas in the crossed case it is the same for each subject.

In the crossed case, the model can be described as a two-way ANOVA model. A subject by occasion interaction could in this case be included in addition to the main effects of each factor. However, since there are no replications for each subject-occasion combination, the interaction term is confounded with the error term ϵ_{ij}. If a random effect is specified for subjects and a fixed effect for occasion, we obtain a so-called mixed-effects two-way ANOVA model. Such a model can be fitted by introducing a dummy variable for the second occasion in the fixed part of the model by using the commands

```
. generate occ2 = occasion==2
. xtmixed wm occ2 || id:, mle
Mixed-effects ML regression                     Number of obs      =        34
Group variable: id                              Number of groups   =        17
                                                Obs per group: min =         2
                                                               avg =       2.0
                                                               max =         2
                                                Wald chi2(1)       =      0.18
Log likelihood = -184.48885                     Prob > chi2        =    0.6714

------------------------------------------------------------------------------
          wm |      Coef.   Std. Err.      z    P>|z|     [95% Conf. Interval]
-------------+----------------------------------------------------------------
        occ2 |   2.882353   6.793485     0.42   0.671    -10.43263    16.19734
       _cons |   452.4706   26.40554    17.14   0.000     400.7167    504.2245
------------------------------------------------------------------------------

------------------------------------------------------------------------------
  Random-effects Parameters  |   Estimate   Std. Err.     [95% Conf. Interval]
-----------------------------+------------------------------------------------
id: Identity                 |
                   sd(_cons) |   107.0561   18.67683      76.05229    150.6991
-----------------------------+------------------------------------------------
                sd(Residual) |   19.80624   3.396742      14.15214    27.71928
------------------------------------------------------------------------------
LR test vs. linear regression: chibar2(01) =    46.44 Prob >= chibar2 = 0.0000
```

or the corresponding `xtreg` and `gllamm` commands

```
xtreg wm occ2, i(id) mle
gllamm wm occ2, i(id) nip(12) adapt
```

We see that there is no evidence for an effect of occasion, which in this example could only be interpreted as a practice effect. Models with crossed random effects are discussed in chapter 11.

2.9 Assigning values to the random intercepts

Remember that the cluster-specific intercepts ζ_j are treated as random variables and not as model parameters in multilevel models. However, having obtained estimates $\widehat{\beta}$, $\widehat{\psi}$, and $\widehat{\theta}$ of the model parameters β, ψ, and θ, we may wish to assign values to the random intercepts ζ_j for individual clusters, analogous to obtaining estimated residuals $\widehat{\epsilon}_i$ in ordinary linear regression.

There are a number of reasons why we may want to obtain values for the random intercepts ζ_j for individual clusters $j = 1, \ldots, J$. For instance, we will use such assigned values for model diagnostics (see sec. 3.9 and 4.8.4), for interpreting and visualizing models (see sec. 4.8.3), and for inference regarding individual clusters (see sec. 4.8.5). An example of the last type would be to estimate subjects' true expiratory flow $\beta + \zeta_j$ from the fallible measurements.

The assigned values are either referred to as predictions or estimates of the random intercepts ζ_j. The term *prediction* is used when ζ_j is regarded as a random variable, whereas the term *estimation* is used when ζ_j is regarded as an unknown fixed parameter. There are two common approaches to assigning values to the random intercepts: maximum likelihood estimation, described in section 2.9.1, and empirical Bayes prediction, described in section 2.9.2.

2.9.1 Maximum likelihood estimation

The estimated model parameters are treated as if they were the true parameter values. For instance, we substitute the parameter estimate $\widehat{\beta}$ into the model giving

$$y_{ij} \;=\; \widehat{\beta} + \underbrace{\zeta_j + \epsilon_{ij}}_{\xi_{ij}}$$

We now view the ζ_j as the only unknown parameters to be estimated. Specifically, for each subject j, we find the value of ζ_j that maximizes the likelihood of the observed responses y_{1j} and y_{2j}

$$\text{Likelihood}(y_{1j}, y_{2j}|\zeta_j)$$

treating the model parameters as known. This approach of treating ζ_j as an unknown (and fixed) parameter contradicts the original model specification, where ζ_j was treated as a random effect.

We can rearrange the above model by subtracting $\widehat{\beta}$ from y_{ij} to obtain estimated total residuals $\widehat{\xi}_{ij}$ and regarding these as the responses

$$\widehat{\xi}_{ij} \;=\; y_{ij} - \widehat{\beta} = \zeta_j + \epsilon_{ij} \tag{2.7}$$

Implementation via OLS regression

The model in (2.7) has a different mean ζ_j for each subject, and we can estimate these means by regressing $\widehat{\xi}_{ij}$ on dummy variables for each of the subjects, excluding the overall intercept because this is now redundant. The ordinary least-squares (OLS) estimates of the regression coefficients for the subject dummies are the required maximum likelihood estimates $\widehat{\zeta}_j^{\text{ML}}$ of the ζ_j.

To obtain these estimates in Stata, we first refit the model using `xtmixed` (with the `quietly` prefix to suppress the output) and then subtract the predicted fixed part $\widehat{\beta}$, obtained using `predict` with the `xb` option, from the responses:

```
. quietly xtmixed wm || id:, mle
. predict pred, xb
. generate res = wm - pred
```

Next we generate dummy variables `subject1`–`subject17` for the subjects by using `tabulate` with the `generate()` option and fit a linear regression model using `regress` with the `noconstant` option to suppress the overall constant:

```
. quietly tabulate id, generate(subject)
. regress res subject1-subject17, noconstant
      Source |       SS       df       MS              Number of obs =      34
-------------+------------------------------           F( 17,    17) =   58.81
       Model |  396343.235    17  23314.308            Prob > F      =  0.0000
    Residual |      6739.5    17  396.441176           R-squared     =  0.9833
-------------+------------------------------           Adj R-squared =  0.9666
       Total |  403082.735    34  11855.3746           Root MSE      =  19.911

------------------------------------------------------------------------------
         res |      Coef.   Std. Err.      t    P>|t|     [95% Conf. Interval]
-------------+----------------------------------------------------------------
    subject1 |   64.58823   14.07908     4.59   0.000     34.88396     94.2925
    subject2 |  -31.41177   14.07908    -2.23   0.039    -61.11604   -1.707504
    subject3 |   60.08823   14.07908     4.27   0.001     30.38396     89.7925
    subject4 |  -17.91177   14.07908    -1.27   0.220    -47.61604     11.7925
    subject5 |   46.08823   14.07908     3.27   0.004     16.38396     75.7925
    subject6 |   158.5882   14.07908    11.26   0.000      128.884    188.2925
    subject7 |  -41.91177   14.07908    -2.98   0.008    -71.61604    -12.2075
    subject8 |  -68.91177   14.07908    -4.89   0.000    -98.61604    -39.2075
    subject9 |   196.0882   14.07908    13.93   0.000      166.384    225.7925
   subject10 |  -15.41177   14.07908    -1.09   0.289    -45.11604     14.2925
   subject11 |  -27.91177   14.07908    -1.98   0.064    -57.61604    1.792496
   subject12 |   161.5882   14.07908    11.48   0.000      131.884    191.2925
   subject13 |  -210.4118   14.07908   -14.94   0.000     -240.116   -180.7075
   subject14 |   18.08823   14.07908     1.28   0.216    -11.61604     47.7925
   subject15 |  -190.4118   14.07908   -13.52   0.000     -220.116   -160.7075
   subject16 |  -93.91177   14.07908    -6.67   0.000     -123.616    -64.2075
   subject17 |  -6.911774   14.07908    -0.49   0.630    -36.61604     22.7925
------------------------------------------------------------------------------
```

From the output, we see that, for instance, $\widehat{\zeta}_1^{\text{ML}} = 64.58823$.

Implementation via the mean total residual

Remember that ζ_j represents the mean residual for subject j for the population of measurement occasions. A natural estimate of ζ_j is therefore the sample mean of the estimated total residual over the n_j occasions (here $n_j = 2$) for which we have data:

$$\widehat{\zeta}_j^{\text{ML}} = \frac{1}{n_j}\sum_{i=1}^{n_j}\widehat{\xi}_{ij} = \frac{1}{2}(\widehat{\xi}_{1j} + \widehat{\xi}_{2j})$$

This is identical to the maximum likelihood or OLS estimator of ζ_j obtained using regression with subject dummies.

The subject-specific means can be calculated using the `egen` command:

```
. egen ml = mean(res), by(id)
```

For the first subject, we get the same result as before:

```
. sort id
. display ml[1]
64.588226
```

In this subsection, we have used all the terminology usually associated with estimating model parameters. However, it is important to remember that ζ_j is not a parameter in the original model. It is only for the purpose of assigning values to ζ_j that we reformulate the problem by treating the original parameters as known constants and the ζ_j as unknown parameters.

2.9.2 Empirical Bayes prediction

Having obtained estimates $\widehat{\beta}$, $\widehat{\psi}$, and $\widehat{\theta}$ of the model parameters and treating them as the true parameter values, we can predict values of the random intercepts ζ_j for individual clusters (subjects in the application). Here we continue to treat ζ_j as a random variable, not as a fixed parameter as in maximum likelihood estimation.

Maximum likelihood estimation of ζ_j uses the responses y_{ij} for subject j as the only information about ζ_j by maximizing the likelihood of observing these particular values

$$\text{Likelihood}(y_{1j}, y_{2j}|\zeta_j)$$

In contrast, empirical Bayes prediction also uses the *prior distribution* of ζ_j, summarizing our knowledge about ζ_j before seeing the data for subject j

$$\text{Prior}(\zeta_j)$$

This prior distribution is just the normal distribution specified for the random intercept with zero mean and estimated variance $\widehat{\psi}$. It represents what we know about ζ_j before we have seen the responses y_{1j} and y_{2j} for subject j. For instance, the most likely value of ζ_j is zero. (Obviously, we have already used all responses to obtain the estimate $\widehat{\psi}$, but we now pretend that ψ is known and not estimated.)

Once we have observed the responses, we can combine the prior distribution with the likelihood to obtain the *posterior distribution* of ζ_j given the observed responses y_{1j} and y_{2j}

$$\text{Posterior}(\zeta_j|y_{1j}, y_{2j}) \;\propto\; \text{Prior}(\zeta_j) \times \text{Likelihood}(y_{1j}, y_{2j}|\zeta_j)$$

where $\propto$ means "proportional to". The posterior of ζ_j represents our updated knowledge regarding ζ_j after seeing the data y_{1j} and y_{2j} for subject j.

The empirical Bayes prediction is just the mean of the posterior distribution with parameter estimates ($\widehat{\beta}$, $\widehat{\psi}$, and $\widehat{\theta}$) plugged in. In a linear model with normal error terms, the posterior is normal and the mean is thus equal to the mode.

Figure 2.9 shows the prior, likelihood, and posterior for a hypothetical example of a subject with $n_j = 2$ responses. In both panels, the estimated total residuals $\widehat{\xi}_{ij}$ are 3 and 5, and the estimated total variance is $\widehat{\psi} + \widehat{\theta} = 5$. In the top panel, 80% of this variance is due to within-subject variability, whereas in the bottom panel, 80% is due to between-subject variability. In both cases, the likelihood (dotted curve) has its maximum at $\zeta_j = 4$, i.e., the mode is 4 (see vertical dotted lines). The maximum likelihood estimate therefore is $\widehat{\zeta}_j^{\text{ML}} = 4$. In contrast, the mode (and mean) of the posterior depends on the relative sizes of the variance components and is 1.33 in the top panel and 3.56 in the bottom panel (see vertical dashed lines). The mean of the posterior lies between the mean of the prior (zero, vertical solid lines) and the mode of the likelihood.

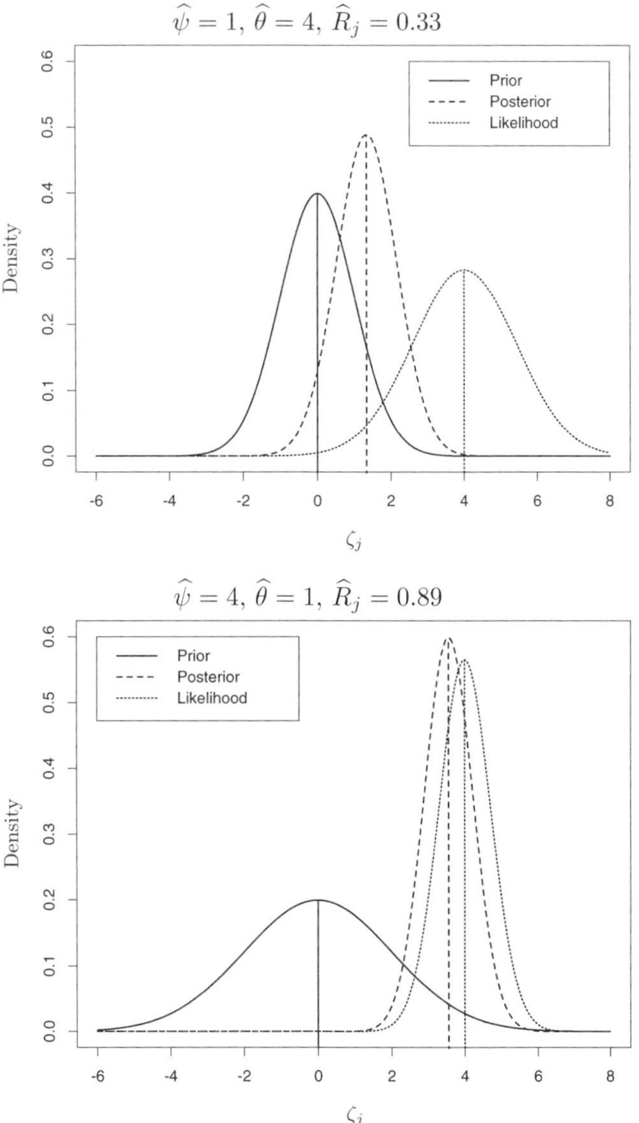

Figure 2.9: Prior, likelihood (normalized), and posterior for a hypothetical subject with $n_j = 2$ responses with total residuals $\widehat{\xi}_{1j} = 3$, $\widehat{\xi}_{2j} = 5$ (the vertical lines represent modes (and means) of the distributions)

In fact, there is a simple formula relating the empirical Bayes prediction $\widetilde{\zeta}_j^{\rm EB}$ to the maximum likelihood estimator $\widehat{\zeta}_j^{\rm ML}$ in linear random-intercept models

$$\widetilde{\zeta}_j^{\rm EB} = \widehat{R}_j \widehat{\zeta}_j^{\rm ML}, \qquad \text{where} \qquad \widehat{R}_j = \frac{\widehat{\psi}}{\widehat{\psi} + \widehat{\theta}/n_j}$$

Here $\widehat{R}_j$ is similar to the estimated intraclass correlation, except that we divide the estimated level-1 variance $\widehat{\theta}$ by the number of responses n_j. $\widehat{R}_j$ therefore represents the *reliability* of the ML estimator of ζ_j, the mean of n_j residuals. $\widehat{R}_j$ is also known as the *shrinkage factor* because $0 < \widehat{R}_j < 1$ so that the empirical Bayes prediction is shrunken toward 0 (the mean of the prior). There will be more shrinkage (i.e., greater influence of the prior) if we have

- a small random-intercept variance $\widehat{\psi}$ (an informative prior)
- a large level-1 residual variance $\widehat{\theta}$ (uninformative data)
- a small cluster size n_j (uninformative cluster)

A nice feature of empirical Bayes prediction is that the prediction error, defined as the difference $\widetilde{\zeta}_j - \zeta_j$ between the prediction and the truth, has zero mean over repeated samples of ζ_j and ϵ_{ij} (or repeated samples of clusters and units from clusters), as well as having the smallest possible variance (treating the model parameters as fixed and known). In linear mixed models, the empirical Bayes predictor is therefore also known as the best linear unbiased predictor (BLUP). However, empirical Bayes predictions are conditionally biased in the sense that their mean over repeated samples of ϵ_{ij} for a given ζ_j (or repeated samples of units from the same cluster) will be too close to zero (due to shrinkage). In contrast, the maximum likelihood estimator is conditionally unbiased but has a greater prediction-error variance. In most applications, shrinkage is desirable because it only affects clusters that provide little information and effectively downplays their influence, "borrowing strength" from other clusters.

For the Mini Wright meter measurements, the shrinkage factor can be calculated from the estimates of $\sqrt{\psi}$ and $\sqrt{\theta}$ in table 2.2:

```
. display 107.05^2/(107.05^2+(19.91^2)/2)
.98299832
```

We can now obtain empirical Bayes predictions in two ways: either by multiplying the maximum likelihood estimates obtained in section 2.9.1 by the shrinkage factor

```
. generate eb1 = .98299832*ml
```

or by using the `predict` command with the `reffects` option after estimation using `xtmixed`:

```
. quietly xtmixed wm || id:, mle
. predict eb2, reffects
```

```
. sort id
. format eb1 eb2 %8.2f
. list id eb1 eb2 if occasion==1, clean noobs
    id       eb1       eb2
     1     63.49     63.49
     2    -30.88    -30.88
     3     59.07     59.07
     4    -17.61    -17.61
     5     45.30     45.30
     6    155.89    155.89
     7    -41.20    -41.20
     8    -67.74    -67.74
     9    192.75    192.75
    10    -15.15    -15.15
    11    -27.44    -27.44
    12    158.84    158.84
    13   -206.83   -206.83
    14     17.78     17.78
    15   -187.17   -187.17
    16    -92.31    -92.31
    17     -6.79     -6.79
```

Both methods give identical results to two decimal places.

There are no simple formulas for empirical Bayes predictions for the models for noncontinuous responses considered in later chapters. For such models, we would have to use the `gllapred` command with the `u` option (after restoring the `gllamm` estimates):

```
estimates restore RI
gllapred eb, u
```

The predictions would then be placed in the variable `ebm1` and would be close to the values in `eb1` and `eb2`.

2.9.3 ❖ Empirical Bayes variances

There are several different kinds of variances for empirical Bayes predictions that can be used to express uncertainty regarding the predictions. However, it is important to remember that these variances do not take into account the uncertainty in the parameter estimates since they are treated as known in empirical Bayes prediction.

Posterior and prediction-error variances

The posterior variance is the variance of the random intercept ζ_j given the observed responses, the variance of the posterior distribution shown as a dashed curve in figure 2.9. For the model considered here, the posterior variance is

$$\text{Var}(\zeta_j|y_{1j}, y_{2j}) \;=\; \frac{\widehat{\theta}/n_j}{\widehat{\psi}+\widehat{\theta}/n_j}\,\widehat{\psi} \;=\; (1-\widehat{R}_j)\,\widehat{\psi}$$

As expected, the posterior variance is smaller than the estimated prior variance $\widehat{\psi}$ due to the information gained regarding the random intercept by knowing the responses y_{1j} and y_{2j}.

In the linear random-intercept model, the posterior variance equals the variance of the prediction errors

$$\text{Var}(\widetilde{\zeta}_j^{\text{EB}} - \zeta_j)$$

over repeated samples of ζ_j and ϵ_{ij} (or repeated samples of clusters j and units i), but with parameters held constant and equal to the estimates. The square root of this variance is sometimes referred to as the *comparative standard error* because it can be used for inferences regarding differences between subjects' true random intercepts.

At the time this text was printed, there was no `xtmixed` postestimation command for estimating posterior variances. We therefore restore the `gllamm` estimates stored under `RI` and use the `gllapred` command with the `u` option:

```
. estimates restore RI
(results RI are active now)
. gllapred eb, u
(means and standard deviations will be stored in ebm1 ebs1)
```

As indicated by the output, the required posterior standard deviations are stored in the variable `ebs1`. These standard deviations are identical for all clusters because the clusters have the same size $n_j = 2$ so that R_j is the same for all clusters. We therefore display only the value for the first observation:

```
. display ebs1[1]
13.963476
```

Marginal sampling variances

For linear random-intercept models, the sampling distribution of the empirical Bayes predictions (over repeated samples of ζ_j and ϵ_{ij}, or of clusters and units from clusters) is normal with mean 0 and variance

$$\text{Var}(\widetilde{\zeta}_j^{\text{EB}}) \;=\; \frac{\widehat{\psi}}{\widehat{\psi} + \widehat{\theta}/n_j}\,\widehat{\psi} \;=\; \widehat{R}_j\,\widehat{\psi}$$

This variance is useful for deciding if the empirical Bayes prediction for a given subject is aberrant. For instance, 95% of predictions should be no larger in absolute value than about two sampling standard deviations. Thus the sampling standard deviation is often called the *diagnostic standard error*. We can compute this standard error from the estimated prior variance $\widehat{\psi}$ and posterior variance $\text{Var}(\zeta_j|y_{1j}, y_{2j})$ obtained earlier from `gllapred`, using the relationship

$$\text{Var}(\widetilde{\zeta}_j^{\text{EB}}) \;=\; \widehat{R}_j\widehat{\psi} = \widehat{\psi} - (1 - \widehat{R}_j)\,\widehat{\psi} = \widehat{\psi} - \text{Var}(\zeta_j|y_{1j}, y_{2j})$$

The corresponding standard deviation can be obtained using

```
. display sqrt(.98299832*107.05^2)
106.13608
```

Conditional sampling variances

The variance of the empirical Bayes prediction over repeated samples of ϵ_{ij} for a given ζ_j (or repeated samples of units from the same cluster) is given by

$$\text{Var}(\widetilde{\zeta}_j^{\text{EB}}|\zeta_j) \;=\; \widehat{R}_j(1-\widehat{R}_j)\,\widehat{\psi}$$

We can compute the conditional sampling standard deviation using

```
. display sqrt(.98299832*(1-.98299832)*107.05^2)
13.839136
```

Comparing the empirical Bayes variances

If $\widehat{R}_j > 0.5$, as is usually the case in practice, we obtain the following relations among the empirical Bayes variances:

$$\text{Var}(\widetilde{\zeta}_j^{\text{EB}}|\zeta_j) < \text{Var}(\zeta_j|y_{1j}, y_{2j}) = \text{Var}(\widetilde{\zeta}_j^{\text{EB}} - \zeta_j) < \text{Var}(\widetilde{\zeta}_j^{\text{EB}})$$

As we would expect, these relations are satisfied for the Mini Wright data since $\widehat{R}_j = 0.98$.

2.10 Summary and further reading

In this chapter, we introduced the idea of decomposing the total variance between responses into variance components, specifically the between-cluster variance ψ and the within-cluster variance θ. This was accomplished by specifying a model that includes corresponding error components, a level-2 random intercept ζ_j for clusters, and a level-1 residual ϵ_{ij} for units within clusters. The random intercept induces correlations among responses for units in the same cluster, known as the *intraclass correlation*.

These concepts underlie all multilevel or hierarchical modeling. By considering the simplest case of a multilevel model, we have provided some insight into estimation of unknown model parameters and prediction of random effects. We have also shown how to conduct hypothesis testing and construct confidence intervals for variance-components models. Although the expressions for estimators and predictors become more complex for the models discussed in later chapters, the basic ideas remain the same.

For further reading about variance-components models, we recommend Snijders and Bosker (1999), as well as many of the books referred to in later chapters. Streiner and Norman (2003), Shavelson and Webb (1991), and Dunn (2004) are excellent books on measurement models.

2.11 Exercises

2.1 Peak-expiratory-flow data

1. Repeat the analysis of section 2.5 for the Wright peak flow meter measurements (data are in `pefr.dta`) using either `xtmixed` or `gllamm`.
2. Compare the estimates with those obtained for the Mini Wright meter (see table 2.2). Does one measurement method appear to be better than the other?
3. Obtain empirical Bayes predictions for both methods and compare the two sets of predictions graphically.
4. ❖ Which method has a smaller prediction-error variance?

2.2 General-health-questionnaire data

Dunn (1992) reported test–retest data for the 12-item version of Goldberg's (1972) General Health Questionnaire (GHQ) designed to measure psychological distress. Twelve clinical psychology students completed the questionnaire on two occasions, three days apart, giving the scores shown in table 2.3.

Table 2.3: GHQ scores for 12 students tested on two occasions

Student	GHQ1	GHQ2
1	12	12
2	8	7
3	22	24
4	10	14
5	10	8
6	6	4
7	8	5
8	4	6
9	14	14
10	6	5
11	2	5
12	22	16

Source: Dunn 1992.

1. Fit the variance-components model in (2.2) for these data using REML.
2. Obtain maximum likelihood estimates and empirical Bayes predictions of ζ_j. Produce a scatterplot of empirical Bayes predictions versus maximum likelihood estimates with a $y = x$ line superimposed. Describe how the graph would change if there were more shrinkage.

3. Extend the model to allow for different means at the two occasions instead of assuming a common mean β as shown in section 2.8. Is there any evidence for a change in mean GHQ scores?

2.3 Twin-neuroticism data

Sham (1998) analyzed data on 522 female monozygotic (identical) twin-pairs and 272 female dizygotic (nonidentical or fraternal) twin-pairs.

Specifically, the dataset (from MacDonald 1996) contains scores for the neuroticism dimension of the Eysenck Personality Questionnaire (EPQ). Such twin data are often used to find out to what degree a trait (here neuroticism) is due to nature (genes) versus nurture (environment). According to the *equal environment assumption*, monozygotic (MZ) and dizygotic (DZ) twins share the same degree of similarity in their environments so that any excess similarity in neuroticism scores for MZ twins must be due to a greater proportion of shared genes (MZ twins share 100% of their genes whereas DZ twins only share 50%); see Sham (1998) for a more detailed discussion.

The dataset `twin.dta` has the following variables:

- `twin1`: neuroticism score for twin 1 (the twin with the higher score)
- `twin1`: neuroticism score for twin 2
- `num2`: the number of twin pairs with a given pair of neuroticism scores
- `dzmz`: a string variable for DZ versus MZ twins (`dz` and `mz`)

1. Create an identifier for twin pairs (e.g., using `generate pair = _n`) and reshape the data to long form, stacking the neuroticism scores into one variable.
2. Fit the variance-components model in (2.2) separately for MZ and DZ twins by maximum likelihood.

   The data are in "collapsed" or aggregated form with `num2` representing the number of twin pairs having a given pair of neuroticism scores. If you are using `gllamm`, you must use this variable as a level-2 (twin-pair-level) frequency weight by using the `gllamm` option `weight(num)`. If you are using `xtreg` or `xtmixed`, you must expand the data using `expand num2` before creating the twin-pair identifier and reshaping the data in step 1.
3. Compare the estimated variance components and total variances between MZ and DZ twins. Do these estimates suggest that there is a genetic contribution to the variability in neuroticism?
4. Obtain the estimated intraclass correlations. Again do these estimates suggest that there is a genetic contribution to the variability in neuroticism?
5. Why should the Pearson correlation not be used for these data?

2.4 Neighborhood-effects data

We now consider data from Garner and Raudenbush (1991), previously analyzed by Raudenbush and Bryk (2002) and Raudenbush et al. (2004).

The dataset `neighborhood.dta` has the following variables:

- `attain`: a measure of end-of-school educational attainment capturing both attainment and length of schooling (based on the number of O-grades and Higher SCE awards at the A–C levels)
- `neighid`: neighborhood identifier
- `schid`: school identifier

Educational attainment (`attain`) is the response variable.

1. Fit a variance-components model for students nested in schools by maximum likelihood using `xtmixed`. Obtain the estimated intraclass correlation.
2. Fit a variance-components model for students nested in neighborhoods by maximum likelihood using `xtmixed`. Obtain the estimated intraclass correlation.
3. Do neighborhoods or schools appear to have a greater influence on educational attainment?

See also exercises 3.2 and 11.5.

2.5 Essay-grading data

Here we consider a subset of data from Johnson and Albert (1999) on grades assigned to 198 essays by 5 experts. The grades are on a 10-point scale with 10 being "excellent".

The dataset `grader1.dta` has the following variables:

- `essay`: identifier for essays
- `grade1`: grade from grader 1 on 10-point scale
- `grade4`: grade from grader 4 on 10-point scale

1. Reshape the data to stack the grades from graders 1 and 4 into one variable.
2. Fit a linear variance-components model for the essay grades with variance components within and between graders. Use maximum likelihood estimation in `xtmixed`.
3. Obtain the estimated intraclass correlation, here interpretable as an inter-rater reliability.
4. Include a dummy variable for grader 4 in the fixed part of the model to allow for bias between the graders as shown in section 2.8. Does one grader appear to be more generous than the other?
5. Obtain empirical Bayes predictions (using `xtmixed` with the `reffects` option) and plot them using a histogram.

2.6 Head-size data

Frets (1921) analyzed data on the adult head sizes of the first two sons of 25 families. Both the length and breadth of each head were measured in millimeters (mm) and the data are provided by Hand et al. (1994).

The variables in the dataset `headsize.dta` are

- `length1`: length of head of first son (mm)
- `breadth1`: breadth of head of first son (mm)
- `length2`: length of head of second son (mm)
- `breadth2`: breadth of head of second son (mm)

1. We will use a variance-components model to estimate the intraclass correlation between the head lengths of sons nested in families. Why wouldn't it make sense to use this approach for obtaining the intraclass correlation between the head lengths and head breadths nested in heads (or men)?
2. Stack the head lengths of both sons into a variable `length` and the breadths into a variable `breadth`.
3. Fit a linear variance-components model for head length using `xtreg` and `xtmixed` with the `mle` option (the results from both commands should be the same).
4. Interpret the estimated intraclass correlation.
5. Extend the model to allow the mean head length to differ between the first-born and second-born sons (see sec. 2.8).
   a. Write down the model.
   b. Fit the model using `xtmixed` with the `mle` option.
   c. Interpret the results.

2.7 Georgian birthweight data

Adams et al. (1997) analyzed a dataset on all live births that occurred in Georgia, U.S.A, from 1980 to 1992 (regardless of maternal state of residence), or that occurred in other states to Georgia residents and for which birth certificates were sent to Georgia. They linked data on births to the same mother using 27 different variables (maternal social security number was often missing or inaccurately recorded; see the paper for details). Following Neuhaus and Kalbfleisch (1998) we will use a subset of the linked data, including only births to mothers for whom five births were identified.

We will use the following variables in `birthwt.dta`:

- `mother`: mother identifier
- `child`: child identifier
- `birthwt`: child's birthweight (in grams)

1. Fit a variance-components model to the birthweights using `xtmixed` with the `mle` option, treating children as level 1 and mothers at level 2.

2. At the 5% level, is there significant between-mother variability in birthweights? Fully report the method and result of the test.
3. Obtain the estimated intraclass correlation and interpret it.
4. Obtain empirical Bayes predictions of the random intercept and plot a histogram of the empirical Bayes predictions.

See also exercise 3.6.

2.8 Reliability and empirical Bayes prediction

In a hypothetical test–retest study, the estimates for the measurement model in (2.2) were as shown in table 2.4.

Table 2.4: Estimates for hypothetical test–retest study

	Est
β	30
ψ	9
θ	8

1. What is the estimated test–retest reliability of these measurements?
2. For a person with measurements 34 and 36, obtain the maximum likelihood estimate and empirical Bayes prediction of ζ_j and of the true score.
3. Two additional measurements of 37 and 33 are made on the same person. Using all four measurements, obtain the maximum likelihood estimate and empirical Bayes prediction of ζ_j and of the true score. Compare the results with the results for step 2, and explain the reason for any differences.
4. What is the empirical Bayes prediction of ζ_j for a person who has not been measured yet?

2.9 ❖ Maximum likelihood and restricted maximum likelihood

For the Mini Wright meter, use the maximum likelihood estimates in table 2.2 to calculate the REML or ANOVA estimate of ψ. Also obtain the corresponding REML or ANOVA estimate of the intraclass correlation.

3 Random-intercept models with covariates

3.1 Introduction

In this chapter, we extend the variance-components models introduced in the previous chapter by including observed explanatory variables or covariates x. Seen from another perspective, we extend the linear regression models discussed in chapter 1 by introducing random intercepts ζ_j.

Although many of the features of the variance-components models persist, new issues arise in estimating regression coefficients. In particular, we discuss the distinction between within-cluster and between-cluster covariate effects and the problem of omitted cluster-level covariates and endogeneity. We also discuss coefficients of determination or measures of variation explained by covariates.

3.2 Does smoking during pregnancy affect birthweight?

Abrevaya (2006) investigates the effect of smoking on birth outcomes using the Natality datasets derived from birth certificates by the U.S. National Center for Health Statistics.

Abrevaya identified multiple births from the same mothers in 9 datasets from 1990–1998 by matching mothers across the datasets. Unlike, for instance, the Nordic countries, a unique person identifier such as a person identification number, social security number, or name is rarely available in U.S. datasets. Perfect matching is thus precluded, and matching must proceed by identifying mothers who have identical values on a set of variables in all datasets. In this study, matching was accomplished by considering mother's state of birth and child's state of birth, as well as mother's county and city of birth, mother's age, race, education, marital status, and, if married, father's age and race. For the matching on mother's and child's states of birth to be useful, the data were restricted to combinations of states that occur rarely.

Here we consider the subset of the matches where the observed interval between births was consistent with the interval since the last birth recorded on the birth certificate. The data are restricted to births with complete data for the variables considered by Abrevaya (2006), singleton births (no twins or other multiple births) and births to mothers for whom at least two births between 1990 and 1998 could be matched and whose race was classified as white or black. We took a 10% random sample of this dataset yielding 8,604 births from 3,978 mothers.

The birth outcome we will concentrate on is birthweight. Abrevaya (2006) motivates his study by citing a report from the U.S. Surgeon General:

> "Infants born to women who smoke during pregnancy have a lower average birthweight and are more likely to be small for gestational age than infants born to women who do not smoke..."
> (*Women and Smoking: A Report of the Surgeon General*, Centers for Disease Control and Prevention, 2001).

We will use the following variables in `smoking.dta`:

- `momid`: mother identifier
- `birwt`: birthweight (in grams)
- `mage`: mother's age at the birth of the child (in years)
- `smoke`: dummy variable for mother smoking during pregnancy (1: smoking; 0: not smoking)
- `male`: dummy variable for baby being male (1: male; 0: female)
- `married`: dummy variable for mother being married (1: married; 0: unmarried)
- `hsgrad`: dummy variable for mother having graduated from high school (1: graduated; 0: did not graduate)
- `somecoll`: dummy variable for mother having some college education, but no degree (1: some college; 0: no college)
- `collgrad`: dummy variable for mother having graduated from college (1: graduated; 0: did not graduate)
- `black`: dummy variable for mother being black (1: black; 0: white)
- `kessner2`: dummy variable for Kessner index = 2, or intermediate prenatal care (1: index=2; 0: otherwise)
- `kessner3`: dummy variable for Kessner index = 3, or inadequate prenatal care (1: index=3; 0: otherwise)
- `novisit`: dummy variable for no prenatal visit (1: no visit; 0: at least 1 visit)
- `pretri2`: dummy variable for first prenatal visit having occurred in 2nd trimester (1: yes; 0: no)
- `pretri3`: dummy variable for first prenatal visit having occurred in 3rd trimester (1: yes; 0: no)

Smoking status was determined from the answer to the question asked on the birth certificate whether there was tobacco use during pregnancy. The dummy variables for mother's education, `hsgrad`, `somecoll`, and `collgrad`, were derived from the years of education given on the birth certificate. The Kessner index is a measure of the adequacy of prenatal care (1: adequate; 2: intermediate; 3: inadequate) based on the timing of the first prenatal visit and the number of prenatal visits taking into account the gestational age of the fetus.

The data have a two-level structure with births (or children or pregnancies) at level 1 and mothers at level 2. In multilevel models, the response variable always varies at the lowest level, taking on different values for different level-1 units within the same level-2 cluster. However, explanatory variables can either vary at level 1 or at level 2. For instance, while `smoke` can change from one pregnancy to the next, `black` is constant between pregnancies. `smoke` is therefore said to be a level-1 variable whereas `black` is a level-2 variable. Among the variables listed above, `black` appears to be the only one that cannot in principle change between pregnancies. However, because of the way the matching was done, the education dummy variables (`hsgrad`, `somecoll`, and `collgrad`) and marital status also remain constant across births for the same mother and are thus level-2 variables.

We start by reading the smoking and birthweight data into Stata using the command

```
. use http://www.stata-press.com/data/mlmus2/smoking
```

A useful Stata command for exploring how much variables vary at level 1 and 2 is `xtsum`:

```
. xtsum birwt smoke black, i(momid)
Variable         |      Mean   Std. Dev.       Min        Max |    Observations
-----------------+--------------------------------------------+----------------
birwt    overall |  3469.931   527.1394        284       5642 |     N =    8604
         between |             451.1943       1361     5183.5 |     n =    3978
         within  |             276.7966   1528.431   5411.431 | T-bar =  2.1629
                 |                                            |
smoke    overall |  .1399349   .3469397          0          1 |     N =    8604
         between |             .3216459          0          1 |     n =    3978
         within  |             .1368006  -.5267318   .8066016 | T-bar =  2.1629
                 |                                            |
black    overall |  .0717108   .2580235          0          1 |     N =    8604
         between |              .257512          0          1 |     n =    3978
         within  |                    0   .0717108   .0717108 | T-bar =  2.1629
```

The total number of observations is $N = 8604$, the number of clusters is $J = 3978$ (`n` in the output), and there are on average about 2.2 births per mother (`T-bar` in the output) in the dataset.

Three different sample standard deviations are given for each variable: the *overall standard deviation* s_{xO}, defined as usual as the square root of the mean squared deviation of observations from the overall mean

$$s_{xO}^2 = \frac{1}{N-1}\sum_{j=1}^{J}\sum_{i=1}^{n_j}(x_{ij} - \overline{x}_{..})^2$$

the *between standard deviation*, defined as the square root of the mean squared deviation of the cluster means from the overall mean

$$s_{xB}^2 = \frac{1}{J-1}\sum_{j=1}^{J}(\overline{x}_{\cdot j} - \overline{x}_{\cdot\cdot})^2$$

and the *within standard deviation*, defined as the square root of the mean squared deviation of observations from the cluster means

$$s_{xW}^2 = \frac{1}{N-1}\sum_{j=1}^{J}\sum_{i=1}^{n_j}(x_{ij} - \overline{x}_{\cdot j})^2$$

We see that birthweight and smoking vary more between mothers than within mothers whereas being black does not vary at all within mothers as expected.

3.3 The linear random-intercept model with covariates

3.3.1 Model specification

An obvious model to consider for the continuous response variable birthweight is a multiple linear regression model (discussed in chapter 1) including smoking status and various other variables as explanatory variables or covariates.

The model for the birthweight y_{ij} of child i with mother j is specified as

$$y_{ij} = \beta_1 + \beta_2 x_{2ij} + \cdots + \beta_p x_{pij} + \xi_{ij} \tag{3.1}$$

where x_{2ij} through x_{pij} are covariates and ξ_{ij} is a residual.

It may be unrealistic to assume that the birthweights of children born to the same mother are independent given the observed covariates, or in other words that the residuals ξ_{ij} and $\xi_{i'j}$ are independent. We can therefore use the idea introduced in the previous chapter to split the total residual or error into two error components:

$$\xi_{ij} \equiv \zeta_j + \epsilon_{ij}$$

Substituting for ξ_{ij} into the multiple-regression model (3.1), we obtain a *linear random-intercept model with covariates*

$$\begin{aligned} y_{ij} &= \beta_1 + \beta_2 x_{2ij} + \cdots + \beta_p x_{pij} + \zeta_j + \epsilon_{ij} \\ &= (\beta_1 + \zeta_j) + \beta_2 x_{2ij} + \cdots + \beta_p x_{pij} + \epsilon_{ij} \end{aligned} \tag{3.2}$$

This model can be viewed as a regression model with a mother-specific intercept $\beta_1 + \zeta_j$. The random intercept ζ_j can be considered a "random parameter" that is not estimated along with the fixed parameters β_1 through β_p, but whose variance ψ is estimated together with the variance θ of the ϵ_{ij}. The linear random-intercept model with covariates is the simplest example of a *linear mixed (effects) model* where there are both fixed and random "effects".

The random intercept or level-2 residual ζ_j is a mother-specific error component, which remains constant across births, whereas the level-1 residual ϵ_{ij} is a child-specific error component, which varies between children i as well as mothers j. The ζ_j are independent over mothers, the ϵ_{ij} are independent over mothers and children, and the two error components are independent of each other.

The mother-specific error component ζ_j represents the combined effects of omitted mother characteristics or unobserved heterogeneity. If ζ_j is positive, the total residuals for mother j, ξ_{ij}, will tend to be positive, leading to heavier babies than predicted by the covariates, and if ζ_j is negative, the total residuals will tend to be negative. Since ζ_j is shared by all responses for the same mother, it induces within-mother dependence among the total residuals ξ_{ij}.

Letting $\mathbf{x}_{ij}=(x_{2ij},\ldots,x_{pij})'$ be the vector consisting of all observed covariates, the exogeneity assumptions are

$$E(\zeta_j|\mathbf{x}_{ij}) = 0 \tag{3.3}$$

and

$$E(\epsilon_{ij}|\mathbf{x}_{ij},\zeta_j) = 0 \tag{3.4}$$

from which it follows that $E(\epsilon_{ij}|\mathbf{x}_{ij}) = 0$. These assumptions ensure that the population-averaged or marginal regression (averaged over ζ_j and ϵ_{ij}, but given $\mathbf{x}_{ij}$) is linear

$$\begin{aligned} E(y_{ij}|\mathbf{x}_{ij}) &= E(\beta_1+\beta_2 x_{2ij}+\cdots+\beta_p x_{pij}) + \underbrace{E(\zeta_j|\mathbf{x}_{ij})}_{0} + \underbrace{E(\epsilon_{ij}|\mathbf{x}_{ij})}_{0} \\ &= \beta_1+\beta_2 x_{2ij}+\cdots+\beta_p x_{pij} \end{aligned} \tag{3.5}$$

and that the cluster-specific or conditional regression (averaged over ϵ_{ij}, but given ζ_j and $\mathbf{x}_{ij}$) is linear

$$\begin{aligned} E(y_{ij}|\mathbf{x}_{ij},\zeta_j) &= E(\beta_1+\beta_2 x_{2ij}+\cdots+\beta_p x_{pij}) + E(\zeta_j|\mathbf{x}_{ij},\zeta_j) + \underbrace{E(\epsilon_{ij}|\mathbf{x}_{ij},\zeta_j)}_{0} \\ &= \beta_1+\beta_2 x_{2ij}+\cdots+\beta_p x_{pij}+\zeta_j \end{aligned} \tag{3.6}$$

It follows from the exogeneity assumptions stated in (3.3) and (3.4) that both ζ_j and ϵ_{ij} are uncorrelated with the covariates. For example, smoking is assumed to be uncorrelated with the random intercept for mother, which represents the effect of omitted mother-specific covariates on birthweight. Endogeneity, or violation of exogeneity, is therefore often discussed in terms of correlations between the error terms and covariates (see sec. 3.7.4).

Regarding distributional assumptions, we specify that

$$\zeta_j|\mathbf{x}_{ij} \sim N(0,\psi)$$

and

$$\epsilon_{ij}|\mathbf{x}_{ij}, \zeta_j \sim N(0, \theta)$$

from which it follows that $\zeta_j \sim N(0, \psi)$ and $\epsilon_{ij} \sim N(0, \theta)$.

A graphical illustration of the random-intercept model with a single covariate x_{ij} for a mother j is given in figure 3.1. Here the solid line is $E(y_{ij}|x_{ij}) = \beta_1 + \beta_2 x_{ij}$,

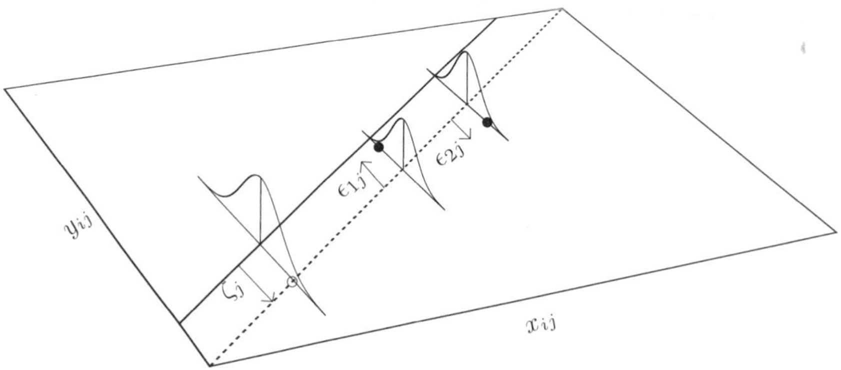

Figure 3.1: Illustration of random-intercept model for one mother

the population-averaged regression line for the population of all mothers j. The normal density curve centered on this line represents the random-intercept distribution and the hollow circle represents a realization ζ_j from this distribution for mother j (this could have been placed anywhere along the line). This negative random intercept ζ_j produces the dotted mother-specific regression line $E(y_{ij}|x_{ij}, \zeta_j) = (\beta_1 + \zeta_j) + \beta_2 x_{ij}$. This line is parallel to and below the population-averaged regression line. For a mother with a positive ζ_j, the mother-specific regression line would be parallel to and above the population-averaged regression line. Two observed responses $y_{ij} = (\beta_1 + \zeta_j) + \beta_2 x_{ij} + \epsilon_{ij}$, $(i = 1, 2)$, are shown where ϵ_{1j} and ϵ_{2j} are random samples from the two normal residual distributions shown on the dotted curve.

3.3.2 Residual variance and intraclass correlation

As shown in section 2.3.2, the total residuals or error terms are homoskedastic (having constant variance),

$$\text{Var}(\xi_{ij}) = \text{Var}(\zeta_j + \epsilon_{ij}) = \psi + \theta$$

where ψ is the variance of ζ_j and θ the variance of ϵ_{ij} as before. It follows that the responses y_{ij}, given the observed covariates $\mathbf{x}_{ij}$, are also homoskedastic

$$\text{Var}(y_{ij}|\mathbf{x}_{ij}) = \psi + \theta$$

As shown in section 2.3, the correlation between the total residuals for any two children i and i' of the same mother j, also called the residual correlation, is

$$\rho \equiv \mathrm{Cor}(\xi_{ij}, \xi_{i'j}) \;=\; \frac{\psi}{\psi + \theta} \tag{3.7}$$

Thus ρ is also the *intraclass correlation* of responses y_{ij} and $y_{i'j}$ for mother j, given the covariates

$$\rho \equiv \mathrm{Cor}(y_{ij}, y_{i'j}|\mathbf{x}_{ij}, \mathbf{x}_{i'j}) \;=\; \frac{\psi}{\psi + \theta}$$

It is important to distinguish between the intraclass correlation in a model not containing any covariates, sometimes called the *unconditional* intraclass correlation, and the *conditional* or *residual* intraclass correlation in a model containing covariates.

3.4 Estimation using Stata

We can use `xtreg`, `xtmixed`, or `gllamm` to fit the models by maximum likelihood. In addition, `xtreg` can be used to obtain generalized least-squares estimates, and `xtmixed` can be used to obtain restricted maximum likelihood estimates. As discussed in chapter 2, `xtreg` is computationally the most efficient, whereas `gllamm` is computationally the least efficient. Unless some special feature of `gllamm` or its prediction command `gllapred` are needed, we do not recommend using `gllamm` for linear models.

3.4.1 Using xtreg

The command for fitting the random-intercept model (3.2) by maximum likelihood using `xtreg` is

(Continued on next page)

```
. xtreg birwt smoke male mage hsgrad somecoll collgrad married black kessner2
> kessner3 novisit pretri2 pretri3, i(momid) mle
Random-effects ML regression                    Number of obs      =      8604
Group variable: momid                           Number of groups   =      3978
Random effects u_i ~ Gaussian                   Obs per group: min =         2
                                                               avg =       2.2
                                                               max =         3
                                                LR chi2(13)        =    659.47
Log likelihood  = -65145.752                    Prob > chi2        =    0.0000

       birwt       Coef.   Std. Err.      z    P>|z|     [95% Conf. Interval]

       smoke   -218.3289   18.20988   -11.99   0.000    -254.0196   -182.6382
        male    120.9375   9.558721    12.65   0.000     102.2027    139.6722
        mage    8.100548   1.347266     6.01   0.000     5.459956    10.74114
      hsgrad    56.84715   25.03538     2.27   0.023     7.778705    105.9156
    somecoll    80.68607   27.30914     2.95   0.003     27.16115     134.211
    collgrad    90.83273   27.99598     3.24   0.001     35.96162    145.7038
     married     49.9202   25.50319     1.96   0.050    -.0651368    99.90554
       black   -211.4138   28.27818    -7.48   0.000     -266.838   -155.9896
    kessner2   -92.91883   19.92624    -4.66   0.000    -131.9736   -53.86411
    kessner3   -150.8759   40.83414    -3.69   0.000    -230.9093   -70.84246
     novisit   -30.03035   65.69213    -0.46   0.648    -158.7846    98.72387
     pretri2     92.8579   23.19258     4.00   0.000     47.40127    138.3145
     pretri3    178.7295   51.64145     3.46   0.001     77.51416    279.9449
       _cons    3117.191   40.97597    76.07   0.000      3036.88    3197.503

    /sigma_u    338.7674   6.296444                      326.6487    351.3358
    /sigma_e    370.6654   3.867707                      363.1618     378.324
         rho    .4551282   .0119411                      .4318152    .4785967

Likelihood-ratio test of sigma_u=0: chibar2(01)= 1108.77 Prob>=chibar2 = 0.000
```

(If the cluster identifier is first defined by the `xtset` command, the `i()` option is not required in the `xtreg` command.)

The estimated regression coefficients are given next to the corresponding covariate name; for instance, the coefficient β_2 of `smoke` is estimated as -218. This means that, according to the fitted model, the expected birthweight is 218 grams lower for a child of a mother who smoked during the pregnancy compared with a child of a mother who did not smoke, controlling or adjusting for the other covariates. The estimated regression coefficients for the other covariates make sense, although the coefficients for the prenatal care variables (`kessner2`, `kessner3`, `novisit`, `pretri2`, `pretri3`) are not straightforward to interpret because their definitions are partly overlapping.

The estimate of the random-intercept standard deviation $\sqrt{\psi}$ is given under `/sigma_u` as 339 grams, and the estimate of the level-1 residual standard deviation $\widehat{\theta}$ is given under `/sigma_e` as 371 grams.

The maximum likelihood estimates for the random-intercept model are also presented in the first columns of table 3.1. The estimated regression coefficients are reported in the "Fixed part" of the table, and the estimated standard deviations for the random intercept and level-1 residual are given in the "Random part".

Table 3.1: Maximum likelihood estimates for smoking data (in grams)

	Full model		Null model		Level-2 cov.	
	Est	(SE)	Est	(SE)	Est	(SE)
Fixed part						
β_1 [_cons]	3,117	(41)	3,468	(7)	3,216	(26)
β_2 [smoke]	−218	(18)				
β_3 [male]	121	(10)				
β_4 [mage]	8	(1)				
β_5 [hsgrad]	57	(25)			131	(25)
β_6 [somecoll]	81	(27)			181	(27)
β_7 [collgrad]	91	(28)			233	(26)
β_8 [married]	50	(26)			115	(25)
β_9 [black]	−211	(28)			−201	(29)
β_{10} [kessner2]	−93	(20)				
β_{11} [kessner3]	−151	(41)				
β_{12} [novisit]	−30	(66)				
β_{13} [pretri2]	93	(23)				
β_{14} [pretri3]	179	(52)				
Random part						
$\sqrt{\psi}$	339		368		348	
$\sqrt{\theta}$	371		378		378	
Derived estimates						
R^2	0.09		0.00		0.05	
ρ	0.46		0.49		0.46	

3.4.2 Using xtmixed

The random-intercept model (3.2) can also be fitted by maximum likelihood using `xtmixed` with the `mle` option:

(*Continued on next page*)

```
. xtmixed birwt smoke male mage hsgrad somecoll collgrad married black
> kessner2 kessner3 novisit pretri2 pretri3, || momid:, mle
Mixed-effects ML regression                     Number of obs      =      8604
Group variable: momid                           Number of groups   =      3978

                                                Obs per group: min =         2
                                                               avg =       2.2
                                                               max =         3

                                                Wald chi2(13)      =    693.74
Log likelihood = -65145.752                     Prob > chi2        =    0.0000

------------------------------------------------------------------------------
       birwt |      Coef.   Std. Err.      z    P>|z|     [95% Conf. Interval]
-------------+----------------------------------------------------------------
       smoke |  -218.3286   18.15946   -12.02   0.000    -253.9205   -182.7368
        male |   120.9375   9.558003    12.65   0.000     102.2042    139.6708
        mage |   8.100566   1.344573     6.02   0.000     5.465251    10.73588
      hsgrad |   56.84716   25.03543     2.27   0.023     7.778611    105.9157
    somecoll |   80.68605   27.30906     2.95   0.003     27.16127    134.2108
    collgrad |   90.83268   27.99498     3.24   0.001     35.96354    145.7018
     married |   49.92022   25.50309     1.96   0.050    -.0649248    99.90537
       black |  -211.4138   28.27764    -7.48   0.000    -266.8369   -155.9906
    kessner2 |  -92.91882   19.92617    -4.66   0.000    -131.9734   -53.86424
    kessner3 |  -150.8758   40.83027    -3.70   0.000    -230.9017   -70.84992
     novisit |   -30.0303   65.69165    -0.46   0.648    -158.7836    98.72298
     pretri2 |   92.85784   23.19067     4.00   0.000     47.40497    138.3107
     pretri3 |   178.7294   51.63677     3.46   0.001      77.5232    279.9356
       _cons |   3117.191   40.88824    76.24   0.000     3037.051     3197.33
------------------------------------------------------------------------------

------------------------------------------------------------------------------
  Random-effects Parameters  |   Estimate   Std. Err.     [95% Conf. Interval]
-----------------------------+------------------------------------------------
momid: Identity              |
                   sd(_cons) |   338.7686   6.296449      326.6499     351.337
-----------------------------+------------------------------------------------
                sd(Residual) |   370.6648   3.867695      363.1613    378.3234
------------------------------------------------------------------------------
LR test vs. linear regression: chibar2(01) =  1108.77 Prob >= chibar2 = 0.0000
```

The estimates are identical to those reported for `xtreg` in table 3.1.

The `reml` option can be used instead of `mle` to obtain restricted maximum likelihood (REML) estimates (see sec. 2.7.1 for the basic idea of REML). When there are many level-2 units J as here, the REML and ML estimates will be almost identical.

3.4.3 Using gllamm

Although we do not recommend using `gllamm` for linear models because it is slower than `xtreg` and `xtmixed` and sometimes less accurate, there are occasions where estimation in `gllamm` may be the only way to obtain certain kinds of postestimation options as we will see in section 3.9.

The `gllamm` command for fitting the model is

```
. gllamm birwt smoke male mage hsgrad somecoll collgrad married black
> kessner2 kessner3 novisit pretri2 pretri3, i(momid) adapt

number of level 1 units = 8604
number of level 2 units = 3978

Condition Number = 8059.0645

gllamm model

log likelihood = -65145.752
```

birwt	Coef.	Std. Err.	z	P>|z|	[95% Conf.	Interval]
smoke	-218.3289	18.20989	-11.99	0.000	-254.0197	-182.6382
male	120.9375	9.558726	12.65	0.000	102.2027	139.6722
mage	8.100549	1.347267	6.01	0.000	5.459954	10.74114
hsgrad	56.84715	25.03541	2.27	0.023	7.77866	105.9156
somecoll	80.68607	27.30916	2.95	0.003	27.1611	134.211
collgrad	90.83273	27.99601	3.24	0.001	35.96157	145.7039
married	49.9202	25.50322	1.96	0.050	-.0651828	99.90559
black	-211.4138	28.27821	-7.48	0.000	-266.8381	-155.9895
kessner2	-92.91883	19.92625	-4.66	0.000	-131.9736	-53.86409
kessner3	-150.8759	40.83416	-3.69	0.000	-230.9094	-70.84242
novisit	-30.03035	65.69217	-0.46	0.648	-158.7846	98.72394
pretri2	92.8579	23.1926	4.00	0.000	47.40125	138.3145
pretri3	178.7295	51.64148	3.46	0.001	77.5141	279.945
_cons	3117.191	40.97601	76.07	0.000	3036.88	3197.503

```
Variance at level 1
------------------------------------------------------------------------------

  137392.81 (2867.2533)

Variances and covariances of random effects
------------------------------------------------------------------------------


***level 2 (momid)

    var(1): 114763.41 (4266.0686)
------------------------------------------------------------------------------
```

The estimates are close to those using `xtreg` and `xtmixed`. Estimation takes a long time for this example, so if there is any chance we may need the estimates again—for instance to perform diagnostics—we should keep the estimates in memory for later use

```
. estimates store gllamm
```

If estimates may be required in a future Stata session, they can also be saved in a file using `estimates save`.

3.5 Coefficients of determination or variance explained

In section 1.5, we motivated the coefficient of determination, or "R-squared", as the proportional reduction in prediction error variance comparing the model without covariates (the "null model") with the model of interest.

In ordinary linear regression without covariates, the predictions are $\widehat{y}_i = \overline{y}$, so the sample prediction error variance is the mean squared error (MSE) for the null model,

$$\text{MSE}_0 = \frac{1}{N-1}\sum_i (y_i - \overline{y})^2 = \widehat{\sigma_0^2}$$

where $\widehat{\sigma_0^2}$ is an estimate of the residual variance in the null model. In the ordinary linear regression model including all covariates, the predictions are $\widehat{y}_i = \widehat{\beta}_1 x_{1i} + \cdots + \widehat{\beta}_p x_{pi}$, and the prediction error variance is the mean squared error in the regression model of interest

$$\text{MSE}_1 = \frac{1}{N-p}\sum_i (y_i - \widehat{y}_i)^2 = \widehat{\sigma_1^2}$$

This is also an estimate of the residual variance σ_1^2 in the model of interest. The coefficient of determination is defined as

$$R^2 = \frac{\sum_i (y_i - \overline{y})^2 - \sum_i (y_i - \widehat{y}_i)^2}{\sum_i (y_i - \overline{y})^2} \approx \frac{\widehat{\sigma_0^2} - \widehat{\sigma_1^2}}{\widehat{\sigma_0^2}}$$

where the approximation improves as N increases.

In a linear random-intercept model, the total residual variance is given by

$$\text{Var}(\zeta_j + \epsilon_{ij}) = \psi + \theta$$

An obvious definition of the coefficient of determination for two-level models, suggested by Snijders and Bosker (1994, 1999), is therefore the proportional reduction in the estimated total residual variance comparing the null model without covariates with the model of interest,

$$R^2 = \frac{\widehat{\psi}_0 + \widehat{\theta}_0 - (\widehat{\psi}_1 + \widehat{\theta}_1)}{\widehat{\psi}_0 + \widehat{\theta}_0}$$

where $\widehat{\psi}_0$ and $\widehat{\theta}_0$ are the estimates for the null model and $\widehat{\psi}_1$ and $\widehat{\theta}_1$ are the estimates for the model of interest.

First, we fit the null model, also often called the *unconditional model*:

```
. xtreg birwt, i(momid) mle
Random-effects ML regression                    Number of obs      =      8604
Group variable: momid                           Number of groups   =      3978
Random effects u_i ~ Gaussian                   Obs per group: min =         2
                                                               avg =       2.2
                                                               max =         3
                                                Wald chi2(0)       =      0.00
Log likelihood  = -65475.486                    Prob > chi2        =         .

------------------------------------------------------------------------------
       birwt |      Coef.   Std. Err.      z    P>|z|     [95% Conf. Interval]
-------------+----------------------------------------------------------------
       _cons |   3467.969   7.137618   485.87   0.000     3453.979    3481.958
-------------+----------------------------------------------------------------
    /sigma_u |   368.2866    6.45442                      355.8509    381.1568
    /sigma_e |   377.6578   3.926794                      370.0393    385.4331
         rho |   .4874391   .0114188                      .4650901    .5098276
------------------------------------------------------------------------------
Likelihood-ratio test of sigma_u=0: chibar2(01)= 1315.66 Prob>=chibar2 = 0.000
```

The estimates for this model are also given under "Null model" in table 3.1. The total variance is estimated as

$$\widehat{\psi}_0 + \widehat{\theta}_0 \;=\; 368.2866^2 + 377.6578^2 = 278260.43$$

For the model including all covariates, whose estimates are given under "Full model" in table 3.1, the total residual variance is estimated as

$$\widehat{\psi}_1 + \widehat{\theta}_1 \;=\; 338.7686^2 + 370.6648^2 = 252156.56$$

It follows that

$$R^2 \;=\; \frac{278260.43 - 252156.56}{278260.43} = 0.09$$

so 9% of the variance is explained by the covariates.

Raudenbush and Bryk (2002) suggest considering the proportional reduction in each of the variance components separately. In our example, the proportion of level-2 variance explained by the covariates is

$$R_2^2 \;=\; \frac{\widehat{\psi}_0 - \widehat{\psi}_1}{\widehat{\psi}_0} = \frac{368.2866^2 - 338.7686^2}{368.2866^2} = 0.15$$

and the proportion of level-1 variance explained is

$$R_1^2 \;=\; \frac{\widehat{\theta}_0 - \widehat{\theta}_1}{\widehat{\theta}_0} = \frac{377.6578^2 - 370.6648^2}{377.6578^2} = 0.04$$

Let us now fit a random-intercept model that includes only the level-2 covariates:

```
. xtreg birwt hsgrad somecoll collgrad married black, i(momid) mle
Random-effects ML regression                    Number of obs      =      8604
Group variable: momid                           Number of groups   =      3978
Random effects u_i ~ Gaussian                   Obs per group: min =         2
                                                               avg =       2.2
                                                               max =         3
                                                LR chi2(5)         =    290.48
Log likelihood  = -65330.247                    Prob > chi2        =    0.0000

------------------------------------------------------------------------------
       birwt |      Coef.   Std. Err.      z    P>|z|     [95% Conf. Interval]
-------------+----------------------------------------------------------------
      hsgrad |   131.4395   24.91149     5.28   0.000     82.61384    180.2651
    somecoll |   180.6879   26.50378     6.82   0.000     128.7414    232.6343
    collgrad |   232.8944   25.58597     9.10   0.000     182.7468    283.0419
     married |    114.765   25.45984     4.51   0.000     64.86465    164.6654
       black |  -201.4773   28.80249    -7.00   0.000    -257.9292   -145.0255
       _cons |   3216.482   25.82479   124.55   0.000     3165.866    3267.097
-------------+----------------------------------------------------------------
    /sigma_u |   348.1441   6.390242                      335.8421    360.8968
    /sigma_e |   377.7638   3.929694                      370.1397    385.5449
         rho |   .4592642   .0118089                       .436201    .4824653
------------------------------------------------------------------------------
Likelihood-ratio test of sigma_u=0: chibar2(01)= 1146.40 Prob>=chibar2 = 0.000
```

In general, and as we can see from comparing the estimates for this model (given under "Level-2 cov." in table 3.1) with the estimates from the null model, adding level-2 covariates will reduce mostly the level-2 variance. However, adding level-1 covariates can reduce both variances as we can see by comparing the estimates for the above model with the full model. Note that the level-2 variance can *increase* when adding level-1 covariates, potentially producing a negative R_2^2.

For the intraclass correlation, we see that the unconditional intraclass correlation for the null model without covariates is estimated as 0.49. This reduces to a conditional or residual intraclass correlation of 0.459 when level-2 covariates are added and to 0.455 when all remaining covariates are added. The conditional intraclass correlation can also be larger than the unconditional intraclass correlation if the estimated level-1 variance decreases more than the level-2 variance when covariates are added.

3.6 Hypothesis tests and confidence intervals

3.6.1 Hypothesis tests for regression coefficients

In ordinary linear regression, we use t tests for testing hypotheses regarding individual regression parameters and F tests for joint hypotheses regarding several regression parameters. Under the null hypothesis, these test statistics have t distributions and F distributions, respectively, with appropriate degrees of freedom in finite samples.

Since finite sample results are not readily available in the multilevel setting, hypothesis testing typically proceeds based on likelihood-ratio or Wald test statistics with asymptotic (large sample) χ^2 null distributions, with the number of restrictions imposed by the null hypothesis as degrees of freedom. The likelihood-ratio and Wald tests (and the less commonly used score or Lagrange multiplier test) are asymptotically equivalent to each other but may produce different conclusions in small samples.

Hypothesis tests for individual regression coefficients

The most commonly used hypothesis test concerns an individual regression parameter, say β_2, with null hypothesis

$$H_0\colon \beta_2 = 0$$

versus the two-sided alternative $H_a\colon \beta_2 \neq 0$.

The Wald statistic for testing the null hypothesis is

$$w = \left(\frac{\widehat{\beta}_2}{\widehat{\mathrm{SE}}(\widehat{\beta}_2)} \right)^2$$

with a χ^2 null distribution with 1 degree of freedom since the null hypothesis imposes one restriction. In practice, the test statistic

$$z = \frac{\widehat{\beta}_2}{\widehat{\mathrm{SE}}(\widehat{\beta}_2)}$$

is usually used, which has a standard normal null distribution (because its square has a χ^2 distribution with 1 degree of freedom).

The z statistic is reported as `z` in the Stata output. For instance, in the output from `xtmixed` on page 100, the z statistic for the regression parameter of smoking is -12.02, which gives a two-sided p-value of less than 0.001.

The likelihood-ratio test statistic is less commonly used for testing individual regression parameters but takes the same form as explained below for several coefficients.

Joint hypothesis tests for several regression coefficients

Consider now the null hypothesis that the regression coefficients of two covariates x_{2ij} and x_{3ij} are both zero,

$$H_0\colon \beta_2 = \beta_3 = 0$$

versus the alternative hypothesis that at least one of the parameters is nonzero. For example, for the smoking and birthweight application, we may want to test the null hypothesis that the quality of prenatal care (as measured by the Kessner index) makes no difference to birthweight (controlling for the other covariates), where the Kessner index is represented by two dummy variables, `kessner2` and `kessner3`.

Let $\widehat{\beta}_2$ and $\widehat{\beta}_3$ be maximum likelihood estimates from the model including the covariates. Both Wald and likelihood-ratio tests have asymptotic χ^2 null distributions with 2 degrees of freedom since the null imposes two restrictions.

The Wald statistic can be expressed as

$$
\begin{aligned}
w &= (\widehat{\beta}_2, \widehat{\beta}_3) \left\{ \begin{array}{cc} \widehat{\text{SE}}(\widehat{\beta}_2)^2 & \widehat{\text{Cor}}(\widehat{\beta}_2, \widehat{\beta}_3)\, \widehat{\text{SE}}(\widehat{\beta}_2)\, \widehat{\text{SE}}(\widehat{\beta}_3) \\ \widehat{\text{Cor}}(\widehat{\beta}_2, \widehat{\beta}_3)\, \widehat{\text{SE}}(\widehat{\beta}_2)\, \widehat{\text{SE}}(\widehat{\beta}_3) & \widehat{\text{SE}}(\widehat{\beta}_3)^2 \end{array} \right\}^{-1} (\widehat{\beta}_2, \widehat{\beta}_3)' \\
&= \frac{1}{1 - \widehat{\text{Cor}}(\widehat{\beta}_2, \widehat{\beta}_3)^2} \left\{ \left(\frac{\widehat{\beta}_2}{\widehat{\text{SE}}(\widehat{\beta}_2)} \right)^2 + \left(\frac{\widehat{\beta}_3}{\widehat{\text{SE}}(\widehat{\beta}_3)} \right)^2 - 2\, \widehat{\text{Cor}}(\widehat{\beta}_2, \widehat{\beta}_3)\, \frac{\widehat{\beta}_2}{\widehat{\text{SE}}(\widehat{\beta}_2)}\, \frac{\widehat{\beta}_3}{\widehat{\text{SE}}(\widehat{\beta}_3)} \right\}
\end{aligned}
$$

The Wald test for the null hypothesis that the Kessner index does not have an effect, i.e., that the coefficients of `kessner2` and `kessner3` are both zero, $\beta_{10} = \beta_{11} = 0$, can be performed by using the `testparm` command:

```
. quietly xtreg birwt smoke male mage hsgrad somecoll collgrad married
> black kessner2 kessner3 novisit pretri2 pretri3, i(momid) mle
. testparm kessner2 kessner3
 ( 1)  [birwt]kessner2 = 0
 ( 2)  [birwt]kessner3 = 0
           chi2(  2) =   26.94
         Prob > chi2 =    0.0000
```

We can reject the null hypothesis at the 5% level with $w = 26.94$, df $= 2$, $p < 0.001$. A more robust version of the test is obtained by specifying robust standard errors in the estimation command using the `vce(robust)` option for `xtreg` (only with the `re` option) or the `robust` option for `gllamm` followed by the `testparm` command.

We can also test the simultaneous hypothesis that three or more regression coefficients are all zero, but the expression for the Wald statistic becomes convoluted unless matrix expressions are used.

The analogous likelihood-ratio test statistic is

$$
L = 2(l_1 - l_0)
$$

where l_1 and l_0 are now the maximized log likelihoods for the models including and excluding both `kessner2` and `kessner3`, respectively. A likelihood-ratio test for the null hypothesis that the Kessner index does not have an effect (given the other covariates) can be performed by using the `lrtest` command:

```
. estimates store full
. quietly xtreg birwt smoke male mage hsgrad somecoll collgrad married black
> novisit pretri2 pretri3, i(momid) mle
. lrtest full .
Likelihood-ratio test                                 LR chi2(2)  =     26.90
(Assumption: . nested in full)                        Prob > chi2 =    0.0000
```

Note that likelihood-ratio tests for regression coefficients cannot be based on "log likelihoods" from restricted maximum likelihood (REML) estimation. When using `xtmixed`, the `mle` option must therefore be specified.

Sometimes it is required to test hypotheses regarding linear combinations of coefficients as demonstrated in section 1.8. In section 3.7.4, we will encounter a special case of this when testing the null hypothesis that two regression coefficients are equal, or in other words that the difference between the coefficients is 0, a simple example of a *contrast*. Wald tests of such hypotheses can be performed in Stata using the `lincom` command.

3.6.2 Predicted means and confidence intervals

We can use the `adjust` command to obtain predicted means for mothers and pregnancies with particular covariate values. This is useful for interpreting the results from regression modeling.

For example, if we want to interpret the effects of smoking and education on birthweight, we can set all other covariates to some meaningful values and produce a table of predicted means by smoking and education. It is useful to first define a categorical variable for level of education and give the categories value labels

```
. generate education = hsgrad*1 + somecoll*2 + collgrad*3
. label define ed 0 "no HS grad" 1 "HS grad" 2 "some Coll" 3 "Coll grad"
. label values education ed
```

After retrieving the estimates of the full model, we then use the `adjust` command, setting all dummy variables either to 0 or 1 and all continuous values to their mean (the latter is accomplished by simply listing the variables). We do not specify values for smoking and the education dummy variables, but specify `smoke` and `education` in the `by()` option instead:

(*Continued on next page*)

```
. estimates restore full

. adjust male=0 mage married=1 black=0 kessner2=0 kessner3=0
> novisit=0 pretri2=0 pretri3=0, by(smoke education) ci format(%4.0f)

------------------------------------------------------------------------------
     Dependent variable: birwt     Equation: birwt     Command: xtreg
   Variables left as is: hsgrad, somecoll, collgrad
  Covariate set to mean: mage = 28.591818
Covariates set to value: male = 0, married = 1, black = 0,
                         kessner2 = 0, kessner3 = 0,
                         novisit = 0, pretri2 = 0, pretri3 = 0
------------------------------------------------------------------------------

----------------------------------------------------------------
          |                        education
    smoke |   no HS grad      HS grad    some Coll    Coll grad
----------+-----------------------------------------------------
Nonsmoker |         3399         3456         3479         3490
          |  [3350,3448]  [3426,3485]  [3450,3509]  [3464,3515]
          |
   Smoker |         3180         3237         3261         3271
          |  [3128,3233]  [3196,3279]  [3218,3304]  [3228,3314]
----------------------------------------------------------------
     Key:  Linear Prediction
           [95% Confidence Interval]
```

The first part of the output states that the mean birthweights are based on the following covariate values: mothers aged 28.59 at the time of birth (the sample mean age when giving birth), female babies, married mothers, white mothers, adequate prenatal care (`kessner2` and `kessner3` equal to zero), and first prenatal visit occurring in the first trimester (`novisit`, `pretri2`, and `pretri3` equal to zero). We have used the `ci` option to obtain 95% confidence intervals—importantly, these are based on the estimated standard errors from the multilevel model estimated previously. The `format()` option rounds the results to the nearest gram.

3.6.3 Hypothesis test for between-cluster variance

Consider testing the null hypotheses that the between-cluster variance is zero

$$H_0\colon \psi = 0 \quad \text{against} \quad H_a\colon \psi > 0$$

This null hypothesis is equivalent to the hypothesis that $\zeta_j = 0$ or that there is no random intercept in the model. If this is true, a multilevel model is not required.

Likelihood-ratio tests are typically used with the test statistic

$$L = 2(l_1 - l_0)$$

where l_1 is the maximized log likelihood for the random-intercept model (which includes ζ_j) and l_0 is the maximized log likelihood for an ordinary regression model (without ζ_j). A correct p-value is obtained by dividing the naive p-value based on the χ^2 with 1 df by 2, as discussed in more detail in section 2.6.2. The result for the correct test procedure is provided in the last row of output from `xtreg` and `xtmixed`, giving $L = 1109$ and $p < 0.001$ for the full model.

Alternative tests for variance parameters were described in section 2.6.2.

3.7 Between and within effects

We now turn to the estimated regression coefficients for the random-intercept model with covariates. For births where the mother smoked during the pregnancy, the population mean birthweight is estimated to be 218g lower than for births where the mother did not smoke, holding all other covariates constant. This estimate represents either a comparison between children of *different* mothers, one of whom smoked during the pregnancy and one of whom did not (holding all other covariates constant), or a comparison between children of the *same* mother where the mother smoked during one pregnancy and not during the other (holding all other covariates constant).

This is neither purely a between-mother comparison (because smoking status can change between pregnancies) nor a purely within-mother comparison (because some mothers either smoke or do not smoke during *all* their pregnancies).

3.7.1 Between-mother effects

If we wanted to obtain purely between-mother effects of the covariates, we could average the response and explanatory variables for each mother j over children i and perform the regression on the resulting means:

$$\frac{1}{n_j}\sum_{i=1}^{n_j} y_{ij} = \frac{1}{n_j}\sum_{i=1}^{n_j}(\beta_1 + \beta_2 x_{2ij} + \cdots + \beta_p x_{pij} + \zeta_j + \epsilon_{ij})$$

or

$$\overline{y}_{\cdot j} = \beta_1 + \beta_2 \overline{x}_{2\cdot j} + \cdots + \beta_p \overline{x}_{p\cdot j} + \zeta_j + \overline{\epsilon}_{\cdot j} \tag{3.8}$$

Here $\overline{y}_{\cdot j}$ is the mean response for mother j, $\overline{x}_{2\cdot j}$ is the mean of the first explanatory variable `smoke` for mother j, etc., and $\overline{\epsilon}_{\cdot j}$ is the mean of the level-1 residuals in the original regression model (3.2). The error term $\zeta_j + \overline{\epsilon}_{\cdot j}$ has population mean zero, $E(\zeta_j + \overline{\epsilon}_{\cdot j}) = 0$, and variance, $\text{Var}(\zeta_j + \overline{\epsilon}_{\cdot j}) = \psi + \theta/n_j$. Any information on the regression coefficients from within-mother variability is eliminated and the coefficients of covariates that do not vary between mothers are absorbed by the intercept.

Ordinary least-squares estimates[1] $\widehat{\boldsymbol{\beta}}^B$ of the parameters $\boldsymbol{\beta}$ in the between model (3.8) whose corresponding covariates vary between mothers (here all covariates) can be obtained using `xtreg` with the `be` (between) option:

1. For unbalanced data ($n_j \neq n$), *weighted* least-squares (WLS) estimates can be obtained using the `wls` option. Then the cluster weights $1/(\widehat{\psi} + \widehat{\theta}/n_j)$ are used, where $\widehat{\psi}$ and $\widehat{\theta}$ are obtained from OLS. OLS still produces consistent albeit inefficient estimators of the regression coefficients but inconsistent estimators of the corresponding standard errors.

```
. xtreg birwt smoke male mage hsgrad somecoll collgrad married black kessner2
> kessner3 novisit pretri2 pretri3, i(momid) be

Between regression (regression on group means)  Number of obs      =      8604
Group variable: momid                           Number of groups   =      3978

R-sq:  within  = 0.0299                         Obs per group: min =         2
       between = 0.1168                                        avg =       2.2
       overall = 0.0949                                        max =         3

                                                F(13,3964)         =     40.31
sd(u_i + avg(e_i.))=  424.7306                  Prob > F           =    0.0000

------------------------------------------------------------------------------
       birwt |      Coef.   Std. Err.      t    P>|t|     [95% Conf. Interval]
-------------+----------------------------------------------------------------
       smoke |  -286.1476   23.22554   -12.32   0.000    -331.6828   -240.6125
        male |   104.9432   19.49531     5.38   0.000     66.72141     143.165
        mage |   4.398704   1.505448     2.92   0.003     1.447179     7.35023
      hsgrad |   58.80977   25.51424     2.30   0.021     8.787497     108.832
    somecoll |   85.07129    28.1348     3.02   0.003     29.91126    140.2313
    collgrad |   99.87509   29.35324     3.40   0.001     42.32622     157.424
     married |   41.91268   26.10719     1.61   0.108    -9.272101    93.09745
       black |  -218.4045   28.57844    -7.64   0.000    -274.4344   -162.3747
    kessner2 |  -101.4931   37.65605    -2.70   0.007    -175.3202   -27.66607
    kessner3 |  -201.9599   79.28821    -2.55   0.011    -357.4094   -46.51042
     novisit |  -51.02733   124.2073    -0.41   0.681    -294.5435    192.4889
     pretri2 |   125.4776   44.72006     2.81   0.005     37.80114    213.1541
     pretri3 |   241.1201   100.6567     2.40   0.017     43.77638    438.4637
       _cons |    3241.45   46.15955    70.22   0.000     3150.951    3331.948
------------------------------------------------------------------------------
```

The estimates of the between-mother effects are also shown under "Between" in table 3.2. The estimated coefficient $\widehat{\beta}_2^B$ of `smoke` is considerably larger, in absolute value, than the maximum likelihood estimate $\widehat{\beta}_2^R$ for the random-intercept model, shown under "MLE random effects" in table 3.2.

Table 3.2: Random-, between-, and within-effects estimates for smoking data (in grams)

	MLE random effects $\widehat{\boldsymbol{\beta}}^R$		Between $\widehat{\boldsymbol{\beta}}^B$		Within $\widehat{\boldsymbol{\beta}}^W$	
	Est	(SE)	Est	(SE)	Est	(SE)
Fixed part						
β_1 [_cons]	3,117	(41)	3,241	(46)	2,768	(86)
β_2 [smoke]	−218	(18)	−286	(23)	−105	(29)
β_3 [male]	121	(10)	105	(19)	126	(11)
β_4 [mage]	8	(1)	4	(2)	23	(3)
β_5 [hsgrad]	57	(25)	59	(26)		
β_6 [somecoll]	81	(27)	85	(28)		
β_7 [collgrad]	91	(28)	100	(29)		
β_8 [married]	50	(26)	42	(26)		
β_9 [black]	−211	(28)	−218	(29)		
β_{10} [kessner2]	−93	(20)	−101	(38)	−91	(23)
β_{11} [kessner3]	−151	(41)	−202	(79)	−128	(48)
β_{12} [novisit]	−30	(66)	−51	(124)	−5	(78)
β_{13} [pretri2]	93	(23)	125	(45)	81	(27)
β_{14} [pretri3]	179	(52)	241	(101)	153	(60)
Random part						
$\sqrt{\psi}$	339				440 [a]	
$\sqrt{\theta}$	371				369 [a]	

[a]Not parameter estimates, but standard deviations of estimates $\widehat{\epsilon}_{ij}$ and $\widehat{\zeta}_j$.

3.7.2 Within-mother effects

If we wanted to obtain purely within-mother effects, we could subtract the between-mother regression (3.8) from the original model (3.2) to obtain the within model

$$y_{ij} - \overline{y}_{\cdot j} \;=\; \beta_2(x_{2ij} - \overline{x}_{2\cdot j}) + \cdots + \beta_p(x_{pij} - \overline{x}_{p\cdot j}) + \epsilon_{ij} - \overline{\epsilon}_{\cdot j} \tag{3.9}$$

Here the response and all covariates have simply been centered around their respective cluster means. The error term $\epsilon_{ij} - \overline{\epsilon}_{\cdot j}$ has population mean zero, $E(\epsilon_{ij} - \overline{\epsilon}_{\cdot j}) = 0$, and variance $\text{Var}(\epsilon_{ij} - \overline{\epsilon}_{\cdot j}) = \theta(1 - 1/n_j)$. Covariates that do not vary within clusters drop out of the equation because the mean-centered covariate is zero. Ordinary least squares can be used to estimate the within-effects in (3.9).

Identical estimates of within-mother effects can be obtained by replacing the random intercept ζ_j for each mother in the original model in (3.2) by a fixed intercept α_j. This could be accomplished by using dummy variables for each mother and omitting the intercept β_1, so that α_j represents the total intercept for mother j, previously represented by $\beta_1 + \zeta_j$. Letting d_{kj} be the dummy variable for the kth mother, ($k = 1, \ldots, 3{,}978$), the *fixed-effects* model can be written as

$$\begin{aligned} y_{ij} &= \beta_2 x_{2ij} + \cdots + \beta_p x_{pij} + \sum_{k=1}^{3,978} d_{kj}\alpha_k + \epsilon_{ij} \\ &= \beta_2 x_{2ij} + \cdots + \beta_p x_{pij} + \alpha_j + \epsilon_{ij} \end{aligned} \tag{3.10}$$

In this model, all mother-specific effects are accommodated by α_j, leaving only within-mother effects to be explained by covariates. Another way of seeing this is by considering the fact that the set of dummy variables is collinear with any between-mother covariates. In practice, it is more convenient to eliminate the intercepts by subtracting the between-mother regression from the random-intercept model as in (3.9) instead of estimating 3,978 intercepts.

The within-estimates $\widehat{\boldsymbol{\beta}}^W$ for the coefficients of covariates that vary within mothers can be obtained using `xtreg` with the `fe` (fixed effects) option[2]:

```
. xtreg birwt smoke male mage hsgrad somecoll collgrad married black kessner2
> kessner3 novisit pretri2 pretri3, i(momid) fe

Fixed-effects (within) regression               Number of obs      =      8604
Group variable: momid                           Number of groups   =      3978

R-sq:  within  = 0.0465                         Obs per group: min =         2
       between = 0.0557                                        avg =       2.2
       overall = 0.0546                                        max =         3

                                                F(8,4618)          =     28.12
corr(u_i, Xb)  = -0.0733                        Prob > F           =    0.0000

------------------------------------------------------------------------------
       birwt |      Coef.   Std. Err.      t    P>|t|     [95% Conf. Interval]
-------------+----------------------------------------------------------------
       smoke |  -104.5494   29.10075    -3.59   0.000    -161.6007   -47.49798
        male |   125.6355   10.92272    11.50   0.000     104.2217    147.0492
        mage |   23.15832   3.006667     7.70   0.000     17.26382    29.05282
      hsgrad |  (dropped)
    somecoll |  (dropped)
    collgrad |  (dropped)
     married |  (dropped)
       black |  (dropped)
    kessner2 |  -91.49483   23.48914    -3.90   0.000    -137.5448    -45.4449
    kessner3 |   -128.091   47.79636    -2.68   0.007    -221.7947   -34.38731
     novisit |  -4.805898    77.7721    -0.06   0.951    -157.2764    147.6646
     pretri2 |   81.29039   27.04974     3.01   0.003     28.25998    134.3208
     pretri3 |    153.059   60.08453     2.55   0.011     35.26462    270.8534
       _cons |   2767.504   86.23602    32.09   0.000      2598.44    2936.567
-------------+----------------------------------------------------------------
     sigma_u |  440.05052
     sigma_e |  368.91787
         rho |  .58725545   (fraction of variance due to u_i)
------------------------------------------------------------------------------
F test that all u_i=0:     F(3977, 4618) =     2.83          Prob > F = 0.0000
```

2. Here the overall constant (next to `_cons`) is obtained by adding the overall means $\overline{y}_{\cdot\cdot}$, $\overline{x}_{2\cdot\cdot}$, etc., and $\overline{\zeta}_{\cdot} + \overline{\epsilon}_{\cdot\cdot}$ back onto the corresponding differences:

$$y_{ij} - \overline{y}_{\cdot j} + \overline{y}_{\cdot\cdot} = \beta_1 + \beta_2(x_{2ij} - \overline{x}_{2\cdot j} + \overline{x}_{2\cdot\cdot}) + \cdots + \beta_p(x_{pij} - \overline{x}_{p\cdot j} + \overline{x}_{p\cdot\cdot}) + \epsilon_{ij} - \overline{\epsilon}_{\cdot j} + \overline{\zeta}_{\cdot} + \overline{\epsilon}_{\cdot\cdot}$$

The estimates of the within-mother effects are also reported under "Within" in table 3.2. Variables that do not vary between pregnancies for the same mother have been dropped because they cancel out in (3.9). The estimated coefficient $\widehat{\beta}_2^W$ for `smoke` is considerably smaller, in absolute value, than the estimate $\widehat{\beta}_2^R$ for the random-intercept model.

In the output `sigma_u` and `sigma_e` are standard deviations of estimated level-2 and level-1 residuals, $\widehat{\zeta}_j = \overline{y}_{\cdot j} - (\widehat{\beta}_1 + \widehat{\beta}_2\overline{x}_{2\cdot j} + \cdots + \widehat{\beta}_p\overline{x}_{p\cdot j})$ and $\widehat{\epsilon}_{ij} = y_{ij} - \widehat{\zeta}_j - (\widehat{\beta}_1 + \widehat{\beta}_2\overline{x}_{2ij} + \cdots + \widehat{\beta}_p\overline{x}_{pij})$, the latter standard deviation adjusted for the number of estimated means.

3.7.3 ❖ Relations among within estimator, between estimator, and estimator for random-intercept model

We will now show that the estimator for random-intercept model can be expressed as a weighted average of the within estimator and the between estimator.

The original random-intercept model in (3.2) implicitly assumes that the between and within effects of the set of covariates that vary both between and within mothers are identical since the between-mother model (3.8) and the within-mother model (3.9) derived from the random-intercepts model (3.2) have the same regression coefficients $\boldsymbol{\beta} = (\beta_1, \beta_2, \ldots, \beta_p)'$.

The estimators for the random-intercept model therefore use both within- and between-mother information. This can be seen explicitly for the generalized least-squares (GLS) estimator (obtained using `xtreg` with the `re` option), which is equivalent to the maximum likelihood estimator in large samples. The relations among the estimators are most transparent for a single covariate x_{ij} with regression coefficient β_2 and balanced data $n_j = n$.

It can be shown that the sampling variance of the between estimator $\widehat{\beta}_2^B$ can be consistently estimated as

$$\widehat{\text{SE}}(\widehat{\beta}_2^B)^2 = \frac{\widehat{\text{Var}}(\zeta_j + \overline{\epsilon}_{\cdot j})}{(J-1)s_{xB}^2}$$

where $\widehat{\text{Var}}(\zeta_j + \overline{\epsilon}_{\cdot j})$ is the mean squared error from the between regression in (3.8) and $s_{xB}^2 = \frac{1}{J-1}\sum_{j=1}^J(\overline{x}_{\cdot j} - \overline{x}_{\cdot\cdot})^2$ is the between variance of x_{ij} given in section 3.2. The sampling variance of the within estimator $\widehat{\beta}_2^W$ can be consistently estimated as

$$\widehat{\text{SE}}(\widehat{\beta}_2^W)^2 = \frac{\frac{Jn-1}{J(n-1)-1}\widehat{\text{Var}}(\epsilon_{ij} - \overline{\epsilon}_{\cdot j})}{(Jn-1)s_{xW}^2} = \frac{\widehat{\text{Var}}(\epsilon_{ij} - \overline{\epsilon}_{\cdot j})}{\{J(n-1)-1\}s_{xW}^2}$$

where $\widehat{\text{Var}}(\epsilon_{ij} - \overline{\epsilon}_{\cdot j})$ is the mean squared error from the within regression in (3.9) and $s_{xW}^2 = \frac{1}{Jn-1}\sum_{j=1}^J\sum_{i=1}^n(x_{ij} - \overline{x}_{\cdot j})^2$ is the within variance of x_{ij} given in section 3.2. The term $\frac{Jn-1}{J(n-1)-1}$ is necessary because of the loss of J degrees of freedom due to mean-centering. The numerator after the first equality is the square of `sigma_e` reported by `xtreg` with the `fe` option.

The generalized least-squares (GLS) estimator $\widehat{\beta}_2^R$ for the random-intercept model can then be written as

$$\widehat{\beta}_2^R = (1-\widehat{\omega})\widehat{\beta}_2^B + \widehat{\omega}\widehat{\beta}_2^W$$

where

$$\widehat{\omega} = \frac{\widehat{\mathrm{SE}}(\widehat{\beta}_2^B)^2}{\widehat{\mathrm{SE}}(\widehat{\beta}_2^B)^2 + \widehat{\mathrm{SE}}(\widehat{\beta}_2^W)^2} \qquad \text{and} \qquad 1-\widehat{\omega} = \frac{\widehat{\mathrm{SE}}(\widehat{\beta}_2^W)^2}{\widehat{\mathrm{SE}}(\widehat{\beta}_2^B)^2 + \widehat{\mathrm{SE}}(\widehat{\beta}_2^W)^2}$$

We see that the estimator $\widehat{\beta}_2^R$ for the random-intercept model approaches the within estimator $\widehat{\beta}_2^W$ when $\widehat{\omega}$ approaches 1, i.e., when the within standard error is much smaller than the between standard error. This happens when n becomes large, or $\widehat{\theta}$ becomes small, or $\widehat{\psi}$ becomes large, or s_{xB} becomes small.

In contrast, $\widehat{\beta}_2^R$ approaches the between estimator $\widehat{\beta}_2^B$ when $\widehat{\omega}$ approaches 0, i.e., when the between standard error is much smaller than the within standard error. This happens when n becomes small, or $\widehat{\theta}$ becomes large, or $\widehat{\psi}$ becomes small, or s_{xW} becomes small.

Although $\widehat{\beta}_2^R$, $\widehat{\beta}_2^B$, and $\widehat{\beta}_2^W$ are all estimators of the same parameter β_2, the generalized least-squares (GLS) estimator $\widehat{\beta}_2^R$ is more efficient (varies less in repeated samples) than the other estimators if the random-intercept model is true because it exploits both within and between-mother information.

3.7.4 Endogeneity and different within- and between-mother effects

The estimated between effect $\widehat{\beta}_2^B$ based on (3.8) may differ from the estimated within-effect $\widehat{\beta}_2^W$ from (3.9) due to omitted mother-specific explanatory variables that affect both $\overline{x}_{2\cdot j}$ and the mother-specific residual ζ_j and hence the mean response $\overline{y}_{\cdot j}$, given the included explanatory variables.

As discussed by Abrevaya (2006), mothers who smoke during their pregnancy may also adopt other behaviors such as drinking and poor nutritional intake. They are also likely to have lower socioeconomic status and be less educated. These variables adversely affect birthweight and have not been adequately controlled for, so that the between effect is likely to be an overestimate of the true effect (in absolute value). We thus have cluster-level confounding or cluster-level omitted-variable bias. In contrast, each mother serves as her own control for the within estimate so all mother-specific explanatory variables have been held constant. Indeed, this was the reason why Abrevaya (2006) constructed the matched dataset: to get closer to the true causal effect of smoking by using within-mother estimates.

To illustrate the idea of different between and within effects, figure 3.2 shows data for hypothetical clusters and a continuous covariate where the between-cluster effect (slope of dashed line) is positive and the within-cluster effect (slope of dotted lines) is negative. Here the hollow circles represent the observed data and the solid circles cluster means.

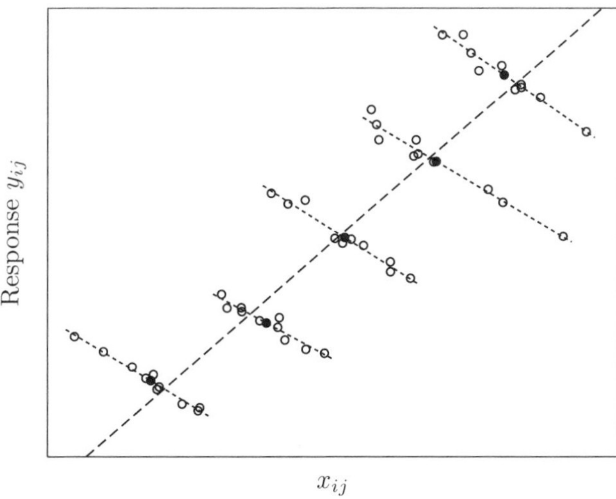

Figure 3.2: Illustration of within-cluster and between-cluster effects

The problem of cluster-level confounding occurs when x_{2ij} is correlated with the error term ζ_j, which represents the effects of omitted cluster-level covariates. This is often referred to as *endogeneity* in econometrics because x_{2ij} should be a response (or endogenous) in a correctly specified model, rather than being exogenous or determined outside the model. Cluster-level confounding can also be responsible for the *ecological fallacy* where between-cluster relationships (ecological relationships) can differ substantially from within-cluster relationships.

We can easily relax the assumption that the between and within effects are the same for a particular explanatory variable, say x_{2ij}, by using the model

$$y_{ij} = \beta_1 + \beta_2^1(x_{2ij} - \overline{x}_{2\cdot j}) + \beta_2^2\overline{x}_{2\cdot j} + \beta_3 x_{3ij} + \cdots + \beta_p x_{pij} + \zeta_j + \epsilon_{ij} \qquad (3.11)$$

which collapses to the original random-intercept model in (3.2) if $\beta_2^1 = \beta_2^2 = \beta_2$. The deviation from the cluster mean of smoking $x_{2ij} - \overline{x}_{2\cdot j}$ is an *instrumental variable* for x_{2ij} because it is correlated with x_{2ij} but uncorrelated with the random intercept ζ_j. We can also view this model as relaxing the assumption that the random intercept is uncorrelated with x_{2ij} if we think of $(\beta_2^2\overline{x}_{2\cdot j} + \zeta_j)$ as the random intercept.

It is important to note that we do not need to subtract the cluster mean $\overline{x}_{2\cdot j}$ from x_{2ij} as long as we include the cluster mean in the model, since

$$\begin{aligned} y_{ij} &= \beta_1 + \beta_2^1(x_{2ij} - \overline{x}_{2\cdot j}) + \beta_2^2\overline{x}_{2\cdot j} + \beta_3 x_{3ij} + \cdots + \beta_p x_{pij} + \zeta_j + \epsilon_{ij} \\ &= \beta_1 + \beta_2^1 x_{2ij} + (\beta_2^2 - \beta_2^1)\overline{x}_{2\cdot j} + \beta_3 x_{3ij} + \cdots + \beta_p x_{pij} + \zeta_j + \epsilon_{ij} \qquad (3.12) \end{aligned}$$

Whether x_{2ij} is cluster-mean centered affects only the interpretation of the coefficient of the cluster mean $\overline{x}_{2\cdot j}$. If x_{2ij} is cluster-mean centered as in the first line of (3.12), the coefficient of the cluster mean represents the between effect, and if not as in the second line of (3.12) it represents the difference in between and within effects.

Figure 3.3 illustrates relationships between within and between effects where $\beta_2^1 = \beta_2^2$ (left panel) and $\beta_2^1 < \beta_2^2$ (right panel). The cluster-specific regression lines are shown for two clusters (solid and dashed lines for clusters 1 and 2, respectively) having the same value of the random intercept ζ_j but nonoverlapping ranges of x_{ij}. The bullets represent the cluster means $\overline{x}_{\cdot j}$ and $\overline{y}_{\cdot j}$. The figure shows that $\beta_2^2 - \beta_2^1$ is the additional increase of the second cluster's mean $\overline{y}_{\cdot j}$ after allowing for the within effect β_2^1 of x_{ij} and the increase in cluster mean $\overline{x}_{\cdot j}$.

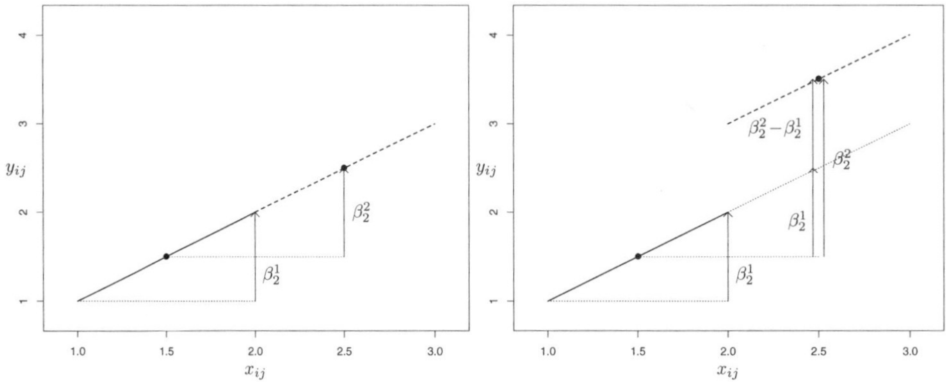

Figure 3.3: Illustration of different within and between effects for two clusters having the same value of ζ_j (β_2^1 is the within effect and β_2^2 is the between effect)

It is common to include only the cluster-mean centered covariate but not the cluster mean itself. However, as shown in figure 3.4, this makes the unrealistic assumption that the additional increase in the second cluster's $\overline{y}_{\cdot j}$, after allowing for the within effect of x_{ij} and the increase in cluster mean $\overline{x}_{\cdot j}$ (which is positive in the left panel), happens to be equal to minus the within effect (as shown in the right panel). It is important to note that the original model that assumed equal between and within effects is no longer a special case if the cluster mean is omitted from the model.

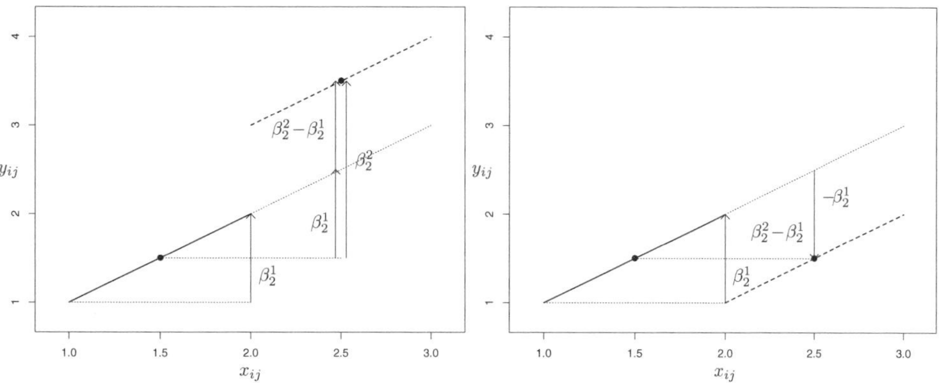

Figure 3.4: Illustration of assuming zero between effect for two clusters having the same value of ζ_j (β_2^1 is the within effect and β_2^2 is the between effect)

Including the cluster-mean centered variable only is likely to lead to cluster-level confounding for other cluster-level variables. For instance, if the clusters are schools, including school-mean centered socioeconomic status (SES) but not the school-mean SES has the consequence that any tendency for the schools with larger mean SES to have larger mean outcomes is attributed to other cluster-level variables that are correlated with mean SES and to the random intercept. Only purely within-school variables, such as perhaps gender, have thus been controlled for SES. The effects of school-level variables such as the curriculum used by the school, have not been controlled for the socioeconomic composition of the school.

We will now fit (3.11) with the cluster mean of `smoke` (the proportion of pregnancies in which the mother smokes), as well as the child-specific deviation from the cluster mean of `smoke` as covariates. These covariates are produced by the commands

```
. egen mn_smok = mean(smoke), by(momid)
. generate dev_smok = smoke - mn_smok
```

and included in the random-intercept model using

(*Continued on next page*)

```
. xtreg birwt dev_smok mn_smok male mage hsgrad somecoll collgrad married
> black kessner2 kessner3 novisit pretri2 pretri3, i(momid) mle

Random-effects ML regression                    Number of obs      =      8604
Group variable: momid                           Number of groups   =      3978

Random effects u_i ~ Gaussian                   Obs per group: min =         2
                                                               avg =       2.2
                                                               max =         3

                                                LR chi2(14)        =    684.32
Log likelihood  = -65133.327                    Prob > chi2        =    0.0000

------------------------------------------------------------------------------
       birwt |      Coef.   Std. Err.      z    P>|z|     [95% Conf. Interval]
-------------+----------------------------------------------------------------
    dev_smok |  -104.2331   29.18164    -3.57   0.000    -161.4281   -47.03815
     mn_smok |   -289.904   23.12917   -12.53   0.000    -335.2364   -244.5717
        male |   121.1497   9.543995    12.69   0.000     102.4438    139.8556
        mage |   8.191991   1.345733     6.09   0.000     5.554404    10.82958
      hsgrad |   43.10052   25.15802     1.71   0.087    -6.208299    92.40934
    somecoll |   62.74526   27.51558     2.28   0.023     8.815722    116.6748
    collgrad |   66.89807   28.37807     2.36   0.018     11.27806    122.5181
     married |   35.21194    25.6433     1.37   0.170    -15.04801    85.47189
       black |  -218.9694   28.28467    -7.74   0.000    -274.4064   -163.5325
    kessner2 |  -92.03346   19.89661    -4.63   0.000    -131.0301   -53.03683
    kessner3 |  -149.2771   40.77281    -3.66   0.000    -229.1903   -69.36383
     novisit |   -23.3923   65.60531    -0.36   0.721    -151.9763    105.1917
     pretri2 |   92.33952   23.15722     3.99   0.000     46.95221    137.7268
     pretri3 |    176.774   51.56358     3.43   0.001      75.7112    277.8367
       _cons |     3154.8   41.57504    75.88   0.000     3073.314    3236.285
-------------+----------------------------------------------------------------
    /sigma_u |   338.4563    6.27314                      326.3818    350.9775
    /sigma_e |   370.0488   3.856608                      362.5666    377.6853
         rho |   .4554982   .0119054                      .4322541    .4788957
------------------------------------------------------------------------------
Likelihood-ratio test of sigma_u=0: chibar2(01)= 1115.22 Prob>=chibar2 = 0.000
```

The estimated between effect of smoking (coefficient of `mn_smok`) is $\widehat{\beta}_2^1 = -290$ and is different from $\widehat{\beta}_2^2 = -104$, the estimated within effect of smoking (coefficient of `dev_smok`). We can formally test the null hypothesis that the corresponding coefficients are the same, H_0: $\beta_2^1 - \beta_2^2 = 0$, using the postestimation command `lincom`:

```
. lincom mn_smok - dev_smok
 ( 1)  - [birwt]dev_smok + [birwt]mn_smok = 0

------------------------------------------------------------------------------
       birwt |      Coef.   Std. Err.      z    P>|z|     [95% Conf. Interval]
-------------+----------------------------------------------------------------
         (1) |  -185.6709   37.21887    -4.99   0.000    -258.6186   -112.7233
------------------------------------------------------------------------------
```

There is strong evidence that the within effect of smoking differs from the between effect.

As stated previously, $x_{2ij} - \overline{x}_{2\cdot j}$ is an instrumental variable for x_{2ij} because it is correlated with x_{2ij} but uncorrelated with the random intercept ζ_j. However, ζ_j may be correlated with another within-mother covariate x_{3ij} and the inconsistency in estimating the corresponding regression coefficient β_3 can be transmitted to the estimator for β_2^1.

To address this problem we can follow Mundlak (1978) and include the cluster means of *all* within-mother covariates. This ensures consistent estimation of all within effects because the deviations from the cluster means are uncorrelated with the cluster means themselves, with any between-mother covariate (such as x_{4j}), and with ζ_j. However, between-mother covariates may still be correlated with ζ_j, producing inconsistent estimates of the between-effects and the random-intercept variance ψ. In contrast, the estimator for the level-1 residual variance θ is consistent.

We start by constructing the cluster means (apart from mn_smok that already exists):

```
. egen mn_male = mean(male), by(momid)
. egen mn_mage = mean(mage), by(momid)
. egen mn_kessner2 = mean(kessner2), by(momid)
. egen mn_kessner3 = mean(kessner3), by(momid)
. egen mn_novisit = mean(novisit), by(momid)
. egen mn_pretri2 = mean(pretri2), by(momid)
. egen mn_pretri3 = mean(pretri3), by(momid)
```

A more elegant way of forming cluster means for such a large number of variables is to use a foreach loop:

```
foreach var of varlist male mage kessner* novisit pretri* {
   egen mn_`var' = mean(`var'), by(momid)
}
```

Here the loop repeats the egen command for each of the variables in the variable list male mage kessner* novisit pretri*. Inside the curly braces, the local macro var (which we could have called whatever we like) evaluates to the current variable name (first male, then mage, etc.), and this variable name is accessed by placing the macro name in single quotes (really a left quote and an apostrophe). We therefore obtain exactly the same commands we previously ran one at a time.

If we include the cluster means of all level-1 covariates but leave the latter variables in the model without cluster-mean centering them as in (3.12), the coefficients for the cluster means represent the differences in the between and within effects for the set of covariates having both between and within variation. We fit the model using

(Continued on next page)

```
. xtreg birwt smok mn_smok male mn_male mage mn_mage hsgrad somecoll
> collgrad married black kessner2 mn_kessner2 kessner3 mn_kessner3 novisit
> mn_novisit pretri2 mn_pretri2 pretri3 mn_pretri3, i(momid) mle
Random-effects ML regression                    Number of obs      =      8604
Group variable: momid                           Number of groups   =      3978

Random effects u_i ~ Gaussian                   Obs per group: min =         2
                                                               avg =       2.2
                                                               max =         3

                                                LR chi2(21)        =    719.28
Log likelihood  = -65115.846                    Prob > chi2        =    0.0000

------------------------------------------------------------------------------
       birwt |      Coef.   Std. Err.      z    P>|z|     [95% Conf. Interval]
-------------+----------------------------------------------------------------
       smoke |  -104.5494    29.1063    -3.59   0.000    -161.5967   -47.50206
     mn_smok |  -183.1657   37.19657    -4.92   0.000    -256.0696   -110.2617
        male |   125.6355    10.9248    11.50   0.000     104.2232    147.0477
     mn_male |  -20.22363   22.31026    -0.91   0.365    -63.95093    23.50367
        mage |   23.15832   3.007241     7.70   0.000     17.26424     29.0524
     mn_mage |  -18.59407   3.360264    -5.53   0.000    -25.18007   -12.00808
      hsgrad |   56.29698   25.38638     2.22   0.027     6.540583    106.0534
    somecoll |   83.07017   27.99083     2.97   0.003     28.20914    137.9312
    collgrad |   98.17599   29.18708     3.36   0.001     40.97037    155.3816
     married |   42.46127   26.03156     1.63   0.103    -8.559647    93.48219
       black |  -219.0013   28.41769    -7.71   0.000     -274.699   -163.3037
    kessner2 |  -91.49483   23.49362    -3.89   0.000    -137.5415   -45.44819
 mn_kessner2 |  -9.050791   44.22205    -0.20   0.838    -95.72442    77.62284
    kessner3 |   -128.091   47.80548    -2.68   0.007     -221.788     -34.394
 mn_kessner3 |  -79.42459   92.26946    -0.86   0.389    -260.2694    101.4202
     novisit |  -4.805899   77.78694    -0.06   0.951    -157.2655    147.6537
  mn_novisit |  -38.11621   146.3218    -0.26   0.794    -324.9017    248.6693
     pretri2 |   81.29039    27.0549     3.00   0.003     28.26376     134.317
  mn_pretri2 |   44.76713   52.06045     0.86   0.390    -57.26948    146.8037
     pretri3 |    153.059   60.09599     2.55   0.011     35.27303     270.845
  mn_pretri3 |   96.07044    116.707     0.82   0.410    -132.6711    324.8119
       _cons |   3238.407   45.98903    70.42   0.000     3148.271    3328.544
-------------+----------------------------------------------------------------
    /sigma_u |   338.4422   6.243867                       326.423    350.9039
    /sigma_e |   368.9882   3.840715                      361.5368    376.5932
         rho |   .4569014    .011855                      .4337527    .4801973
------------------------------------------------------------------------------
Likelihood-ratio test of sigma_u=0: chibar2(01)= 1127.34 Prob>=chibar2 = 0.000
```

A Wald test of the joint null hypothesis that all coefficients for the cluster means in the above model are 0 can be performed using the `testparm` command:

```
. testparm mn_*
 ( 1)  [birwt]mn_male = 0
 ( 2)  [birwt]mn_mage = 0
 ( 3)  [birwt]mn_kessner2 = 0
 ( 4)  [birwt]mn_kessner3 = 0
 ( 5)  [birwt]mn_novisit = 0
 ( 6)  [birwt]mn_pretri2 = 0
 ( 7)  [birwt]mn_pretri3 = 0
 ( 8)  [birwt]mn_smok = 0

           chi2(  8) =   60.04
         Prob > chi2 =    0.0000
```

The Wald statistic is 60.04 with df = 8, so the null hypothesis that the coefficients of the cluster means are all zero is rejected at the 5% level. The above null hypothesis is equivalent to the hypothesis of equal between and within effects (for the covariates having both within and between variation), which is thus also rejected.

An advantage of setting the within and between effects of a covariate equal is that the common effect will be more precisely estimated than separate effects because it pools the within and between information. We may therefore want to set the within and between effects equal when this appears to be appropriate but allow different within and between effects when such equality restrictions are inappropriate. Instead of including the cluster means of all level-1 covariates, we may thus for instance proceed by including cluster means for only the particular covariates where the within and between effects are significantly different at the 5% level.

We therefore consider the model including the cluster means `mn_smoke` and `mn_mage` to allow the within and between effects of smoking and mother's age to be different:

```
. xtreg birwt smok mn_smok male mage mn_mage hsgrad somecoll collgrad
> married black kessner2 kessner3 novisit pretri2 pretri3, i(momid) mle

Random-effects ML regression                    Number of obs      =      8604
Group variable: momid                           Number of groups   =      3978

Random effects u_i ~ Gaussian                   Obs per group: min =         2
                                                               avg =       2.2
                                                               max =         3

                                                LR chi2(15)        =    715.55
Log likelihood  = -65117.711                    Prob > chi2        =    0.0000

------------------------------------------------------------------------------
       birwt |      Coef.   Std. Err.      z    P>|z|     [95% Conf. Interval]
-------------+----------------------------------------------------------------
       smoke |  -105.1819   29.10171    -3.61   0.000    -162.2202   -48.14359
     mn_smok |  -183.0369   37.14847    -4.93   0.000    -255.8466   -110.2272
        male |    120.504   9.523061    12.65   0.000     101.8391    139.1689
        mage |   23.25909    3.00667     7.74   0.000     17.36612    29.15205
     mn_mage |  -18.78885   3.358056    -5.60   0.000    -25.37052   -12.20718
      hsgrad |   55.71939   25.24007     2.21   0.027     6.249765     105.189
    somecoll |   82.18035   27.71151     2.97   0.003      27.8668    136.4939
    collgrad |   96.68248   28.84582     3.35   0.001     40.14571    153.2192
     married |   42.46782   25.65649     1.66   0.098    -7.817982    92.75362
       black |  -220.6388   28.26592    -7.81   0.000     -276.039   -165.2386
    kessner2 |  -94.02909   19.85653    -4.74   0.000    -132.9472     -55.111
    kessner3 |  -151.8727   40.68577    -3.73   0.000    -231.6154   -72.13008
     novisit |  -22.37888   65.46304    -0.34   0.732    -150.6841    105.9263
     pretri2 |   94.18321    23.1086     4.08   0.000      48.8912    139.4752
     pretri3 |    180.753    51.4546     3.51   0.000     79.90383    281.6022
       _cons |     3236.1   43.97523    73.59   0.000     3149.911     3322.29
-------------+----------------------------------------------------------------
    /sigma_u |   338.6018   6.246565                      326.5774    351.0688
    /sigma_e |   369.0286   3.841704                      361.5753    376.6355
         rho |   .4570812   .0118565                      .4339293    .4803797
------------------------------------------------------------------------------
Likelihood-ratio test of sigma_u=0: chibar2(01)= 1127.59 Prob>=chibar2 = 0.000
```

The estimated coefficient of `smoke` is close to that in the previous model where all cluster means were included.

A great advantage of clustered data is that we can investigate and address the endogeneity problem at the cluster level for unit-level covariates (correlation between ζ_j and x_{ij}). However, we cannot handle cluster-level endogeneity for cluster-level covariates (correlation between ζ_j and x_j), which is tackled by an approach suggested by Hausman and Taylor (1981) and implemented in the `xthtaylor` command. However, it is not straightforward to check for endogeneity at the unit level, i.e., to check if ϵ_{ij} is correlated with either cluster-level or unit-level covariates. To correct for this kind of endogeneity, external instrumental variables are usually required.

3.7.5 Hausman endogeneity test

The Hausman test (Hausman 1978), more aptly called the Durbin–Wu–Hausman test, can be used to compare two alternative estimators of $\boldsymbol{\beta}$, both of which are consistent (the estimates approach the true parameter values as the sample size tends to infinity) if the model is true. In its simplest form, one of the estimators is efficient (is the most precise estimator as the sample size tends to infinity) if the model is true, but inconsistent when the model is misspecified. The other estimator is consistent also under misspecification but not efficient when the model is true.

For instance, both the fixed-effects estimator $\widehat{\boldsymbol{\beta}}^W$ and the generalized least-squares estimator $\widehat{\boldsymbol{\beta}}^R$ are consistent, whereas only $\widehat{\boldsymbol{\beta}}^R$ is efficient, if the random-intercept model is correctly specified. However, for certain model violations, such as correlations between the random intercept and any of the covariates, $\widehat{\boldsymbol{\beta}}^R$ becomes inconsistent, whereas $\widehat{\boldsymbol{\beta}}^W$ remains consistent.

Consider first the simple case of a model with a single covariate x_{ij} that varies both between and within clusters. The Hausman test statistic for endogeneity then takes the form

$$h = \frac{(\widehat{\beta}^W - \widehat{\beta}^R)^2}{\widehat{\mathrm{SE}}(\widehat{\beta}^W)^2 - \widehat{\mathrm{SE}}(\widehat{\beta}^R)^2} \tag{3.13}$$

which has a χ^2 null distribution with 1 degree of freedom. The denominator of the test statistic would usually take the form $\widehat{\mathrm{SE}}(\widehat{\beta}^W)^2 + \widehat{\mathrm{SE}}(\widehat{\beta}^R)^2 - 2\,\widehat{\mathrm{Cov}}(\widehat{\beta}^W, \widehat{\beta}^R)$, where the covariance between the within and between estimators would be hard to obtain. However, it can be shown that the denominator simplifies to the one in (3.13) because the random-effects estimator is efficient when the random-intercept model is true.

Consider now the case where there are several covariates that all vary both between and within clusters. Let the fixed effects and GLS estimates be denoted $\widehat{\boldsymbol{\beta}}^W$ and $\widehat{\boldsymbol{\beta}}^R$, respectively, and let the corresponding estimated covariance matrices be denoted $\widehat{\mathrm{Cov}}(\widehat{\boldsymbol{\beta}}^W)$ and $\widehat{\mathrm{Cov}}(\widehat{\boldsymbol{\beta}}^R)$. The Hausman test statistic then takes the form

$$h = (\widehat{\boldsymbol{\beta}}^W - \widehat{\boldsymbol{\beta}}^R)\left\{\widehat{\text{Cov}}(\widehat{\boldsymbol{\beta}}^W) - \widehat{\text{Cov}}(\widehat{\boldsymbol{\beta}}^R)\right\}^{-1}(\widehat{\boldsymbol{\beta}}^W - \widehat{\boldsymbol{\beta}}^R)'$$

The h statistic has a χ^2 null distribution with degrees of freedom given as the number of overlapping estimated regression coefficients from the two approaches, that is the number of covariates with both between- and within-cluster variation.

We can use the **hausman** command to perform the Hausman test in Stata, following estimation of $\widehat{\boldsymbol{\beta}}^W$ using **xtreg** with the **fe** option and $\widehat{\boldsymbol{\beta}}^R$ using **xtreg** with the **re** option:

```
. quietly xtreg birwt smoke male mage hsgrad somecoll collgrad married
> black kessner2 kessner3 novisit pretri2 pretri3, i(momid) fe
. estimates store fixed
. quietly xtreg birwt smoke male mage hsgrad somecoll collgrad married
> black kessner2 kessner3 novisit pretri2 pretri3, i(momid) re
. estimates store random
. hausman fixed random
                 —— Coefficients ——
             |      (b)          (B)            (b-B)     sqrt(diag(V_b-V_B))
             |     fixed        random       Difference          S.E.
-------------+----------------------------------------------------------------
       smoke |   -104.5494    -217.7488        113.1995        22.71343
        male |    125.6355     120.9874        4.648084        5.297981
        mage |    23.15832     8.137158        15.02116        2.687211
    kessner2 |   -91.49483    -92.89604        1.401212        12.44845
    kessner3 |    -128.091    -150.6366        22.54563        24.87574
     novisit |   -4.805898     -29.9223        25.11641        41.66561
     pretri2 |    81.29039     92.73087       -11.44048        13.94097
     pretri3 |     153.059     178.4334       -25.37443        30.76114
------------------------------------------------------------------------------
                           b = consistent under Ho and Ha; obtained from xtreg
            B = inconsistent under Ha, efficient under Ho; obtained from xtreg

    Test:  Ho:  difference in coefficients not systematic

                  chi2(8) = (b-B)'[(V_b-V_B)^(-1)](b-B)
                          =        60.07
                Prob>chi2 =       0.0000
```

There is strong evidence for model misspecification since the Hausman test statistic is 60.07 with df = 8. The Hausman statistic is almost identical to the Wald statistic for the joint null hypothesis that all regression coefficients of the cluster means are 0 shown in the previous section. Indeed, these tests become equivalent in large samples.

A significant Hausman test is often taken to mean that the random-intercept model should be abandoned in favor of a fixed-effects model that only utilizes within information. However, this would preclude estimation of the coefficients of covariates that vary only between clusters (admittedly, these estimates must be interpreted with caution since they may be inconsistent unless the covariates are exogenous). Moreover, if there are covariates having the same within- and between-effects, we obtain more precise estimates of these coefficients by exploiting both within- and between-cluster information. We therefore prefer using a random-intercept model where cluster means are included for some covariates, as demonstrated for smoking and mother's age in the previous section. Indeed, when mn_smok and mn_mage are included in the random-intercept model, the Hausman test is no longer significant at the 5% level.

3.8 Fixed versus random effects revisited

In section 2.4, we discussed whether the effects of clusters should be treated as random or fixed. We argued that this depends on whether inferences are for the population of clusters or only for the clusters included in the sample. In table 3.3, we consider these as the main questions and then ask questions related to each main question. The answers to these questions delineate the main differences between fixed-effects and random-effects approaches.

Table 3.3: Overview of distinguishing features of fixed- and random-effects approaches

	Answers:	
Questions:	**Fixed effects**	**Random effects**
Inference for population of clusters?	No	Yes
Minimum number of clusters required?	Any number	For estimating ψ, at least 10 or 20
What assumptions are required?	None for distribution of fixed intercepts α_j	Random intercepts ζ_j normal, constant variance ψ, exogenous covariates, etc.
Can estimate effects of cluster-level covariates?	No	Yes
Inference for clusters in particular sample?	Yes	No, not for βs, Yes, for ζ_j by using EB
Minimum cluster size required?	Any sizes if many ≥ 2, but large for est. α_j	Any sizes if many ≥ 2
Is the model parsimonious?	No, J parameters α_j but can eliminate α_j	Yes, one variance parameter ψ for all J clusters
Can estimate within-cluster effects of covariates?	Yes	Yes, by including cluster means

Unlike the fixed-effects model, the random-effects model can be used to make inferences regarding the population of clusters, but at the cost of requiring many clusters if inference regarding ψ is desired and making additional assumptions regarding the random-intercept distribution. The additional assumptions include the specification of a normal distribution for ζ_j, a constant variance ψ for the ζ_j, and exogeneity of the observed covariates $\mathbf{x}_{ij}$ with respect to ζ_j. An advantage of the random-effects model is that it can be used to estimate the effects of between-cluster covariates, in contrast to the fixed-effects model.

While the fixed-effects model is designed for making inferences regarding the clusters in the sample, the random-effects model can also to some extent be used for this purpose by predicting the ζ_j using empirical Bayes (EB). However, inferences regarding the regression coefficients from random-effects models, such as estimated standard errors, are for the population of clusters as discussed in section 2.7.2. The fixed-effects approach requires large cluster sizes if we want to estimate the intercepts α_j and is much less parsimonious than the random-intercept model because it includes one parameter α_j for each cluster, whereas the random-intercept model has only one parameter ψ for the variance of the random intercepts ζ_j. Eliminating the α_j by mean centering as shown in section 3.7.2 simplifies the estimation problem but does not make the estimates of the remaining parameters any more efficient. Unlike the random-effects approach, the fixed-effects approach controls for clusters, providing estimates of within-cluster effects of covariates. The random-effects model can provide estimates of within-cluster effects only with extra effort, namely, by including cluster means of those covariates for which the between effect differs from the within effect.

3.9 Residual diagnostics

We now consider residual diagnostics for assessing the normality assumptions for ζ_j and ϵ_{ij}.

In section 2.9.2, we discussed empirical Bayes (EB) prediction of the random intercepts for different clusters j. Such predictions $\widetilde{\zeta}_j$ can be interpreted as estimated level-2 residuals for the mothers. We can obtain corresponding estimated level-1 residuals for birth i of mother j as

$$\widetilde{\epsilon}_{ij} = \widehat{\xi}_{ij} - \widetilde{\zeta}_j$$

where

$$\widehat{\xi}_{ij} = y_{ij} - (\widehat{\beta}_1 + \widehat{\beta}_2 x_{2ij} + \cdots + \widehat{\beta}_p x_{pij})$$

In linear mixed models (but not in the generalized linear mixed models discussed later), these estimated level-2 and level-1 residuals have normal sampling distributions if the model is true. We can therefore use histograms or normal quantile–quantile plots of the estimated residuals to assess the assumptions that the true residuals ζ_j and ϵ_{ij} are normally distributed.

We can also try to find outliers by using standardized residuals and looking for values that are unlikely under the standard normal distribution, e.g., values outside the range ± 4. In section 2.9.3, we discussed the sampling standard deviation, or diagnostic standard error, of the empirical Bayes predictions. A standardized level-2 residual can therefore be obtained as

$$r_j^{(2)} = \frac{\widetilde{\zeta}_j}{\sqrt{\widehat{\text{Var}}(\widetilde{\zeta}_j)}}$$

For the level-1 residuals, we simply divide by the estimated level-1 standard deviation:

$$r_{ij}^{(1)} = \frac{\widetilde{\epsilon}_{ij}}{\sqrt{\widehat{\theta}}}$$

After using `xtmixed`, we can use the `predict` command with the option `reffects` to obtain $\widetilde{\zeta}_j$ and with the option `rstandard` to obtain $r_{ij}^{(1)}$. However, $r_j^{(2)}$ is, at the time of writing this book, not provided by `predict`, so we use the prediction command `gllapred` for `gllamm`, which produces both kinds of standardized residuals $r_{ij}^{(1)}$ and $r_j^{(2)}$.

We begin by retrieving the `gllamm` estimates stored in section 3.4.3. We then use the `gllapred` command with the `ustd` option to obtain $r_j^{(2)}$ and with the `pearson` option to obtain $r_{ij}^{(1)}$:

```
. estimates restore gllamm
. gllapred lev2, ustd
(means and standard deviations will be stored in lev2m1 lev2s1)
. gllapred lev1, pearson
(residuals will be stored in lev1)
```

Histograms of the standardized level-1 residuals $r_{ij}^{(1)}$ and the standardized level-2 residuals $r_j^{(2)}$ can be plotted as follows:

```
. histogram lev1, normal xtitle(Standardized level-1 residuals)
. histogram lev2m1 if idx==1, normal xtitle(Standardized level-2 residuals)
```

These commands produce figures 3.5 and 3.6, respectively.

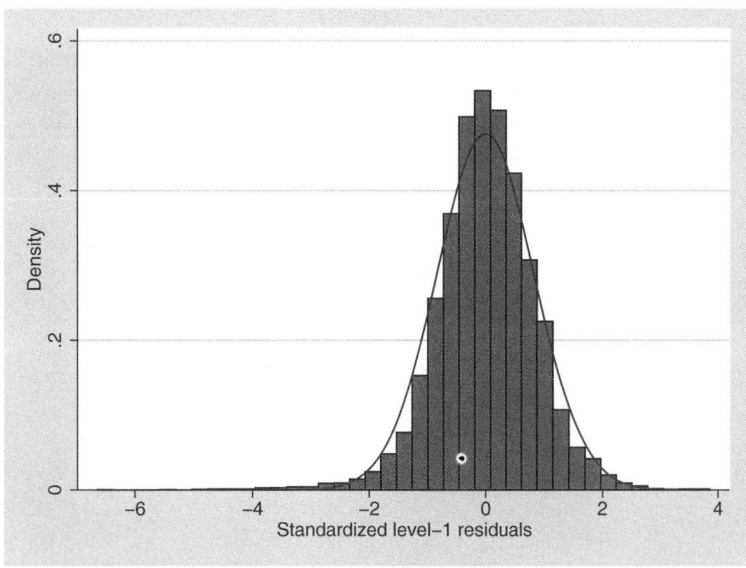

Figure 3.5: Histogram of standardized level-1 residuals

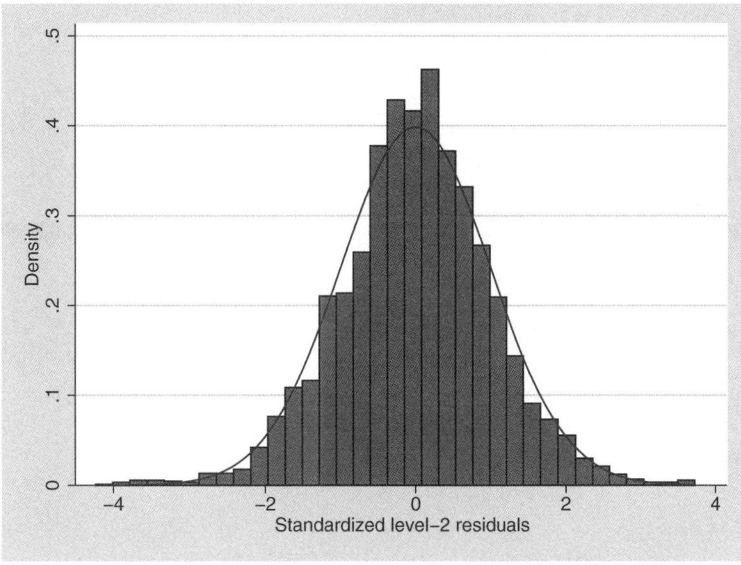

Figure 3.6: Histogram of standardized level-2 residuals

Both histograms look normal, perhaps with thicker tails than the normal distribution. However, there are some extremely small level-1 residuals. The expected number of standardized level-1 residuals less than -4 is 0.27 (=`8604*normal(-4)`). Thus it may be judicious to use robust standard errors based on the sandwich estimator, as they do not rely on the model being correctly specified:

```
. gllamm, robust

number of level 1 units = 8604
number of level 2 units = 3978

Condition Number = 8059.0645

gllamm model

log likelihood = -65145.752

Robust standard errors
```

birwt	Coef.	Std. Err.	z	P>|z|	[95% Conf.	Interval]
smoke	-218.3289	19.23251	-11.35	0.000	-256.0239	-180.6339
male	120.9375	9.675183	12.50	0.000	101.9745	139.9005
mage	8.100549	1.408359	5.75	0.000	5.340216	10.86088
hsgrad	56.84715	25.81829	2.20	0.028	6.244234	107.4501
somecoll	80.68607	28.37876	2.84	0.004	25.06471	136.3074
collgrad	90.83273	29.10669	3.12	0.002	33.78467	147.8808
married	49.9202	26.61475	1.88	0.061	-2.243757	102.0842
black	-211.4138	29.30634	-7.21	0.000	-268.8532	-153.9744
kessner2	-92.91883	21.64216	-4.29	0.000	-135.3367	-50.50098
kessner3	-150.8759	40.94536	-3.68	0.000	-231.1273	-70.62446
novisit	-30.03035	80.90957	-0.37	0.711	-188.6102	128.5495
pretri2	92.8579	24.73624	3.75	0.000	44.37577	141.34
pretri3	178.7295	52.85098	3.38	0.001	75.14353	282.3156
_cons	3117.191	42.73218	72.95	0.000	3033.438	3200.945

```
Variance at level 1
------------------------------------------------------------------------------

  137392.81 (4613.4372)

Variances and covariances of random effects
------------------------------------------------------------------------------


***level 2 (momid)

    var(1): 114763.41 (4726.832)
------------------------------------------------------------------------------
```

Most of the standard errors are larger than before but the basic conclusions remain the same. We could also use `xtreg` with the `vce(robust)` option, but at the time of writing this book, this option was not available together with the `mle` option.

3.10 More on statistical inference for regression coefficients

3.10.1 Consequences of using ordinary regression for clustered data

In this section, we presume that the random-intercept model

$$y_{ij} \;=\; \beta_1 + \beta_2 x_{ij} + \underbrace{\zeta_j + \epsilon_{ij}}_{\xi_{ij}} \tag{3.14}$$

is the *true* model, with the assumptions stated in section 3.3.1 satisfied. However, the regression coefficient β_2 in the above random-intercept model is estimated using ordinary least squares (OLS), based on the assumptions stated in section 1.5, for the linear regression model

$$y_i \;=\; \beta_1 + \beta_2 x_i + \epsilon_i$$

which is *misspecified.*

The crucial difference between the two models is that there is a positive within-cluster correlation between the (total) residuals in the random-intercept model if $\psi > 0$, whereas there is no such within-cluster correlation in the ordinary linear regression model. For simplicity, we assume that the data are balanced ($n_j = n > 1$) and have large $N = Jn$.

The good news is that the OLS estimator is consistent (approaches the unknown parameter value in large samples) when the random-intercept model is true and $\psi > 0$. Indeed, for a given dataset the OLS estimates are usually close to the ML estimates for the correctly specified random-intercept model, $\widehat{\beta}_1^{\rm OLS} \approx \widehat{\beta}_1^{\rm ML}$, $\widehat{\beta}_2^{\rm OLS} \approx \widehat{\beta}_2^{\rm ML}$, and $\widehat{\sigma^2}^{\rm OLS} \approx \widehat{\psi}^{\rm ML} + \widehat{\theta}^{\rm ML}$.

The results are worse when it comes to estimated standard errors and p-values for hypothesis tests. The estimated standard error of the OLS estimator $\widehat{\beta}_2^{\rm OLS}$ for the ordinary linear regression model is

$$\widehat{\rm SE}_{\rm OR}(\widehat{\beta}_2^{\rm OLS}) \;=\; \sqrt{\frac{\widehat{\sigma^2}^{\rm OLS}}{Jn\, s_{xO}^2}} \;\approx\; \sqrt{\frac{\widehat{\psi} + \widehat{\theta}^{\rm ML}}{Jn\, s_{xO}^2}}$$

where s_{xO}^2 is the overall sample variance of x_{ij} given in section 3.2.

For a *between-cluster covariate* (with $x_{ij} = \overline{x}_{\cdot j}$) the within-cluster variance s_{xw}^2 is zero and the estimated standard error of the ML estimator $\widehat{\beta}_2^{\rm ML}$ for the random-intercept model is

$$\widehat{\rm SE}_{\rm RI}(\widehat{\beta}_2^{\rm ML}) \;=\; \sqrt{\frac{n\widehat{\psi}^{\rm ML} + \widehat{\theta}^{\rm ML}}{Jn\, s_{xO}^2}} \;>\; \widehat{\rm SE}_{\rm OR}(\widehat{\beta}_2^{\rm OLS}) \qquad \text{if } n > 1,\ \widehat{\psi} > 0$$

Because the OLS standard error is smaller than it should be, the corresponding p-value is too small.

For a *within-cluster covariate* (with $\overline{x}_{.j} = \overline{x}_{..}$) the between-cluster variance s^2_{xb} is zero and the estimated standard error of $\widehat{\beta}_2^{\rm ML}$ for the random-intercept model is

$$\widehat{\rm SE}_{\rm RI}(\widehat{\beta}_2^{\rm ML}) = \sqrt{\frac{\widehat{\theta}^{\rm ML}}{Jn\, s^2_{xO}}} < \widehat{\rm SE}_{\rm OR}(\widehat{\beta}_2^{\rm OLS}) \qquad \text{if } \widehat{\psi} > 0$$

Because the OLS standard error is larger than it should be, the *p*-value is now too large.

Thus assuming an ordinary regression model when the random-intercept model is true and $\psi > 0$ will produce too small or too large standard errors and *p*-values depending on the nature of the covariate. This is contrary to the common belief that the estimated standard errors are always too small.

An important lesson is that robust standard errors for clustered data should be employed when ordinary regression is used for clustered data. In Stata, such standard errors can be obtained using the `vce(cluster` *clustvar*`)` option as demonstrated in section 2.7.2.

The above expressions for the estimated standard errors also hold if the model includes other covariates that are uncorrelated with x_{ij}.

3.10.2 ❖ Power and sample-size determination

Here we consider power and sample-size determination for the random-intercept model in (3.14) with a single covariate. A typical problem is to determine the sample size to achieve a required power γ at a given significance level α for the two-sided test of the null hypothesis H_0: $\beta_2 = 0$.

From the construction of the z test for the above null hypothesis, it follows that

$$\frac{\beta_2}{{\rm SE}(\widehat{\beta}_2)} = z_{1-\alpha/2} + z_\gamma \tag{3.15}$$

where $z_{1-\alpha/2}$ and z_γ are the $1-\alpha/2$ and γ quantiles of a standard normal distribution, respectively. Often the significance level and power are chosen as $\alpha = 0.05$ and $\gamma = 0.80$, respectively, which gives $z_{1-\alpha/2} + z_\gamma = 1.96 + 0.84 = 2.80$.

For a random-intercept model with a *between-cluster covariate*, the standard error of the maximum likelihood estimate of the coefficient is

$${\rm SE}(\widehat{\beta}_2) = \sqrt{\frac{n\psi + \theta}{Jn\, s^2_{xO}}}$$

We can then substitute this standard error into (3.15) and solve for the total number of clusters J for given cluster size n (or vice versa), with given values of the parameters β_2, ψ, and θ and with a given variance of the covariate s^2_{xO}.

For example, let $n=4$, $\beta_2=1$, $\psi=1$, $\theta=1$, and x_{2j} be a between-cluster treatment dummy variable, equal to 0 for half the clusters and 1 for the other half so that $s_{xO}^2 = 0.5\times(1-0.5) = 0.25$. We then obtain

$$\mathrm{SE}(\widehat{\beta}_2) \;=\; \sqrt{\frac{4\times 1+1}{J\times 4\times 0.25}} = \sqrt{5/J}$$

from which it follows that

$$\frac{\beta_2}{\mathrm{SE}(\widehat{\beta}_2)} = \sqrt{\frac{1}{5/J}} = \sqrt{J/5}$$

Substituting for $\beta_2/\mathrm{SE}(\widehat{\beta}_2)$ and $z_{1-\alpha/2}+z_\gamma$ in (3.15) we obtain the equation

$$\sqrt{J/5} \;=\; 2.80$$

which we solve to get $J = 39.2$. We see that about 40 clusters are needed (20 per group) to achieve 80% power to detect the treatment effect at the 5% significance level.

For a random-intercept model with a *within-cluster covariate*, the standard error of the maximum likelihood estimate of the coefficient is

$$\mathrm{SE}(\widehat{\beta}_2) \;=\; \sqrt{\frac{\theta}{Jn\, s_{xO}^2}}$$

Assuming that x_{2ij} is a treatment dummy variable equal to 0 for half the units in each cluster and 1 for the other half, and keeping all other assumptions the same as in the example above, we obtain

$$\mathrm{SE}(\widehat{\beta}_2) \;=\; \sqrt{\frac{1}{J\times 4\times 0.25}} = \sqrt{1/J}$$

and it follows that

$$\frac{\beta_2}{\mathrm{SE}(\widehat{\beta}_2)} = \sqrt{\frac{1}{1/J}} = \sqrt{J}$$

Substituting in (3.15),

$$\sqrt{J} \;=\; 2.80$$

and solving for J, we see that only about 8 clusters are now needed in total, compared with about 40 if treatment is between clusters.

If treatment is allocated in a randomized experiment, studies with between-cluster treatment allocation are often called cluster randomized trials whereas studies where treatment allocation is stratified within cluster are called multisite studies. Multisite studies have more power and thus require smaller sample sizes, particularly if $n\psi$ is large compared with θ as in our example, but sometimes such studies are not feasible, for instance if the clusters are classrooms and the treatment is a new curriculum.

3.11 Summary and further reading

We have introduced linear random-intercept models, which are important for considering the relationship between a continuous response and a set of covariates when the data have a clustered or hierarchical structure.

An important problem in any regression model is the potential for bias due to omitted covariates. In clustered data, we can identify the problem of omitted cluster-level covariates by comparing within-cluster and between-cluster effects of covariates varying both within and between clusters. Although this problem is emphasized in econometrics, where the Hausman test is used to test for endogeneity, it is not usually considered in other disciplines. However, a nice, relatively nontechnical treatment with applications in biostatistics can be found in Neuhaus and Kalbfleisch (1998). Exercise 3.6 uses a dataset considered in this paper.

Another area where within- and between-cluster effects are contrasted is education (Raudenbush and Bryk 2002). Specifically, interest often concerns the difference in within-school and between-school effects of socioeconomic status (SES); see exercise 3.8. The difference in the within-cluster effect of a covariate (such as students' individual SES) and the between-cluster effect of the aggregated covariate (such as school mean SES) is often called a *contextual effect* in contrast to *compositional effect*.

We have also discussed a number of topics relevant for random-intercept modeling with covariates, including hypothesis testing, different kinds of coefficients of determination, the choice between fixed- versus random-effects approaches, model diagnostics to assess the normality of level-1 and level-2 residuals, consequences of using ordinary regression for clustered data, and power and sample-size determination.

An excellent discussion of linear random-intercept models can be found in Snijders and Bosker (1999), which is also a useful reference on model diagnostics and power analysis. Wooldridge (2002) provides a good but demanding treatment of within and between estimators and endogeneity.

3.12 Exercises

3.1 Tax-preparer data

Frees (2004) analyzed panel or repeated-measurement data on tax returns filed by 258 taxpayers for the years 1982, 1983, 1984, 1986, and 1987. (The data come from the Statistics of Income (SOI) panel of individual returns.)

The dataset `taxprep.dta` has the following variables:

- `subject`: subject identifier
- `time`: identifier for the panel wave or occasion
- `lntax`: natural logarithm of tax liability in 1983 dollars
- `prep`: dummy variable for using a tax preparer for the tax return
- `ms`: dummy variable for being married

- `hh`: dummy variable for being the head of the household
- `depend`: number of dependents claimed by taxpayer
- `age`: dummy variable for being at least 65 years old
- `lntpi`: natural logarithm of the sum of all positive income line items on return in 1983 dollars
- `mr`: marginal tax rate computed on total income less exemptions and the standard deductions
- `emp`: dummy variable for schedule C or F being present on the return, a proxy for self-employment income

1. Fit a random-intercept model for `lntax` with all the subsequent variables given above as covariates and with a random intercept for subjects.
2. Obtain between and within estimates for all the covariates using `xtreg` with the `fe` and `be` options. Compare the estimates for the effect of tax preparer.
3. Perform a Hausman specification test.
4. Obtain histograms for the level-1 and level-2 residuals. Do the normality assumptions appear plausible?

3.2 Neighborhood-effects data

Garner and Raudenbush (1991), Raudenbush and Bryk (2002), and Raudenbush et al. (2004) considered neighborhood effects on educational attainment for young people in one education authority in Scotland who left school between 1984 and 1986.

The dataset `neighborhood.dta` (previously used in exercise 2.4) has the following variables:

- Level 1 (students)
  - `attain`: a measure of end-of-school educational attainment capturing both attainment and length of schooling (based on the number of O-grades and Higher SCE awards at the A–C levels)
  - `p7vrq`: verbal-reasoning quotient (test at age 11–12 in primary school)
  - `p7read`: reading test score (test at age 11–12 in primary school)
  - `dadocc`: father's occupation scaled on the Hope–Goldthorpe scale in conjunction with the Registrar General's social-class index (Willms 1986)
  - `dadunemp`: dummy variable for father being unemployed (1: unemployed; 0: not unemployed)
  - `daded`: dummy variable for father's schooling being past the age of 15
  - `momed`: dummy variable for mother's schooling being past the age of 15
  - `male`: dummy variable for student being male
- Level 2 (neighborhoods)
  - `neighid`: neighborhood identifier
  - `deprive`: social-deprivation score derived from poverty concentration, health, and housing stock of local community

1. Fit a random-intercept model with `attain` as the response variable and without any covariates by maximum likelihood using `xtmixed`. What are the estimated variance components between and within neighborhoods? Obtain the estimated intraclass correlation.
2. Include the covariate `deprive` in the model, and interpret the estimates. Discuss the changes in the estimated standard deviations of the random intercept and level-1 residual.
3. Include the student-level covariates and interpret the estimates. Also comment on how the estimated standard deviations have changed.
4. Obtain the overall coefficient of determination R^2 for the model in step 3.

See also exercise 11.5.

3.3 Grade-point-average data

Hox (2002) analyzed simulated longitudinal or panel data on 200 college students whose grade point average (GPA) was recorded over six successive semesters.

The variables in the dataset `gpa.dta` are

- `gpa1`–`gpa6`: grade point average (GPA) for semesters 1–6
- `student`: student identifier
- `highgpa`: high school GPA
- `job1`–`job6`: amount of time per week spent working for pay (0: not at all; 1: 1 hour; 2: 2 hours; 3: 3 hours; 4: 4 or more hours) in semesters 1–6
- `sex`: sex (1: male; 2: female)

1. Reshape the data to long form, stacking the time-varying variables `gpa1`–`gpa6` and `job1`–`job6` into two new variables `gpa` and `job` and generating a new variable `time` taking the values 1–6 for semesters 1–6.
2. Fit a random-intercept model with covariates `time`, `highgpa`, and `job`, and a dummy variable for males.
3. Assess whether the linearity assumptions for the three continuous covariates appear to be reasonable. You could use graphical methods, include quadratic terms, or use dummy variables for the different values of the covariates (if there are not too many).
4. Test whether there are interactions between each of the two student-level covariates and `time`.
5. For the chosen model, obtain empirical Bayes predictions of the random intercepts and produce graphs to assess their normality.

3.4 Jaw-growth data

In this jaw growth dataset from Pothoff and Roy (1964), eleven boys and sixteen girls had the distance between the center of the pituitary to the pterygomaxillary fissure recorded at ages 8, 10, 12, and 14.

The dataset `growth.dta` has the following variables:

- `idnr`: subject identifier
- `measure`: distance between pituitary and maxillary fissure in millimeters
- `age`: age in years
- `sex`: gender (1: boys; 2: girls)

1. Plot the observed growth trajectories; i.e., plot `measure` against `age`, connecting successive observations on the same subject using `connect(ascending)`. Use the `by()` option to obtain separate graphs by sex.
2. Fit a linear random-intercept model with `measure` as the response variable and with `age` and a dummy variable for girls as explanatory variables.
3. Test whether there is a significant interaction between sex and age at the 5% level.
4. For the chosen model (with or without the interaction), add the predicted mean trajectories for boys and girls to the graph of the observed growth trajectories. (Hint: use `predict, xb`.)

See also exercise 5.3.

3.5 Rat-pups data

Dempster et al. (1984) analyzed data from a reproductive study on rats to assess the effect of an experimental compound on general reproductive performance and pup weights.

30 dams (rat mothers) were randomized to three groups of 10 dams: control, low dose, and high dose of the compound. In the high-dose group, one female did not conceive, one cannibalized her litter, and one delivered one still birth, so that data on only 7 litters were available for that group.

The dataset `pups.dta` has the following variables:

- `dam`: dam (pup's mother) identifier
- `sex`: sex of pup (0: male; 1: female)
- `dose`: dose group (0: controls; 1: low dose; 2: high dose)
- `w`: weight of pup in grams

1. Construct a variable `size` representing the size of each litter.
2. Construct a variable `mnw` representing the mean weight of each litter.
3. Plot `mnw` versus `size`, using different symbols for the three treatment groups. Describe the graph.

4. Fit a random-intercept model for pup weights with `sex`, `dose` (dummy variables for low and high doses), and `size` as covariates and a random intercept for `dam`. Use maximum likelihood estimation.
5. Obtain level-1 and level-2 residuals, and produce graphs to assess the normality assumptions for ζ_j and ϵ_{ij}.

3.6 Georgian birthweight data

Here we use the data described in exercise 2.7.

Following Neuhaus and Kalbfleisch (1998) and Pan (2002), we consider the relationship between the child's birthweight and his or her mother's age at the time of the birth, distinguishing between within-mother and between-mother effects of age.

The variables in `birthwt.dta` are

- `mother`: mother identifier
- `child`: child identifier
- `birthwt`: child's birthweight (in grams)
- `age`: mother's age at the time of the child's birth

1. Fit a random-intercept model with birthweight as the response variable and age as the explanatory variable.
2. Perform a Hausman specification test.
3. Modify the model to estimate both within-mother and between-mother effects of age.
4. Discuss what mother-level omitted covariates might be responsible for the difference between the estimated within-mother and between-mother effects.

We recommend reading either Neuhaus and Kalbfleisch (1998) or Pan (2002), the latter being less technical.

3.7 Wage-panel data

Vella and Verbeek (1998) analyzed panel data for 545 young males taken from the U.S. National Longitudinal Survey (Youth Sample) for the period 1980–1987. We focus on modeling a wage equation for the men.

The dataset `wagepan.dta` was supplied by Wooldridge (2002). The subset of variables considered here is

- `nr`: person identifier (j)
- `year`: 1980 to 1987 (i)
- `lwage`: log of hourly wage in U.S. dollars (y_{ij})
- `educ`: years of schooling (x_{2j})
- `black`: dummy variable for being black (x_{3j})
- `hisp`: dummy variable for being Hispanic (x_{4j})

- exper: labor market experience defined as age−6−educ (x_{5ij})
- expersq: labor market experience squared (x_{6ij})
- married: dummy variable for being married (x_{7ij})
- union: dummy variable for being a member of a union (i.e., wage being set in collective bargaining agreement) (x_{8ij})

You can use the describe command to get a description of the other variables in the file.

1. Ignore the clustered nature of the data, and use the regress command to fit the regression model

$$y_{ij} = \alpha_i + \beta_2 x_{2j} + \beta_3 x_{3j} + \beta_4 x_{4j} + \beta_5 x_{5ij} + \beta_6 x_{6ij} + \beta_7 x_{7ij} + \beta_8 x_{8ij} + \epsilon_{ij}$$

   where α_i are fixed year-specific intercepts and the covariates are defined in the bulleted list above.
2. Refit the above model using the vce(cluster nr) option to get standard errors taking the clustering into account. Compare the estimated standard errors with those from step 1.
3. Fit the random-intercept model

$$y_{ij} = \alpha_i + \beta_2 x_{2j} + \beta_3 x_{3j} + \beta_4 x_{4j} + \beta_5 x_{5ij} + \beta_6 x_{6ij} + \beta_7 x_{7ij} + \beta_8 x_{8ij} + \zeta_j + \epsilon_{ij}$$

   with the usual assumptions. How do the estimates of the βs compare with those from the models ignoring clustering?
4. Modify the random-intercept model to investigate whether the effect of education has increased linearly over time after controlling for the other covariates.

See also chapter 5.

3.8 High-school-and-beyond data

Raudenbush and Bryk (2002) and Raudenbush et al. (2004) analyzed data from the High School and Beyond Survey.

The variables in the dataset hsb.dta that we will use here are

- schoolid: school identifier (j)
- mathach: a measure of mathematics achievement (y_{ij})
- SES: socioeconomic status (SES) based on parental education, occupation, and income (x_{ij})

1. Use xtsum to explore the between-school and within-school variability of SES.
2. Produce a variable mn_ses equal to the schools' mean SES and another variable dev_ses equal to the difference between the students' SES and the mean SES for their school.

3. Use `xtreg` to fit a model for `mathach` with a fixed effect for SES and a random intercept for school.
4. The model above assumes that SES has the same effect within and between schools. Check this by using the covariates `mn_ses` and `dev_ses` instead of `ses` and comparing the coefficients using `lincom`.
5. Interpret the coefficients of `mn_ses` and `dev_ses`.
6. Returning to the model with `ses` as the only predictor, perform a Hausman specification test, and comment on the result.

3.9 Antisocial-behavior data

Allison (2005) considered a sample of 581 children who were interviewed in 1990, 1992, and 1994 as part of the U.S. "National Longitudinal Survey of Youth".

The dataset `antisocial.dta` includes the following variables:

- `id`: child identifier (j)
- `occ`: year of interview (90, 92, 94)
- `anti`: a measure of the child's antisocial behavior y_{ij} (higher values mean more antisocial behavior)
- `pov`: dummy variable for child being from a poor family x_{2ij} (1: poor; 0: otherwise). This covariate varies both between and within children
- `momage`: mother's age at birth of child in years x_{3j}
- `female`: dummy variable for child being female (1: female; 0: male) x_{4j}
- `childage`: child's age in years in 1990 x_{5j}
- `hispanic`: dummy variable for child being Hispanic (1: Hispanic; 0: otherwise) x_{6j}
- `black`: dummy variable for child being black x_{7j} (1: black; 0: otherwise)
- `momwork`: dummy variable for mother being employed in 1990 x_{8j} (1: employed; 0: not employed)
- `married`: dummy variable for mother being married in 1990 x_{9j} (1: married; 0: otherwise)

1. Fit the random-intercept model

$$y_{ij} = \beta_1+\beta_2 x_{2ij}+\beta_3 x_{3j}+\beta_4 x_{4j}+\beta_5 x_{5j}+\beta_6 x_{6j}+\beta_7 x_{7j}+\beta_8 x_{8j}+\beta_9 x_{9j}+\zeta_j+\epsilon_{ij}$$

(with the usual assumptions) and interpret the estimates.

2. Test the null hypothesis that the random-intercept variance is zero.
3. State the expression for the residual intraclass correlation in terms of the model parameters and give the estimate.
4. Replace x_{2ij} in the random-intercept model by the covariates

$$x_{10,j} = \frac{1}{n_j}\sum_{i=1}^{n_j} x_{2ij}$$

where n_j is the number of units in cluster j, and

$$x_{11,ij} = x_{2ij} - x_{10,j}$$

Fit the resulting model and interpret the estimates $\widehat{\beta}_{10}$ and $\widehat{\beta}_{11}$ corresponding to $x_{10,j}$ and $x_{11,ij}$, respectively.

5. Test the null hypothesis $\beta_{10} = \beta_{11}$ against the alternative $\beta_{10} \neq \beta_{11}$. Explain why this test can be interpreted as an endogeneity test.

See also exercise 5.4.

3.10 ❖ Relations among within, between and GLS estimators

For the Georgian birthweight data used in exercise 3.6, the within-mother and between-mother standard deviations of mothers' age x_{ij} at the birth of the child are $s_{xW} = 2.796$ years and $s_{xB} = 3.693$ years, respectively. Consider a random-intercept model for the child's birthweight (with the usual assumptions)

$$y_{ij} \;=\; \beta_1 + \beta_2 x_{ij} + \zeta_j + \epsilon_{ij}$$

The within-mother and between-mother estimates of the effect of age on birthweight are $\widehat{\beta}_2^W = 11.832$ grams/year and $\widehat{\beta}_2^B = 30.355$ grams/year, respectively. There are 878 mothers with $n = 5$ births each, $\widehat{\text{Var}}(\epsilon_{ij} - \overline{\epsilon}_{\cdot j}) = 150,631$ grams^2 and $\widehat{\text{Var}}(\zeta_j + \overline{\epsilon}_{\cdot j}) = 161,457$ grams^2.

Based solely on the above information, obtain

1. The estimated standard error of $\widehat{\beta}_2^W$
2. The estimated standard error of $\widehat{\beta}_2^B$
3. The GLS estimate of β_2 for the random-intercept model

3.11 ❖ Power analysis

1. For a random-intercept model with a single between-cluster covariate, calculate the total sample size Jn required to have 80% power to reject the null hypothesis that $\beta_2 = 0$ using a two-sided test at the 5% level. Assume that $n = 20$, $\beta_2 = 1$, $\psi = 1$, $\theta = 5$, and $s_{xO}^2 = 0.25$.
2. Repeat the calculation in step 1 with the same assumptions except that $\psi = 0$ and $\theta = 6$, i.e., keeping the total variance as before but setting the intraclass correlation to zero.
3. Now consider the general situation where there are two scenarios, each having the same total residual variance $\psi + \theta$ but one having $\psi > 0$ and the other $\psi = 0$. Obtain an expression for the ratio of the required sample sizes Jn for these two scenarios in terms of the intraclass correlation ρ and the cluster size n. (This factor is sometimes called the "design effect" for clustered data.)

4 Random-coefficient models

4.1 Introduction

In the previous chapter, we considered linear random-intercept models where the overall level of the response was allowed to vary over clusters after controlling for covariates.

In this chapter, we include random coefficients or random slopes in addition to random intercepts, thus also allowing the effects of covariates to vary over clusters. Such models involving both random intercepts and random slopes are often called random-coefficient models. In longitudinal settings, where the level-1 units are occasions and the clusters are typically subjects, random-coefficient models are also referred to as growth-curve models. Such models are discussed in chapter 5.

4.2 How effective are different schools?

We start by analyzing a dataset on inner-London schools that accompanies the MLwiN software (Rasbash et al. 2005) and is part of the data analyzed by Goldstein et al. (1993).

At age 16, students took their Graduate Certificate of Secondary Education (GCSE) exams in a number of subjects. A score was derived from the individual exam results. Such scores often form the basis for school comparisons, for instance, to allow parents to choose the best school for their child. However, schools can differ considerably in their intake achievement levels, and it may be argued that what should be compared is the "value added"; the difference in mean GCSE score between schools after controlling for achievement before entering the school. One such measure of prior achievement is the London Reading Test (LRT) taken by these students at age 11.

The dataset `gcse.dta` has the following variables:

- `school`: school identifier
- `student`: student identifier
- `gcse`: Graduate Certificate of Secondary Education (GCSE) score (z score, multiplied by 10)
- `lrt`: London Reading Test (LRT) score (z score, multiplied by 10)
- `girl`: dummy variable for child being a girl (1: girl; 0: boy)
- `schgen`: type of school (1: mixed gender; 2: boys only; 3: girls only)

One purpose of the analysis is to investigate the relationship between GCSE and LRT and how this relationship varies between schools. The model can then be used to address the question of which schools are most effective, taking intake achievement into account.

4.3 Separate linear regressions for each school

Before developing a model for all 65 schools combined, we consider a separate model for each school. For school j, an obvious model for the relationship between GCSE and LRT is a simple regression model

$$y_{ij} = \beta_{1j} + \beta_{2j}x_{ij} + \epsilon_{ij}$$

where y_{ij} is the GCSE score for the ith student in school j, x_{ij} is the corresponding LRT score, β_{1j} the school-specific intercept, β_{2j} the school-specific slope, and ϵ_{ij} is a residual error term with school-specific variance θ_j.

For school 1, OLS estimates of the intercept $\widehat{\beta}_{11}$ and the slope $\widehat{\beta}_{21}$ can be obtained using `regress`:

```
. use http://www.stata-press.com/data/mlmus2/gcse
. regress gcse lrt if school==1
      Source |       SS       df       MS              Number of obs =      73
-------------+------------------------------           F(  1,    71) =   59.44
       Model |  4084.89209     1  4084.89209           Prob > F      =  0.0000
    Residual |  4879.35761    71  68.7233466           R-squared     =  0.4557
-------------+------------------------------           Adj R-squared =  0.4480
       Total |  8964.24969    72  124.503468           Root MSE      =    8.29

------------------------------------------------------------------------------
        gcse |      Coef.   Std. Err.      t    P>|t|     [95% Conf. Interval]
-------------+----------------------------------------------------------------
         lrt |   .7093406   .0920061     7.71   0.000     .5258856    .8927955
       _cons |   3.833302   .9822377     3.90   0.000     1.874776    5.791828
------------------------------------------------------------------------------
```

To assess if this is a reasonable model, we can obtain a scatterplot of the data for school 1 with the corresponding regression line superimposed:

```
. predict p_gcse, xb
. twoway (scatter gcse lrt) (line p_gcse lrt, sort) if school==1,
> xtitle(LRT) ytitle(GCSE)
```

The resulting graph in figure 4.1 suggests that the model assumptions are reasonably met.

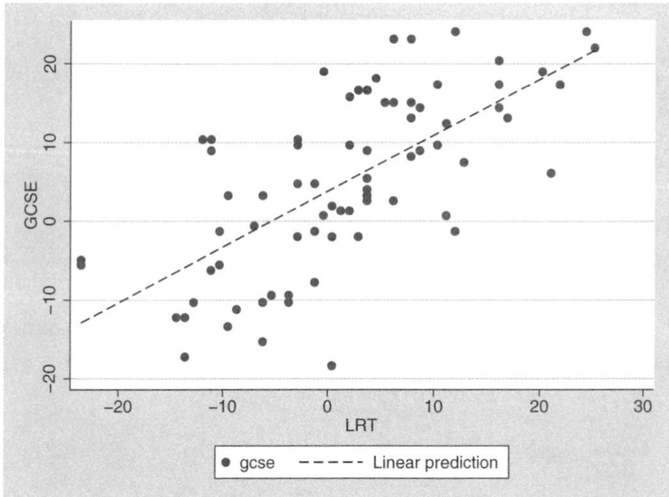

Figure 4.1: Scatterplot of gcse versus lrt for school 1 with regression line

We will now fit a simple linear regression model for each school, which is easily done using Stata's prefix command statsby, and examine the variability in the estimated intercepts and slopes. We first calculate the number of students per school to preclude fitting lines to schools with fewer than 5 students:

```
. egen num = count(gcse), by(school)
```

We then use statsby to create a new dataset ols.dta in the local directory with the variables inter and slope containing estimates of the intercepts (_b[_cons]) and slopes (_b[lrt]) from the command regress gcse lrt if num>4 applied to each school (as well as containing the variable school):

```
. statsby inter=_b[_cons] slope=_b[lrt], by(school) saving(ols):
> regress gcse lrt if num>4
(running regress on estimation sample)

      command:  regress gcse lrt if num>4
        inter:  _b[_cons]
        slope:  _b[lrt]
           by:  school

Statsby groups
----+--- 1 ---+--- 2 ---+--- 3 ---+--- 4 ---+--- 5
..................................................    50
..............
```

We can merge the estimates inter and slope into the gcse dataset using the merge command (after sorting the "master data" by school; the "using data" created by statsby are already sorted by school):

```
. sort school
. merge school using ols
variable school does not uniquely identify observations in the master data
. drop _merge
```

Here we have deleted the variable `_merge` produced by the `merge` command to avoid error messages when we want to run the `merge` command in the future.

A scatterplot is produced using the command

```
. twoway scatter slope inter, xtitle(Intercept) ytitle(Slope)
```

and given in figure 4.2. We see that there is considerable variability between the intercepts and slopes of different schools. To investigate this further, we first create a dummy variable to pick out one observation per school,

```
. egen pickone = tag(school)
```

and then produce summary statistics for the schools by using the `summarize` command:

```
. summarize inter slope if pickone == 1
    Variable |       Obs        Mean    Std. Dev.       Min        Max
-------------+--------------------------------------------------------
       inter |        64   -.1805974    3.291357  -8.519253   6.838716
       slope |        64    .5390514    .1766135   .0380965   1.076979
```

To allow comparison with the parameter estimates obtained from the random-coefficient model considered later on, we also obtain the covariance matrix of the estimated intercepts and slopes:

```
. correlate inter slope if pickone == 1, covariance
(obs=64)
             |    inter     slope
-------------+------------------
       inter |   10.833
       slope |  .208622   .031192
```

The diagonal elements 10.83 and 0.03 are the sample variances of the intercepts and slopes, respectively, whereas the off-diagonal element 0.21 is the sample covariance between the intercepts and slopes, the correlation times the product of the intercept and slope standard deviations.

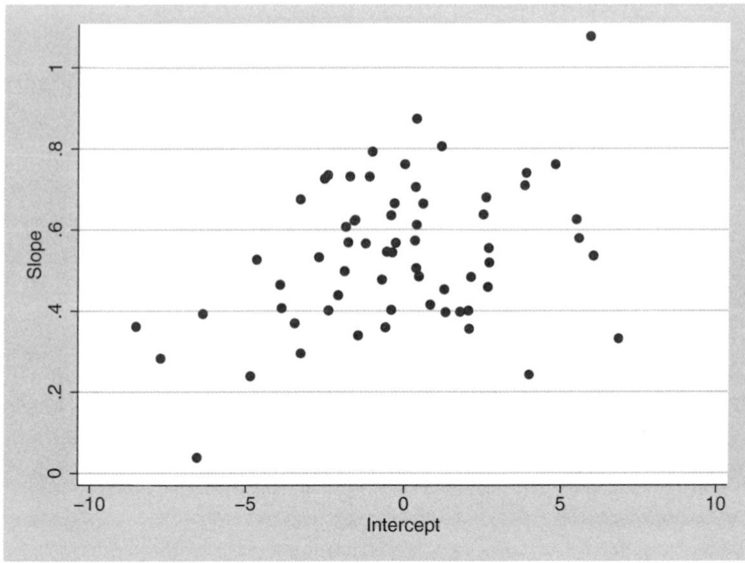

Figure 4.2: Scatterplot of intercepts and slopes for all schools with at least 5 students

We can also plot the predicted school-specific regression lines by first calculating the fitted values $\widehat{y}_{ij} = \widehat{\beta}_{1j} + \widehat{\beta}_{2j}x_{ij}$:

```
. generate pred = inter + slope*lrt
(2 missing values generated)
. sort school lrt
. twoway (line pred lrt, connect(ascending)), xtitle(LRT)
> ytitle(Fitted regression lines)
```

To produce the graph, we first sorted the data so that `lrt` increases within a given school and then jumps to its lowest value for the next school in the dataset. The `connect(ascending)` option is used to connect points only as long as `lrt` is increasing and ensures that only data for the same school are connected. The resulting graph is shown in figure 4.3.

(*Continued on next page*)

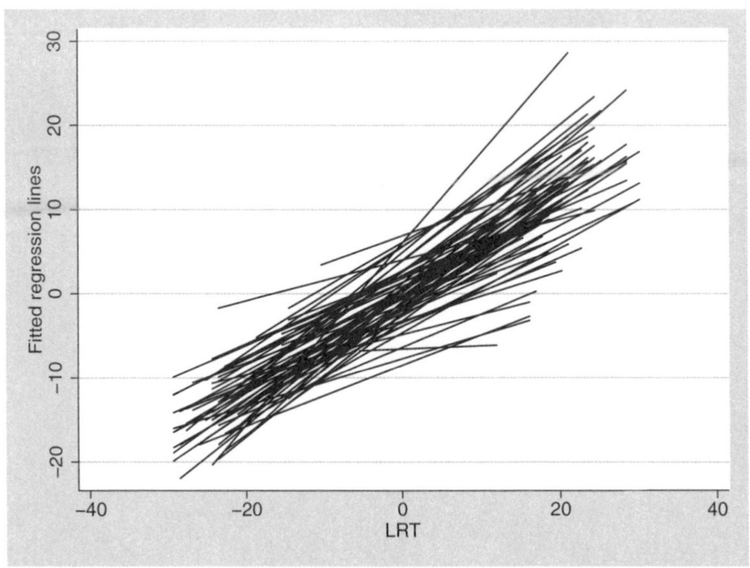

Figure 4.3: Least-squares regression lines for all schools with at least 5 students

4.4 Specification and interpretation of a random-coefficient model

4.4.1 Specification of random-coefficient model

How can we develop a joint model for the relationships between `gcse` and `lrt` in all schools?

One way would be to use dummy variables for all schools (omitting the overall constant) to estimate school-specific intercepts and interactions between these dummy variables and `lrt` to estimate school-specific slopes. The only difference between the resulting model and separate regressions is that a common residual error variance $\theta_j = \theta$ is assumed. However, this model has 130 regression coefficients! Furthermore, if the schools are viewed as a (random) sample of schools from a population of schools, we are not interested in the individual coefficients characterizing each school's regression line. Rather, we would like to estimate the mean intercept and slope as well as the (co)variability of the intercepts and slopes in the population of schools.

A parsimonious model for the relationships between `gcse` and `lrt` can be obtained by specifying a school-specific random intercept ζ_{1j} and a school-specific random slope ζ_{2j} for `lrt` (x_{ij}):

$$\begin{aligned} y_{ij} &= \beta_1 + \beta_2 x_{ij} + \zeta_{1j} + \zeta_{2j} x_{ij} + \epsilon_{ij} \\ &= (\beta_1 + \zeta_{1j}) + (\beta_2 + \zeta_{2j}) x_{ij} + \epsilon_{ij} \end{aligned} \qquad (4.1)$$

We assume that the covariate x_{ij} is exogenous with $E(\zeta_{1j}|x_{ij}) = 0$, $E(\zeta_{2j}|x_{ij}) = 0$, and $E(\epsilon_{ij}|x_{ij}, \zeta_{1j}, \zeta_{2j}) = 0$. Then ζ_{1j} represents the deviation of school j's intercept from the mean intercept β_1, and ζ_{2j} represents the deviation of school j's slope from the mean slope β_2. It follows from the zero expectations that all three random terms are uncorrelated with x_{ij} and that ϵ_{ij} is uncorrelated with both ζ_{1j} and ζ_{2j}. Both the intercepts ζ_{1j} and slopes ζ_{2j} are independent across schools and the level-1 residuals ϵ_{ij} are independent across schools and students.

An illustration of this random-coefficient model with one covariate x_{ij} for a school j is shown in the bottom panel of figure 4.4. A random-intercept model is shown for comparison in the top panel. In each panel, the lower, bold, solid line represents the population-averaged regression line

$$E(y_{ij}|x_{ij}) = \beta_1 + \beta_2 x_{ij}$$

across all schools. The thinner solid line represents the school-specific regression line for school j. For the random-intercept model, this is

$$E(y_{ij}|x_{ij}, \zeta_{1j}) = (\beta_1 + \zeta_{1j}) + \beta_2 x_{ij}$$

which is parallel to the population-averaged line with vertical displacement given by the random intercept ζ_{1j}. In contrast, in the random-coefficient model, the school-specific line

$$E(y_{ij}|x_{ij}, \zeta_{1j}, \zeta_{2j}) = (\beta_1 + \zeta_{1j}) + (\beta_2 + \zeta_{2j})x_{ij}$$

is not parallel to the population-averaged line but has a greater slope because the random slope ζ_{2j} is positive in the illustration. Here the dashed line is parallel to the population-averaged regression line and has the same intercept as school j. The vertical deviation between this dashed line and the line for school j is $\zeta_{2j}x_{ij}$, as shown in the diagram for $x_{ij}=1$. The bottom panel illustrates that the total intercept for school j is $\beta_1 + \zeta_{1j}$ and the total slope is $\beta_2 + \zeta_{2j}$. The arrows from the school-specific regression lines to the responses y_{ij} are the within-school residual error terms ϵ_{ij} (with variance θ). It is clear that $\zeta_j x_{ij}$ represents an *interaction* between the clusters, treated as random, and the covariate x_{ij}.

(*Continued on next page*)

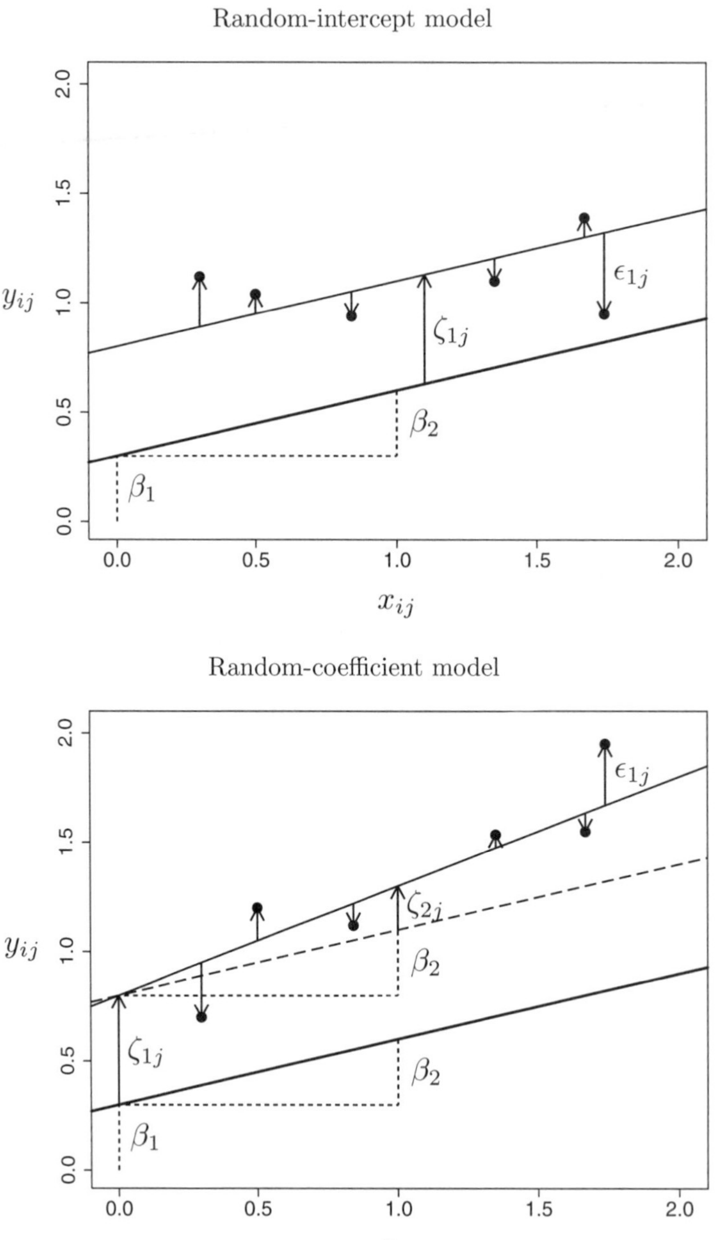

Figure 4.4: Illustration of random-intercept and random-coefficient models

We will, as is usually done, assume that, given x_{ij}, the random intercept and random slope have a bivariate normal distribution with zero mean and covariance matrix:

$$\mathbf{\Psi} = \begin{bmatrix} \psi_{11} & \psi_{12} \\ \psi_{21} & \psi_{22} \end{bmatrix} \equiv \begin{bmatrix} \text{Var}(\zeta_{1j}|x_{ij}) & \text{Cov}(\zeta_{1j},\zeta_{2j}|x_{ij}) \\ \text{Cov}(\zeta_{2j},\zeta_{1j}|x_{ij}) & \text{Var}(\zeta_{2j}|x_{ij}) \end{bmatrix}, \qquad \psi_{21}=\psi_{12}$$

The correlation between the random intercept and slope becomes

$$\rho_{21} = \frac{\psi_{21}}{\sqrt{\psi_{11}\psi_{22}}}$$

An example of a bivariate normal distribution with $\psi_{11}=\psi_{22}=4$ and $\psi_{21}=\psi_{12}=1$ is shown as a perspective plot in figure 4.5 and as a contour plot in figure 4.6. Specifying a bivariate normal distribution implies that the (marginal) univariate distributions of the intercept and slope are also normal.

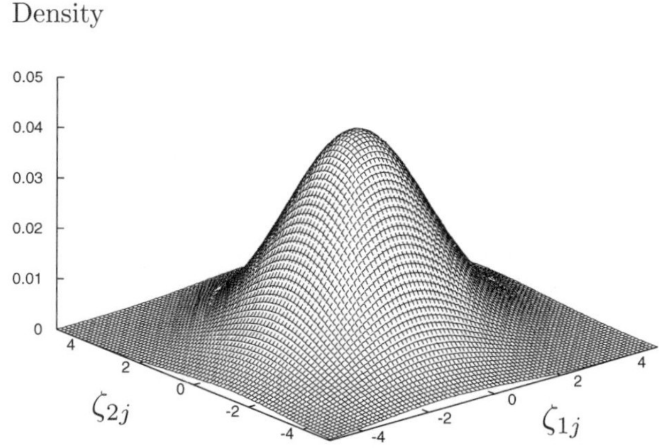

Figure 4.5: Perspective plot of bivariate normal distribution

(Continued on next page)

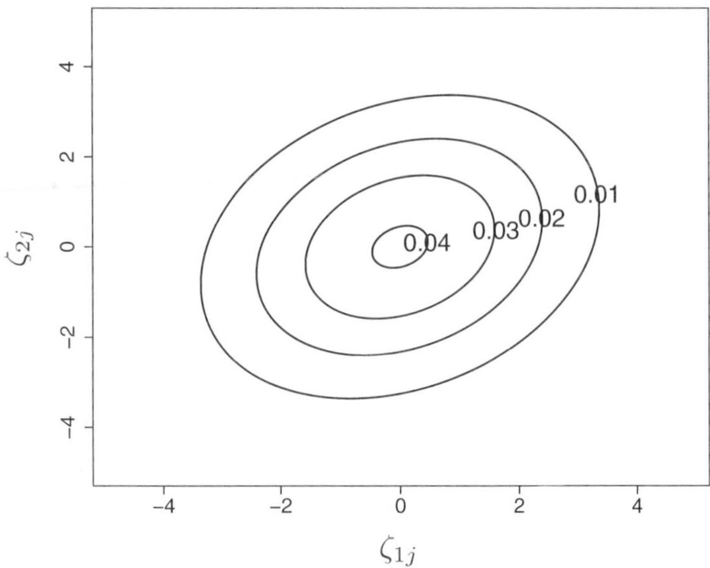

Figure 4.6: Contour plot of bivariate normal distribution

4.4.2 Interpretation of the random-effects variances and covariances

Interpreting the covariance matrix $\mathbf{\Psi}$ of the random effects is not straightforward. First, the random-slope variance ψ_{22} and the covariance between random slope and intercept ψ_{21} depend not just on the scale of the response variable, but also on the scale of the covariate, here `lrt`. Let the units of the response and explanatory variable be denoted as u_y and u_x, respectively. For instance, in the application considered in the next chapter on children's increase in weight, u_y is kilograms, and u_x is years. The units of ψ_{11} are u_y^2, the units of ψ_{21} are u_y^2/u_x, and the units of ψ_{22} are u_y^2/u_x^2. It therefore does not make sense to compare the magnitude of random-intercept and random-slope variances.

A second difficulty is that the total residual variance is no longer constant as in random-intercept models. The total residual is now

$$\xi_{ij} \equiv \zeta_{1j} + \zeta_{2j}x_{ij} + \epsilon_{ij}$$

and the conditional variance of the responses given the covariate, or the variance of the total residual, is

$$\text{Var}(y_{ij}|x_{ij}) = \text{Var}(\xi_{ij}|x_{ij}) = \psi_{11} + 2\psi_{21}x_{ij} + \psi_{22}x_{ij}^2 + \theta \tag{4.2}$$

This variance depends on the value of the covariate x_{ij} and the total residual is therefore *heteroskedastic*. The conditional covariance for two students i and i' in the same school j is

$$
\begin{aligned}
\text{Cov}(y_{ij}, y_{i'j}|x_{ij}, x_{i'j}) &= \text{Cov}(\xi_{ij}, \xi_{i'j}|x_{ij}, x_{i'j}) \\
&= \psi_{11} + \psi_{21}x_{ij} + \psi_{21}x_{i'j} + \psi_{22}x_{ij}x_{i'j} \qquad (4.3)
\end{aligned}
$$

and the conditional intraclass correlation becomes

$$
\text{Cor}(y_{ij}, y_{i'j}|x_{ij}, x_{i'j}) = \frac{\text{Cov}(\xi_{ij}, \xi_{i'j}|x_{ij}, x_{i'j})}{\sqrt{\text{Var}(\xi_{ij}|x_{ij})\text{Var}(\xi_{i'j}|x_{i'j})}}
$$

When $x_{ij} = x_{i'j} = 0$, the expression for the intraclass correlation is the same as for the random-intercept model and represents the correlation of the residuals (from the overall mean regression line) for two students in the same school who both have `lrt` scores equal to 0 (the mean). However, for other pairs of students in the same school, the intraclass correlation is a complicated function of x_{ij} and $x_{i'j}$. Due to the heteroskedastic total residual variance, it is not straightforward to define coefficients of determination, such as R^2, R_2^2, and R_1^2 discussed in section 3.5, for random-coefficient models.

Finally, interpreting the parameters ψ_{11} and ψ_{21} can be difficult because their values depend on the translation of the covariate, or in other words on how much we add or subtract from the covariate. Adding a constant to `lrt` and refitting the model would result in different estimates of ψ_{11} and ψ_{21} (see also exercise 4.7). This is because the intercept variance is the variability in the vertical positions of school-specific regression lines where `lrt`=0 (which changes when `lrt` is translated) and the covariance or correlation is the tendency for regression lines that are higher up where `lrt`=0 to have higher slopes. This lack of invariance of ψ_{11} and ψ_{21} to translation of the explanatory variable x_{ij} is illustrated in figure 4.7. Here identical cluster-specific regression lines are shown in the two panels, but with the explanatory variable $x'_{ij} = x_{ij} - 3.5$ in the lower panel translated relative to the explanatory variable x_{ij} in the upper panel. The intercepts are the intersections of the regression lines with the vertical line at zero. Clearly these intercepts vary more in the upper panel than the lower panel, whereas the correlation between intercepts and slopes is negative in the upper panel and positive in the lower panel.

(*Continued on next page*)

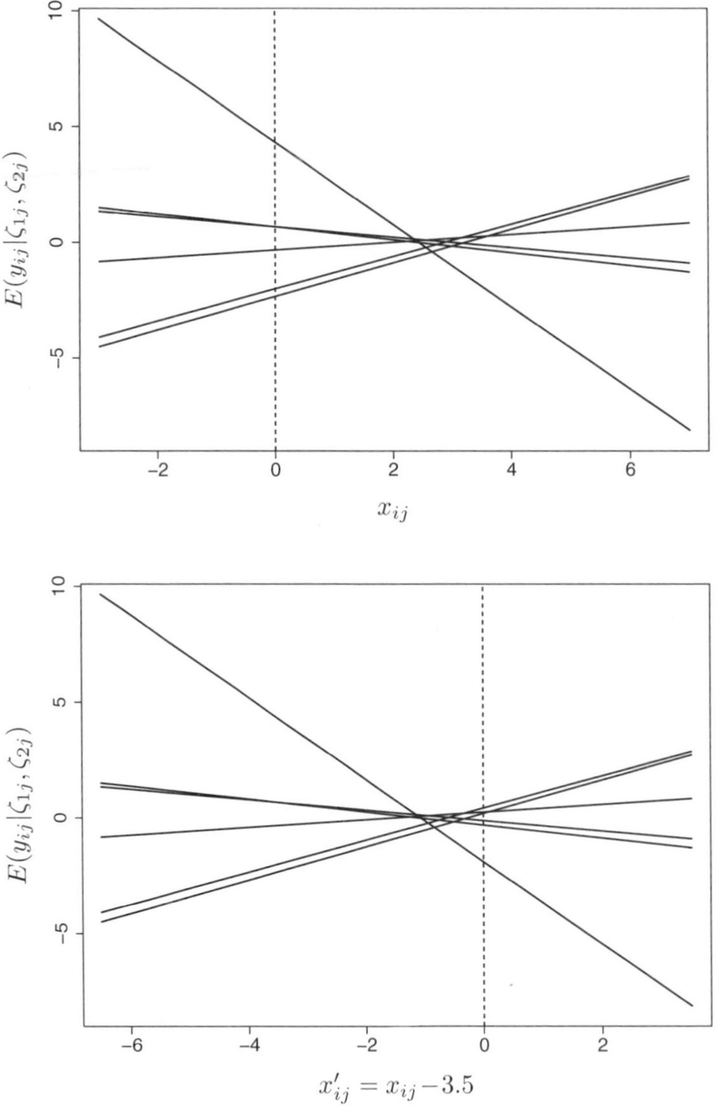

Figure 4.7: Cluster-specific regression lines for random-coefficient model, illustrating lack of invariance under translation of explanatory variable (*Source:* Skrondal and Rabe-Hesketh 2004)

To make ψ_{11} and ψ_{21} interpretable, it makes sense to translate x_{ij} so that $x_{ij} = 0$ is meaningful in some way. Typical choices are either mean centering (as for `lrt`), or, if x_{ij} is time as in the next chapter, defining 0 to be the "initial time" in some sense. Because the magnitude and interpretation of ψ_{21} depend on the translation of x_{ij}, which is often arbitrary, it generally does not make sense to set ψ_{21} to 0 by specifying uncorrelated intercepts and slopes.

A useful way of interpreting the magnitude of the estimated variances $\widehat{\psi}_{11}$ and $\widehat{\psi}_{22}$ is by considering the intervals $\widehat{\beta}_1 \pm 1.96\sqrt{\widehat{\psi}_{11}}$ and $\widehat{\beta}_2 \pm 1.96\sqrt{\widehat{\psi}_{22}}$ that contain about 95% of the intercepts and slopes in the population, respectively. To aid interpretation of the random part of the model, it is also useful to produce plots of school-specific regression lines as discussed in section 4.8.3.

4.5 Estimation using Stata

We will describe two Stata commands for linear random-coefficient models, `xtmixed` and `gllamm`. In general, we recommend using `xtmixed` rather than `gllamm` for linear random-coefficient models because it is computationally more efficient and sometimes more accurate. However, there are certain diagnostics available using the `gllapred` command for `gllamm` that are, at the time of writing this book, not provided by the `predict` command for `xtmixed`.

4.5.1 Using xtmixed

Random-intercept model

We first consider the more familiar random-intercept model

$$y_{ij} = (\beta_1 + \zeta_{1j}) + \beta_2 x_{ij} + \epsilon_{ij} \tag{4.4}$$

discussed in the previous chapter. This model is a special case of the random-coefficient model in (4.1) with $\zeta_{2j} = 0$ or, equivalently, with zero random-slope variance and zero random intercept and slope covariance, $\psi_{22} = \psi_{21} = 0$.

Maximum likelihood estimates for the random-intercept model can be obtained using `xtmixed` with the `mle` option:

(Continued on next page)

```
. xtmixed gcse lrt || school:, mle

Mixed-effects ML regression                     Number of obs      =      4059
Group variable: school                          Number of groups   =        65

                                                Obs per group: min =         2
                                                               avg =      62.4
                                                               max =       198

                                                Wald chi2(1)       =   2042.57
Log likelihood = -14024.799                     Prob > chi2        =    0.0000

------------------------------------------------------------------------------
        gcse |      Coef.   Std. Err.      z    P>|z|     [95% Conf. Interval]
-------------+----------------------------------------------------------------
         lrt |   .5633697   .0124654    45.19   0.000     .5389381    .5878014
       _cons |   .0238706   .4002258     0.06   0.952    -.7605576    .8082987
------------------------------------------------------------------------------

------------------------------------------------------------------------------
  Random-effects Parameters  |   Estimate   Std. Err.     [95% Conf. Interval]
-----------------------------+------------------------------------------------
school: Identity             |
                   sd(_cons) |   3.035271   .3052516      2.492262     3.69659
-----------------------------+------------------------------------------------
                sd(Residual) |   7.521481   .0841759      7.358295    7.688285
------------------------------------------------------------------------------
LR test vs. linear regression: chibar2(01) =   403.27 Prob >= chibar2 = 0.0000
```

To allow later comparison with random-coefficient models using likelihood-ratio tests, we store these estimates using

```
. estimates store ri
```

The random-intercept model assumes that the school-specific regression lines are parallel. The common coefficient or slope β_2 of `lrt`, shared by all schools, is estimated as 0.56 and the mean intercept as 0.02.

Schools vary in their intercepts with an estimated standard deviation of 3.04. Within the schools, the estimated residual standard deviation around the school-specific regression lines is 7.52. The within-school correlation, after controlling for `lrt`, is therefore estimated as

$$\widehat{\rho} = \frac{\widehat{\psi}_{11}}{\widehat{\psi}_{11} + \widehat{\theta}} = \frac{3.035^2}{3.035^2 + 7.521^2} = 0.14$$

The maximum likelihood estimates for the random-intercept model are also given under "Model 1: Random intercept" in table 4.1.

Table 4.1: Maximum likelihood estimates for inner-London school data

	Model 1: Random intercept		Model 2: Random coefficient		Model 3: Rand. coefficient & level-2 covariates		
Parameter	Est	(SE)	Est	(SE)	Est	(SE)	γ_{xx}
Fixed part							
β_1 [_cons]	0.02	(0.40)	−0.12	(0.40)	−1.00	(0.51)	γ_{11}
β_2 [lrt]	0.56	(0.01)	0.56	(0.02)	0.57	(0.03)	γ_{21}
β_3 [boys]					0.85	(1.09)	γ_{12}
β_4 [girls]					2.43	(0.84)	γ_{13}
β_5 [boys_lrt]					−0.02	(0.06)	γ_{22}
β_6 [girls_lrt]					−0.03	(0.04)	γ_{23}
Random part							
xtmixed							
$\sqrt{\psi_{11}}$	3.04		3.01		2.80		
$\sqrt{\psi_{22}}$			0.12		0.12		
ρ_{21}			0.50		0.60		
$\sqrt{\theta}$	7.52		7.44		7.44		
gllamm							
ψ_{11}	9.21		9.04				
ψ_{22}			0.01				
ψ_{21}			0.18				
θ	56.57		55.37				
Log likelihood	$-14,024.80$		$-14,004.61$		$-13,998.83$		

Random-coefficient model

We now relax the assumption that the school-specific regression lines are parallel by introducing random school-specific slopes $\beta_2 + \zeta_{2j}$ of `lrt`:

$$y_{ij} = (\beta_1 + \zeta_{1j}) + (\beta_2 + \zeta_{2j})x_{ij} + \epsilon_{ij}$$

To introduce a random slope for `lrt` using `xtmixed`, we simply add that variable name in the specification of the random part, replacing `school:` with `school: lrt`. We must also specify the `covariance(unstructured)` option (here abbreviated as `cov(unstructured)`) because `xtmixed` will otherwise set the covariance ψ_{21} (and the corresponding correlation) to zero by default. Maximum likelihood estimates for the random-coefficient model are then obtained using

```
. xtmixed gcse lrt || school: lrt, cov(unstructured) mle

Mixed-effects ML regression                     Number of obs      =      4059
Group variable: school                          Number of groups   =        65

                                                Obs per group: min =         2
                                                               avg =      62.4
                                                               max =       198

                                                Wald chi2(1)       =    779.80
Log likelihood = -14004.613                     Prob > chi2        =    0.0000

------------------------------------------------------------------------------
        gcse |      Coef.   Std. Err.      z    P>|z|     [95% Conf. Interval]
-------------+----------------------------------------------------------------
         lrt |   .5567291   .0199367    27.92   0.000     .5176539    .5958043
       _cons |  -.1150841   .3978336    -0.29   0.772    -.8948236    .6646554
------------------------------------------------------------------------------

------------------------------------------------------------------------------
  Random-effects Parameters  |   Estimate   Std. Err.     [95% Conf. Interval]
-----------------------------+------------------------------------------------
school: Unstructured         |
                     sd(lrt) |   .1205631   .0189827      .0885508    .1641483
                   sd(_cons) |   3.007436   .3044138      2.466252    3.667375
            corr(lrt,_cons)  |   .4975474   .1487416      .1572843    .7322131
-----------------------------+------------------------------------------------
                sd(Residual) |   7.440788   .0839482      7.278059    7.607157
------------------------------------------------------------------------------
LR test vs. linear regression:       chi2(3) =   443.64   Prob > chi2 = 0.0000

Note: LR test is conservative and provided only for reference.
```

Because the variance option was not used, the output shows the standard deviations sd(lrt) of the slope and sd(_cons) of the intercept instead of variances, and the correlation between intercepts and slopes corr(lrt,_cons) instead of the covariance. We can obtain the covariance matrix using the postestimation command estat recovariance:

```
. estat recovariance

Random-effects covariance matrix for level school

             |       lrt      _cons
-------------+----------------------
         lrt |  .0145355
       _cons |  .1804036    9.04467
```

The maximum likelihood estimates for the random-coefficient model are also given under "Model 2: Random intercept and slope" in table 4.1. We store the estimates under the name rc for later use:

```
. estimates store rc
```

We could also use restricted maximum likelihood estimation by specifying the reml option.

4.5.2 Using gllamm

Random-intercept model

We start by using gllamm to fit the random-intercept model:

```
. gllamm gcse lrt, i(school) adapt
number of level 1 units = 4059
number of level 2 units = 65

Condition Number = 35.786606

gllamm model

log likelihood = -14024.799
```

gcse	Coef.	Std. Err.	z	P>|z|	[95% Conf.	Interval]
lrt	.5633697	.0124863	45.12	0.000	.538897	.5878425
_cons	.0239115	.4002945	0.06	0.952	-.7606514	.8084744

```
Variance at level 1
------------------------------------------------------------------------------

  56.572669 (1.2662546)

Variances and covariances of random effects
------------------------------------------------------------------------------


***level 2 (school)

    var(1): 9.2127069 (1.8529779)
------------------------------------------------------------------------------
```

We store the estimates under the name rig:

```
. estimates store rig
```

Random-coefficient model

In the previous gllamm command for the random-intercept model, all that was required to specify the random intercept was the i(school) option.

To introduce a random slope ζ_{2j}, we will also need to specify the variable multiplying the random slope in (4.1), i.e., x_{ij} or lrt. This is done by specifying an equation for the slope:

```
. eq slope: lrt
```

We also need an equation for the variable multiplying the random intercept ζ_{1j}, and since it is an intercept, we just specify a variable equal to 1:

```
. generate cons = 1
. eq inter: cons
```

We must also add a new option, `nrf(2)`, standing for "number of random effects is 2" (an intercept and a slope), and specify both equations, `inter` and `slope`, in the `eqs()` option.

To speed up estimation, we use the previous estimates for the random-intercept model as starting values for the regression coefficients and random-intercept variance, and set the starting values for the additional two parameters (for the random-slope variance and the random intercept and slope covariance) to zero:

```
. matrix a = e(b)
. matrix a = (a,0,0)
```

(The order in which the parameters are given in the matrix matters, and when going from a random-intercept to a random-coefficient model, the two new parameters are always at the end.) To use the parameter matrix `a` as starting values, we specify the `from(a)` and `copy` options. Finally, to get good estimates as fast as possible, we use a spherical quadrature rule of degree 15 (see sec. 6.11.2), specified using `ip(m)` and `nip(15)`:

```
. gllamm gcse lrt, i(school) nrf(2) eqs(inter slope) ip(m) nip(15)
> adapt from(a) copy
number of level 1 units = 4059
number of level 2 units = 65

Condition Number = 35.440522

gllamm model

log likelihood = -14004.613
```

gcse	Coef.	Std. Err.	z	P>|z|	[95% Conf.	Interval]
lrt	.556729	.0199969	27.84	0.000	.5175358	.5959222
_cons	-.1150848	.3982896	-0.29	0.773	-.8957181	.6655485

```
Variance at level 1
------------------------------------------------------------------------------

  55.365324 (1.2492818)

Variances and covariances of random effects
------------------------------------------------------------------------------


***level 2 (school)

    var(1): 9.0446842 (1.8310103)
    cov(2,1): .18040306 (.06915204) cor(2,1): .49754322

    var(2): .01453559 (.00457725)
------------------------------------------------------------------------------
```

The gllamm output gives the maximum likelihood estimates of the within-school variance θ under Variance at level 1 and the estimates of the elements ψ_{11}, ψ_{21}, and ψ_{22}, of the covariance matrix of the random intercept and slope under Variances and covariances of random effects. These estimates are also given in table 4.1.

4.6 Testing the slope variance

Before interpreting the parameter estimates, we may want to test whether the random-coefficient model "fits better" than the nested random-intercept model. Specifically, we test the null hypothesis

$$H_0\colon\ \psi_{22} = \psi_{21} = 0$$

which is equivalent to the hypothesis that the random slopes ζ_{2j} are zero. The null hypothesis lies on the boundary of the parameter space since the variance ψ_{22} must be nonnegative. Therefore, as discussed in section 2.6.2, the likelihood-ratio statistic L does not have a chi-squared distribution under the null hypothesis. Fortunately, the correct p-value for testing the slope variance can also simply be obtained by dividing the naive p-value from the likelihood-ratio test by 2.

The naive likelihood-ratio test can be performed using the lrtest command in Stata:

```
. lrtest rc ri
Likelihood-ratio test                                 LR chi2(2)  =     40.37
(Assumption: ri nested in rc)                         Prob > chi2 =    0.0000
Note: LR test is conservative
```

The output correctly states that the test is conservative. We can divide the p-value by 2 but this makes no difference to the conclusion that the random-intercept model is rejected in favor of the random-coefficient model.

4.7 Interpretation of estimates

The population-mean intercept and slope are estimated as -0.12 and 0.56, respectively. These estimates are similar to those for the random-intercept model (see table 4.1) and also similar to the means of the school-specific least-squares regression lines given on page 144. The estimated random-intercept standard deviation and level-1 residual standard deviation are somewhat lower than for the random-intercept model. The latter is due to a better fit of the school-specific regression lines for the random-coefficient model, which relaxes the parallel regression line restriction. The estimated covariance matrix of the intercepts and slopes is similar to the sample covariance matrix of the ordinary least-squares estimates reported on page 144.

The easiest way to interpret the estimated standard deviations of the random intercept and random slope is to form intervals within which 95% of the schools' random intercepts and slopes are expected to lie. It is important to remember that these intervals represent ranges within which 95% of the realizations of a *random variable* are expected to lie, a concept different from confidence intervals that are ranges within which an *unknown parameter* is believed to lie. For the intercepts, we obtain $-0.12 \pm 1.96 \times 3.01$, so 95% of schools have their intercept in the range -6.0 to 5.8. In other words, the school mean GCSE scores for children with average (`lrt`=0) LRT scores vary between -6.0 and 5.8. For the slopes, we obtain $0.56 \pm 1.96 \times 0.12$ giving an interval from 0.32 to 0.80. Ninety-five percent of schools have slopes between 0.32 and 0.80. This exercise of forming intervals is particularly important for slopes because it is useful to know whether the slopes are likely to have different signs for different schools, although that would be odd in the current example. The range from 0.32 to 0.80 is fairly wide and the regression lines of schools may cross: one school could "add more value" (produce higher mean GCSE scores for given LRT scores) than another school for students with low LRT scores and add less value than the other school for students with high LRT scores.

The estimated correlation $\widehat{\rho}_{21} = 0.50$ between intercepts and slopes means that schools with larger mean GCSE scores for students with average LRT scores than other schools also tend to have larger slopes than those other schools. This information, combined with the random-intercept and slope variances and the range of LRT scores determines how much the lines cross, something that is best explored by plotting the predicted regression lines for the schools as demonstrated in section 4.8.3.

The variance of the total residual ξ_{ij} (equal to the conditional variance of the responses y_{ij} given the covariate x_{ij}) was given in (4.2). We can estimate the corresponding standard deviation by plugging in the maximum likelihood estimates:

$$\begin{aligned}\sqrt{\widehat{\mathrm{Var}}(\xi_{ij}|x_{ij})} &= \sqrt{\widehat{\psi}_{11} + 2\widehat{\psi}_{21}x_{ij} + \widehat{\psi}_{22}x_{ij}^2 + \widehat{\theta}} \\ &= \sqrt{9.0447 + 2 \times 0.1804 \times x_{ij} + 0.0145 \times x_{ij}^2 + 55.3653}\end{aligned}$$

A graph of the estimated standard deviation of the total residual against the covariate `lrt` (x_{ij}) can be obtained using the `twoway function` command:

```
. twoway function sqrt(9.0447+2*0.1804*x+0.0145*x^2+55.3653), range(-30 30)
> xtitle(LRT) ytitle(Estimated standard deviation of total residual)
```

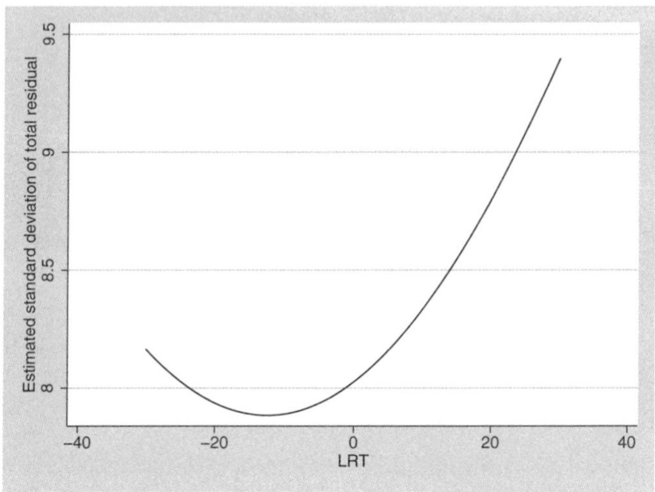

Figure 4.8: Heteroskedasticity of total residual ξ_{ij}

The estimated standard deviation of the total residual varies between just under 8 and just under 9.5.

4.8 Assigning values to the random intercepts and slopes

Having obtained estimated model parameters $\widehat{\beta}_1$, $\widehat{\beta}_2$, $\widehat{\psi}_{11}$, $\widehat{\psi}_{22}$, $\widehat{\psi}_{21}$, and $\widehat{\theta}$, we now assign values to the random intercepts and slopes, treating the estimated parameters as known (see sec. 2.9). This is useful for model visualization, residual diagnostics, and inference for individual clusters as will be demonstrated in sections 4.8.3–4.8.5.

4.8.1 Maximum likelihood estimation

Maximum likelihood estimates of the random intercepts and slopes can be obtained by first estimating the total residuals $\widehat{\xi}_{ij} = y_{ij} - (\widehat{\beta}_1 + \widehat{\beta}_2 x_{ij})$ and then fitting individual regressions of $\widehat{\xi}_{ij}$ on x_{ij} for each school by OLS using the `statsby` prefix command. We first retrieve the `xtmixed estimates` stored under `rc`, then obtain the estimated total residual and use `statsby` to produce the variables `mli` and `mls` containing the maximum likelihood estimates $\widehat{\zeta}_{1j}$ and $\widehat{\zeta}_{2j}$ of the random intercepts and slopes, respectively:

```
. estimates restore rc
(results rc are active now)
. predict fixed, xb
. generate totres = gcse - fixed
```

```
. statsby mli=_b[_cons] mls=_b[lrt], by(school) saving(ols, replace):
> regress totres lrt
(running regress on estimation sample)

      command:  regress totres lrt
          mli:  _b[_cons]
          mls:  _b[lrt]
           by:  school

Statsby groups
----+--- 1 ---+--- 2 ---+--- 3 ---+--- 4 ---+--- 5
..................................................    50
..............
. sort school
. merge school using ols
variable school does not uniquely identify observations in the master data
. drop _merge
```

Maximum likelihood estimates will not be available for schools with only one observation or for schools within which x_{ij} does not vary. There are no such schools in the dataset, but school 48 has only two observations and the maximum likelihood estimates look odd:

```
. list lrt gcse mli mls if school==48, clean noobs

        lrt      gcse       mli         mls
    -4.5541   -1.2908   -32.607   -7.458484
    -3.7276   -6.9951   -32.607   -7.458484
```

Here the fitted line goes through the two points exactly and the intercept and slope take on extreme values, a phenomenon often encountered when using maximum likelihood estimation of random effects for clusters that provide little information. In general, we therefore do not recommend using this method and suggest using empirical Bayes prediction instead.

4.8.2 Empirical Bayes prediction

As discussed for random-intercept models in section 2.9.2, empirical Bayes (EB) predictions have a smaller prediction error variance than maximum likelihood estimates due to shrinkage towards the mean. Furthermore, empirical Bayes predictions are available for schools with only one observation or only one unique value of x_{ij}.

Empirical Bayes predictions of the random intercepts ζ_{1j} and slopes ζ_{2j} can be obtained using the `predict` command with the `reffects` option after estimation with `xtmixed`:

```
. estimates restore rc
. predict ebs ebi, reffects
```

Here we specified the variable names `ebs` and `ebi` for the empirical Bayes predictions $\widetilde{\zeta}_{2j}$ and $\widetilde{\zeta}_{1j}$ of the slopes and intercepts. The intercept variable comes last because `xtmixed` treats the intercept as the last random effect, as reflected by the output. This order is consistent with Stata's convention of treating the fixed intercept as the last regression parameter in estimation commands.

To compare the empirical Bayes predictions with the maximum likelihood estimates, we list one observation per school for schools 1–9 and school 48:

```
. list school mli ebi mls ebs if pickone==1 & (school<10 | school==48), noobs
```

school	mli	ebi	mls	ebs
1	3.948386	3.749336	.1526115	.1249755
2	4.937837	4.702129	.2045584	.1647261
3	5.69259	4.79768	.0222564	.0808666
4	.1526213	.3502505	.2047173	.1271821
5	2.719524	2.462805	.1232875	.0720576
6	6.14715	5.18381	-.0213859	.0586242
7	4.100311	3.640942	-.3144541	-.1488697
8	-.1368859	-.1218861	.010678	.0068854
9	-2.2586	-1.767983	-.1555334	-.0886194
48	-32.607	-.4098185	-7.458484	-.0064854

Most of the time, the empirical Bayes predictions are closer to zero than the maximum likelihood estimates due to shrinkage as discussed for random-intercept models in section 2.9.2. However, for models with several random effects, the relationship between empirical Bayes predictions and maximum likelihood estimates is somewhat more complex than for random-intercept models. The benefit of shrinkage is apparent for school 48 where the empirical Bayes predictions appear more reasonable than the maximum likelihood estimates.

We can see shrinkage more clearly by plotting the empirical Bayes predictions against the maximum likelihood estimates and superimposing a $y = x$ line. For the random intercept, the command is

```
. twoway (scatter ebi mli if pickone==1 & school!=48, mlabel(school))
> (function y=x, range(-10 10)), xtitle(ML estimate)
> ytitle(EB prediction) legend(off)
```

and for the random slope, it is

```
. twoway (scatter ebs mls if pickone==1 & school!=48, mlabel(school))
> (function y=x, range(-0.6 0.6)), xtitle(ML estimate)
> ytitle(EB prediction) legend(off)
```

These commands produce the graphs in figure 4.9. (We excluded school 48 from the graphs because the ML estimates are so extreme.)

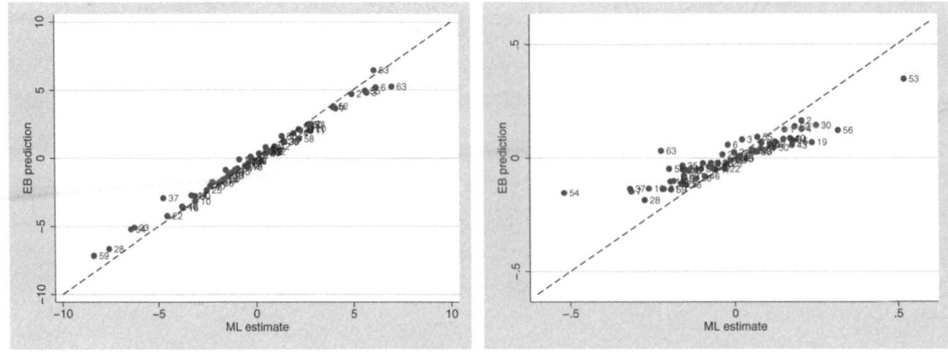

Figure 4.9: Scatterplot of empirical Bayes (EB) predictions versus maximum likelihood (ML) estimates of school-specific intercepts (left) and slopes (right) with equality shown as reference lines

For maximum likelihood estimates above the mean, the empirical Bayes prediction tends to be smaller than the maximum likelihood estimate; the reverse is true for maximum likelihood estimates below the mean.

4.8.3 Model visualization

To better understand the random-intercept and random-coefficient models, and in particular the variability implied by the random part, it is useful to produce graphs of predicted model-implied regression lines for the individual schools. This can be achieved using the `predict` command with the `fitted` option to obtain school-specific fitted regression lines, with maximum likelihood estimates substituted for the regression parameters (β_1 and β_2) and empirical Bayes predictions substituted for the random effects (ζ_{1j} for the random-intercept model and ζ_{1j} and ζ_{2j} for the random-intercept model). For instance, for the random-coefficient model, the predicted regression line for school j is

$$\widehat{y}_{ij} = \widehat{\beta}_1 + \widehat{\beta}_2 x_{ij} + \widetilde{\zeta}_{1j} + \widetilde{\zeta}_{2j} x_{ij}$$

These predictions are obtained and plotted as follows:

```
. predict murc, fitted
. sort school lrt
. twoway (line murc lrt, connect(ascending)), xtitle(LRT)
> ytitle(Empirical Bayes regression lines for model 2)
```

To obtain predictions for the random-intercept model, we must first restore the estimates stored under the name `ri`:

```
. estimates restore ri
(results ri are active now)
. predict muri, fitted
. sort school lrt
```

```
. twoway (line muri lrt, connect(ascending)), xtitle(LRT)
> ytitle(Empirical Bayes regression lines for model 1)
```

The resulting graphs of the school-specific regression lines for both the random-intercept model and the random-coefficient model are given in figure 4.10.

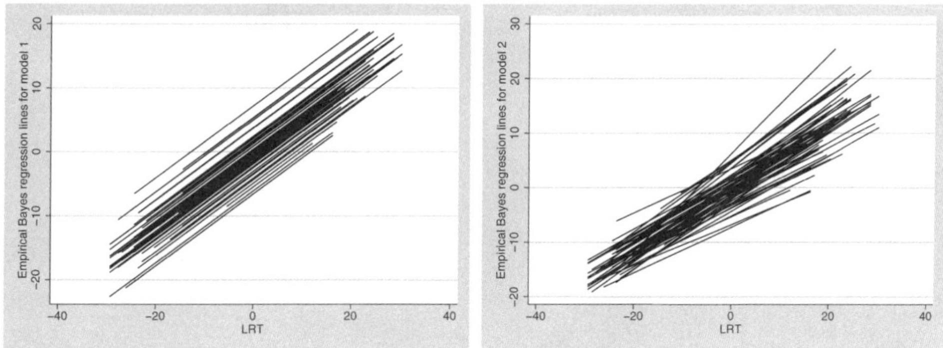

Figure 4.10: Empirical Bayes (EB) predictions of school-specific regression lines for the random-intercept model (left) and the random-intercept and random-slope model (right)

The predicted school-specific regression lines are parallel for the random-intercept model (with vertical shifts given by the $\widetilde{\zeta}_{1j}$) but are not parallel for the random-coefficient model where the slopes $\beta_2 + \widetilde{\zeta}_{2j}$ also vary across schools.

After estimation with `gllamm`, the syntax for obtaining school-specific regression lines would be

```
estimates restore rcg
gllapred murcg, linpred
```

4.8.4 Residual diagnostics

If the normality assumptions for the random intercepts ζ_{1j}, random slopes ζ_{2j}, and level-1 residuals ϵ_{ij} are satisfied, the corresponding empirical Bayes predictions should also have normal distributions. After estimation with `xtmixed`, we obtain the predicted level-1 residuals using

```
. predict res1, residuals
```

To plot the distributions of the predicted random effects, we must pick one prediction per school, and we can accomplish this using the `pickone` variable created earlier. We can now plot all three distributions using

```
. histogram reff1 if pickone==1, normal xtitle(Predicted random slopes)
. histogram reff2 if pickone==1, normal xtitle(Predicted random intercepts)
. histogram res1, normal xtitle(Predicted level-1 residuals)
```

The histograms in figure 4.11 and 4.12 look approximately normal although the one for the slopes is perhaps a little skewed:

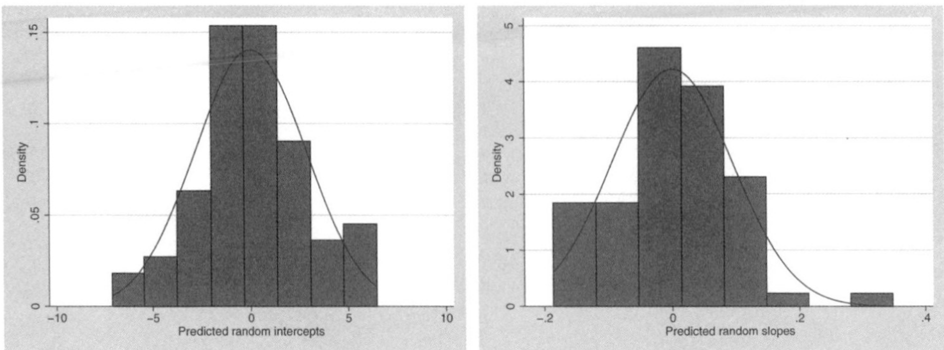

Figure 4.11: Histograms of predicted random intercepts and slopes

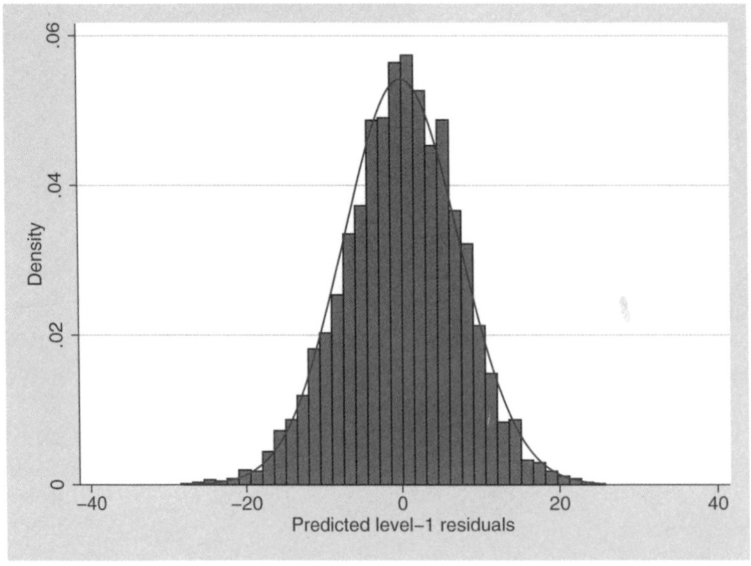

Figure 4.12: Histogram of predicted level-1 residuals

To obtain standardized level-1 residuals, use the `rstandard` option in the `predict` command after estimation using `xtmixed`. After estimation with `gllamm`, standardized level-2 residuals (random intercepts and random slopes) can be obtained using `gllapred` with the `ustd` option and standardized level-1 residuals with the `pearson` option.

4.8.5 Inferences for individual schools

The random-intercept predictions $\widetilde{\zeta}_{1j}$ can be viewed as measures of how much "value" the schools add for children with LRT scores equal to zero (the mean). Therefore, the left panel of figure 4.9 sheds some light on the research question: which schools are most effective? We could also produce similar plots for children with different values x^0 of the LRT scores:

$$\widehat{\beta}_1 + \widehat{\beta}_2 x^0 + \widetilde{\zeta}_{1j} + \widetilde{\zeta}_{2j} x^0$$

For instance, in a similar application, Goldstein et al. (2000) substitute the 10th percentile of the intake measure to compare school effectiveness for poorly performing children. It does not matter whether we add the predicted fixed part of the model since the ranking of schools is not affected by this. Unfortunately, `xtmixed` does not provide standard errors for random-effects predictions at the time of writing. We therefore restore the `gllamm` estimates

```
. estimates restore rcg
```

and use the postestimation command `gllapred` with the `u` option to obtain empirical Bayes predictions and corresponding standard errors for the random intercepts $\widetilde{\zeta}_{1j}$ and random slopes $\widetilde{\zeta}_{2j}$

```
. gllapred reff, u
(means and standard deviations will be stored in reffm1 reffs1 reffm2 reffs2)
```

The first random effect, with predictions stored in `reffm1`, is now the random intercept ζ_{1j}, whereas this was the second random effect in `xtmixed`. The predictions of the random slopes ζ_{2j} are stored in `reffm2`.

Returning to the question of comparing the schools' effectiveness for children with LRT scores equal to 0, we can plot the predicted random effects with approximate 95% confidence intervals (based on the prediction error standard deviations in `reffs1`) using the following commands:

```
. gsort +reffm1 -f
. generate rank = sum(f)
. generate labpos = reffm1 + 1.96*reffs1 + .4
. serrbar reffm1 reffs1 rank if f==1, addplot(scatter labpos rank,
> mlabel(school) msymbol(none) mlabpos(0)) scale(1.96) xtitle(Rank)
> ytitle(Prediction) legend(off)
```

(The school labels were added to the graph by superimposing a scatterplot onto the error bar plot using the `addplot()` option where the vertical positions of the labels are given by the variable `labpos`.) The resulting graph is shown in figure 4.13.

(*Continued on next page*)

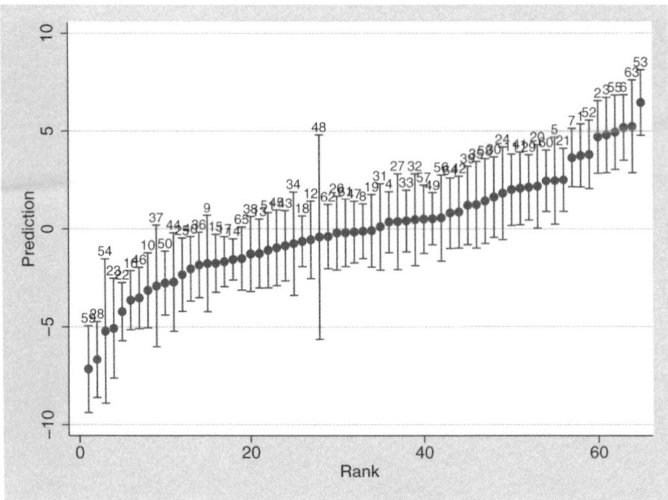

Figure 4.13: Random-intercept predictions and approximate 95% confidence intervals versus ranking (school identifiers shown on top of confidence intervals)

The interval for school 48 is particularly wide because there are only two students from this school in the dataset. It is clear from the large confidence intervals that the rankings are not precise and that perhaps only a coarse classification into poor, medium, and good schools can be justified. (To obtain confidence intervals for different values of x^0 requires posterior correlations that are produced by `gllapred` with the `corr()` option.)

4.9 Two-stage model formulation

In this section, we describe an alternative way of specifying random-coefficient models that is popular in some branches of U.S. social science, such as education (e.g., Raudenbush and Bryk 2002). As shown below, models are specified in two stages (for levels 1 and 2), necessitating a distinction between level-1 and level-2 covariates. Many people find this formulation helpful for interpreting and specifying models. Identical models can be formulated using either the approach discussed up to this point or the two-stage formulation.

To express the random-coefficient model using a two-stage formulation, the *level-1 model* is written as

$$y_{ij} = \eta_{1j} + \eta_{2j}x_{ij} + \epsilon_{ij}$$

where the intercept η_{1j} and slope η_{2j} are school-specific coefficients (the Greek letter η is pronounced eta). The *level-2 models* have these coefficients as responses

$$\begin{aligned} \eta_{1j} &= \gamma_{11} + \zeta_{1j} \\ \eta_{2j} &= \gamma_{21} + \zeta_{2j} \end{aligned} \tag{4.5}$$

Sometimes the first of these level-2 models is referred to as a "means as outcomes" or "intercepts as outcomes" model and the second as a "slopes as outcomes" model. It is assumed that, given the covariate(s), the residuals or disturbances ζ_{1j} and ζ_{2j} in the level-2 model have a bivariate normal distribution with zero mean and covariance matrix:

$$\boldsymbol{\Psi} = \begin{bmatrix} \psi_{11} & \psi_{12} \\ \psi_{21} & \psi_{22} \end{bmatrix}, \qquad \psi_{21} = \psi_{12}$$

It is important to remember that the level-2 models cannot be estimated on their own because the random effects η_{1j} and η_{2j} are not observed. Instead, we must substitute the level-2 models into the level-1 model to obtain the *reduced-form* model for the observed responses y_{ij}

$$\begin{aligned} y_{ij} &= \underbrace{\gamma_{11} + \zeta_{1j}}_{\eta_{1j}} + \underbrace{(\gamma_{21} + \zeta_{2j})}_{\eta_{2j}} x_{ij} + \epsilon_{ij} \\ &= \underbrace{\gamma_{11} + \gamma_{21} x_{ij}}_{\text{fixed}} + \underbrace{\zeta_{1j} + \zeta_{2j} x_{ij} + \epsilon_{ij}}_{\text{random}} \\ &\equiv \beta_1 + \beta_2 x_{ij} + \zeta_{1j} + \zeta_{2j} x_{ij} + \epsilon_{ij} \end{aligned}$$

In the reduced form, the fixed part is usually written first followed by the random part, and we can return to our previous notation by defining $\beta_1 \equiv \gamma_{11}$ and $\beta_2 \equiv \gamma_{21}$. The above model is thus equivalent to the model in (4.1).

Any level-2 covariates (covariates that do not vary at level 1) are included in the level-2 models. For instance, we could include dummy variables for type of school, w_{2j} for boys' schools and w_{3j} for girls' schools with mixed schools as the reference category. If we include these dummy variable in the model for the random intercept

$$\eta_{1j} = \gamma_{11} + \gamma_{12} w_{2j} + \gamma_{13} w_{3j} + \zeta_{1j}$$

the reduced form becomes

$$\begin{aligned} y_{ij} &= \underbrace{\gamma_{11} + \gamma_{12} w_{2j} + \gamma_{13} w_{3j} + \zeta_{1j}}_{\eta_{1j}} + \underbrace{(\gamma_{21} + \zeta_{2j})}_{\eta_{2j}} x_{ij} + \epsilon_{ij} \\ &= \underbrace{\gamma_{11} + \gamma_{12} w_{2j} + \gamma_{13} w_{3j} + \gamma_{21} x_{ij}}_{\text{fixed}} + \underbrace{\zeta_{1j} + \zeta_{2j} x_{ij} + \epsilon_{ij}}_{\text{random}} \end{aligned}$$

If we also include the dummy variables for type of school in the model for the random slope

$$\eta_{2j} = \gamma_{21} + \gamma_{22} w_{2j} + \gamma_{23} w_{3j} + \zeta_{2j}$$

we obtain so-called *cross-level interactions* in the reduced form

$$\begin{aligned} y_{ij} &= \underbrace{\gamma_{11} + \gamma_{12}w_{2j} + \gamma_{13}w_{3j} + \zeta_{1j}}_{\eta_{1j}} + \underbrace{(\gamma_{21} + \gamma_{22}w_{2j} + \gamma_{23}w_{3j} + \zeta_{2j})}_{\eta_{2j}} x_{ij} + \epsilon_{ij} \\ &= \underbrace{\gamma_{11} + \gamma_{12}w_{2j} + \gamma_{13}w_{3j} + \gamma_{21}x_{ij} + \gamma_{22}w_{2j}x_{ij} + \gamma_{23}w_{3j}x_{ij}}_{\text{fixed}} + \underbrace{\zeta_{1j} + \zeta_{2j}x_{ij} + \epsilon_{ij}}_{\text{random}} \end{aligned}$$

The effect of `lrt` now depends on the type of school, with γ_{22} representing the additional effect of `lrt` on `gcse` for boys' schools compared with mixed schools and γ_{23} the additional effect for girls' schools compared with mixed schools.

For estimation in `xtmixed` or `gllamm`, it is necessary to convert the two-stage formulation to the reduced form. For instance, to fit the model above using `xtmixed`, we must generate new variables representing interactions. First, we produce dummy variables for type of school:

```
. tabulate schgen, generate(w)
    schgend |      Freq.     Percent        Cum.
------------+-----------------------------------
          1 |      2,169       53.44       53.44
          2 |        513       12.64       66.08
          3 |      1,377       33.92      100.00
------------+-----------------------------------
      Total |      4,059      100.00
. rename w2 boys
. rename w3 girls
```

Then we create the interaction terms:

```
. generate boys_lrt = boys*lrt
. generate girls_lrt = girls*lrt
```

Using `xtmixed`, we obtain

```
. xtmixed gcse lrt boys girls boys_lrt girls_lrt || school: lrt,
> cov(unstructured) mle
Mixed-effects ML regression                     Number of obs      =      4059
Group variable: school                          Number of groups   =        65

                                                Obs per group: min =         2
                                                               avg =      62.4
                                                               max =       198

                                                Wald chi2(5)       =    803.90
Log likelihood = -13998.825                     Prob > chi2        =    0.0000

------------------------------------------------------------------------------
        gcse |      Coef.   Std. Err.      z    P>|z|     [95% Conf. Interval]
-------------+----------------------------------------------------------------
         lrt |   .5712362   .0271255    21.06   0.000     .5180712    .6244012
        boys |   .8546725   1.085019     0.79   0.431    -1.271926    2.981271
       girls |   2.433411   .8433395     2.89   0.004     .7804959    4.086326
    boys_lrt |  -.0230097   .0573892    -0.40   0.688    -.1354905    .0894711
   girls_lrt |   -.029544   .0447029    -0.66   0.509    -.1171601    .0580721
       _cons |  -.9976069    .506808    -1.97   0.049    -1.990932   -.0042815
------------------------------------------------------------------------------
```

```
------------------------------------------------------------------------------
  Random-effects Parameters  |   Estimate   Std. Err.     [95% Conf. Interval]
-----------------------------+------------------------------------------------
school: Unstructured         |
                     sd(lrt) |   .1199144   .0189129      .0880277    .1633514
                   sd(_cons) |   2.797928   .2886796      2.285666    3.424997
             corr(lrt,_cons) |   .5967756   .1381306      .2614296    .8035692
-----------------------------+------------------------------------------------
                sd(Residual) |   7.441832   .0839662      7.279068    7.608236
------------------------------------------------------------------------------
LR test vs. linear regression:       chi2(3) =   381.45   Prob > chi2 = 0.0000
Note: LR test is conservative and provided only for reference.
```

We see that students from girls' schools perform significantly better at the 5% level than students from mixed schools whereas students from boys' schools do not perform significantly better than students from mixed schools. The effect of `lrt` does not differ significantly between boys' schools and mixed schools or between girls' schools and mixed schools. The estimates and the parameters in the two-stage formulation are given in the last three columns of table 4.1 on page 155.

Although equivalent models can be specified using either the reduced-form (used by `xtmixed` and `gllamm`) or the two-stage formulation (used in the HLM software of Raudenbush et al. 2004), in practice model specification to some extent depends on the approach adopted. For instance, cross-level interactions are easily included using the two-stage specification in the HLM software, whereas same-level interactions must be created outside the program. Papers using this software tend to include more cross-level interactions and more random coefficients in the models (because the level-2 models look odd without residuals) than papers using for instance Stata.

Raudenbush and Bryk (2002) use slightly different notation for the two-stage formulation. For example, they use the subscript 0 instead of 1 for intercepts; see exercise 4.2.

4.10 Some warnings about random-coefficient models

It rarely makes sense to include a random slope if there is no random intercept, just like interactions between two variables usually do not make sense without including the variables themselves in ordinary regression models. Similarly, it is seldom sensible to include a random slope without including the corresponding fixed slope because it is usually strange to allow the slope to vary randomly but constrain its population mean to zero.

It may be tempting to allow many different covariates to have random slopes. However, it should be remembered that the number of parameters for the random part of the model increases rapidly with the number of random slopes because there is a variance parameter for each random effect (intercept or slope) and a covariance parameter for each *pair* of random effects. If there are k random slopes (plus 1 random intercept), then there are $(k+2)(k+1)/2+1$ parameters in the random part (for example, $k=3$ gives 11 parameters). Another problem is that clusters may not provide much information on cluster-specific slopes and hence on the slope variance either if the clusters are small or if x_{ij} does not vary much within clusters, or varies only in a small number of clusters. It should be noted that it does not matter if some of the clusters do not provide information on the slope variance as long as there are an adequate number of clusters that do.

It is generally not a good idea to include a random coefficient for a covariate that does not vary at a lower level than the random coefficient itself. For example, in the inner-London school data, it does not make sense to include a random slope for type of school because type of school does not vary within schools. Since we cannot estimate the effect of type of school for individual schools, it also appears impossible to estimate the variability of the effect of type of school between schools. However, as discussed in section 5.14, a level-2 random coefficient of a level-2 covariate can be used to construct a random-intercept model with a heteroskedastic random intercept.

Sometimes convergence problems will occur because the estimated covariance matrix "tries" to become nonpositive semidefinite, meaning for instance that variances try to become negative or correlations try to be greater than 1 or less than -1. All the commands in Stata force the covariance matrix to be positive semidefinite, and when parameters approach nonpermissible values, convergence can be slow or even fail. It may help to translate and rescale x_{ij} since variances and covariances are not invariant to these transformations. Often a better remedy is to simplify the model by removing some random slopes. The overall message is that random slopes should be included only if strongly suggested by the subject-matter theory related to the application.

Sometimes random-coefficient models are simply not identified. As an important example, consider balanced data with clusters of size $n_j = 2$ and with a covariate x_{ij} taking the same two values t_1 and t_2 for each cluster (an example would be the peak-expiratory-flow data, `pefr.dta`, from chapter 2). The model including a random intercept and a random slope of x_{ij} is not identified. This can be seen by considering the two distinct variances (for $i = 1$ and $i = 2$) and one covariance of the total residuals when $t_1 = 0$ and $t_2 = 1$:

$$
\begin{aligned}
\text{Var}(\xi_{1j}) &= \psi_{11} + \theta \\
\text{Var}(\xi_{2j}) &= \psi_{11} + 2\psi_{21} + \psi_{22} + \theta \\
\text{Cov}(\xi_{1j}, \xi_{2j}) &= \psi_{11} + \psi_{21}
\end{aligned}
$$

The marginal distribution of y_{ij}, given the covariates, is completely characterized by the fixed part of the model and these three model-implied moments (two variances and a covariance). However, the three moments are determined by four parameters of the random part (ψ_{11}, ψ_{22}, ψ_{21}, and θ), so fitting the model-implied moments to the data would effectively involve solving three equations for four unknowns. The model is therefore not identified. We could identify the model by setting $\theta = 0$, which does not impose any restrictions on the covariance matrix. The original model becomes identified if the covariate x_{ij} varies also between clusters because the model-implied covariance matrix of the total residuals then differs between clusters, yielding more equations to solve for the four parameters.

4.11 Summary and further reading

In this chapter, we have introduced the notion of slopes or regression coefficients varying randomly over clusters in linear models. Linear random-coefficient models are parsimonious representations of situations where each cluster has a separate regression model with its own intercept and slope. The linear random-coefficient model was applied to a cross-sectional study of school effectiveness. Here students were nested in schools, and we considered school-specific regressions.

An important consideration when using random-coefficient models is that the interpretation of the covariance matrix of the random effects depends on the scale and translation of the variables having random slopes. One should thus be careful when interpreting the variance and covariance estimates. We briefly demonstrated a two-stage formulation of random-coefficient models that is popular in some fields of social science. This formulation can be used to specify models that are equivalent to models specified using the more common reduced-form formulation used in this book. The utility of empirical Bayes prediction was demonstrated for visualizing the model, making inferences for individual clusters, and for diagnostics.

Random-coefficient models for longitudinal data, often called growth-curve models, are considered in chapter 5. Exercises 5.3, 5.5, 5.7, and 5.8 in that chapter can be viewed as supplementary exercises for the current chapter. Random-coefficient models for noncontinuous responses are discussed in part III of the book.

Introductory books on random-coefficient models include Raudenbush and Bryk (2002), and Snijders and Bosker (1999). A brief overview of the material covered in chapters 2–4 is given in Kreft and de Leeuw (1998).

4.12 Exercises

4.1 ❖ Inner-London schools data

1. Fit the random-coefficient model fitted on page 170.
2. Write down a model with the same covariates as in step 1 that also allows the mean for mixed schools to differ between boys and girls. (The variable `girl` is a dummy variable for the student being a girl.) Write down null hypotheses in terms of linear combinations of regression coefficients for the following research questions: Do girls do better in girls' schools than in mixed schools (after controlling for the other covariates)? Do boys do better in boys' schools than in mixed schools (after controlling for the other covariates)?
3. Fit the model from step 2 and test the null hypotheses from step 2. Discuss whether there is evidence that children of a given gender do better in single-sex schools.

4.2 High-school-and-beyond data

Raudenbush and Bryk (2002) and Raudenbush et al. (2004) analyzed data from the High School and Beyond Survey.

The dataset `hsb.dta` has the following variables:

- Level 1 (student)
  - `mathach`: a measure of mathematics achievement
  - `minority`: dummy variable for student being nonwhite
  - `female`: dummy variable for student being female
  - `ses`: socioeconomic status (SES) based on parental education, occupation, and income
- Level 2 (school)
  - `schoolid`: school identifier
  - `sector`: dummy variable for a school being Catholic
  - `pracad`: proportion of students in the academic track
  - `disclim`: scale measuring disciplinary climate
  - `himinty`: dummy variable for more than 40% minority enrollment

Raudenbush et al. (2004) specify a two-level model. We will use their model and notation here. At level 1, math achievement Y_{ij} is regressed on student's SES, centered around the school mean

$$Y_{ij} = \beta_{0j} + \beta_{1j}(X_{1ij} - \overline{X}_{1.j}) + r_{ij}, \quad r_{ij} \sim N(0, \sigma^2)$$

where X_{1ij} is the student's SES, $\overline{X}_{1.j}$ is the school mean SES, and r_{ij} is a level-1 residual. At level 2, the intercepts and slopes are regressed on the dummy variable W_{1j} for the school being a Catholic school (`sector`) and on the school mean SES

$$\beta_{pj} = \gamma_{p0} + \gamma_{p1} W_{1j} + \gamma_{p2}\overline{X}_{1.j} + u_{pj}, \quad p=0,1, \quad (u_{0j}, u_{1j})' \sim N(\mathbf{0}, \mathbf{T})$$

where u_{pj} is a random effect (a random intercept if $p=0$ and a random slope if $p=1$). The covariance matrix

$$\mathbf{T} = \begin{bmatrix} \tau_{00} & \tau_{01} \\ \tau_{10} & \tau_{11} \end{bmatrix}$$

has three unique elements with $\tau_{10} = \tau_{01}$.

1. Substitute the level-2 models into the level-1 model, and write down the resulting reduced form using the notation of this book.
2. Construct the variables `meanses` equal to the school-mean SES ($\overline{X}_{1.j}$) and `devses` equal to the deviations of the student's SES from their school means ($X_{1ij} - \overline{X}_{1.j}$).
3. Fit the model considered by Raudenbush et al. (2004) using `xtmixed` and interpret the coefficients. In particular, interpret the estimate of γ_{12}.

4. Fit the model that also includes `disclim` in the level-2 models and `minority` in the level-1 model.

4.3 Homework data

Kreft and de Leeuw (1998) consider a subsample of students in eighth grade from the National Education Longitudinal Study of 1988 (NELS–88) collected by the National Center for Educational Statistics of the U.S. Department of Education. The students are viewed as nested in schools.

The data are given in `homework.dta`. In this exercise, we will use the following subset of the variables:

- `schid`: school identifier
- `math`: continuous measure of achievement in mathematics
- `homework`: number of hours of homework done per week
- `white`: student's race (1: white; 0: nonwhite)
- `ratio`: class size as measured by the student-teacher ratio
- `meanses`: school mean SES

1. Write down and state the assumptions of a random-coefficient model with `math` as response variable and `homework`, `white`, and `ratio` as covariates. Let the intercept and the effect of `homework` vary between schools.
2. Fit the model by maximum likelihood and interpret the estimated parameters.
3. Derive an expression for the estimated variance of math achievement conditional on the covariates.
4. How would you extend the model to investigate whether the effect of homework on math achievement depends on the mean socioeconomic status of schools? Write down both the two-stage and the reduced-form formulation of your extended model.
5. Fit the model from step 4.

4.4 Wheat and moisture data

Littell et al. (2006) describe data on ten randomly chosen varieties of winter wheat. Each variety was planted on six randomly selected 1-acre plots of land in a 60-acre field. The amount of moisture in the top 36 inches of soil was determined for each plot before planting the wheat. The response variable is the yield in bushels per acre.

The data `wheat.dta` contains the following variables:

- `variety`: variety (or type) of wheat j
- `plot`: plot (1 acre) on which wheat was planted i
- `yield`: yield in bushels per acre y_{ij}
- `moist`: amount of moisture in top 36 inches of soil prior to planting x_{ij}

In this exercise, variety of wheat will be treated as random.

1. Write down the model for `yield` with a fixed and random intercept and a fixed and random slope of `moist`. State all model assumptions.
2. Fit the random-coefficient model using maximum likelihood estimation.
3. Use a likelihood-ratio test to test the null hypothesis that the true random-coefficient variance is zero.
4. For the chosen model, obtain the predicted yields for each variety (with empirical Bayes predictions substituted for the random effects).
5. Plot lines for predicted yield versus moisture, using the `by()` option to obtain a separate graph for each variety. (Such a graph is sometimes called a "trellis graph".)
6. Produce the same graphs as above, with observed values of yield added as dots.

4.5 ❖ Family birthweight data

Rabe-Hesketh, Skrondal, and Gjessing (2008) analyzed a random subset of the birthweight data from the Medical Birth Registry of Norway described in Magnus et al. (2001). There are 1000 nuclear families each comprising mother, father, and one child (not necessarily the only child in the family).

The data are given in `family.dta`. In this exercise, we will use the following variables:

- `family`: family identifier j
- `member`: family member i (1: mother; 2: father; 3: child)
- `bwt`: birthweight in grams (y_{ij})
- `male`: dummy variable for being male (x_{1ij})
- `first`: dummy variable for being the first child (x_{2ij})
- `midage`: dummy variable for mother of family member being aged 20–35 at time of birth (x_{3ij})
- `highage`: dummy variable for mother of family member being older than 35 at time of birth (x_{4ij})
- `birthyr`: year of birth minus 1967 (1967 was the earliest birth year in the birth registry) (x_{5ij})

In this dataset, family members are nested within families. Due to additive genetic and environmental influences, there will be a particular covariance structure between the members of the same family. Rabe-Hesketh, Skrondal, and Gjessing (2008) show that the following random-coefficient model can be used to induce the required covariance structure (see also exercise 4.6):

$$y_{ij} = \beta_1 + \zeta_{1j}[M_i + K_i/2] + \zeta_{2j}[F_i + K_i/2] + \zeta_{3j}[K_i/\sqrt{2}] + \epsilon_{ij} \qquad (4.6)$$

where M_i is a dummy variable for mothers, F_i is a dummy variable for fathers and K_i is a dummy variable for children. The random coefficients ζ_{1j}, ζ_{2j}, and ζ_{3j} are constrained to have the same variance ψ and to be uncorrelated with each other. As usual, we assume normality with $\zeta_{1j} \sim N(0,\psi)$, $\zeta_{2j} \sim N(0,\psi)$, $\zeta_{3j} \sim N(0,\psi)$, and $\epsilon_{ij} \sim N(0,\theta)$. The variances ψ and θ can be interpreted as genetic and environmental variances, respectively, and the total residual variance is $\psi + \theta$.

1. Produce the required dummy variables M_i, F_i, and K_i.
2. Generate variables equal to the terms in square brackets in (4.6).
3. Which of the correlation structures available in `xtmixed` should be specified for the random coefficients?
4. Fit the model given in (4.6). Note that the model does not include a random intercept.
5. Obtain the estimated proportion of the total variance that is attributable to additive genetic effects.
6. Now fit the model including all of the covariates listed above and having the same random part as the model in step 3.
7. Interpret the estimated coefficients from step 6.
8. Conditional on the covariates, what proportion of the residual variance is estimated to be due to additive genetic effects?

4.6 ❖ Covariance structure for nuclear family data

This exercise concerns family data such as those of exercise 4.5 consisting of a mother, father, and child. Here we consider three types of influences on birthweight: additive genetic effects (due to shared genes), common environmental effects (due to shared environment), and unique environmental effects. These random effects have variances σ_A^2, σ_C^2 and σ_E^2, respectively.

The additive genetic effects have the following properties:

- The parents share no genes by descent, so their additive genetic effects are uncorrelated
- The child shares half its genes with each parent by decent, giving a correlation of 1/2 with each parent
- The additive genetic variance should be the same for each family member

For birth outcomes, no two family members share a common environment since they all developed in different wombs. We therefore cannot distinguish between common and unique environmental effects.

Rabe-Hesketh, Skrondal, and Gjessing (2008) show that we can use the following random-coefficient model to produce the required covariance structure

$$y_{ij} \;=\; \beta_1 + \zeta_{1j}[M_i + K_i/2] + \zeta_{2j}[F_i + K_i/2] + \zeta_{3j}[K_i/\sqrt{2}] + \epsilon_{ij} \qquad (4.7)$$

where M_i, F_i, and M_i are dummy variables for mothers, fathers, and children, respectively. The random coefficients ζ_{1j}, ζ_{2j}, and ζ_{3j} produce the required additive genetic correlations and variances. These random coefficients are constrained to have the same variance $\psi = \sigma_A^2$ and to be uncorrelated with each other. As usual, we assume normality with $\zeta_{1j} \sim N(0, \sigma_A^2)$, $\zeta_{2j} \sim N(0, \sigma_A^2)$, $\zeta_{3j} \sim N(0, \sigma_A^2)$, and $\epsilon_{ij} \sim N(0, \theta)$.

1. By substituting the appropriate numerical values for the dummy variables M_i, F_i, and K_i in (4.7), write down three separate models, one for mothers, one for fathers, and one for children. It is useful to substitute $i = 1$ for mothers, $i = 2$ for fathers, and $i = 3$ for children in these equations.
2. Using the equations from step 1, demonstrate that the total variance is the same for mothers, fathers, and children.
3. Using the equations from step 1, demonstrate that the covariance between mothers and fathers from the same families is zero.
4. Using the equations from step 1, demonstrate that the correlation between the additive genetic components (terms involving ζ_{1j}, ζ_{2j}, or ζ_{3j}) of mothers and their children is 1/2.
5. What is the relationship between θ, σ_C^2, and σ_E^2?

4.7 ❖ Effect of covariate translation on random-effects covariance matrix

Using (4.2) and the estimates for model 2 on page 155, calculate what values $\widehat{\psi}_{11}$ and $\widehat{\psi}_{21}$ would take if you were to subtract 5 from the variable LRT and refit the model.

5 Longitudinal, panel, and growth-curve models

5.1 Introduction

In this chapter, we focus on multilevel and other models for longitudinal data. By longitudinal data we mean prospective data where a sample is followed up over time with information collected at several occasions or time points. Such data are also sometimes referred to as *panel data*, *repeated measures*, or *cross-sectional time-series data* (the latter term explains the `xt` prefix in Stata's commands for longitudinal modeling). Related types of data that are not discussed here include *time-series* data, where one unit is followed over time (usually at many occasions), and time-to-event or survival data, which are discussed in chapter 8 and in section 9.12.

It is useful to distinguish between different types of longitudinal studies. In *panel studies* all subjects are typically followed up at the same occasions (called "panel waves") leading to so-called balanced or "fixed occasion" data, although there may be missing data. In *cohort studies* (as defined in epidemiology), a group of subjects, sometimes of the same age as in a "birth cohort", is often followed up at subject-specific occasions, which produces unbalanced or "variable occasion" data. Intervention studies and clinical trials are special cases of cohort studies.

Longitudinal data can be viewed as two-level or clustered data with occasions nested in subjects so the subjects become the clusters. We use the terms "occasions" i and "subjects" j for level-1 units and level-2 units or clusters, respectively, in this chapter. The Stata documentation uses the indices t for occasions and i for subjects and, somewhat confusingly, uses the term "panel" for subject.

As for all clustered data, the model should accommodate within-cluster dependence. Indeed, we have already used random-intercept models for longitudinal data, such as the smoking and birthweight data, in previous chapters and exercises. In chapter 3 we discussed the problem of cluster-level confounding or endogeneity that can arise in the random effects approach. To avoid this problem, fixed-effects models are often used, particularly in econometrics. Here we will revisit random- and fixed-effects models in the context of longitudinal data.

A special feature of longitudinal data is that the level-1 units or occasions are ordered in time and not exchangeable unlike for example students nested in schools. Perhaps for this reason, there are a variety of different models designed specifically for longitudinal data. Broadly speaking, these models fall into three different categories:

- *Random- and fixed-effects models*, where unobserved between-subject heterogeneity is represented by subject-specific intercepts and possibly subject-specific coefficients.
- *Marginal models*, where within-subject dependence is modeled by allowing a direct specification of the covariance structure across occasions.
- *Autoregressive- or lagged-response models*, where within-subject dependence is modeled by letting responses at a given occasion depend on previous or lagged responses.

In the first part of the chapter, we review models within these categories and discuss other issues specific to longitudinal data, including the choice of time scales and the problem of attrition or dropout.

In the second part of the chapter, we focus on a special case of the random-coefficient models discussed in chapter 4, called *growth-curve models*, where the coefficient of time varies randomly across subjects. Such growth-curve models are applied to a dataset on children's growth.

Although we confine the discussion to longitudinal modeling of continuous responses in this chapter, the basic ideas also apply to other response types. Indeed, examples of longitudinal modeling of dichotomous and ordinal responses and counts are considered in later chapters.

5.2 How and why do wages change over time?

Labor economists are interested in research questions such as how hourly wage depend on union membership, labor market experience and education and how hourly wages change over time.

To address these questions we will use data from the U.S. National Longitudinal Survey of Youth 1979 (NLSY79). The original sample is representative of noninstitutionalized civilian youth who were aged 14–21 on December 31, 1978. Here we consider a subsample of 545 full-time working males who completed schooling by 1980 and who had complete data for 1980–1987. The data were previously analyzed by Vella and Verbeek (1998) and are provided with the book by Wooldridge (2002). (The data were also used in exercise 3.7.)

The variables in the dataset `wagepan.dta` that we will use here are

- `nr`: person identifier (j)
- `lwage`: log of hourly wage in U.S. dollars (y_{ij})
- `black`: dummy variable for being black (x_{2j})

- hisp: dummy variable for being Hispanic (x_{3j})
- union: dummy variable for being a member of a union (i.e., wage being set in collective bargaining agreement) (x_{4ij})
- married: dummy variable for being married (x_{5ij})
- exper: labor market experience, defined as age−6−educ (L_{ij} or x_{6ij})
- year: period or calendar year 1980–1987 (P_i or x_{7i})
- educ: years of schooling (E_j or x_{8j})

5.3 Data structure

5.3.1 Missing data

The data are panel data, with the same set of occasions for each person. If we did not know it already, we might wonder whether lwage was missing at any occasion for any of the subjects. We can investigate this using xtdescribe:

```
. use http://www.stata-press.com/data/mlmus2/wagepan
. xtdescribe if lwage<., i(nr) t(year)
      nr:  13, 17, ..., 12548                                n =        545
    year:  1980, 1981, ..., 1987                             T =          8
           Delta(year) = 1 unit
           Span(year)  = 8 periods
           (nr*year uniquely identifies each observation)
Distribution of T_i:   min      5%      25%       50%       75%     95%     max
                         8       8        8         8         8       8       8
     Freq.  Percent    Cum. |  Pattern
 ---------------------------+----------
      545    100.00  100.00 |  11111111
 ---------------------------+----------
      545    100.00         |  XXXXXXXX
```

Here Pattern could be any sequence of length 8 consisting of "1" (standing for not missing) and "." (standing for missing). The only pattern in this data is "11111111" corresponding to complete data for everyone (see sec. 6.4 for an example with missing data). For such patterns to be meaningful, the occasions must have the same meaning across subjects and must be labeled by a variable specified in the t() option, here year, if not specified previously by xtset.

5.3.2 Time-varying and time-constant variables

Whenever a variable is not constant over time for each subject, it is called *time varying*. The response variable (log wage) y_{ij} for occasion i and subject j is time varying. Some explanatory variables are subject specific or time constant (education E_j and the ethnicity dummies x_{2j} and x_{3j}), some are occasion specific (year P_i), and some are both subject and occasion specific (union membership x_{4ij} and marital status x_{5ij}).

It is useful to investigate the within-subject, between-subject, and total variability of the variables, here shown only for some of the variables:

```
. xtsum lwage union educ year, i(nr)
Variable         |      Mean   Std. Dev.       Min        Max |    Observations
-----------------+--------------------------------------------+----------------
lwage    overall |  1.649147   .5326094  -3.579079    4.05186 |     N =    4360
         between |             .3907468   .3333435   3.174173 |     n =     545
         within  |             .3622636  -2.467201   3.204687 |     T =       8
                 |                                            |
union    overall |  .2440367   .4295639          0          1 |     N =    4360
         between |             .3294467          0          1 |     n =     545
         within  |             .2759787  -.6309633   1.119037 |     T =       8
                 |                                            |
educ     overall |  11.76697   1.746181          3         16 |     N =    4360
         between |             1.747585          3         16 |     n =     545
         within  |                    0   11.76697   11.76697 |     T =       8
                 |                                            |
year     overall |    1983.5   2.291551       1980       1987 |     N =    4360
         between |                    0     1983.5     1983.5 |     n =     545
         within  |             2.291551       1980       1987 |     T =       8
```

As expected, `lwage` and `union` vary both within and between subjects, whereas `year` does not vary between subjects and `educ` does not vary within subjects.

We see from the fact that `T` is given in the output instead of `T-bar` that the number of occasions n_j is constant over subjects j as we already know.

5.4 Time scales in longitudinal data

Consider the longitudinal data on subjects 45 and 847 from the wage-panel data given in table 5.1. We refer to the birth year as "cohort" and to the calendar year as "period" and have calculated "age" and "cohort" from other variables in the data (see (5.1) below for the relationship among the variables).

Table 5.1: Illustration of longitudinal data

Subject j	Occ. i	Cohort C_j	Age A_{ij}	Period P_i	Black x_{2j}	Hispanic x_{3j}	Union x_{4ij}	Log wage y_{ij}
45	1	1960	20	1980	0	0	1	1.89
45	2	1960	21	1981	0	0	1	1.47
45	3	1960	22	1982	0	0	0	1.47
45	4	1960	23	1983	0	0	0	1.74
45	5	1960	24	1984	0	0	0	1.82
45	6	1960	25	1985	0	0	0	1.91
45	7	1960	26	1986	0	0	0	1.74
45	8	1960	27	1987	0	0	0	2.14
847	1	1959	21	1980	1	0	1	1.56
847	2	1959	22	1981	1	0	1	1.66
847	3	1959	23	1982	1	0	0	1.77
847	4	1959	24	1983	1	0	0	1.79
847	5	1959	25	1984	1	0	0	2.00
847	6	1959	26	1985	1	0	1	1.65
847	7	1959	27	1986	1	0	0	2.13
847	8	1959	28	1987	1	0	1	1.69

When investigating change in the response variable using *age-period-cohort* data such as those in table 5.1, any of three different time scales may be of interest: age, period, or cohort. In our example, the definitions of both cohort and age are in terms of the time of birth, and this reference point is important in many investigations. However, for some applications, the reference point could be occurrence of some other event, such as graduation from university. Then a particular cohort may be referred to as "the class of 2007" and the age-like time scale then becomes time since graduation.

The relationship between age, period, and cohort is given by

$$A_{ij} = P_i - C_j$$

as illustrated in figure 5.1 for 4 subjects for the first 5 waves of the wage-panel data. The top two lines represent subjects 847 and 45 in table 5.1, respectively. For example, we see that subject 847 was born in 1959 and hence belongs to the 1959 cohort (referred to as C59 in the figure) and that his age is 21 years in 1980. In 1984 any member of the 1959 cohort is 25 years old.

(*Continued on next page*)

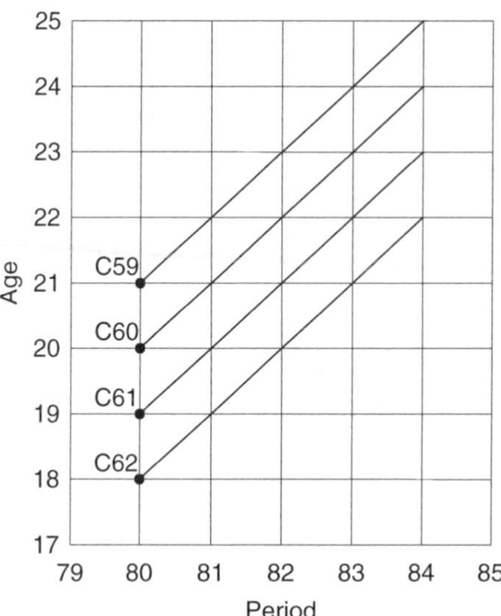

Figure 5.1: Illustration of the relationship between age, period, and cohort

A great advantage of this kind of longitudinal study with several cohorts is that we can investigate the effect of more than one time scale simultaneously.

In contrast, a cross-sectional study provides data for just one period P, which obviously makes it impossible to estimate the effect of period. Furthermore, we cannot separate the effects of age and cohort because each subject's single age A_j is determined by the cohort, $A_j = P - C_j$. For example, there are two competing explanations for older people being more conservative: (1) they are in later stages in life with increased A_j, and/or (2) they were born longer ago (into a different era) with smaller C_j. Cross-sectional data cannot be used to distinguish between these explanations.

A longitudinal study with one cohort C also does not allow us to investigate the effect of more than one time scale. We obviously cannot estimate the effect of cohort and age is determined by period, $A_i = P_i - C$. For instance, we cannot distinguish between two explanations for salary increases: (1) people get more experience as they get older with increased A_i, and/or (2) there is inflation over chronological time P_i.

Longitudinal studies with several cohorts are sometimes said to have a *cohort-sequential design* or *accelerated longitudinal design*, where the term "accelerated" refers to the fact that the range of ages covered exceeds the length of the study. From the relation $A_{ij} = P_i - C_j$, we see that it is possible to estimate the effects of two time scales, but these will be confounded with the third scale. Thus it is useful to pick the time scales that are believed to be most important. For example, conservatism may be viewed as depending on age and cohort (ignoring period), and salary may be viewed as depending on age and period (ignoring cohort).

In the wage-panel data, age A_{ij}, years of education E_j, years of labor market experience L_{ij}, period P_i, and cohort C_j, all potentially have an effect on the response variable log hourly wage. These time scales are defined in terms of three events: birth, entering education (at age 6), and leaving education to enter the labor market, with $A_{ij} = 6 + E_j + L_{ij}$. The relationship between the time scales becomes

$$A_{ij} \;=\; 6 + E_j + L_{ij} \;=\; P_i - C_j \tag{5.1}$$

From the first equality, we see that we have to eliminate one of the collinear variables A_{ij}, E_j, or L_{ij}, for instance A_{ij}. From the second equality, we see that we have to eliminate one of the remaining time scales, for instance C_j, leaving us with three time scales, labor market experience, L_{ij}, period, P_i, and education, E_j.

It is important to emphasize that several time scales can be relevant in longitudinal studies and that the prospect of disentangling the effects of different time scales depends on the research design. Within the limitations of the chosen research design, the time scales that are deemed most relevant for the research problem should be included in the models. These time scales are not necessarily present in the dataset and may have to be constructed from the data.

5.5 Random- and fixed-effects approaches

5.5.1 Correlated residuals

It is not advisable to use ordinary regression to analyze longitudinal data because the data are clustered, and we would expect unobserved between-subject heterogeneity leading to within-subject correlations.

We investigate this for the wage-panel data by inspecting the correlations among the residuals from an ordinary regression model, which includes the three time scales as covariates as well as union membership, race, and being married. First, we translate the time scales to make the intercept more interpretable in later analyses

```
. generate educt = educ - 12
. generate yeart = year - 1980
```

`educt` is the number of years of education beyond the usual time required to complete high school, and `yeart` is the number of years since 1980.

We can now fit a linear regression model and obtain the residuals using

```
. quietly regress lwage black hisp union married exper yeart educt
. predict res, residuals
```

To form the within-subject correlation matrix for pairs of occasions, we must reshape the data to wide form, treating the residuals at times 0–7 as variables `res0`–`res7`. We first preserve and later restore the data because we will require them in long form when fitting models. (If running the subsequent commands from a do-file, all the commands from `preserve`–`restore` must be run in one block.)

```
. preserve
. keep nr res yeart
. reshape wide res, i(nr) j(year)
(note: j = 0 1 2 3 4 5 6 7)
Data                               long   ->   wide
-----------------------------------------------------------------------------
Number of obs.                     4360   ->     545
Number of variables                   3   ->       9
j variable (8 values)             yeart   ->   (dropped)
xij variables:
                                    res   ->   res0 res1 ... res7
-----------------------------------------------------------------------------
```

The variances and correlations of the residuals are then obtained using

```
. tabstat res*, statistics(variance) format(%4.3f)
   stats |     res0     res1     res2     res3     res4     res5     res6     res7
---------+--------------------------------------------------------------------------------
variance |    0.278    0.250    0.210    0.199    0.249    0.235    0.236    0.189
------------------------------------------------------------------------------------------

. correlate res*, wrap
(obs=545)
             |     res0      res1      res2      res3      res4      res5      res6      res7
-------------+------------------------------------------------------------------------------------
        res0 |   1.0000
        res1 |   0.3855    1.0000
        res2 |   0.3673    0.5525    1.0000
        res3 |   0.3298    0.5220    0.6263    1.0000
        res4 |   0.2390    0.4451    0.5740    0.6277    1.0000
        res5 |   0.2673    0.4071    0.5230    0.5690    0.6138    1.0000
        res6 |   0.2084    0.3220    0.4622    0.4710    0.5020    0.5767    1.0000
        res7 |   0.2145    0.4019    0.4232    0.4976    0.5349    0.6254    0.6337    1.0000
. restore
```

We see that there are substantial within-subject correlations among the residuals, ranging from 0.21 to 0.63. These positive correlations could be partly due to between-subject heterogeneity in the intercept, and possibly in the slopes of covariates, which is not accommodated in the ordinary linear regression model.

In the rest of this section, we consider models that handle this within-subject dependence by including either random or fixed subject-specific effects.

5.5.2 Fixed-intercept model

Using xtreg

In the fixed-intercept model, often called the fixed-effects model in econometrics, the subject-specific intercepts are treated as fixed unknown parameters. The model can be written as

$$y_{ij} \;=\; \beta_1 + \alpha_j + \beta_2 x_{2ij} + \cdots + \beta_8 x_{8ij} + \epsilon_{ij}$$

where α_j is the deviation of subject j's intercept from the mean intercept β_1 with $\sum_{j=1}^{J} \alpha_j = 0$. We assume that, for given subjects and covariates, ϵ_{ij} has a normal distribution with mean 0 and variance θ.

The fixed-intercept model can be fitted using `xtreg` with the `fe` option:

```
. xtreg lwage black hisp union married exper yeart educt, i(nr) fe
Fixed-effects (within) regression               Number of obs      =      4360
Group variable: nr                              Number of groups   =       545

R-sq:  within  = 0.1672                         Obs per group: min =         8
       between = 0.0001                                        avg =       8.0
       overall = 0.0513                                        max =         8

                                                F(3,3812)          =    255.03
corr(u_i, Xb)  = -0.1575                        Prob > F           =    0.0000

------------------------------------------------------------------------------
       lwage |      Coef.   Std. Err.      t    P>|t|     [95% Conf. Interval]
-------------+----------------------------------------------------------------
       black |  (dropped)
        hisp |  (dropped)
       union |    .083791    .019414     4.32   0.000      .045728    .1218539
     married |   .0610384   .0182929     3.34   0.001     .0251736    .0969032
       exper |   .0598672   .0025835    23.17   0.000      .054802    .0649325
       yeart |  (dropped)
       educt |  (dropped)
       _cons |   1.211888   .0169244    71.61   0.000     1.178706     1.24507
-------------+----------------------------------------------------------------
     sigma_u |  .40514496
     sigma_e |  .35352815
         rho |  .56772216   (fraction of variance due to u_i)
------------------------------------------------------------------------------
F test that all u_i=0:     F(544, 3812) =     7.78           Prob > F = 0.0000
```

The estimates are also reported under "Fixed intercept" in table 5.2. (`sigma_u` in the output is the sample standard deviation of the estimated intercepts $\widehat{\alpha}_j$.)

(*Continued on next page*)

Table 5.2: Estimates for wage-panel data

	Subject-specific effects						Marginal		Lagged response	
	Fixed intercept		Random intercept		Random coefficient		Residual AR(1)		Response AR(1)	
	Est	(SE)	Est	(SE)	Est	(SE)	Est	(SE)	Est	(SE)
Fixed part:										
β_1 [_cons]	1.21	(0.02)	1.32	(0.04)	1.31	(0.04)	1.30	(0.04)	0.67	(0.03)
β_2 [black]			−0.13	(0.05)	−0.14	(0.05)	−0.13	(0.05)	−0.07	(0.02)
β_3 [hisp]			0.02	(0.04)	0.01	(0.04)	0.02	(0.04)	0.01	(0.02)
β_4 [union]	0.08	(0.02)	0.11	(0.02)	0.11	(0.02)	0.10	(0.02)	0.08	(0.01)
β_5 [married]	0.06	(0.02)	0.08	(0.02)	0.08	(0.02)	0.08	(0.02)	0.05	(0.01)
β_6 [exper]	0.06	(0.00)	0.03	(0.01)	0.04	(0.01)	0.04	(0.01)	0.00	(0.00)
β_7 [yeart]			0.03	(0.01)	0.02	(0.01)	0.03	(0.01)	0.02	(0.01)
β_8 [educt]			0.09	(0.01)	0.10	(0.01)	0.10	(0.01)	0.04	(0.00)
γ									0.56	(0.01)
Random part:										
$\sqrt{\psi_{11}}$			0.33		0.45					
$\sqrt{\psi_{22}}$					0.05					
ρ_{21}					−0.68					
$\sqrt{\theta}$ or res. SD	0.35		0.35		0.33		0.48		0.39	
α							0.58			

As discussed in section 3.7.2, the estimates represent within-subject effects of the covariates. A great advantage of these estimates is that they are not susceptible to bias due to omitted subject-level covariates. Each subject truly serves as their own control in the fixed-effects model.

However, the effects of the time-constant variables `black`, `hisp`, and `educt` cannot be estimated, and these variables are therefore dropped. `yeart` is also dropped because it differs from `exper` only by a subject-specific constant t_j,

$$\texttt{yeart}_i = t_j + \texttt{exper}_{ij}$$

where t_j is the time (in years since 1980) at which the subject entered the labor market. The linear combination `yeart`$-$`exper` is therefore collinear with the dummy variables multiplying the subject-specific intercepts α_j (see sec. 3.7.2 for the dummy variable formulation of the fixed-intercept model).

We see that union membership, being married, and having more experience are all beneficial for wages, according to the fitted model. For instance, each extra year of experience is associated with an estimated increase in the expectation of log hourly wage of 0.06, controlling for the other covariates. If the model assumptions are true, the exponential of this coefficient, 1.06, represents the estimated multiplicative effect of each extra year of experience on the expectation of the hourly wage.[1] In other words, mean hourly wages increase 6% for each extra year of experience. However, this actually represents the combined effect of experience and period, which cannot be disentangled here. In contrast, cohort effects have been controlled for because they are subject specific. Union membership increases mean hourly wages by about 8% and being married by about 6% according to the fitted model, controlling for the other covariates.

We mentioned in section 3.7.2 that subject-specific intercepts can be eliminated by subject-mean centering the responses and covariates (which is what the `xtreg` command with the `fe` option does). Alternatively, we could take first differences,

$$y_{ij} - y_{i-1,j} \;=\; \beta_2(x_{2ij} - x_{2,i-1,j}) + \cdots + \beta_8(x_{8ij} - x_{8,i-1,j}) + \epsilon_{ij} - \epsilon_{i-1,j}$$

to eliminate the intercepts. This approach may be preferable if the residuals in the fixed-effects model are correlated over time.

Using anova

Using the language of experimental design, subjects can be viewed as blocks or plots to which different treatments are applied. In a split-plot design, some treatments are applied to the entire plots (subjects)—these are the whole plot, between-subject or level-2 variables (`black`, `hisp`, and `educt`). The plots are split into subplots (subjects

1. It follows from $\ln(y_{ij}) = \beta_1 + \alpha_j + \beta_2 x_{2ij} + \cdots + \beta_8 x_{8ij} + \epsilon_{ij}$ that $y_{ij} = \exp(\beta_1) \times \exp(\alpha_j) \times \exp(\beta_2)^{x_{2ij}} \times \cdots \times \exp(\beta_8)^{x_{8ij}} \times \exp(\epsilon_{ij})$. If, given the covariates, ϵ_{ij} is normally distributed with zero mean and constant variance θ, then $E(y_{ij}|\mathbf{x}_{ij}) = \exp(\beta_1) \times \exp(\alpha_j) \times \exp(\beta_2)^{x_{2ij}} \times \cdots \times \exp(\beta_8)^{x_{8ij}} \times E\{\exp(\epsilon_i)\} = \exp(\beta_1) \times \exp(\alpha_j) \times \exp(\beta_2)^{x_{2ij}} \times \cdots \times \exp(\beta_8)^{x_{8ij}} \times \exp(\theta/2)$.

at different occasions) to which other treatments can be applied—these are the split-plot, within-subject or level-1 variables (`union`, `married`, `exper`, and `yeart`).

One-way analysis of variance (ANOVA) was briefly discussed in section 1.4, where we showed how the total sum of squares is partitioned into the sum of squares due to a factor (a categorical covariate) and the sum of squared errors. An F statistic is then constructed by dividing the mean squares due to the factor by the mean squared error. In a split-plot design, the important thing to remember is that a different mean squared error is used in the denominator of the F test for testing the whole plot (between-subject) variables than for the split plot (within-subject variables). For the between-subject variables, the denominator is given by the (unique or partial) mean squares due to subjects and for the within-subject variables, the denominator is given by the mean squared error after allowing for fixed effects of subjects. Subject are often viewed as random and an estimator of the between-subject variance can be derived from the mean squares, but the estimators of the effects of within-subject variables are fixed-effects estimators.

To keep it simple, we consider only ethnicity (represented in the dataset by the dummy variables `black` and `hisp`) and calendar year `yeart`. In ANOVA models we use categorical variables directly, not the corresponding dummy variables, and must therefore first construct the variable `ethnicity` (with values 0: white; 1: black; and 2: Hispanic) from the dummy variables `black` and `hisp`:

```
. generate ethnic = black*1 + hisp*2
```

We now fit the following ANOVA model:

```
. anova lwage ethnic / nr|ethnic yeart, continuous(yeart)

                         Number of obs =     4360     R-squared     =  0.6116
                         Root MSE      = .354851      Adj R-squared =  0.5561

                  Source | Partial SS    df       MS           F     Prob > F
              -----------+----------------------------------------------------
                   Model | 756.274864   545   1.3876603      11.02     0.0000
                         |
                  ethnic | 10.6791909     2  5.33959545       4.43     0.0124
               nr|ethnic | 653.797374   542  1.20626822
              -----------+----------------------------------------------------
                   yeart | 91.7982998     1  91.7982998     729.03     0.0000
                         |
                Residual | 480.254778  3814  .125918924
              -----------+----------------------------------------------------
                   Total | 1236.52964  4359  .283672779
```

The model includes the main effects `ethnic`, `nr|ethnic`, and `yeart`, where `nr|ethnic` denotes that subjects are nested within ethnicities in the sense that each subject can have only one ethnicity. The option `continuous(yeart)` specifies that `yeart` should be treated as continuous by assuming the conditional expectation of the log wages to be linearly related to `yeart`. The purpose of the forward slash / is to declare that the denominator for the F test(s) for the preceding term(s) is the mean square for subjects (nested in ethnicity) and not the mean squared error for the entire model, after subtracting the sums of squares due to all terms, including subjects nested within ethnicities. The latter mean squared error is used for the denominator in the F test for `yeart`.

The F test for `ethnic` ($F(2,542) = 4.43$, $p = 0.0124$) is the same as that obtained using `xtreg` with the `be` option:

```
. xi: xtreg lwage i.ethnic, i(nr) be
i.ethnic          _Iethnic_0-2        (naturally coded; _Iethnic_0 omitted)

Between regression (regression on group means)  Number of obs      =      4360
Group variable: nr                              Number of groups   =       545

R-sq:  within  = 0.0000                         Obs per group: min =         8
       between = 0.0161                                        avg =       8.0
       overall = 0.0086                                        max =         8

                                                F(2,542)           =      4.43
sd(u_i + avg(e_i.))=  .3883085                  Prob > F           =    0.0124

------------------------------------------------------------------------------
       lwage |      Coef.   Std. Err.      t    P>|t|     [95% Conf. Interval]
-------------+----------------------------------------------------------------
  _Iethnic_1 |  -.1520501   .0526611    -2.89   0.004     -.255495   -.0486051
  _Iethnic_2 |  -.0537689   .0464083    -1.16   0.247    -.1449311    .0373933
       _cons |    1.67511   .0194886    85.95   0.000     1.636827    1.713392
------------------------------------------------------------------------------
```

The F test for `yeart` ($F(1,3814) = 729.03$, $p < 0.001$) can be obtained from `xtreg` with the `fe` option:

```
. xtreg lwage yeart, i(nr) fe

Fixed-effects (within) regression               Number of obs      =      4360
Group variable: nr                              Number of groups   =       545

R-sq:  within  = 0.1605                         Obs per group: min =         8
       between = 0.0000                                        avg =       8.0
       overall = 0.0742                                        max =         8

                                                F(1,3814)          =    729.03
corr(u_i, Xb)  = -0.0000                        Prob > F           =    0.0000

------------------------------------------------------------------------------
       lwage |      Coef.   Std. Err.      t    P>|t|     [95% Conf. Interval]
-------------+----------------------------------------------------------------
       yeart |   .0633278   .0023454    27.00   0.000     .0587294    .0679262
       _cons |     1.4275   .0098116   145.49   0.000     1.408263    1.446736
-------------+----------------------------------------------------------------
     sigma_u |  .39074676
     sigma_e |  .35485057
         rho |  .54803288   (fraction of variance due to u_i)
------------------------------------------------------------------------------
F test that all u_i=0:     F(544, 3814) =     9.70           Prob > F = 0.0000
```

Here the F statistic is simply the square of the t statistic for `yeart`.

The F statistic for the effects of subjects given at the bottom of the `xtreg, fe` output can also be obtained from the `anova` output: add the sum of squares due to ethnicity to the sum of squares due to subjects nested within ethnicity and divide by the sum of the corresponding degrees of freedom to obtain the mean squares due to subjects. The F statistic is the ratio of the mean squares due to subjects to the mean squared error:

```
. display ((10.6791909+653.797374)/544)/.125918924
9.7004027
```

The ANOVA model assumes that the responses are conditionally independent given the covariates (which include the factor subject). This assumption, together with constant variance, implies *compound symmetry* of the variance–covariance matrix of the responses given the covariates (but not given the factor subject) when subject is treated as random. A less strict assumption that all pairwise differences between responses have the same variance, called *sphericity*, is sufficient for the F test of within-subject variables to be valid. When this assumption is violated, the `repeated()` option can be used in the `anova` command to correct the p-values for within-subject variables. Unfortunately, this option works only for categorical within-subject variables.

The version of repeated measures ANOVA discussed here is sometimes referred to as *univariate* or as applicable to a split-plot design. There is also a multivariate version, called multivariate analysis of variance (MANOVA) and implemented in Stata's `manova` command, that specifies an unstructured covariance matrix for the repeated measures. A great disadvantage of that approach is that it uses listwise deletion, dropping entire subjects if one or more of their responses are missing. Furthermore, the MANOVA approach requires that the within-subject variables take on identical values for all subjects; for instance the variable `yeart` can be used but `union` cannot.

5.5.3 Random-intercept model

As an alternative to the fixed-effects model considered in section 5.5.2, we can specify a random-intercept model

$$y_{ij} = \beta_1 + \beta_2 x_{2ij} + \cdots + \beta_8 x_{8ij} + \zeta_j + \epsilon_{ij}$$

with the usual assumptions that, given the covariates, the random intercept ζ_j, and the level-1 residual ϵ_{ij} are both normally distributed with zero means, independent of one another, with ζ_j independent across subjects and ϵ_{ij} independent across subjects and occasions. In econometrics, the error components ζ_j and ϵ_{ij} are sometimes referred to as "permanent" and "transitory" components, respectively. To be poetic, Crowder and Hand (1990) referred to the fixed part of the model as the "immutable constant of the universe", to ζ_j as the "lasting characteristic of the individual", and to ϵ_{ij} as the "fleeting abberation of the moment".

This model can be fitted by maximum likelihood using `xtmixed`:

```
. xtmixed lwage black hisp union married exper yeart educt || nr:, mle
Mixed-effects ML regression                     Number of obs      =      4360
Group variable: nr                              Number of groups   =       545

                                                Obs per group: min =         8
                                                               avg =       8.0
                                                               max =         8

                                                Wald chi2(7)       =    894.85
Log likelihood = -2214.3572                     Prob > chi2        =    0.0000

------------------------------------------------------------------------------
       lwage |      Coef.   Std. Err.      z    P>|z|     [95% Conf. Interval]
-------------+----------------------------------------------------------------
       black |  -.1338495   .0479549    -2.79   0.005    -.2278395   -.0398595
        hisp |   .0174169   .0428154     0.41   0.684    -.0664998    .1013336
       union |   .1105923   .0179007     6.18   0.000     .0755075    .1456771
     married |   .0753674   .0167345     4.50   0.000     .0425684    .1081664
       exper |   .0331593   .0112023     2.96   0.003     .0112031    .0551154
       yeart |   .0259133   .0114064     2.27   0.023     .0035571    .0482695
       educt |   .0946864   .0107047     8.85   0.000     .0737055    .1156673
       _cons |   1.317175   .0373979    35.22   0.000     1.243877    1.390474
------------------------------------------------------------------------------

------------------------------------------------------------------------------
  Random-effects Parameters  |   Estimate   Std. Err.     [95% Conf. Interval]
-----------------------------+------------------------------------------------
nr: Identity                 |
                   sd(_cons) |   .3271344   .0114153      .3055088    .3502908
-----------------------------+------------------------------------------------
                sd(Residual) |   .3535088   .0040494      .3456606    .3615351
------------------------------------------------------------------------------
LR test vs. linear regression: chibar2(01) =  1547.76 Prob >= chibar2 = 0.0000
```

The estimates, which are also shown under "Random intercept" in table 5.2, are stored for later use

```
. estimates store ri
```

We now have estimates for the effects of all three time scales, but only under the assumption that the random intercept is uncorrelated with the included covariates. We could use the Hausman test, or include subject means of within-subject variables to assess this assumption as discussed in chapter 3. Interestingly, the coefficients of `exper` and `yeart` approximately add up to the coefficient of `exper` in the fixed-effects model where `yeart` was omitted due to collinearity. The estimates suggest that the mere passage of time is associated with a similar effect as each extra year of experience, probably due to inflation.

Each additional year of education is associated with a 10% ($0.10 = \exp(0.0946864) - 1$) increase in expected wages, compared with a 3% increase due to experience, when controlling for other covariates. We also estimate that, given the other covariates, black men's mean wages are 13% lower than white men's ($-0.13 = \exp(-0.1338495) - 1$), although this estimate is particularly prone to subject-level confounding or bias due to omitted subject-level variables. The estimated coefficient of `union` differs substantially from the corresponding estimate for the fixed-intercept model, suggesting that there is some subject-level confounding here. As discussed in section 3.7.4, we could reduce this problem by including the cluster means of the time-varying covariates (although the cluster mean of experience cannot be included because it is collinear with years of education).

From the random part of the model, we see that the between-subject residual standard deviation is estimated as 0.33 compared with an estimate of 0.35 for the within-subject standard deviation. The corresponding estimated residual intraclass correlation is

$$\widehat{\rho} = \frac{\widehat{\psi}}{\widehat{\psi}+\widehat{\theta}} = \frac{0.327^2}{0.327^2+0.354^2} = 0.46$$

Thus 46% of the variance in log wage that is not explained by the covariates is due to time-invariant subject-specific characteristics.

5.5.4 Random-coefficient model

We now consider random-coefficient models for the wage-panel data. We could include random coefficients for any of the time-varying variables to allow the effect of these variables to vary between subjects. The data provide no information on subject-specific effects of time-constant variables and it therefore does not make sense to include random coefficients for these variables (unless we want to model heteroskedasticity as described at the end of section 5.14).

It may well be that different subjects' wages increase at different rates with each extra year of experience. We can investigate this by including a random coefficient of labor market experience L_{ij} in the model

$$y_{ij} = \beta_1 + \beta_2 x_{2ij} + \cdots + \beta_6 L_{ij} + \cdots + \beta_8 x_{8ij} + \zeta_{1j} + \zeta_{2j} L_{ij} + \epsilon_{ij}$$

We fit this random-coefficient model by using

```
. xtmixed lwage black hisp union married exper yeart educt || nr: exper,
> cov(unstructured) mle

Mixed-effects ML regression                     Number of obs      =      4360
Group variable: nr                              Number of groups   =       545

                                                Obs per group: min =         8
                                                               avg =       8.0
                                                               max =         8

                                                Wald chi2(7)       =    573.88
Log likelihood = -2130.4677                     Prob > chi2        =    0.0000

------------------------------------------------------------------------------
       lwage |      Coef.   Std. Err.      z    P>|z|     [95% Conf. Interval]
-------------+----------------------------------------------------------------
       black |   -.139996   .0489058    -2.86   0.004    -.2358496   -.0441423
        hisp |    .009267   .0437623     0.21   0.832    -.0765055    .0950396
       union |   .1098184    .017896     6.14   0.000     .0747429     .144894
     married |   .0757788   .0173732     4.36   0.000      .041728    .1098296
       exper |   .0418495   .0119737     3.50   0.000     .0183815    .0653175
       yeart |   .0171964   .0118898     1.45   0.148    -.0061072    .0405001
       educt |    .097203   .0109324     8.89   0.000     .0757758    .1186302
       _cons |   1.307388   .0404852    32.29   0.000     1.228039    1.386738
------------------------------------------------------------------------------

------------------------------------------------------------------------------
  Random-effects Parameters  |   Estimate   Std. Err.     [95% Conf. Interval]
-----------------------------+------------------------------------------------
nr: Unstructured             |
                   sd(exper) |   .0539497   .0030854       .048229    .0603489
                   sd(_cons) |   .4514402   .0215257       .411162    .4956641
           corr(exper,_cons) |  -.6801072   .0348441     -.7426584   -.6057911
-----------------------------+------------------------------------------------
                sd(Residual) |   .3266336   .0040591       .318774    .3346871
------------------------------------------------------------------------------
LR test vs. linear regression:       chi2(3) =  1715.54   Prob > chi2 = 0.0000
Note: LR test is conservative and provided only for reference.
```

and store the estimates:

```
. estimates store rc
```

The estimates are also shown under "Random coefficient" in table 5.2.

We can perform a likelihood-ratio test comparing the random-intercept and random-coefficient models

```
. lrtest ri rc

Likelihood-ratio test                                 LR chi2(2)  =    167.78
(Assumption: ri nested in rc)                         Prob > chi2 =    0.0000

Note: LR test is conservative
```

The p-value from this test should be divided by 2 (see sec. 4.6), which makes no difference to the conclusion that the random-intercept model is rejected in favor of the random-coefficient model.

We could alternatively have included a random slope for `yeart`. However, if different subjects grow at different rates after 1980 (when `yeart`=0), it makes sense to assume that they have grown at different rates ever since they entered the labor market, which could be before 1980. The variance in wages in the year 1980 must then be larger for those who entered the labor market long before that time (and have grown at different rates for a long time) than for those who entered recently. However, the random-coefficient model would assume a constant variance when `yeart` is zero.

The fixed-effects version of the model is

$$y_{ij} = \beta_1 + \beta_2 x_{2ij} + \cdots + \beta_6 L_{ij} + \cdots + \beta_8 x_{8ij} + \alpha_{1j} + \alpha_{2j} L_{ij} + \epsilon_{ij}$$

where α_{1j} and α_{2j} are fixed subject-specific intercept and slope parameters, respectively. Since L_{ij} increases by the same amount from occasion to occasion for every subject here, the slopes α_{2j} become intercepts after forming differences $(y_{ij} - y_{i-1,j})$, and the original intercepts α_{1j} disappear. The resulting fixed subject-specific intercept model can then be fitted using the approach outlined in section 5.5.2. Alternatively, both α_{1j} and α_{2j} can be eliminated by double differencing, for instance using $(y_{ij} - y_{i-1,j}) - (y_{i-1,j} - y_{i-2,j})$, with estimation simply proceeding by OLS for the resulting regression model.

5.5.5 Marginal mean and covariance structure induced by random effects

So far, we have defined random-effects models in a hierarchical way. First, we specified a model for the response given the random effects, and then combined this model with an assumed distribution for the random effects. Here we consider the implied marginal models.

Marginal mean and covariance structure for random-intercept models

For a random-intercept model, the distribution of y_{ij} given ζ_{1j} (and an observed covariate, time t_{ij}) is specified via

$$y_{ij} = \beta_1 + \beta_2 t_{ij} + \underbrace{\zeta_{1j} + \epsilon_{ij}}_{\xi_{ij}}, \quad \epsilon_{ij}|t_{ij}, \zeta_{1j} \sim N(0, \theta)$$

so that

$$y_{ij}|t_{ij}, \zeta_{1j} \sim N(\beta_1 + \beta_2 t_{ij} + \zeta_{1j}, \theta)$$

The random-intercept distribution is specified as

$$\zeta_{1j}|t_{ij} \sim N(0, \psi_{11})$$

In section 3.3.1, we have also discussed the population-averaged or marginal regression line implied by this model, as well as the marginal variance and covariance and corresponding correlation, called the residual intraclass correlation. Here population averaged and marginal refer to averaging over the random-effects distribution. We are always conditioning on observed covariates, so the term conditional refers to conditioning on the random effects as well, in contrast to marginal where we only condition on observed covariates.

For an overview of the relationship between conditional and marginal (with respect to ζ_{1j}) first- and second-order moments (expectations, variances, and covariances), see table 5.3.

Table 5.3: Conditional and marginal expectation, variance, and covariance of responses for random-intercept model

Conditional or subject-specific			Marginal or population-averaged		
$E(y_{ij}\|t_{ij},\zeta_{1j})$	$=$	$\beta_1+\beta_2 t_{ij}+\zeta_{1j}$	$\beta_1+\beta_2 t_{ij}$	$=$	$E(y_{ij}\|t_{ij})$
$\text{Var}(y_{ij}\|t_{ij},\zeta_{1j})$	$=$	θ	$\psi_{11}+\theta$	$=$	$\text{Var}(y_{ij}\|t_{ij})$
$\text{Cov}(y_{ij},y_{i'j}\|t_{ij},t_{i'j},\zeta_{1j})$	$=$	0	ψ_{11}	$=$	$\text{Cov}(y_{ij},y_{i'j}\|t_{ij},t_{i'j})$

For two occasions $i=1,2$, it follows that we can write the marginal bivariate distribution as

$$\begin{bmatrix} y_{1j} \\ y_{2j} \end{bmatrix} \Bigg| \begin{bmatrix} t_{1j} \\ t_{2j} \end{bmatrix} \sim \text{N}\left(\begin{bmatrix} \beta_1+\beta_2 t_{1j} \\ \beta_1+\beta_2 t_{2j} \end{bmatrix}, \begin{bmatrix} \psi_{11}+\theta & \psi_{11} \\ \psi_{11} & \psi_{11}+\theta \end{bmatrix} \right)$$

The covariance matrix of the responses given the covariates can also be thought of as the covariance matrix of the total residuals $\xi_{ij}=\zeta_{1j}+\epsilon_{ij}$. People often find it easier to interpret correlations than covariances because correlations are unit-free, lying between -1 and $+1$. A correlation between two variables is defined as their covariance divided by the product of their two standard deviations, so the correlation matrix corresponding to the covariance matrix above is

$$\begin{bmatrix} 1 & \frac{\psi_{11}}{\psi_{11}+\theta} \\ \frac{\psi_{11}}{\psi_{11}+\theta} & 1 \end{bmatrix}$$

For the wage-panel data, the residual intraclass correlation is estimated as 0.46 and the total variance is estimated as $0.327^2+0.354^2=0.23$ in section 5.5.3. The random-intercept model therefore implies that all pairwise correlations are equal to 0.46 and all variances equal to 0.23, which are smoothed versions of the estimated residual correlation matrix and variances given on page 186.

Marginal mean and covariance structure for random-coefficient models

Consider now the random-coefficient model

$$y_{ij} = \beta_1 + \beta_2 t_{ij} + \underbrace{\zeta_{1j} + \zeta_{2j} t_{ij} + \epsilon_{ij}}_{\xi_{ij}}$$

For an overview of the relationship between conditional and marginal (with respect to ζ_{1j} and ζ_{2j}) expectations, variances, and covariances for this model, see table 5.4.

Table 5.4: Conditional and marginal expectation, variance, and covariance of responses for random-coefficient model

Conditional or subject-specific		Marginal or population-averaged	
$E(y_{ij}\|t_{ij}, \zeta_{1j}, \zeta_{2j})$	$= \beta_1 + \beta_2 t_{ij} + \zeta_{1j} + \zeta_{2j} t_{ij}$	$\beta_1 + \beta_2 t_{ij} =$	$E(y_{ij}\|t_{ij})$
$\text{Var}(y_{ij}\|t_{ij}, \zeta_{1j}, \zeta_{2j})$	$= \theta$	$\psi_{11} + 2\psi_{21} t_{ij} + \psi_{22} t_{ij}^2 + \theta =$	$\text{Var}(y_{ij}\|t_{ij})$
$\text{Cov}(y_{ij}, y_{i'j}\|t_{ij}, t_{i'j}, \zeta_{1j}, \zeta_{2j})$	$= 0$	$\psi_{11} + \psi_{21}(t_{ij} + t_{i'j}) + \psi_{22} t_{ij} t_{i'j} =$	$\text{Cov}(y_{ij}, y_{i'j}\|t_{ij}, t_{i'j})$

As discussed in section 4.4.2, the variances and covariances (as well as correlations) now depend on the variable having a random coefficient, here t_{ij}.

To get a better idea of the correlation pattern induced by random-coefficient models, it is therefore useful to consider examples of balanced longitudinal data, with say 5 time points, with $t_{ij} = t_i$ given by 0, 1, 2, 3, and 4. Figure 5.2 gives two such examples that differ from each other only in terms of the correlation between intercept and slope, given by 0.2 on the left and −0.8 on the right. As can be seen from the randomly drawn subject-specific regression lines for 10 subjects, the variances increase as a function of time on the left, whereas they change much less, first decreasing and then increasing, on the right. On the left, the correlations between adjacent time points (or "lag-1 correlations") are 0.63 between occasions 1 and 2, 0.76 between occasions 2 and 3, then 0.84, and finally 0.90. Similarly, the lag 2 and lag 3 correlations increase over time. The correlation between occasion 1 and 2 is greater than that between occasion 1 and 3. In general, at a given occasion, the correlations with other occasions decrease as the time lag increases. The correlation matrix on the right follows a less clear pattern, which illustrates that the random-coefficient model can produce a wide range of different correlation patterns.

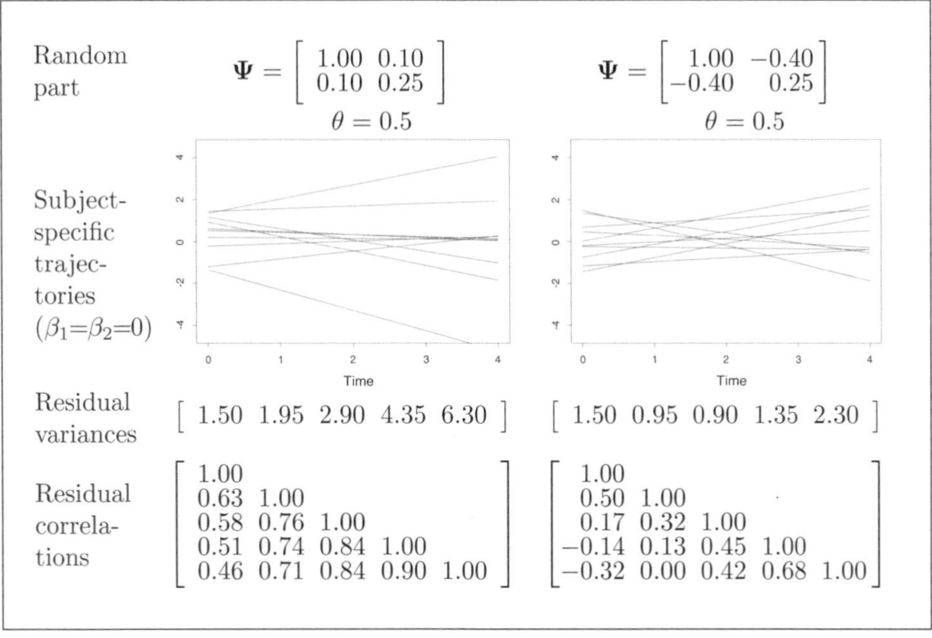

Figure 5.2: Illustration of marginal variances and correlations induced by random-coefficient models

For the random-coefficient model fitted in section 5.5.4, the model-implied residual correlation matrix depends on labor market experience, a covariate that does not take on the same values across occasions for all subjects because subjects entered the labor market at different times.

5.6 Marginal modeling

Instead of specifying multilevel linear models for longitudinal or clustered data in a hierarchical fashion, from which we can derive the marginal expectations and covariances of the responses (averaged over the random effects but conditional on the observed covariates), we can directly specify a model for the marginal expectations and the marginal covariance matrix of the responses given the covariates.

5.6.1 Covariance structures

Some popular marginal covariance structures for marginal modeling are shown in table 5.5 and discussed below.

Table 5.5: Common marginal covariance structures for longitudinal data ($n_j = 3$)

Compound symmetric or exchangeable structure:

$$\begin{bmatrix} \psi_{11}+\theta & & \\ \psi_{11} & \psi_{11}+\theta & \\ \psi_{11} & \psi_{11} & \psi_{11}+\theta \end{bmatrix}$$

Random coefficient structure (t=0,1,2):

$$\begin{bmatrix} \psi_{11}+\theta & & \\ \psi_{11}+\psi_{21} & \psi_{11}+2\psi_{21}+\psi_{22}+\theta & \\ \psi_{11}+2\psi_{21} & \psi_{11}+3\psi_{21}+2\psi_{22} & \psi_{11}+4\psi_{21}+4\psi_{22}+\theta \end{bmatrix}$$

Autoregressive residual structure AR(1):†

$$\frac{\phi}{1-\alpha^2}\begin{bmatrix} 1 & & \\ \alpha & 1 & \\ \alpha^2 & \alpha & 1 \end{bmatrix}$$

Unstructured:

$$\begin{bmatrix} \sigma_{11} & & \\ \sigma_{21} & \sigma_{22} & \\ \sigma_{31} & \sigma_{32} & \sigma_{33} \end{bmatrix}$$

Autoregressive response structure AR(1):†

$$\frac{\sigma^2}{1-\gamma^2}\begin{bmatrix} 1 & & \\ \gamma & 1 & \\ \gamma^2 & \gamma & 1 \end{bmatrix}$$

† Autoregressive residual and autoregressive-response models have the same covariance structure. but different mean structures.

Compound symmetric or exchangeable structure

A simple covariance structure is the *compound symmetric* or *exchangeable* structure where all variances are equal and all covariances (and thus correlations) are also equal. If the common covariance is positive, this is the covariance ψ_{11} structure induced by a random-intercept model as shown in table 5.3, with variance $\psi_{11}+\theta$ and covariance ψ_{11}. Although this dependence structure is parsimonious, it is often unrealistic because responses close in time are expected to be more highly correlated than responses distant in time.

Random-coefficient structure

The *random-coefficient* structure is the marginal covariance structure induced by the random-coefficient model. As shown in table 5.4, this structure is complicated but flexible. The variances change over time and, if t_{ij} is not the same for all subjects, the covariance matrix is also different for different subjects.

When the random-coefficient structure is used in a marginal modeling approach, it can be argued that it is acceptable to relax the usual restriction that the covariance matrix of the random effects is positive semidefinite (i.e., nonnegative variances $0 \leq \psi_{11}$ and $0 \leq \psi_{22}$ and valid correlation $-1 \leq \rho_{21} \leq 1$) and that the level-1 residual variance is nonnegative as long as the *marginal* covariance matrix is positive semidefinite. Relaxing the restriction makes the random-coefficient structure even more flexible, but no longer interpretable as being induced by random intercepts and slopes.

Autoregressive residual structure

In *autoregressive residual* covariance structures the current residual is regressed on the residuals at previous occasions, making correlations between the residuals fall off as the time lag between occasions increases.

A popular special case is the first-order or autoregressive lag-1 structure, AR(1), where the residual is regressed only on the previous residual

$$\epsilon_{ij} = \alpha\epsilon_{i-1,j} + u_{ij}, \qquad u_{ij} \sim \mathrm{N}(0, \phi)$$

with $\epsilon_{i-1,j}$ independent of the errors u_{ij}. If $|\alpha| < 1$, the correlation structure becomes

$$\mathrm{Cor}(y_{ij}, y_{i'j}|\mathbf{x}_{ij}, \mathbf{x}_{i'j}) = \alpha^{|i-i'|}$$

This model makes sense only if the intervals between successive occasions are the same (or similar) across time and for all subjects. When this is restriction violated, a correlation structure of the form

$$\mathrm{Cor}(y_{ij}, y_{i'j}|\mathbf{x}_{ij}, \mathbf{x}_{i'j}) = \alpha^{|t_{ij}-t_{i'j}|}$$

is sometimes specified, where t_{ij} is the time associated with occasion i.

Higher-order autoregressive residual structures extend the AR(1) structure to also regress residuals on lag-2 residuals, lag-3 residuals, and so on.

Unstructured covariance matrix

Finally, an *unstructured* covariance matrix can be used where each variance and unique covariance is freely estimated. This approach may appear to be preferable to the other structures because it is flexible. However, the estimates of the variances and covariances typically become imprecise if there are many time points because there is little information at each occasion. The covariance structure is also not completely unrestricted because it is assumed to be constant across subjects, unlike for instance the random-coefficient structure, when the timing of occasions varies between subjects.

The final covariance structure given in table 5.5, the *autoregressive-response* structure, is discussed in section 5.7.

5.6.2 Marginal modeling using Stata

Stata's `xtgls` command can be used for fitting marginal models in Stata using generalized least squares (GLS). Briefly, this consists of three steps. First, the model is fitted by OLS giving estimated residuals and their sample within-subject covariance matrix. Second, the model-implied covariance matrix is fitted to the sample covariance matrix of the residuals from OLS. Third, the regression coefficients are reestimated by weighted least squares using the inverse of the estimated covariance matrix as a weight matrix. The estimates can be refined using iterated generalized least squares (IGLS), where the residuals are updated using the estimated regression coefficients from the final step of GLS, giving a different fitted residual covariance matrix and new estimated regression coefficients, and so on, until convergence.

For linear marginal models with continuous responses, the `xtgee` command works essentially in the same way as IGLS, except that the covariance matrix is fitted somewhat differently. The estimation method implemented in `xtgee`, called "generalized estimating equations" (GEE), was actually developed for noncontinuous responses such as dichotomous responses and counts (see sec. 6.14.2 and 9.11.3) and is rarely used for continuous responses.

`xtgee` offers all the correlation structures corresponding to the covariance structures discussed in this section except the random-coefficient structure. It also offers higher-order autoregressive structures and many more. `xtgls` offers only the AR(1) covariance structure but accommodates heteroskedastic variances (between subjects) and correlations between subjects. However, neither command allows residual variances to differ between occasions, which is very restrictive. For instance, when modeling children's weights as a function of age, it seems obvious that their weights are less variable at age 0 than they are at age 5.

We briefly demonstrate `xtgee` by fitting a model with an AR(1) residual correlation structure. We specify robust standard errors using the `vce(robust)` option because this is standard practice when using GEE:

```
. xtgee lwage black hisp union married exper yeart educt, i(nr) t(yeart)
> corr(ar 1) vce(robust)
warning: existing time variable is not yeart; assuming delta = 1

GEE population-averaged model                   Number of obs      =      4360
Group and time vars:                nr yeart    Number of groups   =       545
Link:                               identity    Obs per group: min =         8
Family:                             Gaussian                   avg =       8.0
Correlation:                           AR(1)                   max =         8
                                                Wald chi2(7)       =    547.37
Scale parameter:                    .2322738    Prob > chi2        =    0.0000

                                     (Std. Err. adjusted for clustering on nr)

------------------------------------------------------------------------------
             |             Semi-robust
       lwage |      Coef.   Std. Err.      z    P>|z|     [95% Conf. Interval]
-------------+----------------------------------------------------------------
       black |  -.1329348   .0487983    -2.72   0.006    -.2285776   -.0372919
        hisp |   .0195104   .0398453     0.49   0.624     -.058585    .0976058
       union |   .0995969   .0195432     5.10   0.000      .061293    .1379008
     married |   .0813371   .0190996     4.26   0.000     .0439026    .1187715
       exper |   .0351937   .0107019     3.29   0.001     .0142183    .0561691
       yeart |    .025618   .0113301     2.26   0.024     .0034115    .0478246
       educt |   .0952535   .0107629     8.85   0.000     .0741586    .1163485
       _cons |   1.300357   .0381552    34.08   0.000     1.225574     1.37514
------------------------------------------------------------------------------
```

Here a time variable must be specified in the `t()` option unless it has already been declared using `xtset`. This variable is used only to determine the ordering of the time points that are assumed to be equally spaced as indicated by the message **`assuming delta = 1`** in the output. The estimate of the residual variance $\phi/(1-\alpha^2)$ is referred to as the `Scale parameter` and is given by 0.232, so the estimated residual standard deviation is $\sqrt{0.232}=0.48$.

The estimated correlation matrix can be displayed using

```
. estat wcorrelation, format(%4.3f)
Estimated within-nr correlation matrix R:

             c1     c2     c3     c4     c5     c6     c7     c8
--------------------------------------------------------------------
    r1    1.000
    r2    0.575  1.000
    r3    0.330  0.575  1.000
    r4    0.190  0.330  0.575  1.000
    r5    0.109  0.190  0.330  0.575  1.000
    r6    0.063  0.109  0.190  0.330  0.575  1.000
    r7    0.036  0.063  0.109  0.190  0.330  0.575  1.000
    r8    0.021  0.036  0.063  0.109  0.190  0.330  0.575  1.000
```

It follows from the estimated correlations for adjacent occasions that $\widehat{\alpha}=0.575$.

The GEE estimates for the model with an autoregressive lag-1 structure are shown under "Residual AR(1)" in table 5.2.

5.7 Autoregressive- or lagged-response models

A first-order *autoregressive-response* model can be specified by including the previous response $y_{i-1,j}$ as a covariate in a linear regression model for y_{ij}:

$$y_{ij} = \beta_1 + \gamma y_{i-1,j} + \beta_2 x_{2ij} + \cdots + \beta_8 x_{8ij} + \epsilon_{ij}, \qquad \epsilon_{ij}|y_{i-1,j}, \mathbf{x}_{ij} \sim N(0, \sigma^2)$$

The previous response is sometimes referred to as a "lag-1 response" and can be obtained in Stata using

```
. by nr (yeart), sort: generate lag1 = lwage[_n-1]
(545 missing values generated)
```

On the right-hand side of the command, we see that the lag is produced by referring to the previous observation of `lwage` using the row counter `_n` minus 1. The `by` prefix command is used with the subject identifier `nr` for the counter `_n` to be reset to 1 every time `nr` increases. Otherwise, the lag-1 response for the second subject in the data would be the last response of the first subject. For this command to work, we must also sort by `nr` as specified by the `sort` option. `yeart` is also given in parentheses to sort the data by `yeart` within `nr` before defining the counter so that the counter increases in tandem with `yeart`.

To see the result of this command, we list the first nine observations:

```
. sort nr yeart
. list nr yeart lwage lag1 in 1/9, clean noobs
     nr   yeart       lwage        lag1
     13       0     1.19754           .
     13       1     1.85306     1.19754
     13       2    1.344462     1.85306
     13       3    1.433213    1.344462
     13       4    1.568125    1.433213
     13       5    1.699891    1.568125
     13       6   -.7202626    1.699891
     13       7    1.669188   -.7202626
     17       0    1.675962           .
```

Obviously, there is no lag-1 response for the first wave of data, when `yeart` is 0.

We can now fit the autoregressive-response model using the `regress` command:

```
. regress lwage lag1 black hisp union married exper yeart educt

      Source |       SS       df       MS              Number of obs =    3815
-------------+------------------------------           F(  8,  3806) =  379.23
       Model |  455.410767     8  56.9263459           Prob > F      =  0.0000
    Residual |  571.325295  3806  .150111743           R-squared     =  0.4436
-------------+------------------------------           Adj R-squared =  0.4424
       Total |  1026.73606  3814  .269201904           Root MSE      =  .38744

------------------------------------------------------------------------------
       lwage |      Coef.   Std. Err.      t    P>|t|     [95% Conf. Interval]
-------------+----------------------------------------------------------------
        lag1 |   .5572532   .0128673    43.31   0.000     .5320257    .5824807
       black |  -.0721592   .0204291    -3.53   0.000    -.1122123   -.0321062
        hisp |   .0092496   .0179206     0.52   0.606    -.0258854    .0443846
       union |   .0755261   .0149509     5.05   0.000     .0462135    .1048388
     married |   .0482021   .0132629     3.63   0.000      .022199    .0742051
       exper |   .0028574   .0047477     0.60   0.547    -.0064509    .0121657
       yeart |   .0175801   .0056328     3.12   0.002     .0065365    .0286237
       educt |   .0389675   .0046551     8.37   0.000     .0298409    .0480942
       _cons |   .6683362   .0253468    26.37   0.000     .6186417    .7180308
------------------------------------------------------------------------------
```

We see that $\widehat{\gamma} = 0.56$ and that all the other estimated regression coefficients are now closer to zero than for any of the other models. Such a change in estimated effects is not unusual in lagged-response models since the estimates now have a different interpretation, as the estimated effects of the covariates on the response *after controlling for the previous response*. Reexpressing the model as

$$y_{i,j} - \gamma y_{i-1,j} \;=\; \beta_1 + \beta_2 x_{2ij} + \cdots + \beta_8 x_{8ij} + \epsilon_{ij}$$

it is clear that this is similar to a change-score approach since the left-hand side becomes $y_{i,j} - y_{i-1,j}$ if $\gamma = 1$. The lagged response should be used only if it really makes sense to control for the previous response.

The lagged-response model is only sensible if the occasions are equally spaced in time. Otherwise, it would be strange to assume that the lagged response has the same effect on the current response regardless of the time interval between them. It should also be remembered that the sample size is reduced when using a lagged response approach because lags are missing for the first occasion. Furthermore, the problem of missing data becomes exacerbated because not just the missing response itself is discarded but also the subsequent response because its lagged response is missing.

Autoregressive models are sometimes referred to as transition models, Markov models, conditional models, or dynamic models. If $|\gamma| < 1$, the model induces the correlation

$$\mathrm{Cor}(y_{ij}, y_{i'j}|\mathbf{x}_{ij}, \mathbf{x}_{i'j}) = \gamma^{|i-i'|}$$

This correlation structure is given under "Autoregressive-response structure AR(1)" in table 5.5. Higher-order autoregressive processes can also be used where responses are lagged on several previous responses.

5.8 Hybrid approaches

Having discussed fixed-effects, random-effects, marginal, and lagged-response models in their pure forms, we now turn to some of the most common hybrids of these models.

5.8.1 Autoregressive response and random effects

We could specify a random-intercept model that includes lagged responses as covariates. This can be done straightforwardly, using any of the commands for random-intercept models, such as `xtreg` and `xtmixed`, and including the lagged responses as covariates.

A useful feature of such models is that they can be used to distinguish between two competing explanations of within-subject dependence: unobserved heterogeneity (represented by the random effects) or previous experience (represented by the lagged responses). For instance, the within-subject dependence of salaries over time, over and above that explained by observed covariates, may be due to some subjects being especially gifted and thus more valued and/or some subjects having experienced high salaries in the past, which places them in a good bargaining position.

However, a problem with models including both random effects and autoregressive responses is that the lagged responses, which are included as covariates, are correlated with the random intercept. This is because the lagged responses themselves must have been affected by the random intercept. One way of addressing this endogeneity problem is by using Stata's `xtabond` command.

5.8.2 ❖ Autoregressive responses and autoregressive residuals

Consider two simple models: a model with a lagged response and lagged covariate but independent residuals e_{ij}

$$y_{ij} = \gamma\, y_{i-1,j} + \beta_1 x_{ij} + \beta_2 x_{i-1,j} + e_{ij} \tag{5.2}$$

and an autocorrelation model without lagged response or lagged covariate

$$y_{ij} = \beta\, x_{ij} + \epsilon_{ij} \tag{5.3}$$

but residuals ϵ_{ij} having an AR(1) structure. Substituting first for $\epsilon_{ij} = \alpha\epsilon_{i-1,j} + u_{ij}$ and then for $\epsilon_{i-1,j} = y_{i-1,j} - \beta\, x_{i-1,j}$ in (5.3) and reexpressing, the autocorrelation model (5.3) can alternatively be written as

$$y_{ij} = \alpha\, y_{i-1,j} + \beta\, x_{ij} - \alpha\beta\, x_{i-1,j} + u_{ij}$$

This model is equivalent to the lagged-response model (5.2) with the restriction that $\beta_2 = -\gamma\beta_1$. We can thus fit (5.2) and test the nonlinear constraint $\beta_2 = -\gamma\beta_1$ using the `nlcom` or `nltest` command to choose between models with autocorrelated residuals and lagged responses.

5.8.3 Autoregressive residuals and random or fixed effects

An appealing specification is a random-intercept model in which the level-1 residuals have a first-order autoregressive correlation structure. This produces a correlation matrix with serial dependence that does not fall off as rapidly with increasing time lags as the AR(1) structure.

At the time of writing this book, autoregressive residuals cannot be specified in `gllamm` or in Stata's `xtreg` or `xtmixed` commands for multilevel modeling. However, such models can be fitted using Stata's `xtregar` command with the `re` option, which uses generalized least squares. Using `xtregar` with the `fe` option replaces random subject effects by fixed effects.

5.9 Missing data

All the methods discussed so far use data for those subjects j and occasions i where neither the response y_{ij} nor the covariates $\mathbf{x}_{ij}$ are missing. This is in contrast to more old-fashioned approaches to longitudinal data such as MANOVA where subjects with missing responses or covariates are discarded altogether, an approach often referred to as "listwise deletion" or "complete-case analysis". Using all available data does not waste information and is less susceptible to bias.

Using maximum likelihood estimation, as we did for random-intercept and random-coefficient models, has the advantage that consistency (estimates approaching parameter values in large samples) is retained, as long as the missing data are "missing at random" (MAR). This means that the probability of being missing at an occasion i may only depend on the covariates x_{ij} or responses $y_{i'j}$ at previous occasions $i' < i$ (or future occasions, although this seems strange), but not on the responses we would have observed had they not been missing.

5.9.1 ❖ Maximum likelihood estimation under MAR: A simulation

We now use a simulation to investigate how well maximum likelihood estimation works when data are MAR. We first simulate complete data from a random-intercept model (with the usual assumptions)

$$y_{ij} \;=\; 2 + \zeta_j + \epsilon_{ij}, \qquad \zeta_j \sim N(0,1),\ \epsilon_{ij} \sim N(0,1) \tag{5.4}$$

for $J = 10{,}000$ subjects and $n_j = 2$ occasions. It is convenient to simulate the data in wide form, generating a row of data for each subject with variables `y1` and `y2` for the two occasions:

```
. clear
. set obs 10000
. set seed 1323232
. generate zeta = invnormal(uniform())
```

```
. generate y1 = 2 + zeta + invnormal(uniform())
. generate y2 = 2 + zeta + invnormal(uniform())
```

Here the `uniform()` function generates a pseudorandom variable with a uniform distribution on the interval from 0 to (nearly) 1, and this variable is converted to a standard normal pseudorandom variable (as required for ζ_j, ϵ_{1j}, and ϵ_{2j}) using the `invnormal()` function (the inverse standard normal cumulative distribution function). We have set the "seed" of the pseudorandom number generator to an arbitrary number so that you can get the same results as us, but you can also try other seeds if you like.

Among those subjects who have a value of `y1` greater than 2, we now randomly sample about 90% of subjects and replace their `y2` by a missing value:

```
. replace y2 = . if y1>2 & uniform()<0.9
(4485 real changes made, 4485 to missing)
```

Here we made use of the fact that the probability that a (pseudo)random number with a uniform distribution (on the interval from 0 to 1) is less than 0.9 is 0.9. The responses are MAR because the probability of missingness depends only on the previous response.

The sample means at the two occasions are

```
. tabstat y1 y2
```

stats	y1	y2
mean	2.002078	1.513872

We see that the sample mean at the second occasion is 1.51, much lower than the population mean of 2. This is because subjects with larger than average values at occasion 1 ($y_{1j} > 2$) are likely to have larger than average values at occasion 2 as well (due to an intraclass correlation of 0.5), but about 90% of these values are missing.

To see if such a difference in estimated means is also found using maximum likelihood, we fit a slightly more general model than the true model (5.4), which allows the means at the two occasions to be different,

$$y_{ij} = \beta_1 x_{1i} + \beta_2 x_{2i} + \zeta_j + \epsilon_{ij} \tag{5.5}$$

where x_{1i} is a dummy variable for occasion $i = 1$ and x_{2i} is a dummy variable for occasion $i=2$. According to the true model (5.4) used to generate the complete data, the population means at both occasions are $\beta_1 = \beta_2 = 2$.

We must first reshape the data and generate the dummy variables:

```
. generate id = _n

. reshape long y, i(id) j(occasion)
(note: j = 1 2)

Data                               wide   ->   long
-----------------------------------------------------------------------------
Number of obs.                    10000   ->   20000
Number of variables                   4   ->       4
j variable (2 values)                     ->   occasion
xij variables:
                                  y1 y2   ->   y
-----------------------------------------------------------------------------

. quietly tabulate occasion, generate(occ)
```

Now we can use the **xtreg** command to fit the random-intercept model (5.5) using the **noconstant** option because the model does not include an intercept:

```
. xtreg y occ1 occ2, noconstant i(id) mle

Random-effects ML regression                    Number of obs      =     15515
Group variable: id                              Number of groups   =     10000

Random effects u_i ~ Gaussian                   Obs per group: min =         1
                                                               avg =       1.6
                                                               max =         2

                                                Wald chi2(2)       =  22292.14
Log likelihood  =   -26661.4                    Prob > chi2        =    0.0000

------------------------------------------------------------------------------
           y |      Coef.   Std. Err.      z    P>|z|     [95% Conf. Interval]
-------------+----------------------------------------------------------------
        occ1 |   2.002078    .014206   140.93   0.000     1.974235    2.029922
        occ2 |   1.978601   .0209977    94.23   0.000     1.937446    2.019756
-------------+----------------------------------------------------------------
    /sigma_u |   1.006445   .0155622                       .976401    1.037413
    /sigma_e |   1.002578   .0105966                      .9820224    1.023563
         rho |   .5019248   .0116489                      .4791027    .5247407
------------------------------------------------------------------------------
Likelihood-ratio test of sigma_u=0: chibar2(01)=  984.76 Prob>=chibar2 = 0.000
```

We see that all parameter estimates, including $\widehat{\beta}_2$, are close to the true parameter values. The standard error for $\widehat{\beta}_2$ is larger than for $\widehat{\beta}_1$ due to the missing values at the second occasion. The reason these estimates are so good is that the random-intercept model takes the intraclass correlation into account. In contrast, the ordinary least-squares estimates (which are based on assuming zero intraclass correlation) are just the sample means with $\widehat{\beta}_2^{\text{OLS}} = 1.51$.

Of course a serious simulation study would consist of repeatedly simulating data and fitting the model to estimate the bias in the parameter estimates by the average of the difference between estimates and parameter values over repeated samples.

5.10 How do children grow?

We now expand on the random-coefficient models discussed in section 5.5.4 by discussing what is often referred to as growth-curve models.

The dataset considered here is on Asian children in a British community who were weighed on up to four occasions, roughly at ages 6 weeks, and then 8, 12, and 27 months. The dataset `asian.dta` is a 12% random sample, stratified by gender, from the dataset `asian.dat` available from the web page of the Centre for Multilevel Modelling[2]. The full data were previously analyzed by Goldstein (1986) and Prosser, Rasbash, and Goldstein (1991).

The dataset `asian.dta` has the following variables:

- `id`: child identifier
- `weight`: weight in kilograms
- `age`: age in years
- `gender`: gender (1: male; 2: female)

We want to investigate the growth trajectories of childrens' weights as they get older. Both the shape of the trajectories and the degree of variability in growth among the children are of interest.

5.10.1 Observed growth trajectories

We first plot the observed growth trajectories by gender, after defining value labels for `gender` to make them appear on the graph:

```
. use http://www.stata-press.com/data/mlmus2/asian, clear
. label define g 1 "Boy" 2 "Girl"
. label values gender g
. sort id age
. graph twoway (line weight age, connect(ascending)),
> by(gender) xtitle(Age in years) ytitle(Weight in Kg)
```

2. See http://www.cmm.bristol.ac.uk/learning-training/multilevel-m-support/datasets.shtml.

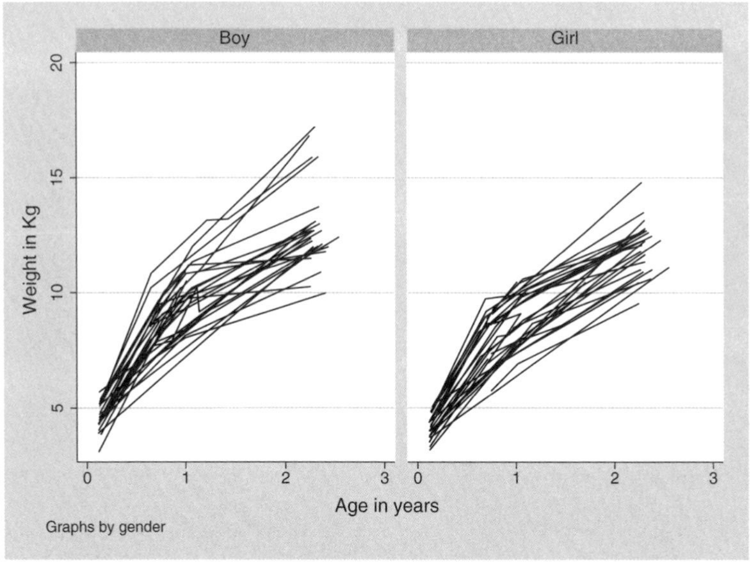

Figure 5.3: Observed growth trajectories for boys and girls

From figure 5.3, it is clear that the growth trajectories are nonlinear; growth is initially fast and then slows down. We will model this nonlinearity by including a quadratic term for `age` in the model. It is also apparent that some children are consistently heavier than others, so a random intercept appears to be warranted.

5.11 Growth-curve modeling

5.11.1 Random-intercept model

We start by considering a random-intercept model (under the standard assumptions)

$$y_{ij} = \beta_1 + \beta_2 x_{ij} + \beta_3 x_{ij}^2 + \zeta_{1j} + \epsilon_{ij}$$

where y_{ij} is the weight of child j at occasion i, x_{ij} is the corresponding age, and ζ_{1j} is a random intercept. The occasion-specific error term ϵ_{ij} allows the responses y_{ij} to deviate from the perfectly quadratic trajectories defined by the first four terms.

(*Continued on next page*)

The model can be fitted in xtmixed using

```
. generate age2 = age^2
. xtmixed weight age age2 || id:, mle
Mixed-effects ML regression                     Number of obs      =       198
Group variable: id                              Number of groups   =        68

                                                Obs per group: min =         1
                                                               avg =       2.9
                                                               max =         5

                                                Wald chi2(2)       =   2623.63
Log likelihood = -276.83266                     Prob > chi2        =    0.0000

------------------------------------------------------------------------------
      weight |      Coef.   Std. Err.      z    P>|z|     [95% Conf. Interval]
-------------+----------------------------------------------------------------
         age |   7.817918   .2896529    26.99   0.000     7.250209    8.385627
        age2 |  -1.705599   .1085984   -15.71   0.000    -1.918448    -1.49275
       _cons |   3.432859   .1810702    18.96   0.000     3.077968     3.78775
------------------------------------------------------------------------------

------------------------------------------------------------------------------
  Random-effects Parameters  |   Estimate   Std. Err.     [95% Conf. Interval]
-----------------------------+------------------------------------------------
id: Identity                 |
                   sd(_cons) |   .9182256   .0973788      .7458965    1.130369
-----------------------------+------------------------------------------------
                sd(Residual) |   .7347063   .0452564      .6511507    .8289837
------------------------------------------------------------------------------
LR test vs. linear regression: chibar2(01) =    78.07 Prob >= chibar2 = 0.0000
```

The coefficients of both age and age2 are significant at the 5% level. There is a considerable estimated random-intercept standard deviation of 0.92 kg. The estimates are shown under "Model 1: Random intercept" in table 5.6.

Table 5.6: Maximum likelihood estimates for children's growth data (in kilograms)

	Model 1: Random intercept		Model 2: Random intercept and slope	
	Est	(SE)	Est	(SE)
Fixed part				
β_1 [_cons]	3.43	(0.18)	3.49	(0.14)
β_2 [age]	7.82	(0.29)	7.70	(0.24)
β_3 [age2]	−1.71	(0.11)	−1.66	(0.09)
Random part				
$\sqrt{\psi_{11}}$	0.92		0.64	
$\sqrt{\psi_{22}}$			0.50	
ρ_{21}			0.27	
$\sqrt{\theta}$	0.73		0.58	
Log likelihood	−276.83		−258.08	

5.11.2 Random-coefficient model

To make the model more realistic, we can specify a random-coefficient model with a random slope ζ_{2j} of age, to allow children to differ in their overall rate of growth

$$y_{ij} = \beta_1 + \beta_2 x_{ij} + \beta_3 x_{ij}^2 + \zeta_{1j} + \zeta_{2j} x_{ij} + \epsilon_{ij} \tag{5.6}$$

(where we make the standard assumptions).

Apart from the quadratic term x_{ij}^2, this model is a traditional linear growth-curve model. The random part of the model $\zeta_{1j} + \zeta_{2j} x_{ij} + \epsilon_{ij}$ can be represented by a path diagram as shown in figure 5.4. The left panel shows the situation where each subject has been observed at the same three occasions, with the corresponding time variable x_i taking the values 0, 1, and 2. The three responses, enclosed in rectangles, are regressed on the intercept ζ_{1j} with regression coefficients set to 1. They are also regressed on the random slope ζ_{2j} with regression coefficients set equal to the times x_i, here 0, 1, and 2. (Lack of an arrow corresponds to a regression coefficient equal to zero.) The short arrows represent the occasion-specific error terms ϵ_{1j}, ϵ_{2j}, and ϵ_{3j}.

The panel on the right is also appropriate for the current application where the time variable can take on different values for different children. Here all variables inside the box labeled "subject j" have a j subscript and vary between subjects. Variables that are also in the box labeled "occasion i" vary between occasions and subjects and have both an i and j subscript. The arrow from x to y therefore represents a regression of y_{ij} on x_{ij}. The latent variable ζ_{2j} modifies this regression or interacts with x_{ij} and therefore represents the random slope.

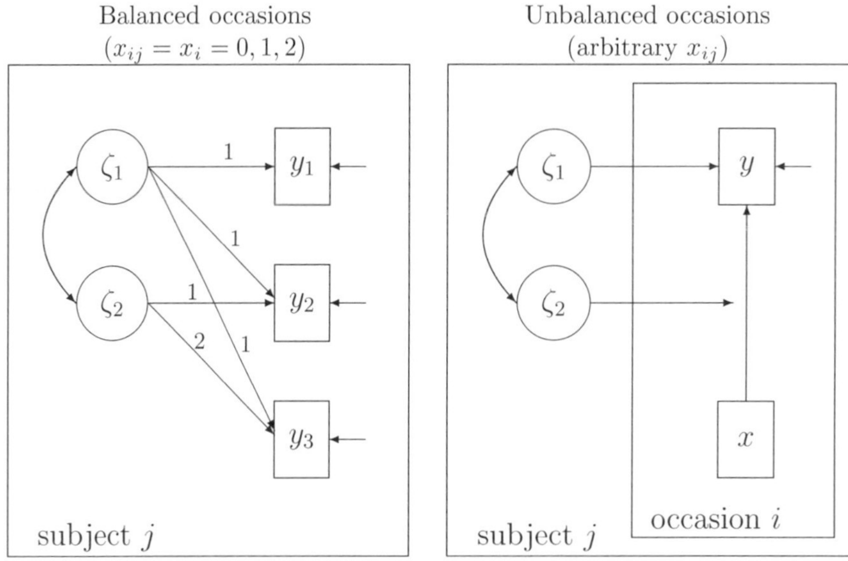

Figure 5.4: Path diagrams for growth-curve models with balanced and unbalanced occasions (*Source:* Skrondal and Rabe-Hesketh 2004)

The model can be fitted as follows:

```
. xtmixed weight age age2 || id: age, cov(unstructured) mle

Mixed-effects ML regression                     Number of obs      =       198
Group variable: id                              Number of groups   =        68

                                                Obs per group: min =         1
                                                               avg =       2.9
                                                               max =         5

                                                Wald chi2(2)       =   1978.20
Log likelihood = -258.07784                     Prob > chi2        =    0.0000

------------------------------------------------------------------------------
      weight |      Coef.   Std. Err.      z    P>|z|     [95% Conf. Interval]
-------------+----------------------------------------------------------------
         age |   7.703998   .2394082    32.18   0.000     7.234767    8.173229
        age2 |  -1.660465   .0885229   -18.76   0.000    -1.833967   -1.486963
       _cons |   3.494512   .1372636    25.46   0.000      3.22548    3.763544
------------------------------------------------------------------------------

------------------------------------------------------------------------------
  Random-effects Parameters  |   Estimate   Std. Err.     [95% Conf. Interval]
-----------------------------+------------------------------------------------
id: Unstructured             |
                     sd(age) |   .5040802   .0879337      .358107    .7095558
                   sd(_cons) |   .6359558   .1293523     .4268684    .9474578
             corr(age,_cons) |   .2747814   .3309063    -.3965135    .7546038
-----------------------------+------------------------------------------------
                sd(Residual) |   .5757751   .0505985     .4846745    .6839993
------------------------------------------------------------------------------
LR test vs. linear regression:       chi2(3) =   115.58   Prob > chi2 = 0.0000

Note: LR test is conservative and provided only for reference
```

The estimates for this model are shown under "Model 2: Random intercept and slope" in table 5.6, together with the estimates for the random-intercept model. The log likelihood has increased by over 18 for two extra parameters, providing evidence that the random slopes are needed. The estimated random-slope standard deviation is 0.50 kg/year. The estimated random-intercept standard deviation has decreased from 0.92 kg to 0.64 kg because it now represents only the variability at age 0 (not the average variability over the ages; see also figure 4.4). However, it is still high. The estimated level-1 residual standard deviation has decreased from 0.73 kg to 0.58 kg reflecting the superior fit of the model-implied growth trajectories for the model including a random slope.

To investigate if there are any systematic differences between boys and girls, we can add a dummy variable w_j for girls to the fixed part of the random-coefficient model

$$y_{ij} = \beta_1 + \beta_2 x_{ij} + \beta_3 x_{ij}^2 + \beta_4 w_j + \zeta_{1j} + \zeta_{2j} x_{ij} + \epsilon_{ij} \tag{5.7}$$

and fit it using the following commands:

```
. generate girl = gender - 1
. xtmixed weight age age2 girl || id: age, cov(unstructured) mle
Mixed-effects ML regression                     Number of obs      =       198
Group variable: id                              Number of groups   =        68

                                                Obs per group: min =         1
                                                               avg =       2.9
                                                               max =         5

                                                Wald chi2(3)       =   1975.44
Log likelihood = -253.86692                     Prob > chi2        =    0.0000

------------------------------------------------------------------------------
      weight |      Coef.   Std. Err.      z    P>|z|     [95% Conf. Interval]
-------------+----------------------------------------------------------------
         age |   7.697967   .2382121    32.32   0.000      7.23108    8.164855
        age2 |  -1.657843   .0880529   -18.83   0.000    -1.830423   -1.485262
        girl |  -.5960093   .1963689    -3.04   0.002    -.9808853   -.2111332
       _cons |   3.794769   .1655053    22.93   0.000     3.470385    4.119153
------------------------------------------------------------------------------

------------------------------------------------------------------------------
  Random-effects Parameters  |   Estimate   Std. Err.     [95% Conf. Interval]
-----------------------------+------------------------------------------------
id: Unstructured             |
                     sd(age) |   .5097089   .0871791      .3645317    .7127039
                   sd(_cons) |    .594731   .1289891      .3887823    .9097762
             corr(age,_cons) |   .1571086   .3240801     -.4564674    .6694143
-----------------------------+------------------------------------------------
                sd(Residual) |   .5723301   .0496274      .4828786    .6783521
------------------------------------------------------------------------------
LR test vs. linear regression:       chi2(3) =   104.17   Prob > chi2 = 0.0000
Note: LR test is conservative and provided only for reference
```

The estimates for this model are shown under "Model 3" in table 5.7, where we have repeated the estimates for Model 2 for comparison. We see that girls at a given age are on average estimated to be 0.60 kg lighter than boys. As expected, including `girl` has also reduced the estimated random-intercept standard deviation somewhat from 0.64 kg to 0.59 kg.

Table 5.7: Maximum likelihood estimates for models including both random intercept and slope for children's growth data in reduced-form notation (in kilograms)

	Model 2		Model 3		Model 4	
	Est	(SE)	Est	(SE)	Est	(SE)
Fixed part						
β_1 [_cons]	3.49	(0.14)	3.79	(0.17)	3.75	(0.17)
β_2 [age]	7.70	(0.24)	7.70	(0.24)	7.81	(0.25)
β_3 [age2]	−1.66	(0.09)	−1.66	(0.09)	−1.66	(0.09)
β_4 [girl]			−0.60	(0.20)	−0.50	(0.21)
β_5 [girl×age]					−0.23	(0.17)
Random part						
$\sqrt{\psi_{11}}$	0.64		0.59		0.59	
$\sqrt{\psi_{22}}$	0.50		0.51		0.50	
ρ_{21}	0.27		0.16		0.19	
$\sqrt{\theta}$	0.58		0.57		0.57	
Log likelihood	−258.08		−253.87		−252.99	

5.11.3 Two-stage model formulation

We now express the random-coefficient model in (5.7) using the two-stage formulation described in section 4.9. The level-1 model is written as

$$y_{ij} = \eta_{1j} + \eta_{2j}x_{ij} + \beta_3 x_{ij}^2 + \epsilon_{ij}$$

where the intercept η_{1j} and slope η_{2j} are child-specific coefficients. The level-2 model has these coefficients as responses

$$\begin{aligned} \eta_{1j} &= \gamma_{11} + \gamma_{12}w_j + \zeta_{1j} \\ \eta_{2j} &= \gamma_{21} + \zeta_{2j} \end{aligned} \tag{5.8}$$

where `girl` (w_j) is a covariate only in the intercept equation. As usual, ζ_{1j} and ζ_{2j} are assumed to have a bivariate normal distribution with zero means and unstructured covariance matrix.

Substituting the level-2 models into the level-1 model, we obtain the *reduced form*

$$\begin{aligned} y_{ij} &= \underbrace{\gamma_{11} + \gamma_{12}w_j + \zeta_{1j}}_{\eta_{1j}} + \underbrace{(\gamma_{21} + \zeta_{2j})}_{\eta_{2j}} x_{ij} + \beta_3 x_{ij}^2 + \epsilon_{ij} \\ &= \gamma_{11} + \gamma_{21}x_{ij} + \beta_3 x_{ij}^2 + \gamma_{12}w_j + \zeta_{1j} + \zeta_{2j}x_{ij} + \epsilon_{ij} \\ &\equiv \beta_1 + \beta_2 x_{ij} + \beta_3 x_{ij}^2 + \beta_4 w_j + \zeta_{1j} + \zeta_{2j}x_{ij} + \epsilon_{ij} \end{aligned}$$

where $\beta_1 \equiv \gamma_{11}$, $\beta_2 \equiv \gamma_{21}$, and $\beta_4 \equiv \gamma_{12}$. This reduced-form model is equivalent to the model in (5.7).

In the two-stage formulation, a natural extension of this model is to include `girl` as a covariate also in the slope equation or level-2 model for η_{2j} by adding the term $\gamma_{22}w_j$. This results in a *cross-level interaction* $\gamma_{22}w_jx_{ij}$ in the reduced form.

The model can be fitted using

```
. generate age_girl = age*girl
. xtmixed weight age age2 girl age_girl|| id: age, cov(unstructured) mle
Mixed-effects ML regression                     Number of obs      =       198
Group variable: id                              Number of groups   =        68

                                                Obs per group: min =         1
                                                               avg =       2.9
                                                               max =         5

                                                Wald chi2(4)       =   2023.53
Log likelihood = -252.99486                     Prob > chi2        =    0.0000

------------------------------------------------------------------------------
      weight |      Coef.   Std. Err.      z    P>|z|     [95% Conf. Interval]
-------------+----------------------------------------------------------------
         age |   7.814711   .2526442    30.93   0.000     7.319538    8.309885
        age2 |  -1.658569   .0879161   -18.87   0.000    -1.830881   -1.486257
        girl |  -.5040552     .20718    -2.43   0.015    -.9101206   -.0979898
    age_girl |  -.2303089   .1731562    -1.33   0.183    -.5696889     .109071
       _cons |   3.748607   .1682409    22.28   0.000     3.418861    4.078353
------------------------------------------------------------------------------

------------------------------------------------------------------------------
  Random-effects Parameters  |   Estimate   Std. Err.     [95% Conf. Interval]
-----------------------------+------------------------------------------------
id: Unstructured             |
                     sd(age) |   .4969465   .0875711      .3518133    .7019514
                   sd(_cons) |   .5890282    .129119      .3833072    .9051597
             corr(age,_cons) |    .187022   .3361449     -.4569598     .702369
-----------------------------+------------------------------------------------
                sd(Residual) |   .5729781   .0498678      .4831207    .6795485
------------------------------------------------------------------------------
LR test vs. linear regression:       chi2(3) =   104.77   Prob > chi2 = 0.0000
Note: LR test is conservative and provided only for reference.
```

and the estimates are shown under "Model 4" in table 5.7. Since the interaction is not significant at the 5% level, we will return to the previous model (Model 3) in the next section.

5.12 Prediction of trajectories for individual children

After estimation with `xtmixed`, the predicted trajectories, based on substituting empirical Bayes predictions for the random intercepts ζ_{1j} and random slopes ζ_{2j}, can be obtained using `predict` with the `fitted` option (after first refitting the model without the cross-level interaction):

```
. quietly xtmixed weight age age2 girl || id: age, cov(unstructured) mle
. predict traj, fitted
```

We can plot these predicted trajectories together with the observed ones using a *trellis graph*, a graph containing a separate two-way plot for each subject. This is accomplished using the `by(id)` option. For girls, we use the command

```
. twoway (scatter weight age) (line traj age, sort) if girl==1,
> by(id, compact legend(off))
```

and similarly for boys. The graphs are shown in figures 5.5 and 5.6 for girls and boys, respectively, and suggest that the model fits reasonably well.

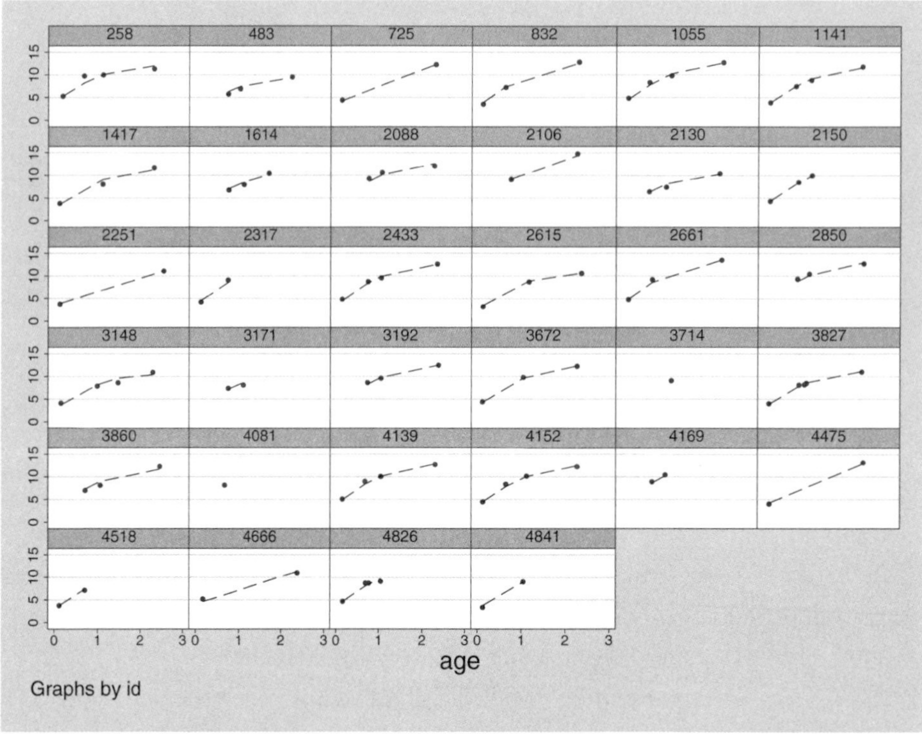

Figure 5.5: Trellis graph of observed responses (dots) and fitted trajectories (dashed lines) for girls

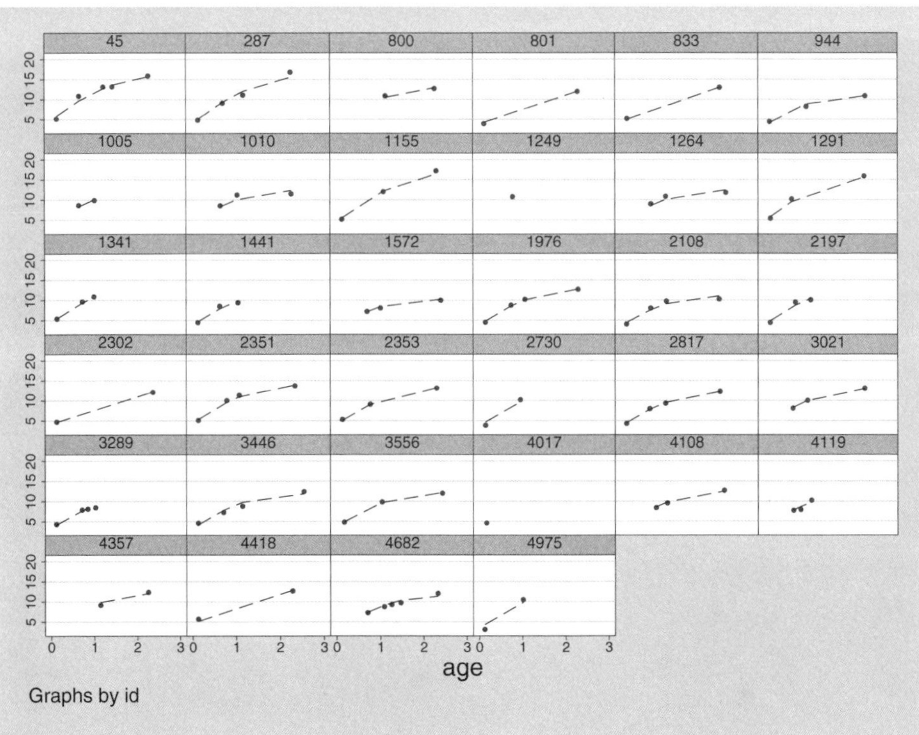

Figure 5.6: Trellis graph of observed responses (dots) and fitted trajectories (dashed lines) for boys

We have not included a random slope for x_i^2 because we only have a few observations per subject, and it would be overly ambitious to allow each subject's growth trajectory to have a different curvature. This model would include seven parameters for the random part instead of four. We can also see from the trellis graph that the model without a random coefficient of x_i^2 appears to fit adequately. In general, it is perfectly reasonable to allow only the lower-order terms of the polynomial used in the fixed part of the model to vary randomly between subjects.

5.13 Prediction of mean growth trajectory and 95% band

We can also plot the estimated population mean trajectory for boys given by the fixed part of the model together with the limits of the band within which 95% of the subject-specific trajectories are expected to lie. For the latter, we must calculate the estimated standard deviation of $\zeta_{1j} + \zeta_{2j}x_{ij}$ given by $\sqrt{\widehat{\psi}_{11} + 2\widehat{\psi}_{21}x_{ij} + \widehat{\psi}_{22}x_{ij}^2}$. We can display the required estimated variances (ψ_{11} and ψ_{22}) and covariance (ψ_{21}) using

```
. estat recovariance

Random-effects covariance matrix for level id

             |       age      _cons
-------------+----------------------
         age |  .2598032
       _cons |  .0476259   .3537049
```

and then use the following `twoway function` command:

```
. twoway (function Weight = _b[_cons] + _b[age]*x + _b[age2]*x^2,
>         range(0.1 2.6) lwidth(medium))
>        (function upper = _b[_cons] + _b[age]*x + _b[age2]*x^2
>         + 2*sqrt(0.354+2*0.0476*x+0.2598*x^2), range(0.1 2.6) lpat(dash))
>        (function lower = _b[_cons] + _b[age]*x + _b[age2]*x^2
>         - 2*sqrt(0.354+2*0.0476*x+0.2598*x^2), range(0.1 2.6) lpat(dash)),
> legend(order(1 "Mean" 2 "Limits of 95% trajectory band"))
> xtitle(Age in years) ytitle(Weight in Kg)
```

This produces figure 5.7 for boys. To obtain the corresponding figure for girls, add `_b[girl]` to the fixed part of the model.

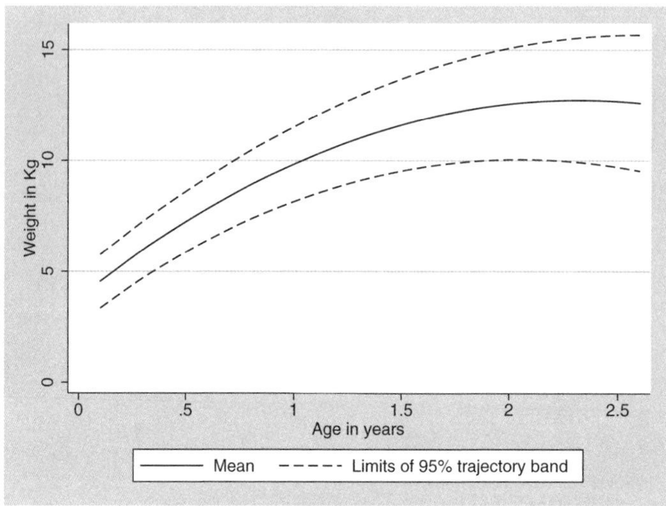

Figure 5.7: Mean trajectory and band within which 95% of subject-specific trajectories lie for boys

5.14 ❖ Complex level-1 variation or heteroskedasticity

In all the models considered so far, we have assumed that the random intercept, random slope, and level-1 residual all have constant variance for all subjects. However, it is sometimes necessary to allow variances to depend on covariates.

Here we consider allowing the level-1 residual variance θ to differ between boys and girls. At the time of writing this book, such complex level-1 variation can be specified only in `gllamm`, by using the `s()` option. Specifically, to allow the variance to depend on two covariates x_{1ij} and x_{2ij}, a linear model for the log standard deviation can be used

$$\ln(\sqrt{\theta_{ij}}) = \ln(\theta_{ij})/2 \;=\; \delta_1 x_{1ij} + \delta_2 x_{2ij}$$

(The Greek letter δ is pronounced delta.) The reason the model is specified for the log of the standard deviation is that this forces the corresponding standard deviation and variance to be positive.

To allow different variances for boys and girls, we can simply include dummy variables for boys and girls in this model. We therefore define the equation:

```
. generate boy = 1 - girl
. eq het: girl boy
```

`gllamm` will not automatically include an intercept in the model for the log standard deviation.

We then define equations for the random intercept and slope:

```
. generate cons = 1
. eq inter: cons
. eq slope: age
```

The equation for the log standard deviation is then passed to `gllamm` using the `s()` option:

```
. gllamm weight age age2 girl, i(id) nrf(2) eqs(inter slope) s(het) adapt
number of level 1 units = 198
number of level 2 units = 68

Condition Number = 9.6099373

gllamm model

log likelihood = -252.40553
```

weight	Coef.	Std. Err.	z	P>|z|	[95% Conf.	Interval]
age	7.629066	.2368424	32.21	0.000	7.164863	8.093268
age2	-1.635112	.0874551	-18.70	0.000	-1.806521	-1.463703
girl	-.6026742	.2049726	-2.94	0.003	-1.004413	-.2009352
_cons	3.829123	.1757714	21.78	0.000	3.484617	4.173629

(*Continued on next page*)

```
Variance at level 1
------------------------------------------------------------------------------

    equation for log standard deviation:

    girl: -.70562727 (.11784508)
    boy: -.44672798 (.1090643)

Variances and covariances of random effects
------------------------------------------------------------------------------


***level 2 (id)

    var(1): .39329069 (.14805678)
    cov(2,1): .04438909 (.08252849) cor(2,1): .145719

    var(2): .23594268 (.08638892)
------------------------------------------------------------------------------
```

The log standard deviations are estimated as −0.706 for girls and −0.447 for boys, corresponding to these variances:

```
. display exp(-.706*2)
.24365548
. display exp(-.447*2)
.40901641
```

The common variance θ for the corresponding model that does not allow the variances to differ between the genders ("Model 3" in table 5.7) was estimated as 0.33, a value between these two estimates.

We see from Model 3 that the log likelihood for the model not allowing for heteroskedasticity at level 1 was −253.87. Using a likelihood-ratio test we obtain

```
. display chi2tail(1,2*(253.87-252.41))
.08748786
```

There is thus little evidence that the level-1 variance differs between boys and girls.

To allow the variance of a random intercept to vary according to some level-2 covariate x_j, we could simply specify a random coefficient of x_j in addition to a random intercept and view the total residual, excluding the level-1 residual ϵ_{ij}, as a heteroskedastic random intercept. The variance of this heteroskedastic intercept $\zeta_{1j}+\zeta_{2j}x_j$ becomes $\psi_{11}+2\psi_{21}x_j+\psi_{22}x_j^2$.

5.15 Summary and further reading

In this chapter, we have discussed different approaches for handling the ubiquitous dependence among responses in longitudinal data. Models described and applied include fixed-effects and repeated measures ANOVA models, random-intercept and random-coefficient (growth curve) models, marginal models, and autoregressive-response models.

Which approach is adopted depends largely on the discipline. In medicine and biostatistics, random-effects and marginal models are most common, whereas random-effects models are most popular in most of the social sciences. In economics, fixed-effects and lagged-response or so-called "dynamic models" appear to be predominant. Two-way error-components models, to be discussed in chapter 11, are also sometimes used. Repeated measures ANOVA is used mostly for experimental designs in areas such as agriculture and psychology, but is increasingly being replaced by random-effects models.

Random-effects or multilevel models have the great advantage of "explaining" the dependence among repeated responses for a subject in terms of "interindividual differences in intraindividual change". In contrast, marginal models treat the dependence as a nuisance, handling it mostly for obtaining correct standard errors for the fixed part of the model. In linear models, both approaches can give nearly identical estimates of fixed regression coefficients and their standard errors. However, as we will see in chapter 6, inferences for regression coefficients can differ substantially between marginal and random-effects approaches in models for dichotomous and other noncontinuous response types. A weakness of both marginal and random-effects approaches is the possibility of subject-level confounding, a problem that can be eliminated in fixed-effects models. As discussed in chapter 3, subject-level confounding can also be eliminated within random-effects models by including subject means of time-varying covariates with the advantage that the effects of between-subject covariates can still be estimated. Lagged-response models should in general be used only if dependence on previous responses is of substantive interest.

Good books on longitudinal data include Wooldridge (2002) in econometrics; Frees (2004) in social science; and Diggle et al. (2002), Fitzmaurice, Laird, and Ware (2004), and Fitzmaurice et al. (2008) in statistics/biostatistics. We also recommend Skrondal and Rabe-Hesketh (2008), which reviews multilevel and related models for longitudinal data.

Introductions to growth-curve models include the books by Singer and Willett (2003) and Bollen and Curran (2006) and the encyclopedia entry by Singer and Willett (2005). The books also show how the models can be expressed as structural equation models. For latent-class or finite-mixture versions of growth-curve models, see Hancock and Samuelsen (2007).

5.16 Exercises

5.1 Wage-panel data

1. Using the same covariates as in this chapter, fit a random-intercept model that contains the lag-1 response as an additional covariate.
   a. Use `xtmixed`.
   b. ❖ Find out how to use `xtabond` and compare the results with those of `xtmixed`.

5.2 Children's growth data

1. Fit the model in (5.7) using `xtmixed`.
2. For the model and estimates from step 1, obtain empirical Bayes (EB) predictions for the random intercepts and slopes.
3. Perform some residual diagnostics.
4. ❖ Also obtain ML (or OLS) estimates of ζ_{1j} and ζ_{2j} by first by subtracting the predicted fixed part of the model and then by using the `statsby` command. Merge these estimates into the data. Compare the ML estimates with the EB predictions using scatterplots by sex.

5.3 Jaw-growth data

1. For the jaw-growth data `growth.dta` of exercise 3.4, extend the model considered there to investigate whether there is significant between-subject variability in the growth rate.
2. For the chosen model (with or without a random slope), plot the fitted growth trajectories by sex and compare them with the corresponding observed growth trajectories.

5.4 Antisocial-behavior data

Consider the data from Allison (2005) that are described in exercise 3.9.

1. Find out if there are any missing data of the response variable `anti`.
2. Calculate a new variable, `time`, taking the values 0, 1, and 2 for the three time points.
3. Fit an ordinary linear regression model for `anti` with `time`, `pov`, `momage`, `female`, `childage`, `hispanic`, `black`, `momwork`, and `married` as covariates. Specify robust standard errors that take the clustering of the data into account.
4. Use `xtgee` to fit models with the same covariates as in step 3 but with the following correlation structures: unstructured, exchangeable, and AR(1). Obtain the corresponding correlation matrices and estimated variances.
5. Fit a random-coefficient model with the same covariates as in step 3 and with a child-specific random intercept and slope of `time`
6. ❖ Derive the three residual variances and correlations implied by the random-coefficient model and compare them with the estimates from step 4.

5.5 Diffusion-of-innovations data

Caudill, Ford, and Kaserman (1995) investigated whether certificate-of-need regulation has an effect in the diffusion of innovations. Specifically, they considered the adoption of hemodialysis (blood filtering) for kidney failure in 50 U.S. states between 1977 and 1990.

Many states implemented certificate-of-need review of dialysis clinics investments in the late 1970s and early 1980s, and many states eliminated such a review in the late 1980s. This change in policy allowed Caudill, Ford, and Kaserman to examine whether certificate-of-need regulation has slowed the rate of diffusion of hemodialysis technology.

They let L_i be the number of dialysis machines in state i in the most recent period in the sample and P_{it} be the number of dialysis machines in state i at time t. Using the transformation $\ln\{P_{it}/(L_i - P_{it})\}$, Caudill, Ford, and Kaserman specified the following model (using their notation)

$$\ln\{P_{it}/(L_i - P_{it})\} \;=\; a_i + c_i T_t + d_i T_t \times \text{Con}_{it} + \epsilon_{it}$$

where T_t is an index of the time period that begins at $T_1 = 1$ in 1977 and Con_{it} is a dummy variable indicating that certificate-of-need regulation was in effect in state i at time t.

The random intercept a_i and the random coefficients c_i and d_i are assumed to be independently normally distributed with means μ_a, μ_c, and μ_d, respectively, and variances σ_a^2, σ_c^2, and σ_d^2, respectively, whereas ϵ_{it} is normally distributed (and independent of a_i, c_i, and d_i) with zero mean and variance σ_ϵ^2.

The dataset `data.cfk` has the following variables:

- `state`: an identifier for the U.S. states
- `T`: the time index T_t
- `TCon`: the interaction $T_t \times \text{Con}_{it}$
- `resp`: the response variable $\ln\{P_{it}/(L_i - P_{it})\}$

1. Read the data using the `infile` command (the variables are in the same order as listed above).
2. Fit the model specified by Caudill, Ford, and Kaserman using the `xtmixed` command. They specified the random effects as uncorrelated. The random effects have nonzero means, whereas `xtmixed` assumes zero means. You can accommodate the means in the fixed part of the model.
3. Do the estimates suggest that certificate-of-need legislation slows the rate of diffusion of hemodialysis technology?
4. If you can get hold of the paper, compare your estimates with the maximum likelihood estimates in table 1 of Caudill, Ford, and Kaserman (1995).
5. Perform a likelihood-ratio test for the null hypothesis

   $$\text{H}_0\text{: } \sigma_c^2 = \sigma_d^2 = 0$$

   The test is conservative because the null hypothesis involves two parameters on the border of the parameter space.
6. Does the conclusion for the effect of certificate-of-need-legislation change when the restricted model (with $\sigma_c^2 = \sigma_d^2 = 0$) is used?
7. Perform residual diagnostics for the selected model, as described in section 3.9.

5.6 Unemployment-claims data

Papke (1994) considered panel data for 1980–1988 from Indiana's enterprise zone program, which provided tax credits for cities with high unemployment and high poverty levels. One of the purposes of the paper was to investigate whether inclusion in an enterprise zone would reduce unemployment claims.

The dataset `ezunem.dta` supplied by Wooldridge (2002) contains the following variables:

- `city`: city identifier (j)
- `year`: year (i)
- `uclms`: number of unemployment claims (y_{ij})
- `ez`: dummy variable for city being in an enterprise zone (x_{2ij})
- `t`: time 1, ..., 9 (x_{3i})

1. What are the benefits of using a panel design for assessing the effects of the enterprise zone program?
2. Use `xtmixed` to fit the random-intercept model

$$\ln(y_{ij}) = \alpha_i + \beta_2 x_{2ij} + \zeta_j + \epsilon_{ij}$$

   where α_i is a fixed year-specific intercept and ζ_j is a normally distributed city-specific random intercept.
3. Interpret the estimated regression coefficients and random-intercept variance. Does the enterprise zone program reduce unemployment claims?
4. Fit the random-coefficient model (with the usual assumptions)

$$\ln(y_{ij}) = \alpha_i + \beta_2 x_{2ij} + \zeta_{1j} + \zeta_{2j} x_{3i} + \epsilon_{ij}$$

5. Interpret the estimated regression coefficients, random-intercept variance, random-slope variance, and correlation between the intercepts and slopes.
6. Perform a likelihood-ratio test to compare random-coefficient and random-intercept models. Which model would you retain at the 5% level?

5.7 Adolescent-alcohol-use data

Singer and Willett (2003) analyzed a dataset from Curran, Hartford, and Muthén (1996). In a longitudinal study, 82 adolescents were interviewed yearly from ages 14–16 and asked about their alcohol consumption during the previous year. Specifically, they were asked to rate the frequency of each of the following on an 8-point scale:

1. drinking wine or beer
2. drinking hard liquor
3. drinking five or more drinks in a row
4. getting drunk

Following Singer and Willett, we will use the square root of the mean of these four items as the response variable. At age 14, the adolescents were also asked what proportion of their peers drink alcohol occasionally and regularly, with each answer scored on a 6-point rating scale. The square root of the mean of these two items will be used as a covariate.

The dataset `alcuse.dta` has the following variables:

- `id`: identifier for the adolescents
- `alcuse`: frequency of alcohol use (square root of mean on 4 alcohol items)
- `age_14`: age − 14, number of years since first interview (t_i)
- `coa`: dummy variable for being a child of an alcoholic (w_{1j})
- `peer`: alcohol use among peers at age 14 (square root of mean of 2 items) w_{2j}

Using Singer and Willett's two-stage formulation (in their notation except that i is occasion and j is subject), consider the following level-1 model

$$Y_{ij} = \pi_{0j} + \pi_{1j} t_i + \epsilon_{ij}$$

and the following level-2 models

$$\begin{aligned} \pi_{0j} &= \gamma_{00} + \gamma_{01} w_{1j} + \gamma_{02} w_{2j} + \zeta_{0j} \\ \pi_{1j} &= \gamma_{10} + \gamma_{11} w_{1j} + \gamma_{12} w_{2j} + \zeta_{1j} \end{aligned}$$

1. Substitute the level-2 models into the level-1 model to obtain the reduced-form model.
2. Interpret each of the parameters in terms of initial status at age 14, π_{0j}, and the rate of growth π_{1j}.
3. Fit the model by restricted maximum likelihood using `xtmixed` and interpret the estimates.
4. Separately for children of alcoholics and other children, plot the fitted trajectories together with the data using trellis graphs.
5. ❖ Obtain the estimated marginal variances and correlation matrix of the total residuals.

5.8 Variance of total residual in random-coefficient model

Consider a random-coefficient model with a random intercept and slope of time, where time takes on the values $t_{1j} = 0$, $t_{2j} = 1$, and $t_{3j} = 2$. The covariance matrix of the intercept and slope is estimated as

$$\widehat{\boldsymbol{\Psi}} = \begin{bmatrix} 4 & 1 \\ 1 & 2 \end{bmatrix}$$

and the level-1 residual variance is estimated as

$$\widehat{\theta} = 1$$

1. Calculate the estimated model-implied variance of the total residual

$$\xi_{ij} = \zeta_{1j} + \zeta_{2j}t_{ij} + \epsilon_{ij}$$

   for the three time points.
2. Calculate the estimated model-implied correlation matrix.

Part III

Two-level generalized linear models

6 Dichotomous or binary responses

6.1 Introduction

Dichotomous, or binary, responses are widespread. Examples include being dead or alive, agreeing or disagreeing with a statement, and succeeding or failing to accomplish something. The responses are usually coded as 1 or 0, where 1 can be interpreted as the answer "yes" and 0 as the answer "no" to some question. For instance, in section 6.2, we will consider the employment status of women where the question is whether the women are employed.

We start by briefly reviewing ordinary logistic and probit regression for dichotomous responses, formulating the models as both generalized linear models, as is common in statistics and biostatistics, and as latent-response models, which is common in econometrics and psychometrics. This prepares the ground for a discussion of various approaches for clustered dichotomous data, with special emphasis on random-intercept models. In this setting, the crucial distinction between conditional or subject-specific effects and marginal or population-averaged effects is highlighted and measures of dependence and heterogeneity are described.

We also discuss special features of statistical inference for random-intercept models with clustered dichotomous responses, including maximum likelihood estimation of model parameters, methods for assigning values to random effects, and how to obtain different kinds of predicted probabilities. Other approaches such as fixed-intercept models (conditional maximum likelihood) and generalized estimating equations (GEE) are briefly discussed.

6.2 Single-level models for dichotomous responses

In this section, we will introduce logit and probit models without random effects that are appropriate for datasets without any kind of clustering. For simplicity, we will start by considering just one covariate x_i for unit (e.g., subject) i. The models can be specified either as generalized linear models or as latent-response models. These two approaches and their relationship are described in sections 6.2.1 and 6.2.2, respectively.

6.2.1 Generalized linear model formulation

As in models for continuous responses, we are interested in the expectation (mean) of the response as a function of the covariate. The expectation of a binary (0 or 1) response is just the probability that the response is 1:

$$E(y_i|x_i) \;=\; \Pr(y_i = 1|x_i)$$

In linear regression, the conditional expectation of the response is modeled as a linear function $E(y_i|x_i) = \beta_1 + \beta_2 x_i$ of the covariate (see sec. 1.5). For dichotomous responses, this approach may be problematic because the probability must lie between 0 and 1, whereas regression lines increase (or decrease) indefinitely as the covariate increases (or decreases). Instead, a nonlinear function is specified in one of two ways:

$$\Pr(y_i = 1|x_i) \;=\; h(\beta_1 + \beta_2 x_i)$$

or

$$g\{\Pr(y_i = 1|x_i)\} \;=\; \beta_1 + \beta_2 x_i = \nu_i$$

where ν_i is referred to as the *linear predictor*. These two formulations are equivalent if the function $h(\cdot)$ is the inverse of the function $g(\cdot)$. Here $g(\cdot)$ is known as the *link function* and $h(\cdot)$ as the *inverse link function*, sometimes written as $g^{-1}(\cdot)$.

We have introduced two components of a generalized linear model: the linear predictor and the link function. The third component is the distribution of the response given the covariates. For dichotomous responses, this is always specified as Bernoulli(π_i) or equivalently binomial($1,\pi_i$), and the responses for different units are assumed to be independent given the covariates.

Typical choices of link function are the logit or probit links. In this section, we focus on the logit link, whereas both links are discussed in section 6.2.2. For the logit link, the model can be written as

$$\Pr(y_i = 1|x_i) \;=\; \text{logit}^{-1}(\beta_1 + \beta_2 x_i) \equiv \frac{\exp(\beta_1 + \beta_2 x_i)}{1 + \exp(\beta_1 + \beta_2 x_i)} \tag{6.1}$$

or

$$\text{logit}\,\{\Pr(y_i = 1|x_i)\} \equiv \ln \underbrace{\left\{ \frac{\Pr(y_i = 1|x_i)}{1 - \Pr(y_i = 1|x_i)} \right\}}_{\text{Odds}(y_i=1|x_i)} \;=\; \beta_1 + \beta_2 x_i \tag{6.2}$$

The term in curly braces in (6.2) represents the odds that $y_i = 1$ given x_i, the expected number of 1 responses for each 0 response. The odds or the expected number of successes for each failure is the standard way of representing the chances of winning in gambling. The natural log (ln) of the odds, or logit function of the probability, is equated to the linear predictor. Correspondingly, (6.1) shows that the probability is given by the inverse logit function (sometimes called logistic function) of the linear predictor.

There is no level-1 residual ϵ_i in (6.2) so that the relationship between the probability and the covariate is deterministic. However, the responses are generated by Bernoulli trials and are therefore random. Including a residual leads to an nonidentified model unless the residual is shared between units in a cluster as in the multilevel models considered later in the chapter.

The logit link is appealing because it produces a linear model for the log of the odds, implying a multiplicative model for the odds themselves. If we add one unit to x_i, we must add β_2 to the log odds or multiply the odds by $\exp(\beta_2)$. This can be seen by considering a 1 unit change in x_i from some value a to $a+1$. The corresponding change in the log odds is

$$\begin{aligned}\ln\{\text{Odds}(y_i = 1|x_i = a+1)\} &- \ln\{\text{Odds}(y_i = 1|x_i = a)\} \\ &= \{\beta_1 + \beta_2(a+1)\} - (\beta_1 + \beta_2 a) = \beta_2\end{aligned}$$

Exponentiating both sides, we obtain the odds ratio (OR):

$$\begin{aligned}\exp\left[\ln\{\text{Odds}(y_i = 1|x_i = a+1)\}\right. &- \left.\ln\{\text{Odds}(y_i = 1|x_i = a)\}\right] \\ &= \frac{\text{Odds}(y_i = 1|x_i = a+1)}{\text{Odds}(y_i = 1|x_i = a)} = \exp(\beta_2)\end{aligned}$$

To illustrate logistic regression, we will consider data on married women from the Canadian Women's Labor Force Participation Dataset used by Fox (1997). The dataset `womenlf.dta` contains women's employment status and two explanatory variables:

- `workstat`: employment status
  (0: not working; 1: employed part-time; 2: employed full time)
- `husbinc`: husband's income in $1,000
- `chilpres`: child present in household (dummy variable)

Fox (1997) considered a multiple logistic regression model for a woman being employed (full or part time) versus not working with covariates `husbinc` and `chilpres`

$$\text{logit}\{\Pr(y_i=1|\mathbf{x}_i)\} = \beta_1 + \beta_2 x_{2i} + \beta_3 x_{3i}$$

where $y_i=1$ denotes employment and $y_i=0$ not working, x_{2i} is `husbinc`, x_{3i} is `chilpres`, and $\mathbf{x}_i = (x_{2i}, x_{3i})'$ is a vector containing both covariates.

(*Continued on next page*)

We can fit the model by maximum likelihood using Stata's `logit` command:

```
. use http://www.stata-press.com/data/mlmus2/womenlf
. recode workstat 2=1
. logit workstat husbinc chilpres
Logistic regression                               Number of obs   =        263
                                                  LR chi2(2)      =      36.42
                                                  Prob > chi2     =     0.0000
Log likelihood = -159.86627                       Pseudo R2       =     0.1023

------------------------------------------------------------------------------
    workstat |      Coef.   Std. Err.      z    P>|z|     [95% Conf. Interval]
-------------+----------------------------------------------------------------
     husbinc |  -.0423084   .0197801    -2.14   0.032    -.0810767   -.0035401
    chilpres |  -1.575648   .2922629    -5.39   0.000    -2.148473   -1.002824
       _cons |    1.33583   .3837632     3.48   0.000     .5836677    2.087992
------------------------------------------------------------------------------
```

The estimated coefficients are negative, so the estimated log odds of employment are lower if the husband earns more and if there is a child in the household. At the 5% significance level, we can reject the null hypotheses that the individual coefficients β_2 and β_3 are zero. These estimated coefficients and their estimated standard errors are also given in table 6.1.

Table 6.1: Maximum likelihood estimates for logistic regression model for women's labor force participation

	Est	(SE)	OR=exp(β)	(95% CI)
β_1 [_cons]	1.34	(0.38)		
β_2 [husbinc]	−0.04	(0.02)	0.96	(0.92, 1.00)
β_3 [chilpres]	−1.58	(0.29)	0.21	(0.12, 0.37)

Instead of considering changes in log odds, it is more informative to obtain odds ratios, the exponentiated regression coefficients. This can be achieved by using the `logit` command with the `or` option:

```
. logit workstat husbinc chilpres, or
Logistic regression                               Number of obs   =        263
                                                  LR chi2(2)      =      36.42
                                                  Prob > chi2     =     0.0000
Log likelihood = -159.86627                       Pseudo R2       =     0.1023

------------------------------------------------------------------------------
    workstat | Odds Ratio   Std. Err.      z    P>|z|     [95% Conf. Interval]
-------------+----------------------------------------------------------------
     husbinc |   .9585741   .0189607    -2.14   0.032     .9221229    .9964661
    chilpres |   .2068734   .0604614    -5.39   0.000     .1166622    .3668421
------------------------------------------------------------------------------
```

Comparing women with and without a child at home, whose husbands have the same income, the odds of working are about 5 ($\approx$1/.2068734) times as high for the women who do not have a child at home as for women who do. Within these two groups of women, each \$1,000 increase in husband's income reduces the odds of working by about 4% $\{\approx(1-.9585741)\times 100\%\}$. Although this odds ratio looks less important than the one for `chilpres`, remember that we cannot directly compare the magnitude of the two odds ratios. The odds ratio for `chilpres` represents a comparison of two distinct groups of women, whereas the odds ratio for `husbinc` merely expresses the effect of a \$1,000 increase in the husband's income. A \$10,000 increase would be associated with an odds ratio of $0.959^{10} = 0.66$.

In an attempt to make effects directly comparable and assess the "relative importance" of covariates, some researchers standardize all covariates to have standard deviation 1, thereby comparing the effects of a standard deviation change in each covariate. As discussed in section 1.5, there are many problems with such an approach, one of them being the meaningless notion of a standard deviation change in a dummy variable such as `chilpres`.

The standard errors of exponentiated estimated regression coefficients should generally not be used for confidence intervals or hypothesis tests. Instead, the 95% confidence intervals in the above output were computed by taking the exponentials of the confidence limits for the regression coefficients β:

$$\exp\{\widehat{\beta} \pm 1.96\times \text{SE}(\widehat{\beta})\}$$

In table 6.1, we therefore report estimated odds ratios with 95% confidence intervals instead of standard errors.

To visualize the model, we can produce a plot of the predicted probabilities versus `husbinc`, with separate curves for women with and without children at home. Because the logit link gives the log of the odds [see (6.2)], the predicted odds of working are given by

$$\widehat{\text{Odds}}(y_i = 1|x_{2i}, x_{3i}) \;=\; \exp(\widehat{\beta}_1 + \widehat{\beta}_2 x_{2i} + \widehat{\beta}_3 x_{3i})$$

Using the relationship between odds and probabilities

$$\text{Odds} = \frac{\text{Pr}}{1-\text{Pr}} \qquad \text{and} \qquad \text{Pr} = \frac{\text{Odds}}{1+\text{Odds}}$$

we get the following predicted probability for woman i, often denoted $\widehat{\pi}_i$:

$$\widehat{\pi}_i \;=\; \frac{\exp(\widehat{\beta}_1 + \widehat{\beta}_2 x_{2i} + \widehat{\beta}_3 x_{3i})}{1+\exp(\widehat{\beta}_1 + \widehat{\beta}_2 x_{2i} + \widehat{\beta}_3 x_{3i})} \;=\; \text{logit}^{-1}(\widehat{\beta}_1 + \widehat{\beta}_2 x_{2i} + \widehat{\beta}_3 x_{3i}) \qquad (6.3)$$

The predicted probability is therefore the *inverse logit* of the estimated linear predictor and can be obtained for the women in the dataset using the `predict` command with the `pr` option:

```
. predict prob, pr
. twoway (line prob husbinc if chilpres==0, sort)
> (line prob husbinc if chilpres==1, sort lpatt(dash)),
> legend(order(1 "No child" 2 "Child"))
> xtitle("Husband's income/$1000") ytitle("Probability that wife works")
```

The resulting graph is shown in figure 6.1.

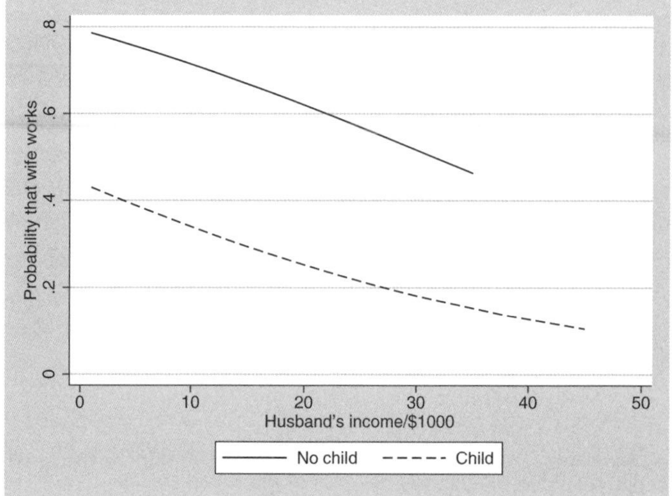

Figure 6.1: Predicted probabilities from logistic regression model (for range of `husbinc` in dataset)

The graph is similar to the graph of the predicted means from an analysis of covariance model (a linear regression model with a continuous and a dichotomous covariate; see section 1.7) except that the curves are not exactly straight. The curves have been plotted for the range of values of `husbinc` observed for the two groups of women, and for these ranges the predicted probabilities are nearly linear functions of `husbinc`.

We will now plot the predicted probabilities for a widely extended range of values of `husbinc` (including negative values, although this does not make sense) to see what the predicted curves look like. This could be accomplished by inventing additional observations with more extreme values of `husbinc` and then using the `predict` command again. More conveniently, we can also use Stata's useful `twoway` plot type, `function`

```
. twoway (function y=invlogit(_b[husbinc]*x+_b[_cons]), range(-100 100))
> (function y=invlogit(_b[husbinc]*x+_b[chilpres]+_b[_cons]),
> range(-100 100) lpatt(dash)),
> xtitle("Husband's income/$1000") ytitle("Probability that wife works")
> legend(order(1 "No child" 2 "Child")) xline(1) xline(45)
```

The estimated regression coefficients are referred to as `_b[husbinc]`, `_b[chilpres]`, and `_b[_cons]`, and we have used Stata's `invlogit()` function to obtain the predicted probabilities as shown in (6.3). The resulting graph is shown in figure 6.2.

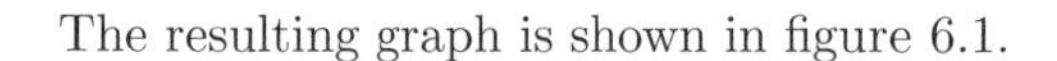

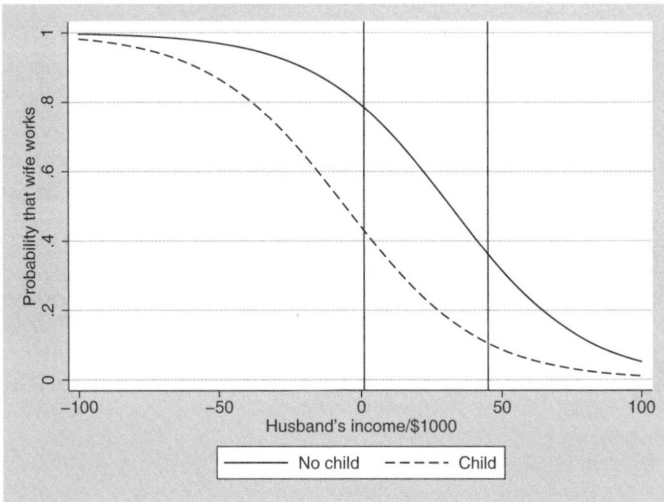

Figure 6.2: Predicted probabilities from logistic regression model (extrapolating beyond the range of `husbinc` in the data)

The range of `husbinc` actually observed in the data lies approximately between the two vertical lines. It would not be safe to rely on predicted probabilities extrapolated outside this range. The curves are approximately linear in the region where the linear predictor is close to zero (and the predicted probability is close to 0.5) and then flatten as the linear predictor becomes extreme. This flattening ensures that the predicted probabilities remain in the permitted interval from 0 to 1.

We can fit the same model using the `glm` command for generalized linear models. The syntax is the same as that of the `logit` command, except that we must specify the logit link function in the `link()` option and the binomial distribution in the `family()` option:

```
. glm workstat husbinc chilpres, link(logit) family(binom)
Generalized linear models                          No. of obs      =        263
Optimization     : ML                              Residual df     =        260
                                                   Scale parameter =          1
Deviance         =  319.7325378                    (1/df) Deviance =   1.229741
Pearson          =  265.9615312                    (1/df) Pearson  =   1.022929
Variance function: V(u) = u*(1-u)                  [Bernoulli]
Link function    : g(u) = ln(u/(1-u))              [Logit]
                                                   AIC             =   1.238527
Log likelihood   = -159.8662689                    BIC             =  -1129.028

------------------------------------------------------------------------------
             |                 OIM
    workstat |      Coef.   Std. Err.      z    P>|z|     [95% Conf. Interval]
-------------+----------------------------------------------------------------
     husbinc |  -.0423084   .0197801    -2.14   0.032    -.0810768   -.0035401
    chilpres |  -1.575648   .2922629    -5.39   0.000    -2.148473   -1.002824
       _cons |    1.33583   .3837634     3.48   0.000     .5836674    2.087992
------------------------------------------------------------------------------
```

To obtain estimated odds ratios, we use the `eform` option (for "exponentiated form"), and to fit a probit model, we simply change the `link(logit)` option to `link(probit)`.

6.2.2 Latent-response formulation

The logistic regression model and other models for dichotomous responses can also be viewed as latent-response models. Underlying the observed dichotomous response y_i (whether the woman works or not), there is an unobserved or latent continuous response y_i^*, representing the propensity to work or the excess utility of working as compared with not working. If this latent response is greater than 0, the observed response is 1:

$$y_i = \begin{cases} 1 & \text{if } y_i^* > 0 \\ 0 & \text{otherwise} \end{cases}$$

For simplicity, we will assume that there is one covariate x_i. A linear regression model is then specified for the latent response y_i^*

$$y_i^* = \beta_1 + \beta_2 x_i + \epsilon_i$$

where ϵ_i is a residual error term with $E(\epsilon_i|x_i) = 0$ and the error terms of different women i are independent.

The latent-response formulation has been used in various disciplines and applications. In genetics, where y_i is often a phenotype or qualitative trait, y_i^* is called a *liability*. In discrete-choice settings, y_i^* is the difference in utilities between alternatives. For attitudes measured by agreement or disagreement with statements, the latent response can be thought of as a "sentiment" in favor of the statement.

Logistic regression

In logistic regression, ϵ_i is assumed to have a logistic cumulative density function given x_i,

$$\Pr(\epsilon_i < \tau|x_i) = \frac{\exp(\tau)}{1+\exp(\tau)}$$

which has mean zero and variance $\pi^2/3 \approx 3.29$ (note that π here represents the famous mathematical constant 'pi').

Figure 6.3 illustrates the equivalence between the latent-response formulation shown in the lower graph and the generalized linear model formulation in terms of a logistic curve for the probability that $y_i = 1$ in the upper graph. The regression line in the lower graph represents the conditional expectation of y_i^* given x_i as a function of x_i, and the density curves represent the conditional distributions of y_i^* given x_i. The dotted horizontal line at $y_i^* = 0$ represents the threshold so that $y_i = 1$ if y_i^* exceeds the threshold and $y_i = 0$ otherwise. Therefore, the areas under the parts of the density curves that lie above the dotted line, here shaded grey, represent the probabilities that $y_i = 1$ given x_i. For the value of x_i indicated by the vertical dotted line, the mean of y_i^* is 0, and therefore half the area of the density lies above the threshold, and the probability curve equals 0.5 at that point.

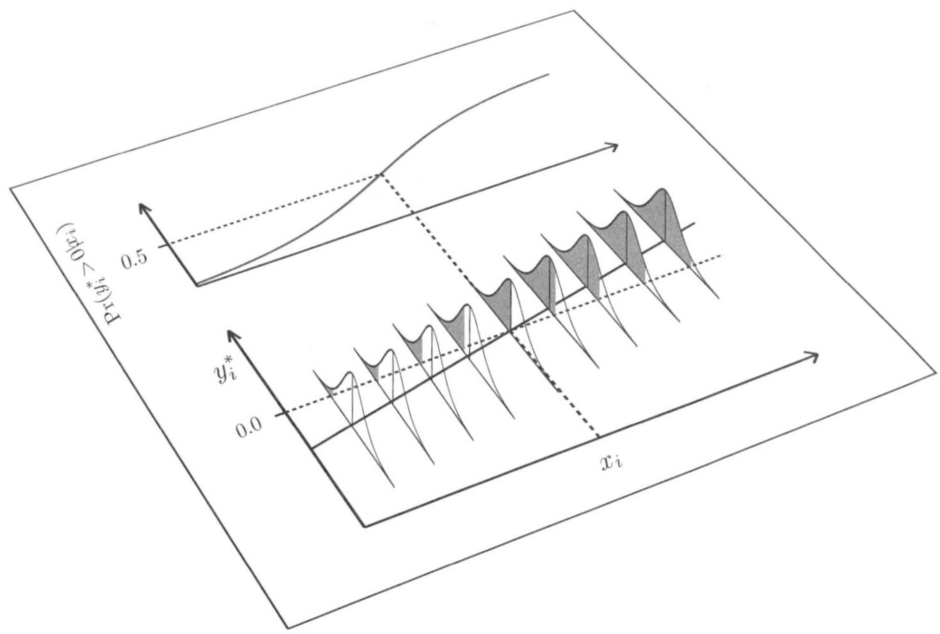

Figure 6.3: Illustration of equivalence of latent-response and generalized linear model formulations for logistic regression

Probit regression

When a latent-response formulation is used, it seems natural to assume that ϵ_i has a normal distribution given x_i, as is usually done in linear regression. If a standard (mean 0 and variance 1) normal distribution is assumed, the model becomes a probit model

$$\begin{aligned}\Pr(y_i\!=\!1|x_i) &= \Pr(y_i^* > 0|x_i) = \Pr(\beta_1 + \beta_2 x_i + \epsilon_i > 0) \\ &= \Pr\{\epsilon_i > -(\beta_1 + \beta_2 x_i)\} = \Pr(-\epsilon_i \le \beta_1 + \beta_2 x_i) \\ &= \Pr(\epsilon_i \le \beta_1 + \beta_2 x_i) = \Phi(\beta_1 + \beta_2 x_i) \qquad (6.4)\end{aligned}$$

Here $\Phi(\cdot)$ is the standard normal cumulative distribution function, the probability that a standard normally distributed random variable (here ϵ_i) is less than the argument. $\Phi(\cdot)$ is the inverse link function $h(\cdot)$, whereas the link function $g(\cdot)$ is $\Phi^{-1}(\cdot)$, the inverse standard normal cumulative distribution function, called the *probit link* function. The penultimate equality in (6.4) exploits the symmetry of the normal distribution.

To understand why a *standard* normal distribution is specified for ϵ_i, with the variance θ fixed at 1, consider the graph in figure 6.4. On the left, the standard deviation is 1, whereas the standard deviation on the right is 2. However, by doubling the slope of the regression line for y_i^* on the right (without changing the point where it intersects the threshold 0), we obtain the same curve for the probability that $y_i=1$. Since we can obtain equivalent models by increasing both the standard deviation and the slope by the same multiplicative factor, the model with a freely estimated standard deviation is not identified.

This lack of identification is also evident from inspecting the expression for the probability if the variance θ were not fixed at 1 [from (6.4)],

$$\Pr(y_i=1|x_i) \;=\; \Pr(\epsilon_i \le \beta_1+\beta_2 x_i) \;=\; \Pr\left(\frac{\epsilon_i}{\sqrt{\theta}} \le \frac{\beta_1+\beta_2 x_i}{\sqrt{\theta}}\right) = \Phi\left(\frac{\beta_1}{\sqrt{\theta}} + \frac{\beta_2}{\sqrt{\theta}} x_i\right)$$

where we see that multiplication of the regression coefficients by a constant can be counteracted by multiplying $\sqrt{\theta}$ by the same constant. This is the reason for fixing the standard deviation in probit models to 1 (see also exercise 6.11). The variance of ϵ_i in logistic regression is also fixed, but to a larger value $\pi^2/3$.

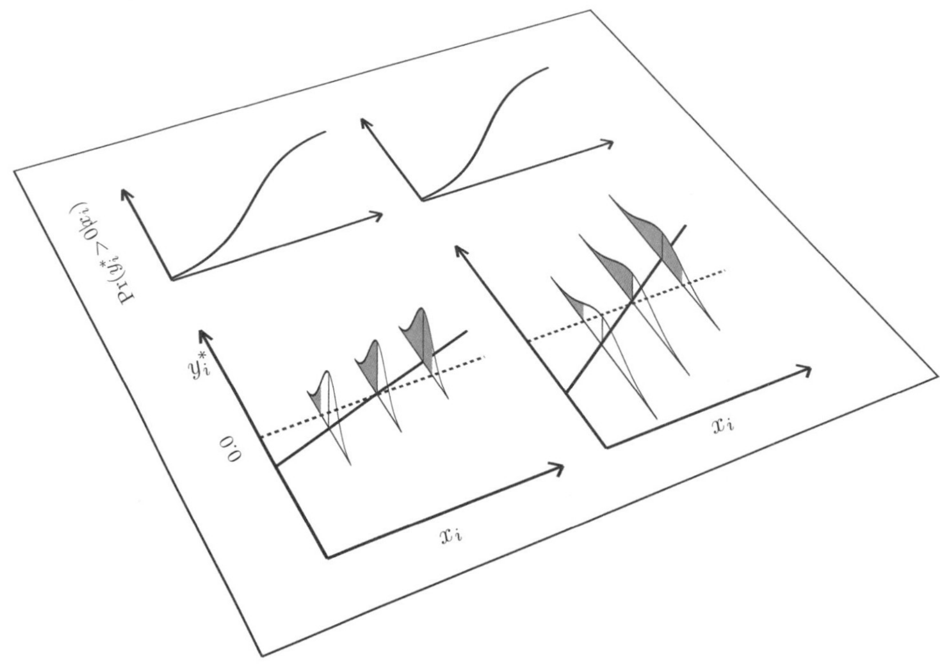

Figure 6.4: Illustration of equivalence between probit models with change in residual standard deviation counteracted by change in slope

A probit model can be fitted to the women's employment data in Stata by using the `probit` command:

```
. probit workstat husbinc chilpres
Probit regression                               Number of obs   =        263
                                                LR chi2(2)      =      36.19
                                                Prob > chi2     =     0.0000
Log likelihood = -159.97986                     Pseudo R2       =     0.1016

    workstat       Coef.   Std. Err.      z    P>|z|     [95% Conf. Interval]

     husbinc   -.0242081   .0114252    -2.12   0.034    -.0466011    -.001815
    chilpres   -.9706164   .1769051    -5.49   0.000    -1.317344   -.6238887
       _cons    .7981508   .2240082     3.56   0.000     .3591028    1.237199
```

These estimates are closer to zero than those reported for the logit model in table 6.1 because the standard deviation of ϵ_i is 1 for the probit model and $\pi/\sqrt{3} \approx 1.81$ for the logit model. Therefore, as we have already seen in figure 6.4, the regression coefficients in logit models must be larger in absolute value to produce nearly the same curve for the probability that $y_i = 1$. Here we say "nearly the same" because the shapes of the probit and logit curves are similar yet not identical. To visualize the subtle difference in shape, we can plot the predicted probabilities for women without children at home from both the logit and probit models:

```
. twoway (function y=invlogit(1.3358-0.0423*x)), range(-100 100))
> (function y=normal(0.7982-0.0242*x), range(-100 100) lpatt(dash)),
> xtitle("Husband's income/$1000") ytitle("Probability that wife works")
> legend(order(1 "Logit link" 2 "Probit link")) xline(1) xline(45)
```

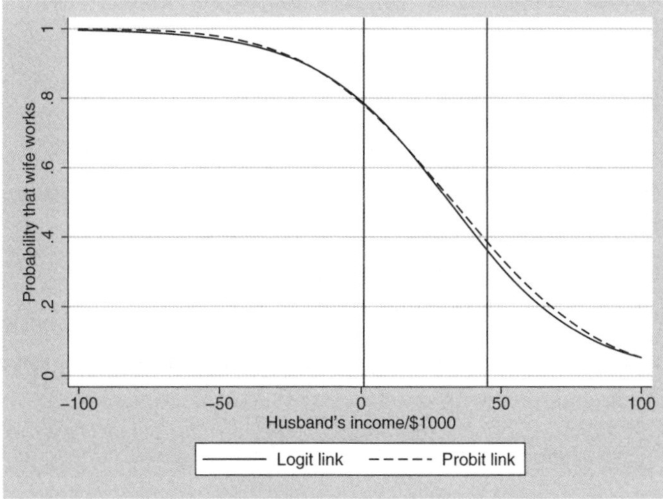

Figure 6.5: Predicted probabilities from logistic and probit regression models for women without children at home

Here the predictions from the models coincide nearly perfectly in the region where most of the data are concentrated and are very similar elsewhere. It is thus futile to attempt to empirically distinguish between the logit and probit links unless one has a huge sample.

6.3 Which treatment is best for toenail infection?

In the previous section, we described conventional modeling of dichotomous responses where it is assumed that the responses are conditionally independent given the covariates $\mathbf{x}_i$. We are now ready to consider multilevel models for *clustered* dichotomous responses, which are dependent even after conditioning on the covariates.

Lesaffre and Spiessens (2001) analyzed data provided by De Backer et al. (1998) from a randomized, double-blind trial of treatments for toenail infection (dermatophyte onychomycosis). Toenail infection is common with a prevalence of about 2% to 3% in the United States, and a much higher prevalence among diabetics and the elderly. The infection is caused by a fungus and does not only disfigure the nails but can also cause physical pain and impair the ability to work.

In this clinical trial, 378 patients were randomly allocated into two oral antifungal treatments (250 mg/day terbinafine and 200 mg/day itraconazole) and evaluated at seven visits, at weeks 0, 4, 8, 12, 24, 36, and 48. One outcome is onycholysis, the degree of separation of the nail plate from the nail bed, which has been dichotomized ("moderate or severe" versus "none or mild") and is available for 294 patients.

The dataset `toenail.dta` contains the following variables:

- `patient`: patient identifier
- `outcome`: onycholysis (separation of nail plate from nail bed) (0: none or mild; 1: moderate or severe)
- `treatment`: treatment group (0: itraconazole; 1: terbinafine)
- `visit`: visit number (1, 2, ..., 7)
- `month`: exact timing of visit in months

The main research question is whether the treatments differ in their efficacy. In other words, do patients receiving one treatment experience a greater decrease in their probability of having onycholysis than those receiving the other treatment?

6.4 Longitudinal data structure

Before investigating the research question, we look at the longitudinal structure of the toenail data. We can use the `xtdescribe` command introduced in chapter 5 because the data were intended to be balanced with seven visits planned for the same set of weeks for each patient (although the exact timing of the visits varied between patients):

```
. use http://www.stata-press.com/data/mlmus2/toenail, clear
. xtdescribe if outcome < ., i(patient) t(visit)
 patient:  1, 2, ..., 383                                    n =        294
   visit:  1, 2, ..., 7                                      T =          7
           Delta(visit) = 1; (7-1)+1 = 7
           (patient*visit uniquely identifies each observation)
Distribution of T_i:   min      5%     25%       50%       75%     95%     max
                         1       3       7         7         7       7       7
     Freq.  Percent    Cum. |  Pattern
 ---------------------------+---------
      224     76.19   76.19 |  1111111
       21      7.14   83.33 |  11111.1
       10      3.40   86.73 |  1111.11
        6      2.04   88.78 |  111....
        5      1.70   90.48 |  1......
        5      1.70   92.18 |  11111..
        4      1.36   93.54 |  1111...
        3      1.02   94.56 |  11.....
        3      1.02   95.58 |  111.111
       13      4.42  100.00 | (other patterns)
 ---------------------------+---------
      294    100.00         |  XXXXXXX
```

The dataset is not balanced since all patients did not attend all planned visits. Specifically, 224 patients have complete data (the pattern "`1111111`"), 21 patients missed the 6th visit ("`11111.1`"), 10 patients missed the 5th visit ("`1111.11`"), and most other patients dropped out at some point, never returning after missing a visit. The latter pattern is sometimes referred to as *monotone missingness* in contrast to *intermittent missingness*, which follows no particular pattern.

As discussed in section 5.9, a nice feature of maximum likelihood estimation for incomplete data such as these is that all information is used. Thus not only patients who attended all visits, but also patients with missing visits contribute information. If the model is correctly specified, maximum likelihood estimates are consistent when the responses are missing at random (MAR).

6.5 Population-averaged or marginal probabilities

A useful graphical display of the data is a bar plot showing the proportion of patients with onycholysis at each visit by treatment group. The following Stata commands can be used to produce the graph shown in figure 6.6:

```
. label define tr 0 "Itraconazole" 1 "Terbinafine"
. label values treatment tr
. graph bar (mean) proportion= outcome, over(visit) by(treatment)
> ytitle(Proportion with onycholysis)
```

Here we defined value labels for `treatment` to make them appear on the graph.

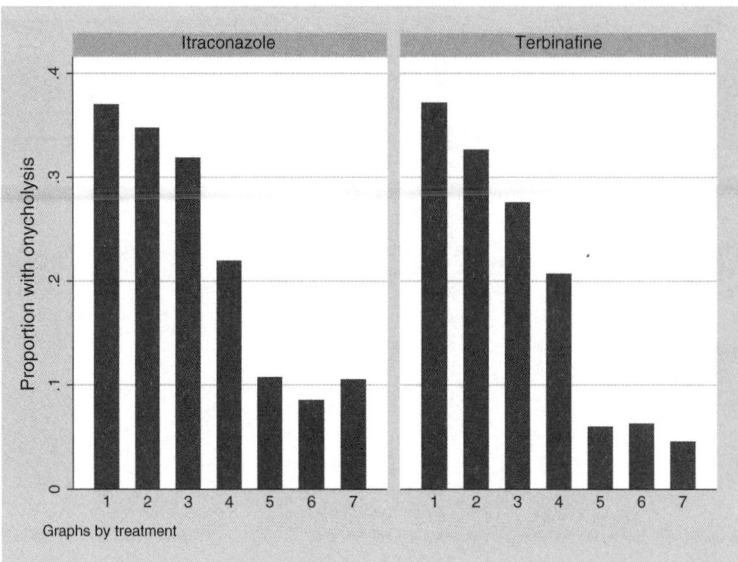

Figure 6.6: Bar plot of proportion of patients with toenail infection by visit and treatment group

We used the visit number `visit` to define the bars instead of the the exact timing of the visit `month` because there would generally not be enough patients with the same timing to reliably estimate the proportions. An alternative display is a line graph, plotting the observed proportions at each visit against time. For this graph, it is better to use the average time associated with each visit for the x axis than using visit number since the visits were not equally spaced. Both the proportions and average times for each visit in each treatment group can be obtained by using the `egen` command with the `mean()` function:

```
. egen prop = mean(outcome), by(treatment visit)
. egen mn_month = mean(month), by(treatment visit)
. twoway line prop mn_month, by(treatment) sort
> xtitle(Time in months) ytitle(Proportion of onycholysis)
```

The resulting graph is shown in figure 6.7.

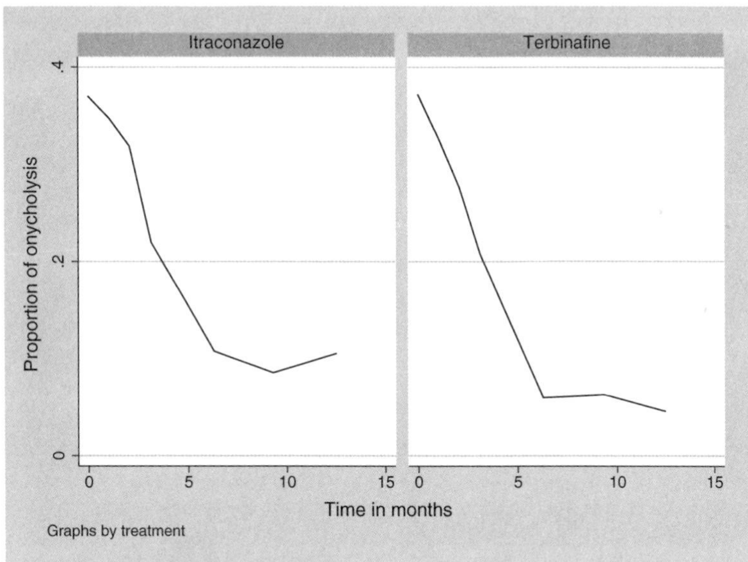

Figure 6.7: Line plot of proportion of patients with toenail infection by visit and treatment group

The proportions shown in figure 6.7 represent the estimated average (or marginal) probabilities of onycholysis given the two covariates: time since randomization and treatment group. We are not attempting to estimate individual patients' personal probabilities, which may vary substantially, but are considering the population averages, given the covariates.

Instead of estimating the probabilities for each combination of `visit` and `treatment`, we can attempt to obtain smooth curves of the estimated probability as a function of time. We then no longer have to group observations for the same visit number together but can directly use the exact timing of the visits. One way to accomplish this is by using a logistic regression model with `month`, `treatment`, and their interaction as covariates. This model for the dichotomous outcome y_{ij} at visit i for patient j can be written as

$$\text{logit}\{\Pr(y_{ij}=1|\mathbf{x}_{ij})\} = \beta_1 + \beta_2 x_{2j} + \beta_3 x_{3ij} + \beta_4 x_{2j} x_{3ij} \tag{6.5}$$

where x_{2j} represents `treatment`, x_{3ij} represents `month`, and $\mathbf{x}_{ij} = (x_{2j}, x_{3ij})'$ is a vector containing both covariates. This model allows for a difference between groups at baseline β_2, and linear changes in the log odds of onycholysis over time with slope β_3 in the itraconazole group and slope $\beta_3 + \beta_4$ in the terbinafine group. Therefore, β_4, the difference in the rate of improvement (on the log odds scale) between treatment groups, can be viewed as the treatment effect (terbinafine versus itraconazole). This model makes the unrealistic assumption that the responses for a given patient are conditionally independent after controlling for the included covariates. We will relax this assumption in the next section.

We now check how well predicted probabilities from the logistic regression model correspond to the observed proportions in figure 6.7. The predicted probabilities are obtained as follows:

```
. generate trt_month = treatment*month
. logit outcome treatment month trt_month, or
Logistic regression                               Number of obs   =       1908
                                                  LR chi2(3)      =     164.47
                                                  Prob > chi2     =     0.0000
Log likelihood = -908.00747                       Pseudo R2       =     0.0830

------------------------------------------------------------------------------
     outcome | Odds Ratio   Std. Err.      z    P>|z|     [95% Conf. Interval]
-------------+----------------------------------------------------------------
   treatment |   .9994185   .1560558    -0.00   0.997     .7359277    1.357249
       month |   .8434052   .0199212    -7.21   0.000     .8052504    .8833679
   trt_month |    .934988   .0350845    -1.79   0.073     .8686913    1.006344
------------------------------------------------------------------------------
. predict prob, p
```

Plotting these together with the observed proportions using the command

```
. twoway (line prop mn_month, sort) (line prob month, sort lpatt(dash)),
> by(treatment) legend(order(1 "Observed proportions" 2 "Fitted probabilities"))
> xtitle(Time in months) ytitle(Probability of onycholysis)
```

results in figure 6.8.

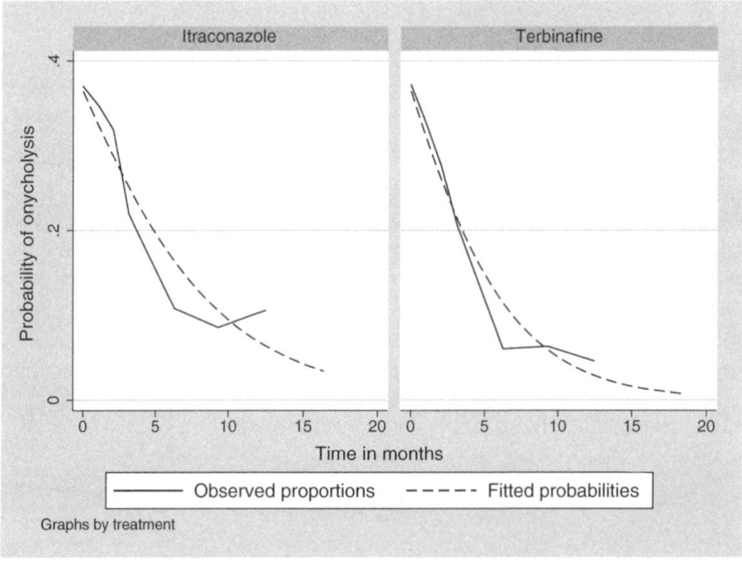

Figure 6.8: Proportions and fitted probabilities using ordinary logistic regression

The marginal probabilities predicted by the model fit the observed proportions well. However, the model assumes that the responses for the same patient are conditionally independent given the covariates, which is likely to be violated. As discussed for linear models in section 3.10.1, the standard errors from ordinary logistic regression are therefore probably not trustworthy. A model for longitudinal or clustered data should capture both the *mean structure* (here the marginal probabilities) as well as the *dependence structure*. So far, we have neglected the second aspect but address it now.

6.6 Random-intercept logistic regression

To relax the assumption of conditional independence among the responses for the same patient given the covariates, we can include a patient-specific random intercept ζ_j in the linear predictor

$$\text{logit}\{\Pr(y_{ij}=1|\mathbf{x}_{ij},\zeta_j)\} = \beta_1 + \beta_2 x_{2j} + \beta_3 x_{3ij} + \beta_4 x_{2j} x_{3ij} + \zeta_j \qquad (6.6)$$

with $\zeta_j|\mathbf{x}_{ij} \sim N(0,\psi)$ and ζ_j independent across patients j, giving a random-intercept logistic regression model. This is a simple example of a *generalized linear mixed model* because it is a generalized linear model with both fixed effects β_1 to β_4 and a random effect ζ_j. The random intercept can be thought of as the combined effect of omitted patient-specific (time-constant) covariates that cause some patients to be more prone to onycholysis than others. It is appealing to model this unobserved heterogeneity in the same way as observed heterogeneity by simply adding the random intercept to the linear predictor.

Model specification is completed by assuming that, given $\pi_{ij} \equiv \Pr(y_{ij}|\mathbf{x}_{ij},\zeta_j)$, y_{ij} are independently distributed as

$$y_{ij}|\pi_{ij} \sim \text{binomial}(1,\pi_{ij})$$

Within a two-stage formulation, Raudenbush and Bryk (2002) refer to this part of the model as the level-1 sampling model and to (6.6) as the structural model (right-hand side) and link function (left-hand side).

Using the latent-response formulation, the model can equivalently be written as

$$y^*_{ij} = \beta_1 + \beta_2 x_{2j} + \beta_3 x_{3ij} + \beta_4 x_{2j} x_{3ij} + \zeta_j + \epsilon_{ij}$$

where $\zeta_j|\mathbf{x}_{ij} \sim N(0,\psi)$ and $\epsilon_{ij}|\mathbf{x}_{ij},\zeta_j$ has a logistic distribution. The ϵ_{ij} are independent across both occasions and patients and independent of ζ_j, and ζ_j is independent across patients.

Confusingly, logistic random-effects models are sometimes written as $y_{ij} = \pi_{ij} + e_{ij}$, where e_{ij} is a heteroscedastic normally distributed level-1 residual. This formulation is inconsistent with the random-intercept logistic regression model and should be avoided. (See Skrondal and Rabe-Hesketh 2007.)

6.7 Estimation of logistic random-intercept models

As of Stata 10, there are three commands for fitting the model in Stata; `xtlogit`, `xtmelogit`, and `gllamm`. All three commands provide maximum likelihood estimation using adaptive quadrature to approximate the integrals involved (see sec. 6.11.1 for more information). `xtlogit` follows essentially the same syntax as `xtreg`, `xtmelogit` follows essentially the same syntax as `xtmixed`, and `gllamm` uses essentially the same syntax for linear, logistic, and other types of models.

All three commands are relatively slow because they use numerical integration, but for random-intercept models `xtlogit` is much faster than `xtmelogit`, which is often faster than `gllamm`. However, the rank ordering is reversed when it comes to the usefulness of the commands for predicting random effects and various types of probabilities as we will see in sections 6.12 and 6.13.

We do not discuss random-coefficient logistic regression in this chapter, but such models can be fitted using `xtmelogit` and `gllamm` (but not using `xtlogit`), using essentially the same syntax as for linear random-coefficient models discussed in section 4.5.1. Also, the probit version of the random-intercept model is available using `xtprobit` or `gllamm`, but random-coefficient probit models are available in `gllamm` only.

6.7.1 Using xtlogit

The `xtlogit` command for fitting the random-intercept model is similar to the `xtreg` command for fitting the corresponding linear model, except that we add the `quad(30)` option to ensure accurate estimates (see sec. 6.11.1):

```
. xtlogit outcome treatment month trt_month, i(patient) quad(30)
Random-effects logistic regression              Number of obs      =      1908
Group variable: patient                         Number of groups   =       294
Random effects u_i ~ Gaussian                   Obs per group: min =         1
                                                               avg =       6.5
                                                               max =         7
                                                Wald chi2(3)       =    150.65
Log likelihood  = -625.38558                    Prob > chi2        =    0.0000

     outcome       Coef.   Std. Err.      z    P>|z|     [95% Conf. Interval]

   treatment   -.160608   .5796716    -0.28   0.782    -1.296744    .9755275
       month   -.390956   .0443707    -8.81   0.000    -.4779209   -.3039911
   trt_month  -.1367758   .0679947    -2.01   0.044     -.270043   -.0035085
       _cons  -1.618795   .4303891    -3.76   0.000    -2.462342   -.7752477

    /lnsig2u   2.775749   .1890237                      2.405269    3.146228

     sigma_u   4.006325   .3786451                      3.328876    4.821641
         rho   .8298976    .026684                      .7710804    .8760322

Likelihood-ratio test of rho=0: chibar2(01) =   565.24 Prob >= chibar2 = 0.000
```

The estimated regression coefficients are given in the usual format. The value next to `sigma_u` represents the estimated standard deviation $\sqrt{\widehat{\psi}}$ of the random intercept and the value next to `rho` represents the estimated residual intraclass correlation of the latent responses (see sec. 6.10.1).

In the itraconazole group (`treatment`=0), the estimated log odds of onycholysis decrease by 0.39 per month. The log odds for the terbinafine group decrease an extra 0.14 per month, giving a downward slope of 0.53. The estimated difference in the slopes of time between the two groups (the coefficient of `trt_month`) can be interpreted as the treatment effect, and this is significant at the 5% level.

We can use the `or` option to obtain exponentiated regression coefficients, which are interpreted as odds ratios here. Instead of refitting the model, we can simply change the way the results are displayed using the following short `xtlogit` command (known as "replaying the estimation results" in Stata parlance):

```
. xtlogit, or

Random-effects logistic regression              Number of obs      =      1908
Group variable: patient                         Number of groups   =       294

Random effects u_i ~ Gaussian                   Obs per group: min =         1
                                                               avg =       6.5
                                                               max =         7

                                                Wald chi2(3)       =    150.65
Log likelihood  = -625.38558                    Prob > chi2        =    0.0000

------------------------------------------------------------------------------
     outcome |         OR   Std. Err.      z    P>|z|     [95% Conf. Interval]
-------------+----------------------------------------------------------------
   treatment |   .8516258   .4936633    -0.28   0.782     .2734207    2.652566
       month |   .6764099   .0300128    -8.81   0.000     .6200712    .7378675
   trt_month |   .8721658   .0593027    -2.01   0.044     .7633467    .9964976
-------------+----------------------------------------------------------------
    /lnsig2u |   2.775749   .1890237                      2.405269    3.146228
-------------+----------------------------------------------------------------
     sigma_u |   4.006325   .3786451                      3.328876    4.821641
         rho |   .8298976    .026684                      .7710804    .8760322
------------------------------------------------------------------------------
Likelihood-ratio test of rho=0: chibar2(01) =   565.24 Prob >= chibar2 = 0.000
```

The estimated odds ratios and their 95% confidence intervals are also given in table 6.2. We see that the estimated odds for a subject in the itraconazole group are multiplied by 0.68 every month and the odds for a subject in the terbinafine group are multiplied by 0.59 (= $.6764099 \times .872165$) every month. In terms of percentage decreases in estimated odds, $100\%(1 - \widehat{\mathrm{OR}})$, the odds decrease 32% per month in the itraconazole group and 41% per month in the terbinafine group.

Table 6.2: Estimates for toenail data

	Marginal effects		Conditional effects	
	Ordinary logistic	GEE logistic	Random int. logistic	Conditional logistic
Parameter	OR (95% CI)	OR (95% CI)†	OR (95% CI)	OR (95% CI)
Fixed part				
$\exp(\beta_2)$ [treatment]	1.00 (0.74, 1.36)	1.01 (0.61, 1.68)	0.85 (0.27, 2.65)	
$\exp(\beta_3)$ [month]	0.84 (0.81, 0.88)	0.84 (0.79, 0.89)	0.68 (0.62, 0.74)	0.68 (0.62, 0.75)
$\exp(\beta_4)$ [trt_month]	0.93 (0.87, 1.01)	0.93 (0.83, 1.03)	0.87 (0.76, 1.00)	0.91 (0.78, 1.05)
Random part				
ψ			16.08	
ρ			0.83	
Log likelihood	−908.01		−625.39	−188.94

†Based on the sandwich estimator

6.7.2 Using xtmelogit

The syntax for `xtmelogit` is similar to that for `xtmixed` except that we also specify the number of quadrature points, or integration points, using the `intpoints()` option:

```
. xtmelogit outcome treatment month trt_month || patient:, intpoints(30)
Mixed-effects logistic regression               Number of obs      =      1908
Group variable: patient                         Number of groups   =       294

                                                Obs per group: min =         1
                                                               avg =       6.5
                                                               max =         7

Integration points =  30                        Wald chi2(3)       =    150.52
Log likelihood = -625.39709                     Prob > chi2        =    0.0000

------------------------------------------------------------------------------
     outcome |      Coef.   Std. Err.      z    P>|z|     [95% Conf. Interval]
-------------+----------------------------------------------------------------
   treatment |  -.1609377   .5842082    -0.28   0.783    -1.305965    .9840893
       month |  -.3910604   .0443958    -8.81   0.000    -.4780744   -.3040463
   trt_month |  -.1368073   .0680236    -2.01   0.044     -.270131   -.0034836
       _cons |  -1.618961   .4347773    -3.72   0.000    -2.471109   -.7668132
------------------------------------------------------------------------------

------------------------------------------------------------------------------
  Random-effects Parameters  |   Estimate   Std. Err.     [95% Conf. Interval]
-----------------------------+------------------------------------------------
patient: Identity            |
                   sd(_cons) |   4.008165   .3813919      3.326217    4.829927
------------------------------------------------------------------------------
LR test vs. logistic regression: chibar2(01) =   565.22 Prob>=chibar2 = 0.0000
```

The results are similar but not identical to those from `xtlogit` because the commands use slightly different versions of adaptive quadrature (see sec. 6.11.1). Since the estimates took some time to obtain, we store them for later use

```
. estimates store xtmelogit
```

Estimated odds ratios can be obtained using the `or` option. `xtmelogit` can also be used with one integration point, which is equivalent to the so-called Laplace approximation. See section 6.11.2 for the results obtained for the toenail data using this method.

6.7.3 Using gllamm

Using `gllamm` for the random-intercept logistic regression model requires that we specify a logit link and binomial distribution using the `link()` and `family()` options (exactly as for the `glm` command). We also use the `nip()` option (for the number of integration points) to request that 30 integration points be used:

```
. gllamm outcome treatment month trt_month, i(patient) link(logit) family(binom)
> nip(30) adapt
number of level 1 units = 1908
number of level 2 units = 294

Condition Number = 23.076299

gllamm model

log likelihood = -625.38558
```

outcome	Coef.	Std. Err.	z	P>|z|	[95% Conf.	Interval]
treatment	-.1608751	.5802054	-0.28	0.782	-1.298057	.9763065
month	-.3911055	.0443906	-8.81	0.000	-.4781095	-.3041015
trt_month	-.136829	.0680213	-2.01	0.044	-.2701484	-.0035097
_cons	-1.620364	.4322408	-3.75	0.000	-2.46754	-.7731873

```
Variances and covariances of random effects
------------------------------------------------------------------------------

***level 2 (patient)

    var(1): 16.084107 (3.0626223)
------------------------------------------------------------------------------
```

The estimates are again similar to those from `xtlogit` and `xtmelogit`. The estimated random-intercept variance is given next to `var(1)` instead of the random-intercept standard deviation reported by `xtlogit` and `xtmelogit`, unless the `variance` option is used for the latter. We store the `gllamm` estimates for later use:

```
. estimates store gllamm
```

We can use the `eform` option to obtain estimated odds ratios or alternatively use the command

```
gllamm, eform
```

after having already fitted the model.

6.8 Inference for logistic random-intercept models

As discussed earlier, we can interpret the regression coefficient β as the difference in log-odds associated with a unit change in the corresponding covariate and the exponentiated regression coefficient as an odds ratio, $\text{OR} = \exp(\beta)$. The relevant null hypothesis for odds ratios usually is H_0: $\text{OR} = 1$, and this corresponds directly to the null hypothesis that the corresponding regression coefficient is zero, H_0: $\beta = 0$.

Wald tests and z tests can be used for regression coefficients just as described in section 3.6.1 for linear models. 95% Wald confidence intervals for individual regression coefficients are obtained using

$$\widehat{\beta} \pm z_{.975}\,\widehat{\text{SE}}(\widehat{\beta})$$

where $z_{.975} = 1.96$ is the 97.5th percentile of the standard normal distribution. The corresponding confidence interval for the odds ratio is obtained by exponentiating both limits of the confidence interval

$$\exp\{\widehat{\beta} - z_{.975}\,\widehat{\text{SE}}(\widehat{\beta})\} \quad \text{to} \quad \exp\{\widehat{\beta} + z_{.975}\,\widehat{\text{SE}}(\widehat{\beta})\}$$

Wald tests for linear combinations of regression coefficients can be used to test the corresponding multiplicative relationships among odds for different covariate values. For instance, for the toenail data, we may want to obtain the odds ratio comparing the treatment groups after 20 months. The corresponding difference in log odds after 20 months is a linear combination of regression coefficients, namely, $\beta_2 + \beta_4 \times 20$ (see sec. 1.8 if this is not clear). We can test the null hypothesis that the difference in log odds is 0 and hence that the odds ratio is 1 using the `lincom` command:

```
. lincom treatment + trt_month*20
 ( 1)  [outcome]treatment + 20 [outcome]trt_month = 0
```

outcome	Coef.	Std. Err.	z	P>|z|	[95% Conf.	Interval]
(1)	-2.897456	1.310367	-2.21	0.027	-5.465727	-.3291841

If we require a confidence interval for the odds ratio after 20 months, we can repeat the `lincom` command but this time with the `or` option, which gives exponentials of the limits of the confidence interval above:

```
. lincom treatment + trt_month*20, or
 ( 1)  [outcome]treatment + 20 [outcome]trt_month = 0
```

outcome	Odds Ratio	Std. Err.	z	P>|z|	[95% Conf.	Interval]
(1)	.0551634	.0722843	-2.21	0.027	.0042293	.7195106

After 20 months of treatment, the odds ratio comparing terbinafine (`treatment`=1) to itraconazole is estimated as 0.055. Such small numbers are difficult to interpret, so we can switch the groups around by taking the reciprocal of the odds ratio, 18 (= 1/0.055), which represents the odds ratio comparing itraconazole to terbinafine. Alternatively, we can always switch the comparison around by simply changing the sign of the corresponding difference in log odds in the `lincom` command

```
lincom -(treatment + trt_month*20), or
```

Multivariate Wald tests can be performed by using `testparm`. Wald tests and confidence intervals can be used with robust standard errors produced by `gllamm` with the `robust` option.

Null hypotheses about individual regression coefficients or several regression coefficients can also be tested using likelihood-ratio tests. Although likelihood-ratio and

Wald tests are asymptotically equivalent, the tests statistics are not identical in finite samples. If the statistics are different, there may be a sparseness problem, for instance with mostly "1" or mostly "0" responses in one of the groups.

Both `xtlogit` and `xtmelogit` provide likelihood-ratio tests for the null hypothesis that the between-cluster variance ψ is zero in the last line of the output. The p-values are based on the correct asymptotic sampling distribution (not the naive χ^2) as described for linear models in section 2.6.2. For the toenail data, the likelihood-ratio statistic is 565.2 giving $p < 0.001$, which suggests that a multilevel model is required.

6.9 Subject-specific vs. population-averaged relationships

The estimated regression coefficients for the random-intercept logistic regression model are more extreme (different from 0) than those for the ordinary logistic regression model (see table 6.2). Correspondingly, the estimated odds ratios are more extreme (different from 1) than those for the ordinary logistic regression model. The reason for this discrepancy is that ordinary logistic regression is fitting overall *population-averaged* probabilities, whereas random-effects logistic regression fits *subject-specific* probabilities for the individual patients.

This important distinction can be seen in the way the two models are written in (6.5) and (6.6). Whereas the former is for the overall or population-averaged probability, conditioning only on covariates, the latter is for the subject-specific probability, given the subject-specific random intercept ζ_j and the covariates. Odds ratios derived from these models can be referred to as population-averaged (although the averaging is applied to the probabilities) or subject-specific odds ratios, respectively. For instance, in the random-intercept model, we can interpret the estimated subject-specific odds ratio of 0.68 for `month` as the odds ratio for each patient in the itraconazole group: each patient's odds decrease 32% per month. In contrast, the estimated population-averaged odds ratio of 0.84 for `month` means that the the odds of having onycholysis *among the patients* in the itraconazole group, decrease 16% per month. Other commonly used terms for population-averaged and subject-specific are *marginal* and *conditional*, respectively

The population-averaged probabilities implied by the random-intercept model can be obtained by averaging the subject-specific probabilities over the random-intercept distribution. Since the random intercepts are continuous, this averaging is accomplished by integration:

$$
\begin{aligned}
\Pr(y_{ij} = 1|x_{2j}, x_{3ij}) \\
&= \int \Pr(y_{ij} = 1|x_{2j}, x_{3ij}, \zeta_j)\phi(\zeta_j; 0, \widehat{\psi})\, d\zeta_j \\
&= \int \frac{\exp(\beta_1 + \beta_2 x_{2j} + \beta_3 x_{3ij} + \beta_4 x_{2j}x_{3ij} + \zeta_j)}{1 + \exp(\beta_1 + \beta_2 x_{2j} + \beta_3 x_{3ij} + \beta_4 x_{2j}x_{3ij} + \zeta_j)} \phi(\zeta_j; 0, \widehat{\psi})\, d\zeta_j \\
&\neq \frac{\exp(\beta_1 + \beta_2 x_{2j} + \beta_3 x_{3ij} + \beta_4 x_{2j}x_{3ij})}{1 + \exp(\beta_1 + \beta_2 x_{2j} + \beta_3 x_{3ij} + \beta_4 x_{2j}x_{3ij})} \qquad (6.7)
\end{aligned}
$$

where $\phi(\zeta_j; 0, \widehat{\psi})$ is the normal density function with mean zero and variance $\widehat{\psi}$.

The difference between population-averaged and subject-specific effects is due to the fact that the average of a nonlinear function is not the same as the nonlinear function of the average. In the present context, the average of the inverse logit of the linear predictor $\beta_1 + \beta_2 x_{2j} + \beta_3 x_{3ij} + \beta_4 x_{2j} x_{3ij} + \zeta_j$ is not the same as the inverse logit of the average of the linear predictor, which is $\beta_1 + \beta_2 x_{2j} + \beta_3 x_{3ij} + \beta_4 x_{2j} x_{3ij}$. We can see this by comparing the simple average of the logits of 1 and 2 with the logit of the average of 1 and 2:

```
. display (invlogit(1) + invlogit(2))/2
.80592783
. display invlogit((1+2)/1)
.81757448
```

We can also see this in figure 6.9, where the individual dotted curves represent subject-specific logistic curves with randomly varying intercepts, whereas the solid, shallower curve represents the average of these curves for each value of x.

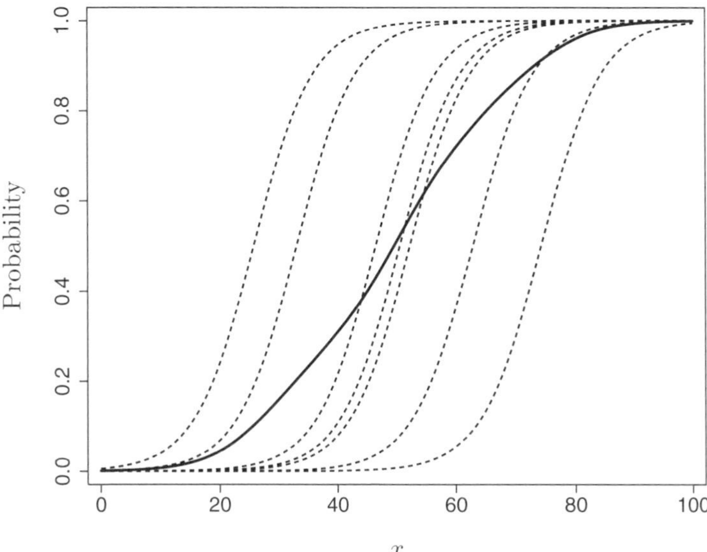

Figure 6.9: Subject-specific versus population-averaged logistic regression

The average curve has a different shape than the individual curves. Specifically, the effect of x on the average curve is smaller than the effect of x on the subject-specific curves. However, the population *median* probability is the same as the subject-specific probability evaluated at the median of ζ_j because the inverse logit function is a strictly increasing function.

Another way of understanding why the subject-specific effects are more extreme than the population-averaged effects is by writing the random-intercept logistic regression model as a latent-response model:

$$y_{ij}^* = \beta_1 + \beta_2 x_{2j} + \beta_3 x_{3ij} + \beta_4 x_{2j} x_{3ij} + \underbrace{\zeta_j + \epsilon_{ij}}_{\xi_{ij}}$$

The total residual variance is

$$\text{Var}(\xi_{ij}) = \psi + \pi^2/3$$

estimated as $\widehat{\psi} + \pi^2/3 = 16.08 + 3.29 = 19.37$, which is much greater than the residual variance of about 3.29 for an ordinary logistic regression model. As we have already seen in figure 6.4 for probit models, the slope in the model for y_i^* has to increase when the residual standard deviation increases to produce an equivalent curve for the marginal probability that the observed response is 1. Therefore, the regression coefficients of the random-intercept model (representing subject-specific effects) must be larger than those of the ordinary logistic regression model (representing population-averaged effects) to obtain a good fit of the model-implied marginal probabilities to the corresponding sample proportions. See section 6.13 for the predicted subject-specific and population-averaged probabilities for the toenail data.

Having described subject-specific and population-averaged probabilities or expectations of y_{ij}, for given covariate values, we now consider the corresponding variances. The subject-specific or conditional variance is

$$\text{Var}(y_{ij}|\mathbf{x}_{ij}, \zeta_j) = \Pr(y_{ij} = 1|\mathbf{x}_{ij}, \zeta_j)\{1 - \Pr(y_{ij} = 1|\mathbf{x}_{ij}, \zeta_j)\}$$

and the population-averaged or marginal variance is

$$\text{Var}(y_{ij}|\mathbf{x}_{ij}) = \Pr(y_{ij} = 1|\mathbf{x}_{ij})\{1 - \Pr(y_{ij} = 1|\mathbf{x}_{ij})\}$$

The random-intercept variance ψ does not affect the relationship between the marginal variance and the marginal mean. This is in contrast to models for counts described in chapter 9 where a random intercept (with $\psi > 0$) produces so-called overdispersion, with a larger marginal variance for a given marginal mean than the model without a random intercept ($\psi = 0$). It is important to note that, contrary to common belief, overdispersion is impossible for dichotomous responses (Skrondal and Rabe-Hesketh 2007).

6.10 Measures of dependence and heterogeneity

6.10.1 Conditional or residual intraclass correlation of the latent responses

Returning to the latent-response formulation, the dependence among the dichotomous responses for the same subject (or the between-subject heterogeneity) can be quantified by the *conditional intraclass correlation* or *residual intraclass correlation* ρ of the latent responses y_{ij}^* given the covariates

$$\rho \equiv \text{Cor}(y_{ij}^*, y_{i'j}^*|\mathbf{x}_{ij}, \mathbf{x}_{i'j}) = \text{Cor}(\xi_{ij}, \xi_{i'j}) = \frac{\psi}{\psi + \pi^2/3}$$

Substituting the estimated variance $\widehat{\psi} = 16.08$, we obtain an estimated conditional intraclass correlation of 0.83, which is large even for longitudinal data. The estimated intraclass correlation is also reported next to `rho` by `xtlogit`.

The reason why the degree of dependence is often expressed this way, in terms of the intraclass correlation for the latent responses y_{ij}^*, is that the intraclass correlation for the observed responses y_{ij} varies according to the values of the covariates.

For probit models, the expression for the intraclass correlations is as above with $\pi^2/3$ replaced by 1.

6.10.2 Median odds ratio

Larsen et al. (2000) and Larsen and Merlo (2005) suggest an alternative measure of heterogeneity. They consider repeatedly sampling two subjects with the same covariate values and forming the odds ratio comparing the subject with the larger random intercept with the other subject. For a given pair of subjects j and j', this odds ratio is given by $\exp(|\zeta_j - \zeta_{j'}|)$ and heterogeneity is expressed as the median of these odds ratios across repeated samples.

The median and other percentiles $a > 1$ can be obtained from the cumulative distribution function

$$\Pr\{\exp(|\zeta_j - \zeta_{j'}|) \le a\} \;=\; \Pr\left\{\frac{|\zeta_j - \zeta_{j'}|}{\sqrt{2\psi}} \le \frac{\ln(a)}{\sqrt{2\psi}}\right\} \;=\; 2\,\Phi\left\{\frac{\ln(a)}{\sqrt{2\psi}}\right\} - 1$$

If the cumulative probability is set to 1/2, a is the median odds ratio $\mathrm{OR}_{\mathrm{median}}$

$$2\,\Phi\left\{\frac{\ln(\mathrm{OR}_{\mathrm{median}})}{\sqrt{2\psi}}\right\} - 1 \;=\; 1/2$$

Solving this equation gives

$$\mathrm{OR}_{\mathrm{median}} \;=\; \exp\{\sqrt{2\psi}\Phi^{-1}(3/4)\}$$

Plugging in the parameter estimates, we obtain $\widehat{\mathrm{OR}}_{\mathrm{median}}$:

```
. display exp(sqrt(2*16.08)*invnormal(3/4))
45.833581
```

When two subjects are chosen at random at a given time point from the same treatment group, the odds ratio comparing the subject with the larger odds to the subject with the smaller odds will exceed 45.83 half the time, which is a very large odds ratio. For comparison, the estimated odds ratio comparing two subjects at 20 months, who had the same value of the random intercept but one of whom received received itraconazole (`treatment`=0) and the other of whom received terbinafine (`treatment`=1), is about 18 $\{= 1/\exp(-.161 + 20 \times -.137)\}$.

6.11 Maximum likelihood estimation

6.11.1 ❖ Adaptive quadrature

The marginal likelihood is the joint probability of all observed responses given the observed covariates. For linear mixed models, this marginal likelihood can be evaluated and maximized easily (see section 2.7). However, in generalized linear mixed models, the marginal likelihood does not have a closed form and must be evaluated by approximate methods.

To see this, we will now construct this marginal likelihood step by step for a random-intercept logistic regression model with one covariate x_j. The responses are conditionally independent given the random intercept ζ_j and the covariate x_j. Therefore, the joint probability of all the responses y_{ij} $(i = 1, \ldots, n_j)$ for cluster j, given the random intercept and covariate, is simply the product of the probabilities of the individual responses

$$\Pr(y_{1j}, \ldots, y_{n_j j}|x_j, \zeta_j) \;=\; \prod_{i=1}^{n_j} \Pr(y_{ij}|x_j, \zeta_j) \;=\; \prod_{i=1}^{n_j} \frac{\exp(\beta_1 + \beta_2 x_j + \zeta_j)^{y_{ij}}}{1 + \exp(\beta_1 + \beta_2 x_j + \zeta_j)} \tag{6.8}$$

In the last term

$$\frac{\exp(\beta_1 + \beta_2 x_j + \zeta_j)^{y_{ij}}}{1 + \exp(\beta_1 + \beta_2 x_j + \zeta_j)} \;=\; \begin{cases} \frac{\exp(\beta_1+\beta_2 x_j+\zeta_j)}{1+\exp(\beta_1+\beta_2 x_j+\zeta_j)} & \text{if } y_{ij} = 1 \\ \frac{1}{1+\exp(\beta_1+\beta_2 x_j+\zeta_j)} & \text{if } y_{ij} = 0 \end{cases}$$

as specified by the logistic regression model.

To obtain the marginal joint probability of the responses, not conditioning on the random intercept ζ_j (but still on the covariate x_j), we integrate out the random intercept

$$\Pr(y_{1j}, \ldots, y_{n_j j}|x_j) \;=\; \int \Pr(y_{1j}, \ldots, y_{n_j j}|x_j, \zeta_j)\, \phi(\zeta_j; 0, \psi)\, d\zeta_j \tag{6.9}$$

where $\phi(\zeta_j, 0, \psi)$ is the normal density of ζ_j with mean 0 and variance ψ. Unfortunately, this integral does not have a closed-form expression.

The marginal likelihood is just the joint probability of all responses for all clusters. Since the clusters are mutually independent, this is given by the product of the marginal joint probabilities of the responses for the individual clusters

$$L(\beta_1, \beta_2, \psi) \;=\; \prod_{j=1}^{N} \Pr(y_{1j}, \ldots, y_{n_j j}|x_j)$$

This marginal likelihood is viewed as a function of the parameters β_1, β_2, and ψ (with the observed responses treated as given). The parameters are estimated by finding the values of β_1, β_2, and ψ that yield the largest likelihood. The search for the maximum is iterative, beginning with some initial guesses or starting values for the parameters and updating these step by step until the maximum is reached, typically using a Newton–Raphson or expectation-maximization (EM) algorithm.

The integral over ζ_j in (6.9) can be approximated by a sum of R terms with e_r substituted for ζ_j and the normal density replaced by a weight w_r for the rth term $r = 1, \ldots, R$

$$\Pr(y_{1j}, \ldots, y_{n_j j}|x_j) \approx \sum_{r=1}^{R} \Pr(y_{1j}, \ldots, y_{n_j j}|x_j, \zeta_j = e_r)\, w_r$$

where e_r and w_r are called Gauss–Hermite quadrature locations and weights, respectively. This approximation can be viewed as replacing the continuous density of ζ_j by a discrete distribution with R possible values of ζ_j having probabilities $\Pr(\zeta_j = e_r)$. The Gauss–Hermite approximation is illustrated for $R = 5$ in figure 6.10. Obviously, the approximation improves when R increases.

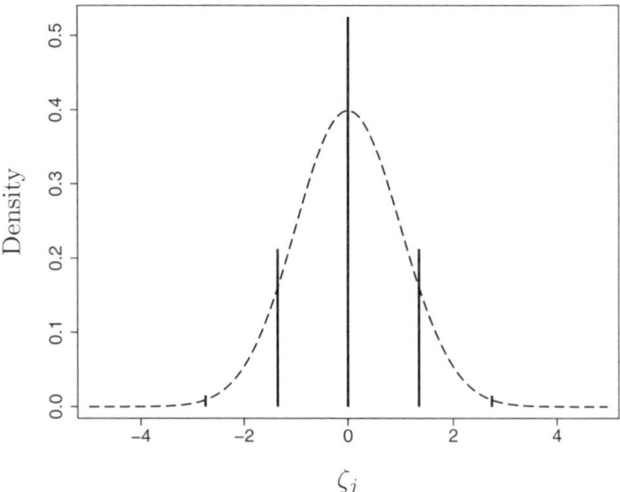

Figure 6.10: Gauss–Hermite quadrature: Approximating continuous density (dashed curve) by discrete distribution (bars)

The ordinary quadrature approximation described above can perform poorly if the function being integrated, called the *integrand*, has a sharp peak, as discussed in Rabe-Hesketh, Skrondal, and Pickles (2002, 2005). Sharp peaks can occur when the clusters are very large so that many functions (the individual response probabilities as functions of ζ_j) are multiplied to yield $\Pr(y_{1j}, \ldots, y_{n_j j}|x_j, \zeta_j)$. Similarly, if the responses are counts or continuous responses, even a few terms can result in a highly peaked function. Another potential problem is a high intraclass correlation. Here the functions being multiplied coincide with each other more closely, due to the greater similarity of responses within clusters, yielding a sharper peak. In fact, the toenail data analyzed in this chapter, which has an estimated conditional intraclass correlation for the latent responses of 0.83, poses real problems for estimation using ordinary quadrature, as pointed out by Lesaffre and Spiessens (2001).

The top panel in figure 6.11 shows the same 5-point quadrature approximation and density of ζ_j as in figure 6.10. The solid curve is proportional to the integrand for a hypothetical cluster. Here the quadrature approximation works poorly because the peak falls between adjacent quadrature points.

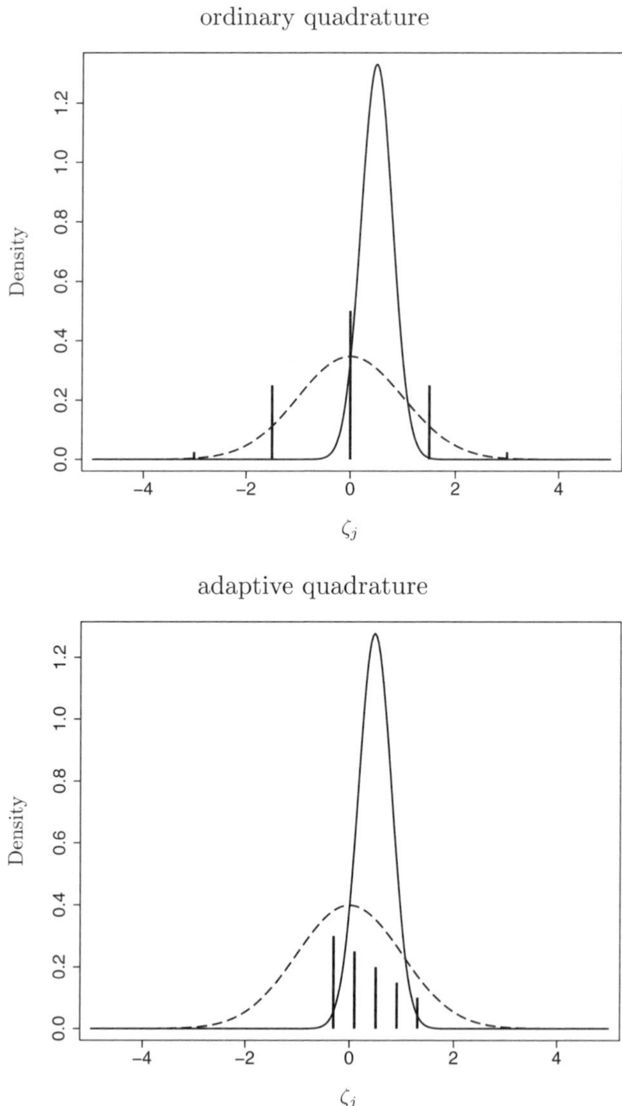

Figure 6.11: Density of ζ_j (dashed curve), normalized integrand (solid curve), and quadrature weights (bars) for ordinary quadrature and adaptive quadrature (*Source:* Rabe-Hesketh, Skrondal, and Pickles 2002)

The bottom panel of figure 6.11 shows an improved approximation, known as *adaptive quadrature*, where the locations are rescaled and translated

$$e_{rj} = a_j + b_j e_r \tag{6.10}$$

to fall under the peak of the integrand, where a_j and b_j are cluster-specific constants. This transformation of the locations is accompanied by a transformation of the weights w_r that also depends on a_j and b_j. The method is called *adaptive* because the quadrature locations and weights are adapted to the data for the individual clusters.

To maximize the likelihood, we start with a set of initial or starting values of the parameters and then keep updating the parameters until the likelihood is maximized. The quantities a_j and b_j needed to evaluate the likelihood are functions of the parameters (as well as the data) and must therefore be updated or "readapted" when the parameters are updated.

There are two different implementations of adaptive quadrature in Stata that differ in the values used for a_j and b_j in (6.10). The method implemented in `gllamm` and the default method in `xtlogit` (as of Stata 10) use the posterior mean of ζ_j for a_j and the posterior standard deviation for b_j. However, obtaining the posterior mean and standard deviation requires numerical integration so adaptive quadrature sometimes does not work when there are too few quadrature points (e.g., fewer than 5). Details of the algorithm are given in Rabe-Hesketh, Skrondal, and Pickles (2002, 2005) and Skrondal and Rabe-Hesketh (2004).

The method implemented in `xtmelogit`, and available in `xtlogit` with the option `method(aghermite)`, uses the posterior mode of ζ_j for a_j and for b_j the standard deviation of the normal density whose logarithm has the same curvature as the log posterior of ζ_j at the mode. An advantage of this approach is that it does not rely on numerical integration and can therefore be implemented even with one quadrature point. With one quadrature point, this version of adaptive quadrature becomes a Laplace approximation.

6.11.2 Some speed considerations

As discussed in section 6.11.1, the likelihood involves integrals that are evaluated by numerical integration. Even with the best approach, adaptive quadrature, the likelihood itself, as well as the maximum likelihood estimates, are therefore only approximate. We can assess whether the approximation is adequate in a given situation by repeating the analysis with a larger number of quadrature points. If we get essentially the same result, the lower number of quadrature points was adequate. Such checking should always be done before estimates are taken at face value.

Also due to numerical integration, estimation can be slow, especially if there are many random effects. The time it takes to fit a model is approximately proportional to the product of the number of quadrature points for all random effects (although this seems to be more true for `gllamm` than for `xtmelogit`). For example, if there are two random effects at level 2 (a random intercept and slope) and 8 quadrature points are used for each random effect, the time will be approximately proportional to 64.

Therefore, using 4 quadrature points for each random effect will take only about one-fourth (16/64) as long as using 8. The time is also approximately proportional to the number of observations, and for programs using numerical differentiation (gllamm and xtmelogit), to the square of the number of parameters. (For xtlogit, computation time increases less dramatically when the number of parameters increases because it uses analytical derivatives.)

For large problems, it may be advisable to estimate how long estimation will take before starting work on a project. In this case, we recommend fitting a similar model with fewer random effects, fewer parameters (e.g., fewer covariates), or fewer observations, and using the above approximate proportionality factors to estimate the time that will be required for the larger problem.

For random-intercept models, by far the fastest command is xtlogit (because it uses analytical derivatives). However, xtlogit cannot fit random-coefficient models, or higher-level models introduced in chapter 10. For such models, xtmelogit or gllamm must be used. The quickest way of obtaining results here is using xtmelogit with one integration point, corresponding to the Laplace approximation. Although this method sometimes works well, it can produce severely biased estimates, especially if the clusters are small and the (true) random-intercept variance is large as for the toenail data. For these data, we obtain the following:

```
. xtmelogit outcome treatment month trt_month || patient:, intpoints(1)

Mixed-effects logistic regression               Number of obs      =      1908
Group variable: patient                         Number of groups   =       294

                                                Obs per group: min =         1
                                                               avg =       6.5
                                                               max =         7

Integration points =   1                        Wald chi2(3)       =    131.96
Log likelihood = -627.80894                     Prob > chi2        =    0.0000

------------------------------------------------------------------------------
     outcome |      Coef.   Std. Err.      z    P>|z|     [95% Conf. Interval]
-------------+----------------------------------------------------------------
   treatment |  -.3070179   .6899612    -0.44   0.656    -1.659317    1.045281
       month |  -.4000919    .047059    -8.50   0.000    -.4923258    -.307858
   trt_month |  -.1372598   .0695865    -1.97   0.049    -.2736469   -.0008728
       _cons |  -2.523352    .788292    -3.20   0.001    -4.068376   -.9783276
------------------------------------------------------------------------------

------------------------------------------------------------------------------
  Random-effects Parameters  |   Estimate   Std. Err.     [95% Conf. Interval]
-----------------------------+------------------------------------------------
patient: Identity            |
                   sd(_cons) |   4.570918   .7199338      3.356885     6.22401
------------------------------------------------------------------------------
LR test vs. logistic regression: chibar2(01) =   560.40 Prob>=chibar2 = 0.0000
Note: log-likelihood calculations are based on the Laplacian approximation.
```

We see that the estimated intercept and coefficient of treatment are different from the estimates in section 6.7.1 using adaptive quadrature with 30 quadrature points. As mentioned in the previous section, gllamm cannot be used with only 1 quadrature point, and adaptive quadrature typically requires at least 5 quadrature points.

Advice for speeding up gllamm

To speed up estimation using `gllamm`, we recommend using good starting values whenever they are available. For instance, when increasing the number of quadrature points or adding or dropping covariates, use the previous estimates as starting values. This can be done by using the `from()` option to specify a row matrix of starting values. This option should be combined with `skip` if the new model contains fewer parameters than supplied. You can also use the `copy` option if your parameters are supplied in the correct order yet are not necessarily labeled correctly. Use of these options is demonstrated throughout this book.

For some datasets and models, you can also represent the data using fewer rows than there are observations, thus speeding up estimation. For example, if the response is dichotomous and we are using one dichotomous covariate in a two-level dataset, we can use one row of data for each combination of covariate and response (00, 01, 10, 11) for each cluster, leading to at most four rows per cluster. We can then specify a variable containing level-1 frequency weights equal to the number of observations, or level-1 units, in each cluster having each combination of the covariate and response values. Level-2 weights can be used if several clusters have the same level-2 covariates and the same number of level-1 units with the same response and level-1 covariate pattern. The `weight()` option in `gllamm` is designed for specifying frequency weights at the different levels. See exercise 6.8 for an example with level-1 weights and exercises 2.3 and 6.3 for examples with level-2 weights. If the dataset is large, starting values could be obtained using a random sample of the data.

For models involving several random effects at the same level, such as two-level random-coefficient models with a random intercept and slope, the multivariate integral can be evaluated more efficiently using *spherical quadrature* instead of the default Cartesian-product quadrature. For the random intercept and slope example, Cartesian-product quadrature consists of evaluating the function being integrated on the rectangular grid of quadrature points consisting of all combinations of $\zeta_{1j} = e_1, \ldots, e_R$ and $\zeta_{2j} = e_1, \ldots, e_R$, giving R^2 terms. In contrast, spherical quadrature consists of evaluating ζ_{1j} and ζ_{2j} at values falling on concentric circles (spheres in more dimensions). The important thing is that the same accuracy can now be achieved with fewer than R^2 points. For example, when $R = 8$, Cartesian-product quadrature requires 64 evaluations and spherical quadrature requires only 44 evaluations, taking nearly 30% less time to achieve the same accuracy. Here accuracy is expressed in terms of the "degree" of the approximation given by $d = 2R - 1$, $d = 15$ in the current example. To use spherical quadrature, specify the `ip(m)` in `gllamm` and give the degree d of the approximation, `nip(15)` here. Unfortunately, spherical integration is available only for certain combinations of numbers of dimensions (or numbers of random effects) and degrees of accuracy, d: For two dimensions, d can be 5, 7, 9, 11, or 15 and for more than two dimensions, d can be 5 or 7. See Rabe-Hesketh, Skrondal, and Pickles (2005) for more information.

6.12 Assigning values to random effects

Having estimated the model parameters, we may want to assign values to the random intercepts ζ_j for individual clusters j. The ζ_j are not model parameters, but as for linear models, we can treat the estimated parameters as known and then either estimate or predict ζ_j.

Unfortunately, this is both harder and less useful than for linear models. The estimated or predicted values of ζ_j should not be used for model diagnostics because their distribution is not known if the model is true. In general, the values should also not be used to obtain cluster-specific predicted probabilities (see sec. 6.13.2). However, they can be used to obtain cluster-specific log-odds and hence a ranking of clusters (the ranking of the probabilities is the same as the ranking of the log odds).

6.12.1 Maximum likelihood estimation

As discussed for linear models in section 2.9.1, we can estimate the intercepts ζ_j by treating them as the only unknown parameters, after estimates have been plugged in for the model parameters,

$$\text{logit}\{\Pr(y_{ij}=1|\mathbf{x}_{ij},\zeta_j)\} \quad = \quad \underbrace{\text{offset}_{ij}}_{\widehat{\beta}_1+\widehat{\beta}_2 x_{2ij}+\cdots} \quad + \quad \zeta_j$$

This is a logistic regression model for cluster j with offset (a term with regression coefficient set to 1) given by the estimated fixed part of the linear predictor from the logistic random-intercept model and with a cluster-specific intercept ζ_j.

We then maximize the corresponding likelihood for cluster j

$$\text{Likelihood}(y_{1j}, y_{2j}, \ldots, y_{n_j,j}|\mathbf{X}_j, \zeta_j)$$

with respect to ζ_j, where $\mathbf{X}_j$ is a matrix containing all covariates for cluster j. This can be accomplished by fitting logistic regression models to the individual clusters. First, obtain the offset from the `xtmelogit` estimates

```
. estimates restore xtmelogit
(results xtmelogit are active now)
. predict offset, xb
```

Then use the `statsby` command to fit the individual logistic regression models, specifying an offset:

```
. statsby mlest=_b[_cons], by(patient) saving(ml): logit outcome, offset(offset)
(running logit on estimation sample)

      command:  logit outcome, offset(offset)
        mlest:  _b[_cons]
           by:  patient

Statsby groups
----+--- 1 ---+--- 2 ---+--- 3 ---+--- 4 ---+--- 5
.....xx......xx..xxx...x.x...xxxxx.xx...xxx.xxxx      50
xx.xxxxxxxxx.xxxx..xxxxxxxxx.x..xx..x.xxx.xxx.x...   100
xx.xxxxxxxxxxx.xxx.x.x...x.xx.xxxxx.xx....xxx.x.xx   150
.x..x.xxxx..xxxxx.xx..xxxxx..xxx.x.xxxxx.x.x.xxx...  200
.xxxxx.xx.xx..x.xxx...xx.x..xxxxx.x..x.x..x..xxxxx   250
x.xx.x..xxxxxx..x..x..xxx.x..xxxxxxxx.x.x...
```

We have saved the estimates under the variable name `mlest` in a file called `ml.dta` in the local directory. The `x`'s in the output indicate that the `logit` command did not converge for many clusters. For these clusters, the variable `mlest` is missing. This happens for clusters where all responses are 0 or all responses are 1 because the maximum likelihood estimate then is $-\infty$ and $+\infty$, respectively.

We now merge the estimates with the data for later use:

```
. sort patient
. merge patient using ml
variable patient does not uniquely identify observations in the master data
. drop _merge
```

6.12.2 Empirical Bayes prediction

The ideas behind empirical Bayes prediction discussed in section 2.9.2 for linear variance-components models also apply to other generalized linear mixed models. Instead of basing inference completely on the likelihood of the responses for a cluster given the random effect, we combine this information with the prior, which is just the density of the random effect, to obtain the posterior density

$$\text{Posterior}(\zeta_j|y_{1j},\ldots,y_{n_jj},\mathbf{X}_j) \;\propto\; \text{Prior}(\zeta_j)\times\text{Likelihood}(y_{1j},\ldots,y_{n_jj}|\mathbf{X}_j,\zeta_j)$$

The product on the right is proportional, but not equal, to the posterior density. Obtaining the posterior density requires dividing this product by a normalizing constant that can only be obtained by numerical integration. The resulting posterior density is no longer normal as for linear models, and hence its mode does not equal its mean. There are therefore two different types of predictions we could consider: the mean of the posterior and its mode. The first is undoubtedly the most common and is referred to as empirical Bayes prediction, whereas the second is sometimes called empirical Bayes modal prediction.

The empirical Bayes prediction of the random intercept for a cluster j is the mean of the posterior distribution of the random intercept, with model parameter estimates plugged in. This can be obtained as

$$\widetilde{\zeta}_j = \int \zeta_j \, \text{Posterior}(\zeta_j | y_{1j}, \ldots, y_{n_j j}, \mathbf{X}_j) \, d\zeta_j \tag{6.11}$$

using numerical integration.

At the time of writing this book, the only Stata command that provides empirical Bayes predictions for generalized linear mixed models is the postestimation command `gllapred` for `gllamm` with the `u` option:

```
. estimates restore gllamm
. gllapred eb, u
(means and standard deviations will be stored in ebm1 ebs1)
```

The empirical Bayes predictions are close to the maximum likelihood estimates whenever the latter exist:

```
. compare mlest ebm1
```

	count	difference minimum	difference average	difference maximum
mlest>ebm1	767	.0884503	.2652569	.76276
jointly defined	767	.0884503	.2652569	.76276
mlest missing only	1141			
total	1908			

Here the maximum likelihood estimates are always positive and always larger than the empirical Bayes predictions. When both are positive, this corresponds to shrinkage, which is minimal here (giving a difference of at most 0.76) because the prior distribution is flat with a large standard deviation of about 4.

We mentioned in previous chapters on linear models that the posterior standard deviations produced by `gllapred` with the `u` option were the same as the prediction-error standard deviations. However, this is not true for generalized linear mixed models not having an identity link, such as the random-intercept logistic model discussed here. There is also no longer an easy way to obtain the sampling standard deviation or "diagnostic" standard error. The `ustd` option for standardized level-2 residuals therefore divides the empirical Bayes predictions by an approximation for this standard deviation (see Skrondal and Rabe-Hesketh 2004, 231–232, for details).

6.12.3 Empirical Bayes modal prediction

Empirical Bayes modal predictions are easy to obtain using the `predict` command with the `reffects` option after estimation using `xtmelogit`:

```
estimates restore xtmelogit
predict ebmodal, reffects
```

6.13 Different kinds of predicted probabilities

6.13.1 Predicted population-averaged probabilities

At the time of writing this book, population-averaged or marginal probabilities can be predicted for random-intercept logistic regression models only by using `gllapred` after estimation using `gllamm`. To obtain estimated marginal probabilities using `gllapred`, specify the options `mu` (for the mean response, here a probability) and `marginal`[1] (for integrating over the random-intercept distribution):

```
. estimates restore gllamm
. gllapred margprob, mu marginal
(mu will be stored in margprob)
```

We now compare predictions of population-averaged or marginal probabilities from the ordinary logit (previously obtained under the variable name `prob`) and the random-intercept logit model

```
. twoway (line prob month, sort) (line margprob month, sort lpatt(dash)),
> by(treatment) legend(order(1 "Ordinary logit" 2 "Random-intercept logit"))
> xtitle(Time in months) ytitle(Fitted marginal probabilities of onycholysis)
```

giving figure 6.12. The predictions are nearly identical.

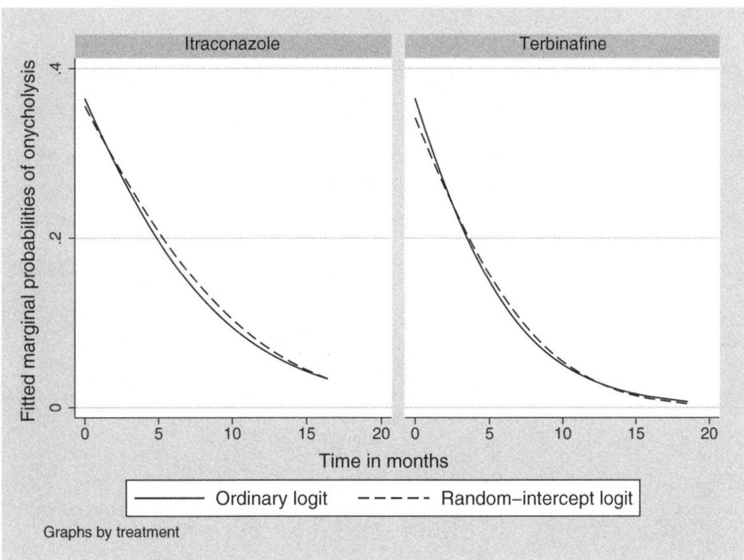

Figure 6.12: Fitted marginal probabilities using ordinary and random-intercept logistic regression

1. *Marginal* is used in a different way here than sometimes used in econometrics, where it represents the derivative of the probability with respect to a covariate.

6.13.2 Predicted subject-specific probabilities

Predictions for hypothetical subjects

Subject-specific curves for different values of ζ_j can be produced using **gllapred** with the `mu` and `us(`*varname*`)` options, where *varname*1 is the name of the variable containing the value of the first (here the only) random effect. We now produce predicted probabilities for ζ_j equal to 0, -4, 4, -2, and 2:

```
. generate zeta1 = 0
. gllapred condprob0, mu us(zeta)
(mu will be stored in condprob0)
. generate lower1 = -4
. gllapred condprobm4, mu us(lower)
(mu will be stored in condprobm4)
. generate upper1 = 4
. gllapred condprob4, mu us(upper)
(mu will be stored in condprob4)
. replace lower1 = -2
. gllapred condprobm2, mu us(lower)
(mu will be stored in condprobm2)
. replace upper1 = 2
. gllapred condprob2, mu us(upper)
(mu will be stored in condprob2)
```

Plotting all these conditional probabilities together with the observed proportions and marginal probabilities

```
. twoway (line prop mn_month, sort)
> (line margprob month, sort lpatt(dash))
> (line condprob0 month, sort lpatt(shortdash_dot))
> (line condprob4 month, sort lpatt(shortdash))
> (line condprobm4 month, sort lpatt(shortdash))
> (line condprob2 month, sort lpatt(shortdash))
> (line condprobm2 month, sort lpatt(shortdash)),
> by(treatment)
> legend(order(1 "Observed proportion" 2 "Marginal probability"
>              3 "Median probability" 4 "Conditional probabilities"))
> xtitle(Time in months) ytitle(Probabilities of onycholysis)
```

produces figure 6.13.

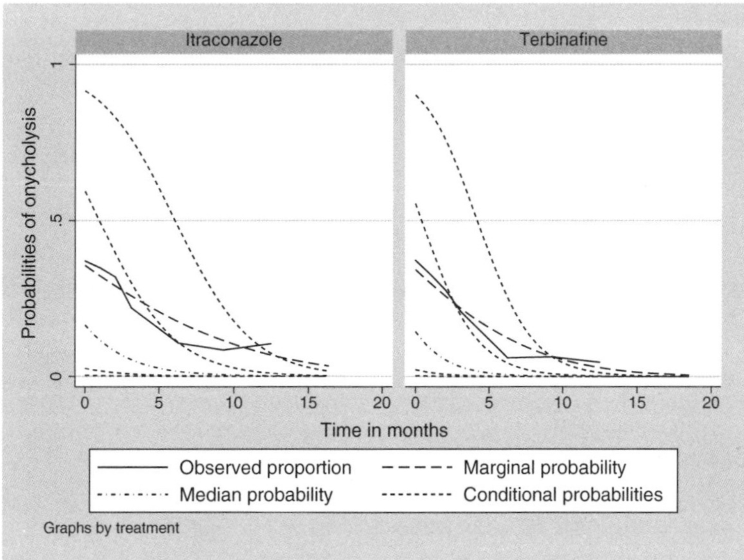

Figure 6.13: Conditional and marginal probabilities for the random-intercept logistic regression model

Clearly the subject-specific curves have steeper downward slopes than the population-averaged curve. The subject-specific curve represented by a "dash-dot" line is for $\zeta_j = 0$ and hence represents the population *median* curve.

Predictions for the subjects in the sample

We may also want to predict the probability that $y_{ij} = 1$ for a given cluster j. The predicted conditional probability, given the unknown random intercept ζ_j is

$$\widehat{\Pr}(y_{ij} = 1|\mathbf{x}_{ij}, \zeta_j) = \frac{\exp(\widehat{\beta}_1 + \widehat{\beta}_2 x_{ij} + \zeta_j)}{1 + \exp(\widehat{\beta}_1 + \widehat{\beta}_2 x_{ij} + \zeta_j)}$$

Because our knowledge about ζ_j for subject j is represented by the posterior distribution, a good prediction $\widetilde{\Pr}(y_{ij}|\mathbf{x}_{ij})$ of the unconditional probability is obtained by integrating over the posterior distribution:

$$\begin{aligned} \widetilde{\Pr}(y_{ij}|\mathbf{x}_{ij}) &\equiv \int \widehat{\Pr}(y_{ij} = 1|\mathbf{x}_{ij}, \zeta_j) \times \text{Posterior}(\zeta_j|y_{1j}, \ldots, y_{n_j j}, \mathbf{X}_j)\, d\zeta_j \\ &\neq \widehat{\Pr}(y_{ij} = 1|\mathbf{x}_{ij}, \widetilde{\zeta}_j) \end{aligned} \tag{6.12}$$

Notice that we cannot simply plug in the posterior mean of the random intercept $\widetilde{\zeta}_j$ for ζ_j in generalized linear mixed models, such as the random-intercept logit model. The reason is that the mean of a given nonlinear function of ζ_j does not in general equal the same function evaluated at the mean of ζ_j.

The posterior means of the predicted probabilities as defined in (9.13) can be obtained using `gllapred` with the `mu` option (and not the `marginal` option) after estimation using `gllamm`:

```
. gllapred cmu, mu
(mu will be stored in cmu)
```

To give you some idea of the relationship between observed responses and empirical Bayes predictions, we reshape the data so that we can list all the responses for a given patient. We first preserve the data and drop patients from the itraconazole group:

```
. preserve
. drop if treatment==0
. keep patient outcome visit ebm1 ebs1 cmu
. reshape wide outcome cmu, i(patient) j(visit)
```

To look at the results for a range of different patients, we pick out patients with the 15th, 45th, 75th, 105th, and 135th largest values of the empirical Bayes predictions of the random intercepts, corresponding approximately to the 10th, 30th, 50th, 70th, and 90th percentiles:

```
. sort ebm1
. keep if _n==15|_n==45|_n==75|_n==105|_n==135
```

The empirical Bayes predictions and outcomes can be listed compactly by first formatting the corresponding variables to a display format consisting of three digits and by abbreviating the variable names to three characters:

```
. format ebm1 cmu* %3.2f
. list ebm1 ebs1 outcome*, clean noobs abbreviate(3)
     ebm1   ebs1   out~1   out~2   out~3   out~4   out~5   out~6   out~7
    -2.70   2.70       0       0       0       0       0       0       0
    -2.69   2.71       0       0       0       0       0       .       .
    -2.66   2.71       0       0       0       0       0       0       0
     2.48   1.11       1       1       0       0       .       0       0
     4.44   1.09       1       1       1       1       0       0       0
```

For clusters where the response is always zero, the empirical Bayes predictions of ζ_j are large negative values and have larger posterior standard deviations (`ebs1`) than for the other response patterns. Maximum likelihood estimates of ζ_j do not exist when all responses are zero. Large positive empirical Bayes predictions of ζ_j occur when the initial responses are 1.

We can also list the posterior mean response probabilities together with the empirical Bayes predictions of the random intercept:

```
. list ebm1 cmu*, clean noobs

    ebm1   cmu1   cmu2   cmu3   cmu4   cmu5   cmu6   cmu7
   -2.70   0.07   0.05   0.03   0.02   0.01   0.00   0.00
   -2.69   0.07   0.05   0.03   0.02   0.00      .      .
   -2.66   0.07   0.04   0.03   0.02   0.00   0.00   0.00
    2.48   0.64   0.52   0.39   0.28      .   0.02   0.00
    4.44   0.90   0.86   0.79   0.69   0.36   0.09   0.02
```

These predicted probabilities reflect the decline in the odds of onycholysis over time, even when this is not reflected in the observed responses for a subject. At the time of writing this book, `gllapred` cannot compute predicted probabilities for observations where the response is missing.

After estimation with `xtmelogit`, the `predict` command with the `mu` option gives the posterior mode of the predicted conditional probability $\widehat{\Pr}(y_{ij}|\mathbf{x}_{ij},\zeta_j)$ instead of the posterior mean. This is achieved by substituting the posterior mode of ζ_j into the expression for the conditional probability. [The mode of a strictly increasing function of ζ_j (here an inverse logit), is the same function evaluated at the mode of ζ_j.]

We now restore the data so we can use them in the following sections:

```
. restore
```

6.14 Other approaches to clustered dichotomous data

6.14.1 Conditional logistic regression

Instead of using random intercepts for clusters (patients in the toenail application), it would be tempting to use fixed intercepts by including a dummy variable for each patient (and omitting the overall intercept). This would be analogous to the fixed-effects estimator of within-patient effects discussed for linear models in section 3.7.2. However, in logistic regression, this approach would lead to inconsistent estimates of the within-patient effects due to the so-called *incidental parameter problem*, which becomes less severe for large clusters. Obviously, we also cannot eliminate the random intercepts by simply cluster-mean-centering the responses and covariates, as in (3.9).

Instead, we can eliminate the patient-specific intercepts by constructing a likelihood that is conditional on the number of responses that take the value 1 (a sufficient statistic for the patient-specific intercept). Indeed, in the linear case, assuming normality, least-squares estimation of the cluster-mean-centered model is equivalent to conditional maximum likelihood estimation. In logistic regression, conditional maximum likelihood estimation is more involved and is known as *conditional logistic regression*. Importantly, this method estimates conditional or subject-specific effects like random-intercept logistic regression. Conditional maximum likelihood can only estimate the effects of within-patient or time-varying effects. Patient-specific covariates such as `treatment` cannot be included. However, interactions between patient-specific and time-varying variables, such as `treatment` by `month`, can be estimated.

Conditional logistic regression can be performed using Stata's **xtlogit** command with the **fe** option or using the **clogit** command (with the **or** option to obtain odds ratios):

```
. clogit outcome month trt_month, group(patient) or
note: multiple positive outcomes within groups encountered.
note: 179 groups (1141 obs) dropped due to all positive or
      all negative outcomes.
Conditional (fixed-effects) logistic regression   Number of obs   =        767
                                                  LR chi2(2)      =     290.97
                                                  Prob > chi2     =     0.0000
Log likelihood = -188.94377                       Pseudo R2       =     0.4350

     outcome | Odds Ratio   Std. Err.      z    P>|z|     [95% Conf. Interval]
-------------+----------------------------------------------------------------
       month |   .6827717   .0321547    -8.10   0.000     .6225707     .748794
   trt_month |   .9065404   .0667426    -1.33   0.183     .7847274    1.047262
```

The odds ratio for the treatment effect is now estimated as 0.91 and is no longer significant at the 5% level. However, both this estimate and the estimate for **month**, also given in the last column of table 6.2, are similar to the estimates for the random-intercept model.

As we discussed in section 3.7.4, within-patient estimates cannot be confounded with omitted between-patient covariates and are in that sense less sensitive to model misspecification than estimates based on the random-intercept model. A further advantage of conditional maximum likelihood is that it does not make any assumptions regarding the distribution of the patient-specific effect. Therefore, it is reassuring that the conditional maximum likelihood estimates are similar to the maximum likelihood estimates for the random-intercept model. If the random-intercept model is true, the latter estimator is more efficient and tends to yield smaller standard errors leading to smaller p-values, as we can see for the treatment by time interaction. Here the conditional logistic regression method is inefficient since, as noted in the output, 179 subjects whose responses were all 0 or all 1 cannot contribute to the analysis. This is because the conditional probabilities of these response patterns, conditioning on the total response across time, are 1 regardless of the covariates (e.g., if the total is zero, all responses must be zero) and the conditional probabilities therefore do not provide any information on covariate effects.

The above model is sometimes referred to as the Chamberlain fixed-effects logit model in econometrics and is used for matched case–control studies in epidemiology. The same trick of conditioning is also used for the Rasch model in psychometrics. Unfortunately, the conditional fixed-effects approach cannot be generalized to more complex settings with subject-specific intercepts and slopes. There is also no counterpart to conditional logistic regression for probit models. Finally, conditional logistic regression gives inconsistent estimates if a lagged response is included as a covariate.

6.14.2 Generalized estimating equations (GEE)

Generalized estimating equations (GEE), first introduced in section 5.6, can be used to estimate marginal or population-averaged effects taking into account the dependence among units nested in clusters. This is usually done by specifying "working correlations" for the observed responses (conditional on covariates), typically one of the following structures:

- Independence:
  Same as ordinary logistic regression
- Exchangeable:
  Same correlation for all pairs of units—the default in Stata
- Autoregressive lag-1 [AR(1)]:
  Correlation declines exponentially with the time lag—only makes sense for longitudinal data
- Unstructured:
  A different correlation for each pair of responses—should not be used for large clusters

The advantage of GEE over ordinary logistic regression is that more efficient estimates are obtained if the working correlation structure resembles the true dependence structure. However, the true marginal (over the random effects) correlation of the responses is in general not constant as specified in GEE but varies according to values of the observed covariates. Using Pearson correlations for dichotomous responses is also somewhat peculiar because the odds ratio is the measure of association in logistic regression.

A nice feature of GEE (and indeed ordinary logistic regression) is that marginal effects can be consistently estimated, even if the dependence among units in clusters is not properly modeled. The standard errors are usually based on the sandwich estimator, which takes the dependence into account. However, note that GEE estimates of marginal effects are not robust against misspecified regression structures, such as omitted covariates. We will not describe the estimation method here apart from noting that it can be viewed as a weighted logistic regression.

GEE estimates of odds ratios using the default exchangeable correlation structure, with "robust" standard errors from the sandwich estimator, are obtained using `xtgee` with the `vce(robust)` and `eform` options:

(*Continued on next page*)

```
. xtgee outcome treatment month trt_month, i(patient) link(logit)
> family(binom) correlation(exchangeable) vce(robust) eform

GEE population-averaged model                   Number of obs      =      1908
Group variable:                    patient      Number of groups   =       294
Link:                                logit      Obs per group: min =         1
Family:                           binomial                     avg =       6.5
Correlation:                  exchangeable                     max =         7
                                                Wald chi2(3)       =     63.44
Scale parameter:                         1      Prob > chi2        =    0.0000

                             (Std. Err. adjusted for clustering on patient)

------------------------------------------------------------------------------
             |             Semi-robust
     outcome | Odds Ratio   Std. Err.      z    P>|z|     [95% Conf. Interval]
-------------+----------------------------------------------------------------
   treatment |   1.007207   .2618022     0.03   0.978     .6051549    1.676373
       month |   .8425856   .0253208    -5.70   0.000     .7943911     .893704
   trt_month |   .9252113   .0501514    -1.43   0.152     .8319576    1.028918
------------------------------------------------------------------------------
```

These estimates are given under "GEE" in table 6.2 and can alternatively be obtained using `xtlogit` with the `pa` option.

It is important to note that GEE is an *estimation method* for marginal effects. It does not require the specification of a statistical model, which is in stark contrast to multilevel modeling where statistical models are explicitly specified and different kinds of conditional effects are estimated.

6.15 Summary and further reading

We have described various approaches to modeling clustered dichotomous data, focusing on random-intercept models for longitudinal data. Alternatives to multilevel modeling such as conditional maximum likelihood estimation and generalized estimating equations have also been briefly discussed. The important distinction between conditional or subject-specific effects and marginal or population-averaged effects has been emphasized.

For random-intercept models, we described adaptive quadrature for maximum likelihood estimation and pointed out that you need to make sure that a sufficient number of quadrature points have been used for a given model and application. We have demonstrated the use of a variety of predictions, either cluster-specific predictions, based on empirical Bayes, or population-averaged predictions. Diagnostics for generalized linear mixed models are still being developed; see chapter 8 of Skrondal and Rabe-Hesketh (2004) for an overview.

We have discussed the most common link functions for dichotomous responses, namely logit and probit links. A third link that is sometimes used is the complementary log-log link, which is introduced in section 8.6. Dichotomous responses are sometimes aggregated into counts, giving the number of successes y_i in n_i trials for unit i. In this situation, it is usually assumed that y_i has a binomial distribution. `xtmelogit` can then be used as for dichotomous responses, but with the `binomial()` option to specify the variable containing the values n_i. Similarly, `gllamm` can be used with the binomial distribution and any of the link functions together with the `denom()` option to specify the variable containing n_i.

Good introductions to single-level logistic regression include Collett (2003a), Long (1997), and Hosmer and Lemeshow (2000). Logistic and other types of regression using Stata are discussed by Long and Freese (2006), primarily with examples from social science, and by Vittinghoff et al. (2006), with examples from medicine.

Generalized linear mixed models are described in the books by Skrondal and Rabe-Hesketh (2004), Molenberghs and Verbeke (2005), Hedeker and Gibbons (2006), and McCulloch and Searle (2001). See also Goldstein (2003) and Raudenbush and Bryk (2002). Several examples with dichotomous responses are discussed in Skrondal and Rabe-Hesketh (2004, chap. 9). Guo and Zhao (2000) is a good introductory paper on multilevel modeling for binary data with applications in social science. We also recommend the book chapter by Rabe-Hesketh and Skrondal (2008), the article by Agresti et al. (2000), and the encyclopedia entry by Hedeker (2005) for overviews of generalized linear mixed models.

Detailed accounts of generalized estimating equations are given in Hardin and Hilbe (2003) and Diggle et al. (2002).

6.16 Exercises

6.1 Toenail data

1. Fit the probit version of the random-intercept model in (6.6) using `gllamm`. How many quadrature points appear to be needed using adaptive quadrature?
2. Estimate the residual intraclass correlation for the latent responses.
3. Obtain empirical Bayes predictions using both the random-intercept logit and probit models and estimate the approximate constant of proportionality between these.
4. ❖ By considering the residual standard deviations of the latent response for the logit and probit models, work out what you think the constant of proportionality should be for the logit and probit empirical Bayes predictions. How does this compare with the constant estimated in step 3?

6.2 Ohio wheeze data

In this exercise, we use data from the Six Cities Study (Ware et al. 1984), previously analyzed by Fitzmaurice (1998), among others. The dataset includes 537 children from Steubenville, Ohio, who were examined annually four times from age 7 to age 10 to ascertain their wheezing status. The smoking status of the mother was also determined at the beginning of the study to investigate whether maternal smoking increases the risk of wheezing in children. The mother's smoking status is treated as time-constant, although it may have changed for some mothers over time.

The dataset `wheeze.dta` has the following variables:

- `id`: child identifier (j)
- `age`: number of years since 9th birthday (x_{2ij})
- `smoking`: mother smokes regularly (1: yes; 0: no) (x_{3j})
- `y`: wheeze status (1: yes; 0: no) (y_{ij})

1. Fit the following transition model considered by Fitzmaurice (1998):

$$\text{logit}\{\Pr(y_{ij}=1|\mathbf{x}_{ij}, y_{i-1,j})\} = \beta_1 + \beta_2 x_{2ij} + \beta_3 x_{3j} + \gamma y_{i-1,j}, \quad i = 2, 3, 4$$

   where x_{2ij} is `age` and x_{3j} is `smoking`. (The lagged responses can be obtained using `by id (age), sort: generate lag = y[_n-1]`.)
2. Fit the following random-intercept model considered by Fitzmaurice (1998):

$$\text{logit}\{\Pr(y_{ij}=1|\mathbf{x}_{ij}, \zeta_j)\} = \beta_1 + \beta_2 x_{2ij} + \beta_3 x_{3j} + \zeta_j, \quad i = 1, 2, 3, 4$$

   where $\zeta_j|\mathbf{x}_{ij} \sim N(0, \psi)$.
3. Interpret the estimated effects of mother's smoking status for the models in steps 1 and 2.

6.3 Vaginal-bleeding data

Fitzmaurice, Laird, and Ware (2004) analyzed data from a trial reported by Machin et al. (1988). Women were randomized to receive an injection of either 100 mg or 150 mg of the long-lasting injectable contraception depot medroxyprogesterone acetate (DMPA) at the start of the trial and at three successive 90-day intervals. In addition, the women were followed up 90 days after the final injection. Throughout the study, each woman completed a menstrual diary that recorded any vaginal bleeding pattern disturbances. The diary data were used to determine whether a woman experienced amenorrhea, defined as the absence of menstrual bleeding for at least 80 consecutive days.

The response variable for each of the four 90-day intervals is whether the woman experienced amenorrhea during the interval. Data are available on 1,151 women for the first interval, but there was considerable dropout after that.

The dataset `amenorrhea.dta` has the following variables:

- `dose`: high dose (1: yes; 0: no)
- `y1`–`y4`: responses for intervals 1–4 (1: amenorrhea; 0: no amenorrhea)
- `wt2`: number of women with the same dose level and response pattern

1. Produce an identifier variable for women, and reshape the data to long form, stacking the responses `y1`–`y4` into one variable and creating a new variable `occasion` taking the values 1–4 for each woman.

2. Fit the following model considered by Fitzmaurice, Laird, and Ware (2004):

$$\text{logit}\{\Pr(y_{ij} = 1|x_j, t_{ij}, \zeta_j)\} = \beta_1 + \beta_2 t_{ij} + \beta_3 t_{ij}^2 + \beta_4 x_j t_{ij} + \beta_5 x_j t_{ij}^2 + \zeta_j$$

   where $t_{ij} = 1, 2, 3, 4$ is the time interval, x_j is `dose`, and $\zeta_j | x_j, t_{ij} \sim N(0, \psi)$. Use `gllamm` with the `weight(wt)` option to specify that `wt2` are level-2 weights.
3. Interpret the estimated coefficients.
4. Plot marginal predicted probabilities as a function of time, separately for women in the two treatment groups.

6.4 Verbal-aggression data

De Boeck and Wilson (2004) discuss a dataset from Vansteelandt (2000) where 316 participants were asked to imagine the following four frustrating situations where either another or oneself is to blame:

- Bus: a bus fails to stop for me (another to blame)
- Train: I miss a train because a clerk gave me faulty information (another to blame)
- Store: the grocery store closes just as I am about to enter (self to blame)
- Operator: the operator disconnects me when I have used up my last 10 cents for a call (self to blame)

For each situation, the participant was asked if it is was true (yes, perhaps, or no) that

- I would (want to) curse
- I would (want to) scold
- I would (want to) shout

For each of the three behaviors above, the words "want to" were both included and omitted, yielding six statements with a 3×2 factorial design (3 behaviors in 2 modes) combined with the four situations. Thus there were 24 items in total.

The dataset `aggression.dta` contains the following variables:

- `person`: subject identifier
- `item`: item (or question) identifier
- `description`: item description (mode: do/want; situation: bus/train/store/operator; behavior: curse/scold/shout)
- `i1`–`i24`: dummy variables for the items, e.g., `i5` equals 1 when `item` equals 5 and zero otherwise
- `y`: ordinal response (0: no; 1: perhaps; 2: yes)

- Person characteristics:
  - `anger`: trait anger score (STAXI, Spielberger [1988]) (w_{1j})
  - `gender`: dummy variable for being male (1: male; 0: female) (w_{2j})
- Item characteristics:
  - `do_want`: dummy variable for mode being "do" (i.e., omitting words *want to*) versus "want" (x_{2ij})
  - `other_self`: dummy variable for others to blame versus self to blame (x_{3ij})
  - `blame`: variable equal to 0.5 for blaming behaviors curse and scold and -1 for shout (x_{4ij})
  - `express`: variable equal to 0.5 for expressive behaviors curse and shout and -1 for scold (x_{5ij})

1. Recode the ordinal response variable `y` so that either a "2" or a "1" for the original variable becomes a "1" for the recoded variable.
2. De Boeck and Wilson (2004, sec. 2.5) consider the following "explanatory item-response model" for the dichotomous response

$$\text{logit}\{\Pr(y_{ij}=1|\mathbf{x}_{ij},\zeta_j)\} \;=\; \beta_1+\beta_2 x_{2ij}+\beta_3 x_{3ij}+\beta_4 x_{4ij}+\beta_5 x_{5ij}+\zeta_j$$

   where $\zeta_j|\mathbf{x}_{ij} \sim N(0,\psi)$ can be interpreted as the latent trait "verbal aggressiveness". Fit this model using `xtlogit`, and interpret the estimated coefficients. In De Boeck and Wilson (2004), the first five terms are negative so that their estimated coefficients have the opposite sign.
3. De Boeck and Wilson (2004, sec. 2.6) extend the above model by including a latent regression, allowing verbal aggressiveness (now denoted η_j instead of ζ_j) to depend on the personal characteristics w_{1j} and w_{2j}:

$$\text{logit}\{\Pr(y_{ij}=1|\mathbf{x}_{ij},\eta_j)\} \;=\; \beta_1+\beta_2 x_{2ij}+\beta_3 x_{3ij}+\beta_4 x_{4ij}+\beta_5 x_{5ij}+\eta_j$$

$$\eta_j \;=\; \gamma_1 w_{1j}+\gamma_2 w_{2j}+\zeta_j$$

   Substitute the level-2 model for η_j into the level-1 model for the item responses, and fit the model using `xtlogit`.
4. Use `xtlogit` to fit the "descriptive item-response model" considered by De Boeck and Wilson (2004, sec. 2.3):

$$\text{logit}\{\Pr(y_{ij}=1|d_{1i},\ldots,d_{24,i},\zeta_j)\} \;=\; \sum_{m=1}^{24}\beta_m d_{mi}+\zeta_j$$

   where d_{mi} is a dummy variable for item i, with $d_{mi}=1$ if $m=i$ and 0 otherwise. In De Boeck and Wilson (2004), the first term is negative so that their β_m coefficients have the opposite sign; see also their page 53.

5. The model above is known as a one-parameter item-response model because there is one parameter β_m for each item. The negative of these item-specific parameters $-\beta_m$ can be interpreted as "difficulties"; the larger $-\beta_m$, the larger the latent trait (here verbal aggressiveness, but often ability) has to be to yield a given probability (e.g., 0.5) of a 1 response.

   Sort the items in increasing order of the estimated difficulties. For the least and most difficult items, look up the variable `description`, and discuss whether it makes sense that these items are particularly easy and hard to endorse (requiring little and a lot of verbal aggressiveness), respectively.

See also exercise 7.2.

6.5 Women's employment data

The "Social Change and Economic Life Initiative", described in Davies, Elias, and Penn (1992) and Davies (1993), followed the employment status of wives from their marriage to the end of the survey in 1987. Here we consider a subsample from Rochdale, one of the six localities studied. These data are analyzed further in Skrondal and Rabe-Hesketh (2004, sec. 9.6).

The dataset `wemp.dta` has the following variables:

- `case`: identifier for wives
- `wemp`: wife's employment status
- `husunemp`: dummy variable taking the value 1 if the wife's husband is unemployed and 0 otherwise
- `time`: year of observation − 1975
- `under1`: dummy variable for the wife having children under the age of 1
- `under5`: dummy variable for the wife having children under the age of 5
- `age`: wife's age in years − 35

1. First, consider a transition or lagged-response model where current employment status depends on employment status in the previous year, as well as the covariates

$$\text{logit}\{\Pr(y_{ij}=1|\mathbf{x}_{ij},y_{i-1,j})\} = \beta_1+\beta_2 x_{2ij}+\cdots+\beta_6 x_{6ij}+\beta_7 y_{i-1,j}$$

   Fit this model using the `logit` command. (The lagged response can be obtained using `by case (age), sort: generate lag=wemp[_n-1]`.)

2. The model above somewhat unrealistically assumes that all the dependence among the responses for the same woman, after controlling for the covariates $\mathbf{x}_{ij}$, is due to "state dependence" on $y_{i-1,j}$. To investigate if the apparent state dependence is spurious due to omitting time-invariant covariates, we can include a random intercept (or permanent component) $\zeta_j|\mathbf{x}_{ij} \sim N(0,\psi)$,

$$\text{logit}\{\Pr(y_{ij}=1|\mathbf{x}_{ij},y_{i-1,j},\zeta_j)\} = \beta_1+\beta_2 x_{2ij}+\cdots+\beta_6 x_{6ij}+\beta_7 y_{i-1,j}+\zeta_j$$

   Fit this model using `xtlogit`. Is there any evidence for spurious state dependence?

6.6 Dairy-cow data

Dohoo et al. (2001) and Dohoo, Martin, and Stryhn (2003) analyzed data on dairy cows from Reunion Island. One outcome considered was the "risk" of conception at the first insemination attempt (first service) since the previous calving. This outcome was available for several lactations (calvings) per cow.

The variables in the dataset `dairy.dta` used here are

- `cow`: cow identifier
- `herd`: herd identifier
- `region`: geographic region
- `fscr`: first service conception risk (dummy variable for cow becoming pregnant)
- `lncfs`: log of time interval (in log days) between calving and first service (insemination attempt)
- `ai`: dummy variable for artificial insemination's being used (versus natural) at first service
- `heifer`: dummy variable for being a young cow that has only calved once

1. Fit a two-level random-intercept logistic regression model for the response variable `fscr`, an indicator for conception at the first insemination attempt (first service). Include a random intercept for cow and the covariates `lncfs`, `ai`, and `heifer`. (Use either `xtlogit`, `xtmelogit`, or `gllamm`.)
2. Obtain estimated odds ratios with 95% confidence intervals for the covariates and interpret them.
3. Obtain the estimated residual intraclass correlation between the latent responses for two observations on the same cow. Is there much variability in the cows' fertility?
4. Obtain the estimated median odds ratio for two randomly chosen cows with the same covariates, comparing the cow with the larger random intercept to the cow with the smaller random intercept.

See also exercises 10.4 and 10.5.

6.7 Union membership data

Vella and Verbeek (1998) analyzed panel data on 545 young males taken from the U.S. National Longitudinal Survey (Youth Sample) for the period 1980–1987. In this exercise, we will focus on modeling whether the men were members of unions or not.

The dataset `wagepan.dta` was provided by Wooldridge (2002) and was previously used in exercise 3.7 and chapter 5. The subset of variables considered here is

- `nr`: person identifier (j)
- `year`: 1980–1987 (i)
- `union`: dummy variable for being a member of a union (i.e., wage being set in collective bargaining agreement) (y_{ij})
- `educ`: years of schooling (x_{2j})
- `black`: dummy variable for being black (x_{3j})
- `hisp`: dummy variable for being Hispanic (x_{4j})
- `exper`: labor market experience, defined as age−6−`educ` (x_{5ij})
- `married`: dummy variable for being married (x_{6ij})
- `poorhlth`: dummy variable for having health disability (x_{7ij})
- `rur`: dummy variable for living in a rural area (x_{8ij})
- `nrtheast`: dummy variable for living in Northeast (x_{9ij})
- `nrthcen`: dummy variable for living in Northern Central ($x_{10,ij}$)
- `south`: dummy variable for living in South ($x_{11,ij}$)

You can use the `describe` command to get a description of the other variables in the file.

1. Fit the random-intercept logistic regression model

$$\text{logit}\{\Pr(y_{ij} = 1|\mathbf{x}_{ij}, \zeta_j)\} = \beta_1 + \beta_2 x_{2j} + \cdots + \beta_{11} x_{11,ij} + \zeta_j$$

   where $\zeta_j|\mathbf{x}_{ij} \sim N(0, \psi)$. Interpret the effects of the covariates in terms of estimated odds ratios. Use `xtlogit` because it is considerably faster than the other commands here.

2. Fit the random-intercept logistic regression model

$$\begin{aligned}\text{logit}\{\Pr(y_{ij} = & \ 1|\, y_{i-1,j}, \mathbf{x}_{ij}, \zeta_j)\} \\ & = \beta_1 + \beta_2 x_{2j} + \cdots + \beta_{11} x_{11,ij} + \beta_{12} y_{i-1,j} + \zeta_j\end{aligned}$$

   where β_{12} is the effect of union membership in the previous year $y_{i-1,j}$. You can create this lagged response using the command `by nr (year), sort: generate lag = union[_n-1]`. How do the estimates of the regression coefficients change after including the lagged effect?

3. Fit probit versions of the previous models using `xtprobit`. Which type of model do you find easiest to interpret?

6.8 School retention in Thailand data

A national survey of primary education was conducted in Thailand in 1988. The data were previously analyzed by Raudenbush and Bhumirat (1992) and are distributed with the HLM software (Raudenbush et al. 2004). Here we will model the probability that a child repeats a grade any time during primary school.

The dataset `thailand.dta` has the following variables:

- `rep`: dummy variable for child having repeated a grade during primary school (y_{ij})
- `schoolid`: school identifier (j)
- `pped`: dummy variable for child having preprimary experience (x_{2ij})
- `male`: dummy variable for child being male (x_{3ij})
- `mses`: school mean socioeconomic status (SES) (x_{4j})
- `wt1`: number of children in the school having a given set of values of `rep`, `pped`, and `male` (level-1 frequency weights).

1. Fit the model

$$\text{logit}\{\Pr(y_{ij}=1|\mathbf{x}_{ij},\zeta_j)\} = \beta_1+\beta_2 x_{2ij}+\beta_3 x_{3ij}+\beta_4 x_{4j}+\zeta_j$$

   where $\zeta_j|\mathbf{x}_{ij} \sim N(0,\psi)$. Use `gllamm` with the `weight(wt)` option to specify that each row in the data represents `wt1` children (level-1 units).
2. Obtain and interpret the estimated odds ratios and the estimated residual intraschool correlation of the latent responses.
3. Use `gllapred` to obtain empirical Bayes predictions of the probability of repeating a grade. These probabilities will be specific to the schools, as well as depending on the student-level predictors.
   a. List the values of `male`, `pped`, `rep`, `wt1` and the predicted probabilities for the school with `schoolid` equal to 10104. Explain why the predicted probabilities are greater than 0, although none of the children in the sample from that school have been retained. For comparison, list the same variables for the school with `schoolid` equal to 10105.
   b. Produce box plots of the predicted probabilities for each school by `male` and `pped` (for instance, using `by(male)` and `over(pped)`). To ensure that each school contributes no more than four probabilities to the graph (one for each combination of the student-level covariates), use only responses where `rep` is 0 (i.e., `if rep==0`). Do the schools appear to be variable in their retention probabilities?

6.9 PISA data

Here we consider data from the 2000 Program for International Student Assessment (PISA) conducted by the Organization for Economic Cooperation and Development (OECD 2000). The survey assessed educational attainment of 15 year olds in 43 countries in various areas, with an emphasis on reading. Following Rabe-Hesketh and Skrondal (2006), we will analyze reading proficiency, treated as dichotomous (1: proficient; 0: not proficient), for the U.S. sample.

The variables in the dataset `pisaUSA2000.dta` are

- `id_school`: school identifier
- `pass_read`: dummy variable for being proficient in reading
- `female`: dummy variable for student being female
- `isei`: international socioeconomic index
- `high_school`: dummy variable for highest education level by either parent being high school
- `college`: dummy variable for highest education level by either parent being college
- `test_lang`: dummy variable for test language (English) being spoken at home
- `one_for`: dummy variable for one parent being foreign born
- `both_for`: dummy variable for both parents being foreign born
- `w_fstuwt`: student-level or level-1 survey weights
- `wnrschbq`: school-level or level-2 survey weights

1. Fit a logistic regression model with `pass_read` as response variable and the variables `female`–`both_for` above as covariates and with a random intercept for schools using `gllamm`. (Use the default 8 quadrature points.)
2. Fit the model from step 1 with the school mean of `isei` as an additional covariate. (Use the estimates from step 1 as starting values.)
3. Interpret the estimated coefficients of `isei` and school mean `isei` and comment on the change in the other parameter estimates due to adding school mean `isei`.
4. From the estimates in step 2, obtain an estimate of the between-school effect of socioeconomic status.
5. Obtain robust standard errors using the command `gllamm, robust` and compare them with the model-based standard errors.
6. ❖ In this survey, schools were sampled with unequal probabilities, π_j, and given that a school was sampled, students were sampled from the school with unequal probabilities $\pi_{i|j}$. The reciprocals of these probabilities are given as school and student-level survey weights, `wnrschbg` ($w_j = 1/\pi_j$) and `w_fstuwt` ($w_{i|j} = 1/\pi_{i|j}$), respectively. As discussed in Rabe-Hesketh and Skrondal (2006), incorporating survey weights in multilevel models using a so-called *pseudolikelihood* approach, can lead to biased estimates, particularly if the level-1 weights $w_{i|j}$ are different from 1 and if the cluster sizes are small. Neither of these issues arise here, so implement pseudo maximum likelihood estimation as follows:

   a. Rescale the student-level weights by dividing them by their cluster means [this is scaling method 2 in Rabe-Hesketh and Skrondal (2006)].
   b. Rename the level-2 weights and rescaled level-1 weights to `wt2` and `wt1`, respectively.

c. Run the `gllamm` command from step 2 above with the additional option `pweight(wt)` (Only the stub of the weight variables is specified; `gllamm` will look for the level-1 weights under `wt1` and the level-2 weights under `wt2`.) Use the estimates from step 2 as starting values.
d. Compare the estimates with those from step 2. Robust standard errors are computed by `gllamm` because model-based standard errors are not appropriate with survey weights.

6.10 Wine tasting data

Fahrmeir and Tutz (2001) analyzed data on the bitterness of white wines.

The dataset `wine.dta` has the following variables:

- `bitter`: dummy variable for bottle being classified as bitter (y_{ij})
- `judge`: judge identifier (j)
- `temp`: temperature (low=1; high=0) x_{2ij}
- `contact`: skin contact when pressing the grapes (yes=1; no=0) x_{3ij}
- `repl`: replication

Interest concerns whether conditions that can be controlled while pressing the grapes such as temperature and skin contact influence the bitterness. For each combination of temperature and skin contact, two bottles of white wine were randomly chosen and the bitterness of each bottle was classified as "bitter" or "nonbitter" by the same nine professional judges.

To allow the judgment of bitterness to vary between judges, the following model was specified:

$$\ln\left\{\frac{\Pr(y_{ij}=1|x_{2ij},x_{3ij},\zeta_j)}{\Pr(y_{ij}=0|x_{2ij},x_{3ij},\zeta_j)}\right\} = \beta_1+\beta_2 x_{2ij}+\beta_3 x_{3ij}+\zeta_j,$$

where $\zeta_j|x_{2ij},x_{3ij} \sim N(0,\psi)$ and ζ_j is independent across judges. Maximum likelihood estimates and estimated standard errors for the model are given in the table below.

1. Interpret the estimated effects of the covariates as odds ratios.
2. State the expression for the residual intraclass correlation of the latent responses for the above model and estimate this intraclass correlation.
3. Consider two bottles characterized by the same covariates and judged by two randomly chosen judges. Estimate the median odds ratio comparing the judge with the larger random intercept to the judge with the smaller random intercept.
4. ❖ Based on the estimates given in the table, provide an approximate estimate of ψ if a probit model is used instead of a logit model. Assume that the estimated residual intraclass correlation of the latent responses is the same as for the logit model.

Table 6.3: Maximum likelihood estimates for bitterness model

	Estimate	Standard error
Fixed part		
β_1	−1.50	(0.90)
β_2	4.26	(1.30)
β_3	2.63	(1.00)
Random part		
ψ	2.80	
Log likelihood	−25.86	

5. ❖ Based on the estimates given in the table, provide approximate estimates for the marginal effects of x_{2ij} and x_{3ij} in an ordinary logistic regression model (without any random effects).

6.11 ❖ Random-intercept logistic model

In a hypothetical study, an ordinary probit model was fitted for students clustered in schools. The response was whether students gave the right answer to a question, and the single covariate was socioeconomic status (SES). The intercept and regression coefficient of SES were estimated as $\widehat{\beta}_1 = 0.2$ and $\widehat{\beta}_2 = 1.6$, respectively. The analysis was then repeated, this time including a normally distributed random intercept for school with variance estimated as $\widehat{\psi} = 0.15$.

1. Guess the values of the estimated school-specific regression coefficients for the random-intercept probit model.
2. Obtain the corresponding residual intraclass correlation for the latent responses.

7 Ordinal responses

7.1 Introduction

An ordinal variable is a categorical variable with ordered categories. An example from sociology would be the response to an attitude statement such as "Murderers should be executed" in one of the ordered categories "disagree strongly", "disagree", "agree", or "agree strongly". A medical example would be neurologists diagnosing multiple sclerosis using the ordered categories "certain", "probable", "possible", "doubtful", "unlikely", or "definitely not".

In this chapter, we generalize the multilevel models for dichotomous responses that were discussed in the previous chapter to handle ordinal responses. We start by introducing single-level cumulative logit and probit models for ordinal responses before extending these models to the multilevel and longitudinal setting by including random effects in the linear predictor.

Many of the issues that were discussed in the previous chapter, such as the distinction between conditional and marginal effects and the use of numerical integration for maximum likelihood estimation and empirical Bayes prediction, persist for multilevel modeling of ordinal responses. Indeed, the main difference between the logistic models considered in this and the previous chapter is that odds ratios must now be interpreted in terms of the odds of exceeding a given category.

7.2 Single-level cumulative models for ordinal responses

For simplicity, we start by introducing ordinal regression models for a single covariate x_i in the single-level case where units have index i. Just as for the dichotomous responses discussed in chapter 6, we can specify models for ordinal responses using either a generalized linear model formulation or a latent response formulation.

7.2.1 Generalized linear model formulation

Consider an ordinal response variable y_i with S ordinal categories denoted s ($s = 1, \ldots, S$). One way of specifying regression models for ordinal responses y_i is to let the cumulative probability that a response is in a higher category than s, given a covariate x_i, be structured as

$$\Pr(y_i > s|x_i) = F(\beta_2 x_i - \kappa_s) \qquad s=1,\ldots,S-1 \tag{7.1}$$

where $F(\cdot)$ is a cumulative distribution function (CDF): a standard normal CDF for the ordinal probit model and a logistic CDF for the ordinal logit model.

This is a generalized linear model if $F(\cdot)$ is considered the inverse link function, denoted $h(\cdot)$ or $g^{-1}(\cdot)$, and if the category-specific linear predictor ν_{is} is

$$\nu_{is} = \beta_2 x_i - \kappa_s$$

Here the κ_s are category-specific parameters, often called thresholds. Notice that we have omitted the intercept β_1 because it is not identified if all the $S-1$ thresholds κ_s are free parameters (see sec. 7.2.4). The covariate effect β_2 is constant across categories, a property sometimes referred to as the parallel-regressions assumption because the linear predictors for different categories s are parallel.

It follows from the cumulative probabilities in (7.1) that the probability for a specific category s can be obtained as

$$\begin{aligned}\Pr(y_i = s|x_i) &= \Pr(y_i > s-1|x_i) - \Pr(y_i > s|x_i)\\ &= F(\beta_2 x_i - \kappa_{s-1}) - F(\beta_2 x_i - \kappa_s)\end{aligned}$$

For this reason, cumulative models for ordinal responses are sometimes called *difference models*.

Having specified the link function and linear predictor, the final ingredient for a generalized linear model is the conditional distribution of the responses. For ordinal responses, this is a multinomial distribution with category-specific probabilities as given above. A simple example of a process following a multinomial distribution is rolling a die, where the probabilities for all categories $s=1,\ldots,6$ are equal to $1/6$.

Cumulative models for ordinal responses differ from generalized linear models for other response types in one important way: the inverse-link function of the linear predictor $g^{-1}(\nu_{is})$ does not represent the expectation or mean of the response (given the covariates), but a cumulative probability. Perhaps for this reason, Stata's `glm` command cannot be used for these models; instead, these models are fitted using the specialty commands `ologit` and `oprobit`.

7.2.2 Latent-response formulation

Cumulative models for ordinal responses can alternatively be viewed as linear regression models for latent (unobserved) continuous responses y_i^*,

$$y_i^* = \beta_2 x_i + \epsilon_i$$

where the conditional distribution of ϵ_i, given the observed covariate x_i, is standard normal for the ordinal probit and logistic for the ordinal logit.

Observed ordinal responses y_i are generated from the latent continuous responses y_i^* via a threshold model:

$$y_i = \begin{cases} 1 & \text{if} \quad y_i^* \le \kappa_1 \\ 2 & \text{if} \quad \kappa_1 < y_i^* \le \kappa_2 \\ \vdots & \vdots \quad\quad \vdots \\ S & \text{if} \quad \kappa_{S-1} < y_i^* \end{cases}$$

The threshold model is illustrated for an ordinal probit model without covariates, $y_i^* \sim N(0,1)$, with three categories ($S=3$) in figure 7.1.

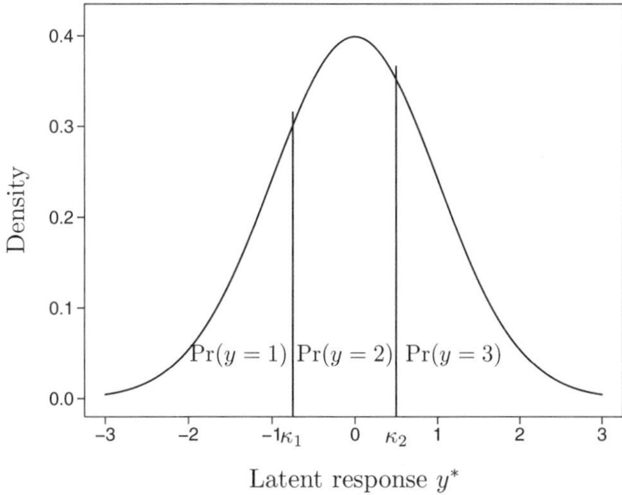

Figure 7.1: Illustration of threshold model for three-category ordinal probit

The observed response is 1 if the latent response y_i^* is less than κ_1 with corresponding probability $\Pr(y_i=1)$ given by the area under the (standard) normal density to the left of κ_1. The response is 2 if y_i^* is larger than κ_1 but less than κ_2 with $\Pr(y_i=2)$ given by the area under the density between these thresholds. Finally, the response is 3 if y_i^* is larger than κ_2 with $\Pr(y_i=3)$ given by the area under the density to the right of κ_2. Similar graphs are also shown in figure 7.4.

The latent-response and generalized linear model formulations are equivalent for ordinal responses, just as previously shown for dichotomous responses in section 6.2.2. This can be seen by considering the cumulative probabilities in (7.1):

$$\begin{aligned} \Pr(y_i > s|x_i) &= \Pr(y_i^* > \kappa_s|x_i) = \Pr(\beta_2 x_i + \epsilon_i > \kappa_s|x_i) \\ &= \Pr(-\epsilon_i \le \beta_2 x_i - \kappa_s|x_i) = F(\beta_2 x_i - \kappa_s) \end{aligned}$$

Figure 7.2 illustrates the equivalence of the formulations for an ordinal logit model with one covariate x_i. In the lower part of the figure, we present the (solid) regression line for the regression of a latent response y_i^* on a covariate x_i:

$$E(y_i^*|x_i) = \beta_2 x_i$$

For selected values of x_i, we have plotted the conditional logistic densities of the latent responses y_i^*. The thresholds κ_1 and κ_2 are represented as dashed and dotted lines, respectively. For a given value of x_i, the probability of observing a category above 1 is given by the grey (combined light and dark) area under the corresponding logistic density above κ_1. The probability of a category above 2 is the dark grey area under the logistic density above κ_2. These probabilities are plotted in the upper part of the figure as dashed and dotted lines, respectively.

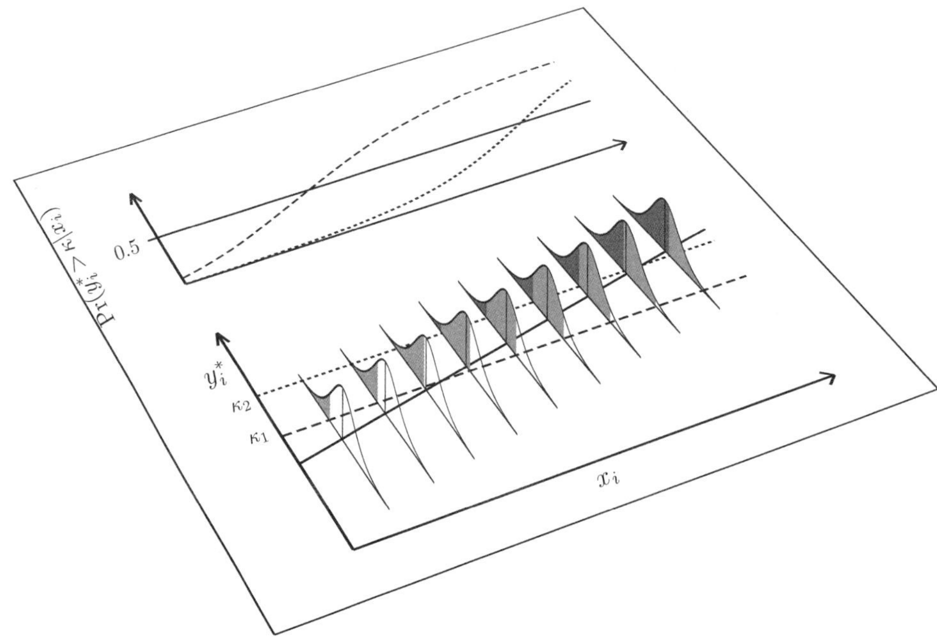

Figure 7.2: Illustration of equivalence of latent-response and generalized linear model formulation for ordinal logistic regression

The top panel of figure 7.3 shows the same cumulative probability curves as figure 7.2:

$$\begin{aligned}\Pr(y_i > 1|x_i) &= \Pr(y_i = 2|x_i) + \Pr(y_i = 3|x_i) \\ \Pr(y_i > 2|x_i) &= \Pr(y_i = 3|x_i)\end{aligned}$$

The corresponding probabilities of individual categories s are given by

$$\begin{aligned}\Pr(y_i = 3|x_i) &= \Pr(y_i > 2|x_i) \\ \Pr(y_i = 2|x_i) &= \Pr(y_i > 1|x_i) - \Pr(y_i > 2|x_i) \\ \Pr(y_i = 1|x_i) &= 1 - \Pr(y_i > 1|x_i)\end{aligned}$$

and are shown in the bottom panel of figure 7.3. The probabilities of the lowest and highest response categories 1 and S, which are categories 1 and 3 in the current example, are monotonic (strictly increasing or decreasing) functions of the covariates in cumulative models for ordinal responses. In contrast, the probability of an intermediate category is a unimodal or single-peaked function, as shown for $\Pr(y_i = 2|x_i)$ in the bottom panel.

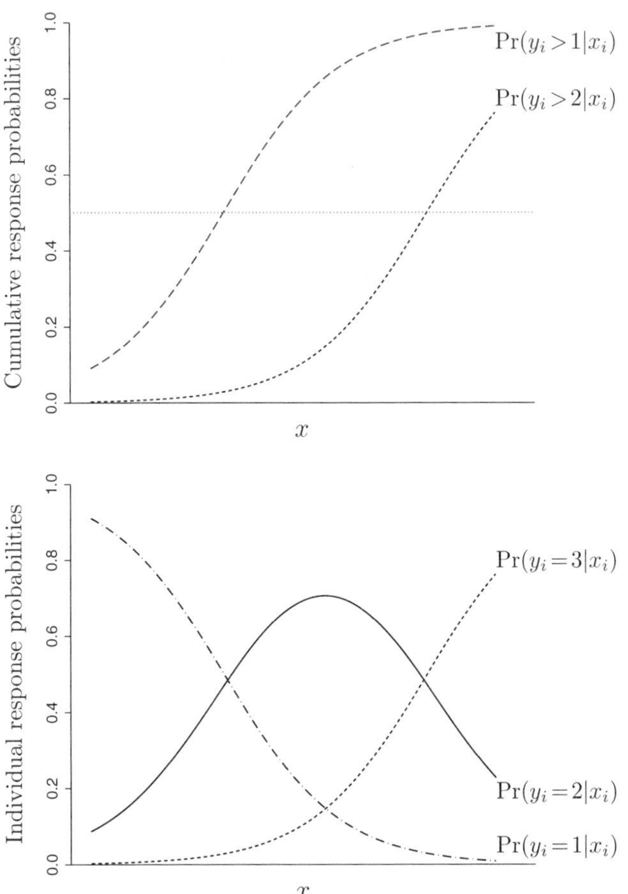

Figure 7.3: Illustration of cumulative and individual or category-specific response probabilities

(*Continued on next page*)

7.2.3 Proportional odds

If we specify a logit link or equivalently assume a logistic distribution for the residual ϵ_i in the latent-response model, the cumulative probabilities become

$$\Pr(y_i > s|x_i) = \frac{\exp(\beta_2 x_i - \kappa_s)}{1+\exp(\beta_2 x_i - \kappa_s)}$$

and the corresponding log-odds are

$$\ln\left\{\frac{\Pr(y_i > s|x_i)}{1-\Pr(y_i > s|x_i)}\right\} = \beta_2 x_i - \kappa_s$$

It follows that the ratios of the odds that the response exceeds s for two covariate values x_i and $x_{i'}$ becomes

$$\frac{\Pr(y_i > s|x_i)/\{1-\Pr(y_i > s|x_i)\}}{\Pr(y_{i'} > s|x_{i'})/\{1-\Pr(y_{i'} > s|x_{i'})\}} = \exp\{\beta_2(x_i - x_{i'})\} = \exp(\beta_2)^{(x_i - x_{i'})}$$

We see that for unit increase in the covariate, $x_i - x_{i'} = 1$, the odds ratio is $\exp(\beta_2)$. Indeed, for dummy variables the only possible increase is a unit increase from 0 to 1. In contrast, for a continuous covariate, we can also consider for instance a 10-unit change. Then we must raise $\exp(\beta_2)$ to the 10th power because the model for the odds is multiplicative.

The category-specific parameter κ_s cancels out from the odds ratio, so the odds ratio is the same whatever category s is considered. For instance, for a three-category response, the odds ratio of 3 versus 1 or 2 is the same as the odds ratio for 3 or 2 versus 1. This property, called *proportional odds*, makes it evident that these models are highly structured. It may hence come as no surprise that several models have been suggested that relax this property (see section 7.12).

If the proportional odds property holds, we can merge or collapse categories—or in the extreme case, dichotomize an ordinal variable—and still get similar estimates. For instance, in the case of three categories, we could redefine the categories as $1 = \{2,3\}$ and $0 = \{1\}$ and fit a binary logistic regression model. The interpretation of the odds ratios for this model would then correspond to one of the interpretations of the odds ratios for the original proportional odds model: the odds ratios for categories 2 or 3 versus 1. However, if the proportional-odds assumption is violated, the estimates may be quite different.

7.2.4 ❖ Identification

Consider now for simplicity a model for y_i^* without covariates, but with an intercept β_1 and a variance parameter θ,

$$y_i^* = \beta_1 + \epsilon_i, \qquad \epsilon_i \sim N(0,\theta)$$

which can alternatively be written as

$$y_i^* \sim N(\beta_1, \theta)$$

It follows from this model that the probability of observing a category larger than s becomes

$$\Pr(y_i > s) = \Pr(y_i^* > \kappa_s) = \Pr\left(\frac{y_i^* - \beta_1}{\sqrt{\theta}} > \frac{\kappa_s - \beta_1}{\sqrt{\theta}}\right) = \Phi\left(\frac{\beta_1 - \kappa_s}{\sqrt{\theta}}\right)$$

where $\Phi(\cdot)$ is the cumulative standard normal density function.

We see from this expression that the cumulative probabilities $\Pr(y_i > s)$ do not change if we add some constant a to both the latent response y_i^*, and therefore to β_1, and to the thresholds κ_s, since the constant cancels out in the numerator of the argument. Similarly, we see that the cumulative probabilities do not change if we multiply y_i^* by a constant b, so that β_1 and $\sqrt{\theta}$ are multiplied by b; and if we multiply κ_s by the same constant b, since we are in effect multiplying both the numerator and the denominator by the same constant. It follows that the location and scale of the latent responses y_i^* are not identified if the thresholds are free parameters and thus cannot be estimated from the data.

This invariance is also illustrated in figure 7.4. It is evident from the left panel (top and bottom) of the figure that response probabilities are invariant to translation of the latent response y_i^*; adding 2 to y_i^* (or to the intercept β_1) can be counteracted by adding 2 to the thresholds κ_s. From the right panel (top and bottom), we see that the response probabilities are invariant to rescaling of y_i^*; multiplying y_i^* by 2 (or multiplying β_1 and $\sqrt{\theta}$ by 2) can be counteracted by multiplying the thresholds κ_s by 2. Thus the location and scale of latent responses y_i^* are identified only relative to the location and spacing of the thresholds κ_s.

(*Continued on next page*)

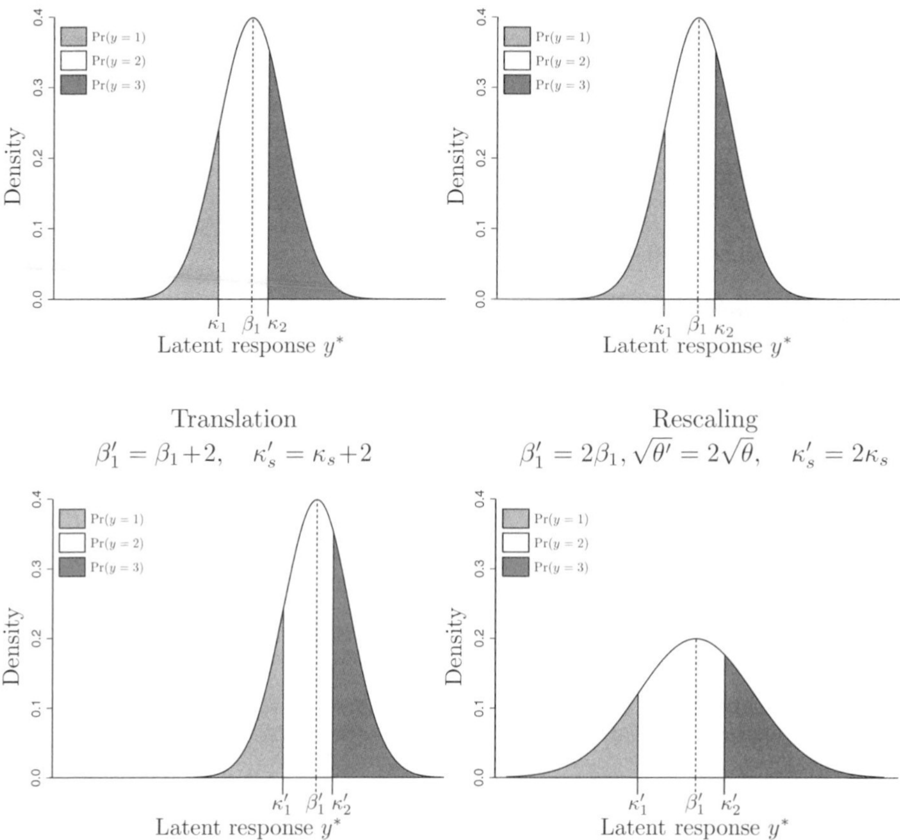

Figure 7.4: Illustration of scale and translation invariance ($\beta_1 = 0$)

To identify ordinal regression models, the location of the latent response is usually fixed by setting the intercept β_1 to zero, which is what we have done so far in this chapter. The scale is usually fixed by setting $\theta = 1$ for probit models and $\theta = \pi^2/3$ for logit models. These two conventions are most common for ordinal models and are used by Stata's `ologit` and `oprobit` commands.

An alternative to fixing the location of y_i^* by setting $\beta_1 = 0$ is to fix the location of the thresholds, typically by setting $\kappa_1 = 0$ to identify the intercept β_1. This parameterization is common for dichotomous responses and is used in Stata's `logit`, `probit`, and `glm` commands. Less commonly, we could fix both the location and the scale of the thresholds by setting $\kappa_1 = 0$ and $\kappa_2 = 1$ so that both β_1 and θ can be estimated.

7.3 Are antipsychotic drugs effective for patients with schizophrenia?

Schizophrenia is a mental illness characterized by impairments in the perception or expression of reality, most commonly in terms of auditory hallucinations and paranoid or bizarre delusions. According to the U.S. National Institute of Mental Health (NIMH), the prevalence of schizophrenia is about 1% in the U.S. population aged 18 and older.

We now use data from the NIMH Schizophrenia Collaborative Study on treatment-related changes in overall severity of schizophrenia. The data have previously been analyzed by Hedeker and Gibbons (1996), Gibbons et al. (1988), Gibbons and Hedeker (1994), and Hedeker and Gibbons (2006).

Patients were randomly assigned to receive one of four treatments: placebo (control), chlorpromazine, fluphenazine, or thioridazine. In the present analysis, we will not distinguish between the three antipsychotic drugs and call the combined group the treatment group. The outcome considered is "severity of illness" as measured by item 79 of the Inpatient Multidimensional Psychiatric Scale (IMPS) of Lorr and Klett (1966). The patients were examined weekly for up to 6 weeks.

The dataset `schiz.dta` has the following variables:

- `id`: subject identifier
- `week`: week of assessment since randomization (0, 1, ..., 6)
- `imps`: item 79 of IMPS
- `treatment`: dummy variable for being in treatment group (1: treatment (drug); 0: control)

Following Hedeker and Gibbons (1996), we recode item 79 of the IMPS into an ordinal severity of illness variable `impso` with four categories (1: normal or borderline mentally ill; 2: mildly or moderately ill; 3: markedly ill; 4: severely or among the most extremely ill). This can be accomplished using the `recode` command:

```
. use http://www.stata-press.com/data/mlmus2/schiz
. generate impso = imps
. recode impso -9=. 1/2.4=1 2.5/4.4=2 4.5/5.4=3 5.5/7=4
```

7.4 Longitudinal data structure and graphs

Before we start modeling the severity of schizophrenia over time, let us look at the dataset, using both descriptive statistics and graphs. This may provide insights that can be used in subsequent model specification.

7.4.1 Longitudinal data structure

We first use the `xtdescribe` command to describe the participation pattern in the dataset:

```
. xtdescribe if impso<., i(id) t(week)
      id:  1103, 1104, ..., 9316                             n =        437
    week:  0, 1, ..., 6                                      T =          7
           Delta(week) = 1; (6-0)+1 = 7
           (id*week uniquely identifies each observation)
Distribution of T_i:   min      5%      25%       50%       75%     95%     max
                         2       2        4         4         4       4       5
     Freq.  Percent    Cum.    Pattern
 ---------------------------+---------
      308     70.48   70.48 |  11.1..1
       41      9.38   79.86 |  11.1...
       37      8.47   88.33 |  11.....
        8      1.83   90.16 |  11....1
        8      1.83   91.99 |  111....
        6      1.37   93.36 |  11.1.1.
        5      1.14   94.51 |  1..1..1
        5      1.14   95.65 |  11.11..
        3      0.69   96.34 |  .1.1..1
       16      3.66  100.00 | (other patterns)
 ---------------------------+---------
      437    100.00         |  XXXXXXX
```

Clearly the data are unbalanced. For instance, all patterns shown have missing assessments for at least three occasions. The predominant pattern "`11.1..1`" (308 subjects) has assessments only at weeks 0, 1, 3, and 6. We can tabulate the number of observations in each week for the treatment and control groups using the `table` command:

```
. table week treatment, contents(count impso) col

----------+-----------------------
          |       treatment
     week |      0       1   Total
----------+-----------------------
        0 |    107     327     434
        1 |    105     321     426
        2 |      5       9      14
        3 |     87     287     374
        4 |      2       9      11
        5 |      2       7       9
        6 |     70     265     335
----------+-----------------------
```

We see that few assessments took place in weeks 2, 4, and 5.

7.4.2 Plotting cumulative proportions

When considering cumulative models, it is natural to first inspect the cumulative proportions in the dataset. We therefore calculate the proportions of patients having responses above 1, 2, and 3 at each occasion and by treatment group. These proportions can be calculated conveniently using Stata's `egen` command with the `mean()` function:

```
. egen propg1 = mean(impso>1), by(week treatment)
. egen propg2 = mean(impso>2), by(week treatment)
. egen propg3 = mean(impso>3), by(week treatment)
```

Since only a few assessments were made in weeks 2, 4, and 5, we will plot cumulative proportions only for weeks 0, 1, 3, and 6. To simplify later commands, we define a dummy variable `nonrare` for these weeks using `egen` with the `anymatch()` function:

```
. egen nonrare = anymatch(week), values(0,1,3,6)
```

We then define a label for `treatment` and produce a graph for the cumulative proportions against week:

```
. label define t 0 "Control" 1 "Treatment", modify
. label values treatment t
. sort treatment id week
. twoway (line propg1 week, sort)
> (line propg2 week, sort lpatt(vshortdash))
> (line propg3 week, sort lpatt(dash)) if nonrare==1, by(treatment)
> legend(order(1 "Prop(y>1)" 2 "Prop(y>2)" 3 "Prop(y>3)")) xtitle("Week")
```

The resulting graph, shown in figure 7.5, suggests that antipsychotic drugs have a beneficial effect since the proportions in the higher, more severe, categories decline more rapidly in the treatment group. We observe that the cumulative proportions also decline somewhat in the control group.

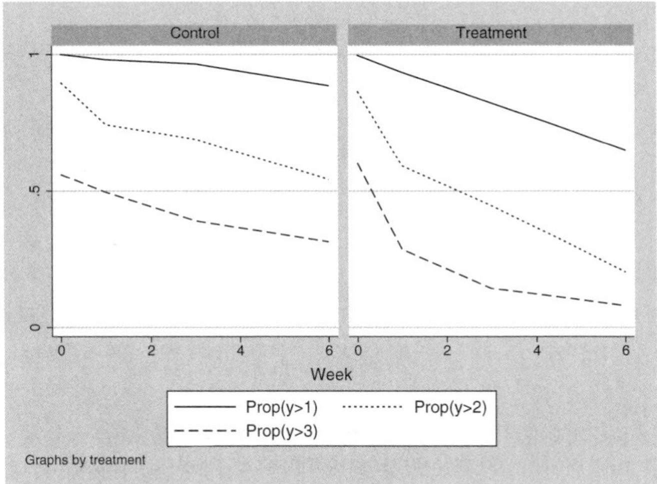

Figure 7.5: Cumulative proportions versus week

7.4.3 Plotting estimated cumulative logits and transforming the time scale

In ordinal logistic regression, the cumulative log odds or logits, and not the cumulative proportions, are specified as a linear function of covariates. Before specifying a model for the time trend, it is therefore useful to plot the estimated cumulative logits against `week`:

```
. generate logodds1 = ln(propg1/(1-propg1))
. generate logodds2 = ln(propg2/(1-propg2))
. generate logodds3 = ln(propg3/(1-propg3))
. twoway (line logodds1 week, sort)
> (line logodds2 week, sort lpatt(vshortdash))
> (line logodds3 week, sort lpatt(dash)) if nonrare==1, by(treatment)
> legend(order(1 "Log Odds(y>1)" 2 "Log Odds(y>2)" 3 "Log Odds(y>3)"))
> xtitle("Week")
```

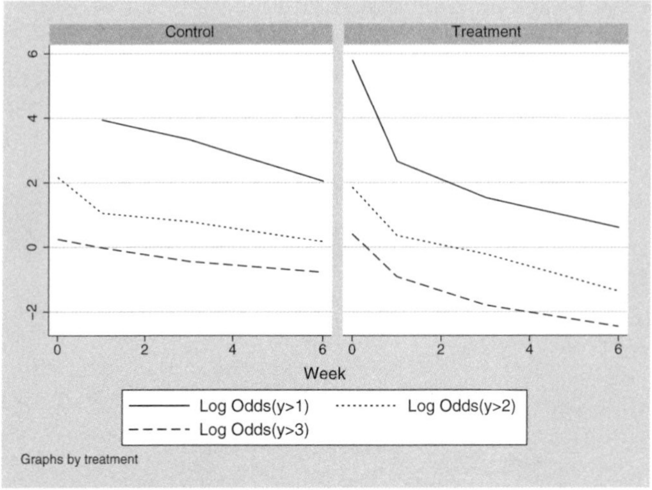

Figure 7.6: Estimated cumulative logits versus week

The missing value for the log odds of being in categories above 1 in week 0 for the control group is due to the corresponding proportion being equal to 1 (see figure 7.5), giving an odds of ∞.

The estimated cumulative logits in figure 7.6 would be poorly approximated by a linear function of `week` in the treatment group. One way to address this problem would be to depart from a linear trend for the cumulative logits and instead consider some polynomial function, such as a quadratic or cubic. Another approach, adopted by Hedeker and Gibbons (1996) and pursued here, is to retain a linear trend for the logits but transform the time scale by taking the square root of `week`:

```
. generate weeksqrt = sqrt(week)
```

Plots for the estimated cumulative logits against the transformed time variable `weeksqrt` are produced by the command

```
. twoway (line logodds1 weeksqrt, sort)
> (line logodds2 weeksqrt, sort lpatt(vshortdash))
> (line logodds3 weeksqrt, sort lpatt(dash)) if nonrare==1, by(treatment)
> legend(order(1 "Log Odds(y>1)" 2 "Log Odds(y>2)" 3 "Log Odds(y>3)"))
> xtitle("Square root of week")
```

and presented in figure 7.7.

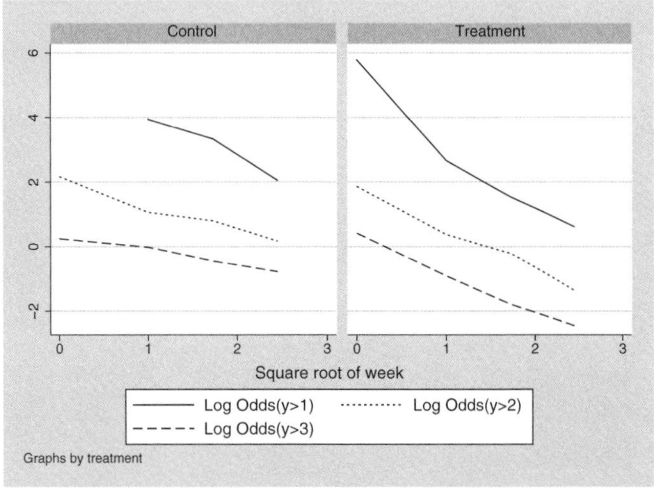

Figure 7.7: Estimated cumulative logits versus square root of week

As desired, the curves generally appear to be more linear after the square-root transformation of time.

7.5 A single-level proportional odds model

We start modeling the schizophrenia data by fitting a single-level ordinal logistic regression model.

7.5.1 Model specification

Based on figure 7.7 it seems reasonable to assume a linear relationship between the cumulative logits and the square root of week, allowing intercepts and slopes to differ between the treatment and control groups,

$$\begin{aligned}\text{logit}\{\Pr(y_{ij} > s|x_{2ij}, x_{3j})\} &= \beta_2 x_{2ij} + \beta_3 x_{3j} + \beta_4 x_{2ij} x_{3j} - \kappa_s \\ &= (\beta_2 + \beta_4 x_{3j}) x_{2ij} + \beta_3 x_{3j} - \kappa_s \qquad (7.2)\end{aligned}$$

where x_{2ij} represents **weeksqrt** and x_{3j} represents **treatment**. Here β_2 is the slope of time (in square root of week) for the control group, β_3 the difference between treatment and control groups at week 0, and β_4 the difference in the slopes of time between treatment and control groups. From the figure, we would expect β_4, which we can interpret as the treatment effect, to be negative, and the corresponding odds ratio, given by $\exp(\beta_4)$, to be less than 1.

To interpret the odds ratio $\exp(\beta_4)$, consider the odds ratio for a unit increase in time x_{2ij}:

$$\exp(\beta_2 + \beta_4 x_{3j}) = \begin{cases} \exp(\beta_2) & \text{if } x_{3j} = 0 \quad \text{(control group)} \\ \exp(\beta_2)\exp(\beta_4) & \text{if } x_{3j} = 1 \quad \text{(treatment group)} \end{cases}$$

Thus $\exp(\beta_4)$ is the odds ratio of time for the treatment group divided by the the odds ratio of time for the control group.

7.5.2 Estimation using Stata

The Stata commands for fitting the single-level proportional odds model (7.2) using the `ologit` command are

```
. generate interact = weeksqrt*treatment
. ologit impso weeksqrt treatment interact, vce(cluster id) or
Ordered logistic regression                     Number of obs   =       1603
                                                Wald chi2(3)    =     440.17
                                                Prob > chi2     =     0.0000
Log pseudolikelihood = -1878.0969               Pseudo R2       =     0.1177
                               (Std. Err. adjusted for 437 clusters in id)
```

impso	Odds Ratio	Robust Std. Err.	z	P>|z|	[95% Conf.	Interval]
weeksqrt	.5847056	.0591797	-5.30	0.000	.4794958	.7130004
treatment	.9993959	.2042595	-0.00	0.998	.6695244	1.491793
interact	.4719089	.0568135	-6.24	0.000	.3727189	.5974961
/cut1	-3.807279	.1956796			-4.190804	-3.423754
/cut2	-1.760167	.1811041			-2.115125	-1.40521
/cut3	-.4221112	.1795596			-.7740415	-.0701808

Here we have used the `vce(cluster id)` option to obtain robust standard errors (based on the sandwich estimator), taking into account the clustered nature of the data, and the `or` option to obtain estimated odds ratios $\exp(\widehat{\beta})$ with corresponding 95% confidence intervals. The output looks just like output from binary logistic regression (using `logit` with the `or` option), except for the additional estimates of the thresholds κ_1–κ_3, labeled /cut1–/cut3. Maximum likelihood estimates for the proportional odds model are presented under the heading "POM" in table 7.1.

Table 7.1: Maximum likelihood estimates and 95% CIs for proportional odds model (POM), random-intercept proportional odds model (RI-POM), and random-coefficient proportional odds model (RC-POM)

	POM		RI-POM		RC-POM	
	Est	(95% CI)†	Est	(95% CI)	Est	(95% CI)
Fixed part: Odds ratios						
$\exp(\beta_2)$ [weeksqrt]	0.58	(0.48, 0.71)	0.46	(0.36, 0.60)	0.41	(0.27, 0.63)
$\exp(\beta_3)$ [treatment]	1.00	(0.67, 1.49)	0.94	(0.51, 1.75)	1.06	(0.49, 2.28)
$\exp(\beta_4)$ [interact]	0.47	(0.37, 0.60)	0.30	(0.22, 0.40)	0.18	(0.11, 0.30)
Fixed part: Thresholds						
κ_1	−3.81		−5.86		−7.32	
κ_2	−1.76		−2.83		−3.42	
κ_3	−0.42		−0.71		−0.81	
Random part: Variances and covariance						
ψ_{11}			3.77		6.99	
ψ_{22}					2.01	
ψ_{21}					−1.51	
Log likelihood	−1,878.10		−1,701.38		−1,662.76	

†Based on the sandwich estimator

The odds ratios for this model represent marginal or population-averaged effects. We see from table 7.1 that the odds ratio of more severe illness per unit of time (on the transformed scale) is estimated as 0.58 (95% CI from 0.48 to 0.71) for the control group. The corresponding odds ratio for the treatment group is estimated as $0.58 \times 0.47 = 0.28$. To obtain a confidence interval for the latter odds ratio, we can use the `lincom` command with the `or` (or `eform`) option:

```
. lincom weeksqrt+interact, or
 ( 1)  [impso]weeksqrt + [impso]interact = 0

       impso   Odds Ratio   Std. Err.      z    P>|z|     [95% Conf. Interval]

         (1)     .2759278    .0184819   -19.22   0.000      .241981    .3146369
```

The 95% CI for the odds ratio per square root of week ranges from 0.24 to 0.31 in the treatment group. All these odds ratios are ratios of the odds of more severe versus less severe illness regardless of where we cut the ordinal scale to define more versus less.

An advantage of estimating marginal relationships is that we can easily assess the fit by comparing model-implied probabilities with observed proportions. Model-implied probabilities for the individual response categories can be obtained by using the `predict` command with the `pr` option:

```
. predict prob1-prob4, pr
```

The corresponding cumulative probabilities of exceeding categories 3, 2, and 1 are obtained by the following commands:

```
. generate probg3 = prob4
. generate probg2 = prob3 + probg3
. generate probg1 = prob2 + probg2
```

We now plot these marginal probabilities together with the corresponding marginal proportions:

```
. twoway (line propg1 week if nonrare==1, sort lpatt(solid))
> (line propg2 week if nonrare==1, sort lpatt(vshortdash))
> (line propg3 week if nonrare==1, sort lpatt(dash))
> (line probg1 week, sort lpatt(solid))
> (line probg2 week, sort lpatt(vshortdash))
> (line probg3 week, sort lpatt(dash)), by(treatment)
> legend(order(1 "Prob(y>1)" 2 "Prob(y>2)" 3 "Prob(y>3)")) xtitle("Week")
```

The resulting graphs shown in figure 7.8 suggest that the marginal proportions are well recovered by the model.

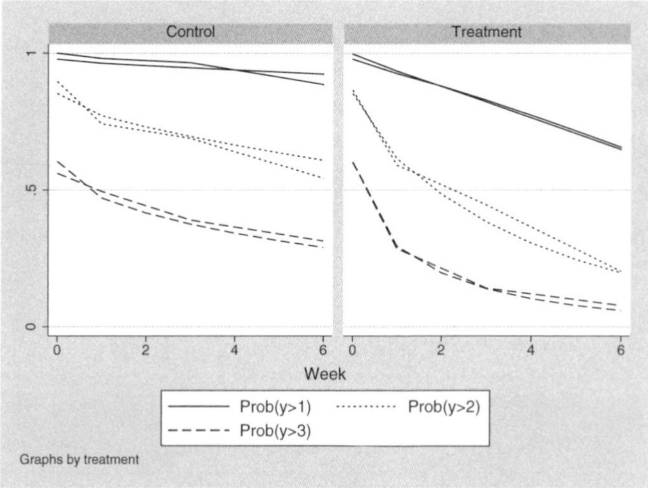

Figure 7.8: Cumulative proportions and predicted cumulative probabilities from ordinal logistic regression versus week

However, the single-level ordinal proportional odds model (7.2) unrealistically assumes that the responses on the same patient are conditionally independent given the covariates.

7.6 A random-intercept proportional odds model

7.6.1 Model specification

We now account for longitudinal dependence by including a patient-specific random intercept ζ_{1j} in the proportional odds model:

$$\text{logit}\{\Pr(y_{ij} > s|\mathbf{x}_{ij}, \zeta_{1j})\} \;=\; \beta_2 x_{2ij} + \beta_3 x_{3j} + \beta_4 x_{2j} x_{3ij} + \zeta_{1j} - \kappa_s \qquad (7.3)$$

The overall level or intercept of the cumulative logits is ζ_{1j} and is hence varying over patients j. As usual, we assume that $\zeta_{1j}|\mathbf{x}_{ij} \sim N(0, \psi)$ and that the ζ_{1j} are independent across patients.

The model can alternatively be written in terms of a latent-response formulation. Here we specify a linear random-intercept model for the latent response y_{ij}^*,

$$y_{ij}^* \;=\; \beta_2 x_{2ij} + \beta_3 x_{3j} + \beta_4 x_{2j} x_{3ij} + \zeta_{1j} + \epsilon_{ij}$$

where the $\epsilon_{ij}|\mathbf{x}_{ij}, \zeta_{1j}$ have logistic distributions and are independent across patients and occasions. The continuous latent responses y_{ij}^* are related to the ordinal severity of illness variable y_{ij} via the threshold model

$$y_{ij} = \begin{cases} 1 & \text{if} \quad y_{ij}^* \le \kappa_1 \\ 2 & \text{if} \quad \kappa_1 < y_{ij}^* \le \kappa_2 \\ 3 & \text{if} \quad \kappa_2 < y_{ij}^* \le \kappa_3 \\ 4 & \text{if} \quad \kappa_3 < y_{ij}^* \end{cases}$$

7.6.2 Estimation using Stata

At the time of writing this book, the only Stata command for fitting multilevel models with ordinal responses is `gllamm`. Ordinal logit and probit models can be fitted in `gllamm` by using the `link(ologit)` and `link(oprobit)` options, respectively. To fit the random-intercept proportional odds model in (7.3), we use the options `i(id)` to specify that the responses are nested in patients with identifier `id`, `adapt` to request adaptive quadrature, and `eform` to obtain exponentiated estimates or odds ratios:

(Continued on next page)

```
. gllamm impso weeksqrt treatment interact, i(id) link(ologit) adapt eform
number of level 1 units = 1603
number of level 2 units = 437

Condition Number = 15.409532

gllamm model

log likelihood = -1701.3807
```

	exp(b)	Std. Err.	z	P>|z|	[95% Conf.	Interval]
impso						
weeksqrt	.4649525	.0608031	-5.86	0.000	.3598277	.6007899
treatment	.9439404	.2962807	-0.18	0.854	.5102375	1.746292
interact	.2993646	.0457031	-7.90	0.000	.2219474	.4037855
_cut11						
_cons	-5.858453	.331792	-17.66	0.000	-6.508753	-5.208153
_cut12						
_cons	-2.825669	.2900513	-9.74	0.000	-3.394159	-2.257179
_cut13						
_cons	-.7077072	.2750904	-2.57	0.010	-1.246875	-.1685399

```
Variances and covariances of random effects
------------------------------------------------------------------------------

***level 2 (id)

    var(1): 3.7733416 (.46496877)
------------------------------------------------------------------------------
. estimates store model1
```

The estimated odds ratios with 95% confidence intervals from the random-intercept proportional odds model are presented under "RI-POM" in table 7.1. As expected, these estimates of conditional or subject-specific effects are further from 1 than the marginal or population-averaged counterparts from model (7.2). The subject-specific odds ratio of time is estimated as 0.46 in the control group and as $0.46 \times 0.30 = 0.14$ in the treatment group.

The random-intercept variance is estimated as $\widehat{\psi}_{11} = 3.77$, implying an estimated residual intraclass correlation for the latent responses y^*_{ij} of

$$\widehat{\rho} = \frac{\widehat{\psi}_{11}}{\widehat{\psi}_{11} + \pi^2/3} = \frac{3.773}{3.773 + \pi^2/3} = 0.53$$

7.7 A random-coefficient proportional odds model

7.7.1 Model specification

To allow the slope of the time variable `weeksqrt` to vary randomly between patients within the two groups, we now include a random slope ζ_{2j} in addition to the random intercept ζ_{1j}:

$$\begin{aligned}\text{logit}\{\Pr(y_{ij} > s|\mathbf{x}_{ij}, \zeta_{1j}, \zeta_{2j})\} & \\ &= \beta_2 x_{2ij} + \beta_3 x_{3j} + \beta_4 x_{2j} x_{3ij} + \zeta_{1j} + \zeta_{2j} x_{2ij} - \kappa_s \\ &= \zeta_{1j} + (\beta_2 + \zeta_{2j}) x_{2ij} + \beta_3 x_{3j} + \beta_4 x_{2j} x_{3ij} - \kappa_s \qquad (7.4)\end{aligned}$$

In this model, not only the intercept but also the slope $\beta_2 + \zeta_{2j}$ of `weeksqrt` (x_{2ij}) varies over patients j. We assume that, given $\mathbf{x}_{ij}$, the random intercept and slope have a bivariate normal distribution with zero mean and covariance matrix

$$\mathbf{\Psi} = \begin{bmatrix} \psi_{11} & \psi_{12} \\ \psi_{21} & \psi_{22} \end{bmatrix}, \qquad \psi_{21} = \psi_{12}$$

and that both the random intercepts and the random slopes are independent across patients.

The random-coefficient proportional odds model can also be specified using a latent-response formulation as shown for the random-intercept proportional odds model in section 7.6.1.

7.7.2 Estimation using gllamm

To fit the random-coefficient proportional odds model using `gllamm`, we must first define equations for the intercept and slope:

```
. generate cons = 1
. eq inter: cons
. eq slope: weeksqrt
```

It is usually a good idea to use the estimates for the random-intercept model as starting values for the random-coefficient model. We can obtain the estimates for the random-intercept model by using

```
. matrix a = e(b)
```

since this model was the last one fitted. However, the random-coefficient model requires two additional parameters, which happen to be located at the end of the parameter vector. We arbitrarily specify values 0.1 and 0 for these and put them in the augmented matrix

```
. matrix a = (a,.1,0)
```

We are now ready to fit the random-coefficient model, using the `from(a)` and `copy` options to specify that the augmented matrix `a` contains the starting values in the correct order:

```
. gllamm impso weeksqrt treatment interact, i(id) nrf(2) eqs(inter slope)
> link(ologit) adapt from(a) copy eform
number of level 1 units = 1603
number of level 2 units = 437

Condition Number = 13.095801

gllamm model

log likelihood = -1662.7601
```

	exp(b)	Std. Err.	z	P>|z|	[95% Conf.	Interval]
impso						
weeksqrt	.413751	.0901232	-4.05	0.000	.2699793	.6340851
treatment	1.058938	.4152881	0.15	0.884	.4909661	2.283967
interact	.1837922	.0463305	-6.72	0.000	.1121387	.3012304
_cut11						
_cons	-7.317755	.4732883	-15.46	0.000	-8.245383	-6.390127
_cut12						
_cons	-3.417135	.3869274	-8.83	0.000	-4.175499	-2.658771
_cut13						
_cons	-.8119554	.3526935	-2.30	0.021	-1.503222	-.1206888

```
Variances and covariances of random effects
------------------------------------------------------------------------------

***level 2 (id)

    var(1): 6.9860366 (1.3149406)
    cov(2,1): -1.5060781 (.53159615) cor(2,1): -.40211944

    var(2): 2.0079551 (.41889393)
------------------------------------------------------------------------------
. estimates store model2
```

We can compare the random-coefficient proportional odds model (7.4) with the random-intercept proportional odds model (7.3) using

```
. lrtest model1 model2
(log likelihoods of null models cannot be compared)
Likelihood-ratio test                                 LR chi2(2)  =     77.24
(Assumption: model1 nested in model2)                 Prob > chi2 =    0.0000
```

Although the test is conservative (since the null hypothesis is on the boundary of the parameter space), we get a p-value that is close to zero. The random-intercept model is thus rejected in favor of the random-coefficient model.

The maximum likelihood estimates for the random-coefficient proportional odds model are presented under "RC-POM" in table 7.1. In this model, the subject-specific odds ratio per unit of time (in square root of week) is estimated as 0.41 in the control group and as $0.41 \times 0.18 = 0.07$ in the treatment group. The random-intercept variance is estimated as $\widehat{\psi}_{11} = 6.99$ and the random-slope variance as $\widehat{\psi}_{22} = 2.01$. The covariance between the random intercept and slope is estimated as $\widehat{\psi}_{21} = -1.51$, corresponding to a correlation of -0.40. This means that patients having severe schizophrenia at the onset of the study (week= 0) tend to have a greater decline in severity than those with less severe schizophrenia in both the control and treatment groups.

7.8 Different kinds of predicted probabilities

7.8.1 Predicted population-averaged probabilities

As shown for binary logistic regression models in section 6.13.1, we can obtain the population-averaged or marginal probabilities implied by the fitted random-coefficient proportional odds model using gllapred with the mu and marginal options. Since there are several response categories in ordinal models, what gllamm actually provides is cumulative probabilities that the response is above category s,

$$\widehat{\Pr}(y_{ij} > s|\mathbf{x}_{ij})$$

where s is specified using the above() options:

```
. gllapred mprobg1, mu marginal above(1)
(mu will be stored in mprobg1)
. gllapred mprobg2, mu marginal above(2)
(mu will be stored in mprobg2)
. gllapred mprobg3, mu marginal above(3)
(mu will be stored in mprobg3)
```

These marginal cumulative probabilities can be plotted together with the proportions against week (we revert to week since we find it easier to interpret this natural time scale than weeksqrt):

```
. twoway (line propg1 week if nonrare==1, sort lpatt(solid))
> (line propg2 week if nonrare==1, sort lpatt(vshortdash))
> (line propg3 week if nonrare==1, sort lpatt(dash))
> (line mprobg1 week, sort lpatt(solid))
> (line mprobg2 week, sort lpatt(vshortdash))
> (line mprobg3 week, sort lpatt(dash)), by(treatment)
> legend(order(1 "Prob(y>1)" 2 "Prob(y>2)" 3 "Prob(y>3)")) xtitle("Week")
```

The resulting graphs are shown in figure 7.9.

(Continued on next page)

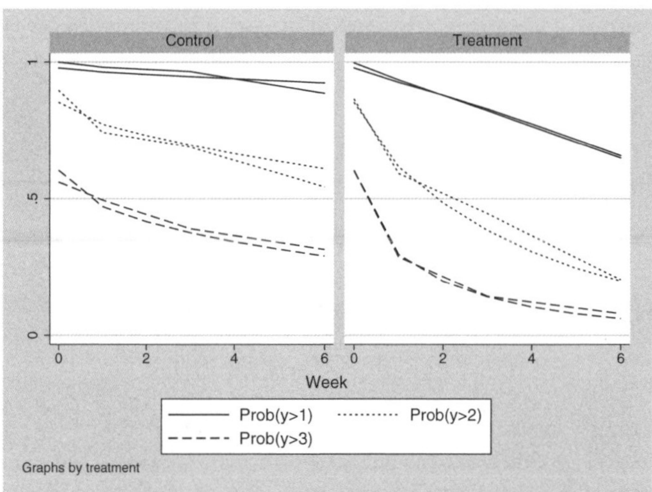

Figure 7.9: Cumulative proportions and cumulative predicted marginal probabilities from random-coefficient proportional odds model versus week

In addition to modeling the dependence of repeated measurements within patients, we see that the random-coefficient proportional odds model fits the cumulative proportions equally well as the single-level proportional odds model in figure 7.8.

The corresponding marginal probabilities for the individual response categories s can be obtained using the relationship

$$\widehat{\Pr}(y_{ij} = s|\mathbf{x}_{ij}) \;=\; \widehat{\Pr}(y_{ij} > s-1|\mathbf{x}_{ij}) - \widehat{\Pr}(y_{ij} > s|\mathbf{x}_{ij})$$

where $\widehat{\Pr}(y_{ij} > 0|\mathbf{x}_{ij}) = 1$. The Stata commands for calculating these predicted probabilities are

```
. generate pr1 = 1 - mprobg1
. generate pr2 = mprobg1 - mprobg2
. generate pr3 = mprobg2 - mprobg3
. generate pr4 = mprobg3
```

Graphs of the marginal probabilities against week are produced by the command

```
. twoway (line pr1 week, sort lpatt(solid))
> (line pr2 week, sort lpatt(vshortdash))
> (line pr3 week, sort lpatt(dash_dot))
> (line pr4 week, sort lpatt(dash)), by(treatment)
> legend(order(1 "Prob(y=1)" 2 "Prob(y=2)" 3 "Prob(y=3)" 4 "Prob(y=4)"))
> xtitle("Week")
```

giving the graph in figure 7.10.

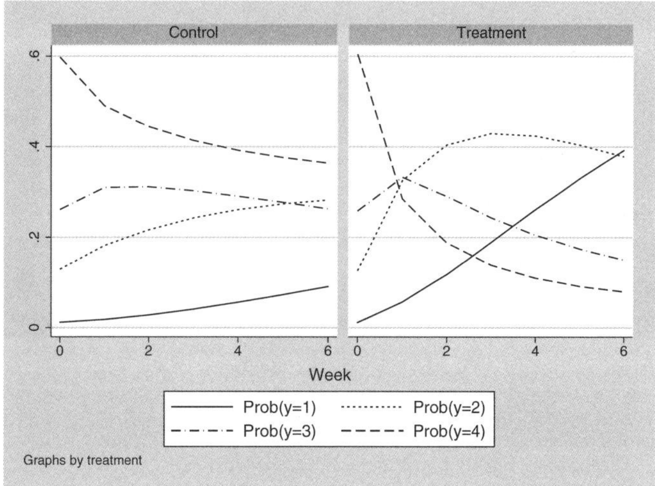

Figure 7.10: Marginal category probabilities from random-coefficient proportional odds model versus week

In the treatment group, the predicted marginal probability of being at most borderline ill (category 1) increases from nearly 0 to about 0.4, whereas the predicted marginal probability of being severely ill or among the most extremely ill (category 4) decreases from about 0.6 to about 0.1. These improvements are much less marked in the control group.

7.8.2 Predicted patient-specific probabilities

As shown for binary logistic regression in section 6.13.2, we can obtain empirical Bayes predictions of subject-specific probabilities. The corresponding predicted cumulative probabilities are

$$\widetilde{\Pr}(y_{ij} > s|\mathbf{x}_{ij}) \;\equiv\; \int \widehat{\Pr}(y_{ij} > s|\mathbf{x}_{ij}, \zeta_j) \;\times\; \text{Posterior}(\zeta_j|y_{1j}, \ldots, y_{n_j j}, \mathbf{X}_j)\, d\zeta_j$$

and the individual response probabilities can be obtained as the differences

$$\widetilde{\Pr}(y_{ij} = s|\mathbf{x}_{ij}) \;=\; \widetilde{\Pr}(y_{ij} > s-1|\mathbf{x}_{ij}) - \widetilde{\Pr}(y_{ij} > s|\mathbf{x}_{ij})$$

with $\widetilde{\Pr}(y_{ij} > 0|\mathbf{x}_{ij}) = 1$.

The patient-specific cumulative probabilities can be obtained by using the `mu` and `above()` options (but not the `marginal` option):

```
. gllapred pprobg1, mu above(1)
(mu will be stored in pprobg1)
. gllapred pprobg2, mu above(2)
(mu will be stored in pprobg2)
. gllapred pprobg3, mu above(3)
(mu will be stored in pprobg3)
```

We first define a new identifier `newid` that numbers patients within each group from 1:

```
. generate week0 = week==0
. sort treatment id week
. by treatment: generate newid = sum(week0)
```

We then use this identifier to plot the curves for twelve subjects in each treatment group:

```
. twoway (line pprobg1 week, sort lpatt(solid))
> (line pprobg2 week, sort lpatt(vshortdash))
> (line pprobg3 week, sort lpatt(dash)) if newid<13 & treatment==0, by(newid)
> legend(order(1 "Prob(y>1)" 2 "Prob(y>2)" 3 "Prob(y>3)")) xtitle("Week")
. twoway (line pprobg1 week, sort lpatt(solid))
> (line pprobg2 week, sort lpatt(vshortdash))
> (line pprobg3 week, sort lpatt(dash)) if newid<13 & treatment==1, by(newid)
> legend(order(1 "Prob(y>1)" 2 "Prob(y>2)" 3 "Prob(y>3)")) xtitle("Week")
```

The resulting graphs shown in figure 7.11 illustrate the considerable variability in the course of schizophrenia over time between patients within the treatment groups.

Control group

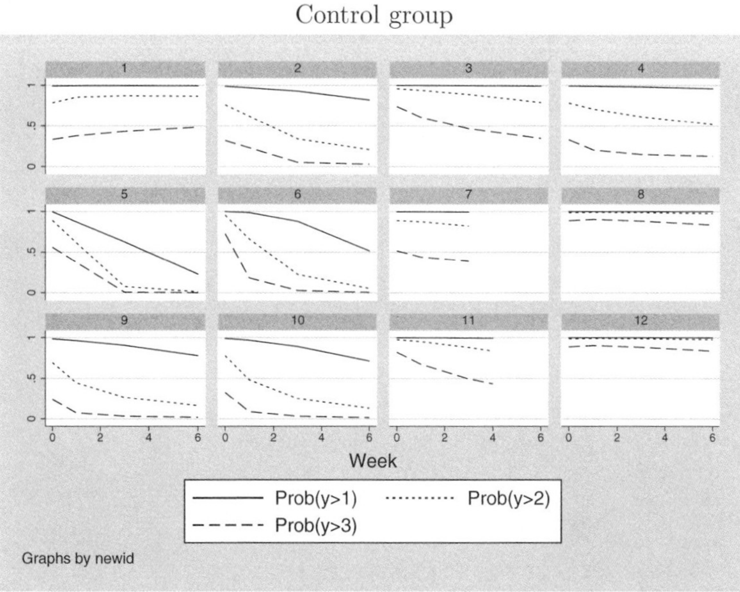

Treatment group

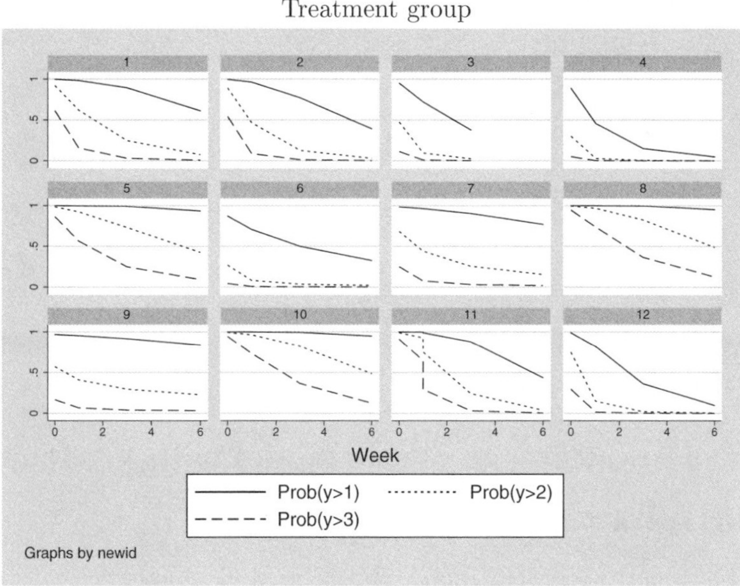

Figure 7.11: Cumulative posterior probabilities for four patients in each treatment group versus week

7.9 Do experts differ in their grading of student essays?

We now consider data from Johnson and Albert (1999) on grades assigned to 198 essays by 5 experts. The grades are on a 10-point scale, with 10 being "excellent".

The variables in the dataset `essays.dta` used here are

- `essay`: identifier for essays j
- `grade`: grade on 10-point scale
- `grader`: identifier of expert grader i

The data can be considered as interrater reliability data. They are similar to the peak-expiratory-flow data considered in chapter 2 except that we have five experts instead of two methods and no replicate measurements for a given expert. The grades are also on an ordinal scale and not an interval scale.

It is useful to obtain a frequency table of the response:

```
. use http://www.stata-press.com/data/mlmus2/essays, clear
. tabulate grade
```

grade	Freq.	Percent	Cum.
1	106	10.71	10.71
2	145	14.65	25.35
3	115	11.62	36.97
4	127	12.83	49.80
5	135	13.64	63.43
6	100	10.10	73.54
7	103	10.40	83.94
8	159	16.06	100.00
Total	990	100.00	

This output shows that the top grades of 9 and 10 were never used and that the remaining 8 grades all occur at similar frequencies.

7.10 A random-intercept probit model with grader bias

7.10.1 Model specification

We consider a cumulative ordinal probit model with a random intercept $\zeta_j \sim N(0, \psi)$. The initial model for the grade y_{ij} assigned by grader i to essay j is

$$\Pr(y_{ij} > s \,|\, \zeta_j) \;=\; \Phi(\zeta_j - \kappa_s)$$

where $\Phi(\cdot)$ is the standard normal cumulative density function.

This model can also be written using the latent-response formulation, with the latent regression model and the threshold model specified as

$$\begin{aligned} y_{ij}^* &= \zeta_j + \epsilon_{ij}, \qquad \zeta_j \sim N(0,\psi), \quad \epsilon_{ij}|\zeta_j \sim N(0,\theta) & (7.5) \\ y_{ij} &= s \qquad\qquad \text{if } \kappa_{s-1} < y_{ij}^* \le \kappa_s & (7.6) \end{aligned}$$

respectively. When the threshold model is written in this compact form, we must also state that $\kappa_0 = -\infty$ and $\kappa_S = \infty$. It is evident that this model assumes that all graders i measure the same truth ζ_j with the same measurement error variance θ and assign grades using the same thresholds κ_s $(s=1,\ldots,S-1)$.

At the least, we should allow for grader bias by including grader-specific intercepts β_i. However, one of the intercepts must be set to zero to identify all thresholds. Retaining the threshold model (7.6), we extend the latent regression model (7.5) to

$$y_{ij}^* = \zeta_j + \beta_2 x_{2i} + \beta_3 x_{3i} + \beta_4 x_{4i} + \beta_5 x_{5i} + \epsilon_{ij} \qquad (7.7)$$

where x_{2i} to x_{5i} are dummy variables for graders 2–5. The corresponding regression coefficients β_2–β_5 represent how much more generous or lenient graders 2–5 are than the first grader.

7.10.2 Estimation

We start by creating dummy variables for the graders (called grad1–grad5) using

```
. quietly tabulate grader, generate(grad)
```

The random-intercept probit model with grader-specific bias can then be fitted by maximum likelihood using gllamm with the link(oprobit) link:

(Continued on next page)

```
. gllamm grade grad2-grad5, i(essay) link(oprobit) adapt

number of level 1 units = 990
number of level 2 units = 198

Condition Number = 13.267348

gllamm model

log likelihood = -1808.0929
```

	Coef.	Std. Err.	z	P>|z|	[95% Conf.	Interval]
grade						
grad2	-1.127551	.1123199	-10.04	0.000	-1.347694	-.9074085
grad3	-.6301196	.1102998	-5.71	0.000	-.8463033	-.4139358
grad4	-.673948	.1087182	-6.20	0.000	-.8870318	-.4608642
grad5	-1.283113	.1128503	-11.37	0.000	-1.504295	-1.06193
_cut11						
_cons	-2.760615	.145463	-18.98	0.000	-3.045717	-2.475512
_cut12						
_cons	-1.831267	.1308097	-14.00	0.000	-2.087649	-1.574885
_cut13						
_cons	-1.275763	.1246812	-10.23	0.000	-1.520133	-1.031392
_cut14						
_cons	-.7281207	.1203823	-6.05	0.000	-.9640657	-.4921757
_cut15						
_cons	-.1590002	.1183832	-1.34	0.179	-.3910269	.0730265
_cut16						
_cons	.3013533	.118667	2.54	0.011	.0687702	.5339364
_cut17						
_cons	.8585616	.1224718	7.01	0.000	.6185212	1.098602

```
Variances and covariances of random effects
------------------------------------------------------------------------------

***level 2 (essay)

    var(1): 1.4066991 (.18918365)
------------------------------------------------------------------------------

. estimates store model1
```

We see that grader 1 is the most generous since the estimated coefficients of the dummies for graders 2–5 are all negative. These graders are all significantly more stringent at, say, the 5% level.

7.11 ❖ Including grader-specific measurement error variances

7.11.1 Model specification

Although the above model accommodates grader bias, it is relatively restrictive because it still assumes that all graders have the same measurement error variance θ. We can relax this homoskedasticity assumption by retaining the previous model (7.6) and (7.7), except that we now also allow each grader to have a grader-specific residual variance or measurement error variance θ_i:

$$\epsilon_{ij}|\mathbf{x}_{ij}, \zeta_j \sim N(0, \theta_i)$$

In gllamm, this can be accomplished by specifying a linear model for the log standard deviation of the measurement errors (see also section 5.14):

$$\ln(\sqrt{\theta_i}) = \ln(\theta_i)/2 \;=\; \delta_2 x_{2i} + \delta_3 x_{3i} + \delta_4 x_{4i} + \delta_5 x_{5i} \qquad (7.8)$$

In this model for level-1 heteroskedasticity, we have again omitted the dummy variable for the first grader, which amounts to setting the standard deviation of the measurement error for this grader to 1 since $\exp(0) = 1$. A constraint like this is necessary to identify the model since all thresholds κ_s $(s = 1, \ldots, S)$ are freely estimated (see sec. 7.2.4). In terms of the above parametrization, the measurement error variance θ_i for grader i (apart from grader 1) becomes $\exp(2\delta_i)$.

In this model, each grader has his or her own mean and variance

$$y^*_{ij}|\zeta_j \sim N(\zeta_j + \beta_i, \theta_i), \quad \beta_1 = 0$$

but applies the same thresholds to the latent responses to generate the observed ratings. The cumulative probabilities are

$$\begin{aligned}\Pr(y_{ij} > s\,|\,\zeta_j) \;=\; \Pr(y^*_{ij} > \kappa_s\,|\,\zeta_j) \;&=\; \Pr\left(\frac{y^*_{ij} - \zeta_j - \beta_i}{\sqrt{\theta_i}} > \frac{\kappa_s - \zeta_j - \beta_i}{\sqrt{\theta_i}}\right)\\ &=\; \Phi\left(\frac{\zeta_j + \beta_i - \kappa_s}{\sqrt{\theta_i}}\right)\end{aligned}$$

This model can be thought of as a generalized linear model with a *scaled probit link*, where the scale parameter $\sqrt{\theta_i}$ differs between graders i.

7.11.2 Estimation

To specify model (7.8) for the log standard deviations of the measurement errors in gllamm, we need to first define a corresponding equation

```
. eq het: grad2-grad5
```

(`gllamm` will not add an intercept to this equation for heteroskedasticity). Then pass this equation on to `gllamm` using the `s()` option, and change the link to `soprobit` (for scaled ordinal probit):

```
. matrix a = e(b)
. gllamm grade grad2-grad5, i(essay) link(soprobit) s(het) from(a) adapt

number of level 1 units = 990
number of level 2 units = 198

Condition Number = 16.599694

gllamm model

log likelihood = -1767.6284
```

	Coef.	Std. Err.	z	P>|z|	[95% Conf.	Interval]
grade						
grad2	-1.372239	.1625285	-8.44	0.000	-1.690789	-1.053689
grad3	-.6924764	.183522	-3.77	0.000	-1.052173	-.33278
grad4	-.8134375	.1183584	-6.87	0.000	-1.045416	-.5814594
grad5	-1.552205	.1632283	-9.51	0.000	-1.872127	-1.232284
_cut11						
_cons	-3.443449	.3145287	-10.95	0.000	-4.059914	-2.826984
_cut12						
_cons	-2.248858	.2234508	-10.06	0.000	-2.686814	-1.810902
_cut13						
_cons	-1.547513	.1784254	-8.67	0.000	-1.897221	-1.197806
_cut14						
_cons	-.8606409	.1468759	-5.86	0.000	-1.148512	-.5727694
_cut15						
_cons	-.1507223	.1338984	-1.13	0.260	-.4131584	.1117138
_cut16						
_cons	.420217	.1403798	2.99	0.003	.1450776	.6953565
_cut17						
_cons	1.120534	.1679034	6.67	0.000	.7914498	1.449619

```
Variance at level 1
------------------------------------------------------------------------------

    equation for log standard deviation:

    grad2: .24143082 (.11558242)
    grad3: .72935227 (.10636421)
    grad4: -.09545464 (.13094184)
    grad5: .04306859 (.12477309)
```

```
Variances and covariances of random effects
------------------------------------------------------------------------------

***level 2 (essay)

    var(1): 2.0587449 (.41200889)
------------------------------------------------------------------------------
```

We see that grader 3 has the largest estimated measurement error variance of $\exp(0.729)^2 = 4.30$.

We can use a likelihood-ratio test to test the null hypothesis that the measurement error variances are identical for the graders, H_0: $\theta_i = \theta$, against the alternative that the measurement error variances are different for at least two graders:

```
. estimates store model2
. lrtest model1 model2
(log likelihoods of null models cannot be compared)
Likelihood-ratio test                                 LR chi2(4)  =     80.93
(Assumption: model1 nested in model2)                 Prob > chi2 =    0.0000
```

There is strong evidence to suggest that the graders differ in their measurement error variances.

7.12 ❖ Including grader-specific thresholds

7.12.1 Model specification

A model allowing for grader-bias by including grader-specific intercepts β_i can equivalently be specified by omitting the β_i from the latent regression model in (7.7) and instead defining grader-specific thresholds:

$$\kappa_{si} \;=\; \kappa_s - \beta_i$$

Here all the thresholds of a given grader are translated relative to the thresholds of another grader.

A final model extension would be to allow different graders to apply entirely different thresholds that are not just shifted or translated by a constant:

$$\kappa_{si} \;=\; \alpha_{s1} + \alpha_{s2}x_{2i} + \cdots + \alpha_{s5}x_{5i} \tag{7.9}$$

The model now becomes

$$\Pr(y_{ij} > s \,|\, \zeta_j) \;=\; \Phi\left(\frac{\zeta_j - \kappa_{si}}{\sqrt{\theta_i}}\right) = \Phi\left(\frac{\zeta_j}{\sqrt{\theta_i}} - \frac{\kappa_{si}}{\sqrt{\theta_i}}\right)$$

Here $\sqrt{\theta_i}$ is identified because it determines the effect of ζ_j on the response probabilities for the individual raters. In fact, $1/\sqrt{\theta_i}$ can be interpreted as a discrimination parameter or factor loading, and the model is equivalent to Samejima's graded-response model (see Embretson and Reise 2000 for a discussion of this model). Since the thresholds are free parameters, dividing them by $\sqrt{\theta_i}$ does not modify the threshold model.

7.12.2 Estimation

In `gllamm`, we can specify model (7.9) for the thresholds by first defining an equation:

```
. eq thr: grad2-grad5
```

The intercept α_{s1} will be added by `gllamm` to the equation for the thresholds and does not need to be specified. This equation is then passed to `gllamm` using the `thresh()` option (beware that it takes a long time to fit the model):

```
. gllamm grade, i(essay) link(soprobit) s(het) thresh(thr) from(a) adapt skip
number of level 1 units = 990
number of level 2 units = 198

Condition Number = 32.297224

gllamm model

log likelihood = -1746.814
```

grade	Coef.	Std. Err.	z	P>|z|	[95% Conf.	Interval]
_cut11						
grad2	1.728075	.2924703	5.91	0.000	1.154844	2.301307
grad3	1.06136	.3765595	2.82	0.005	.3233166	1.799403
grad4	.6380834	.3413259	1.87	0.062	-.0309031	1.30707
grad5	1.277386	.3128459	4.08	0.000	.6642195	1.890553
_cons	-3.170991	.3363765	-9.43	0.000	-3.830277	-2.511706
_cut12						
grad2	1.654291	.2124781	7.79	0.000	1.237842	2.070741
grad3	1.237556	.2612321	4.74	0.000	.7255511	1.749562
grad4	.982663	.2164324	4.54	0.000	.5584633	1.406863
grad5	1.532188	.2193326	6.99	0.000	1.102305	1.962072
_cons	-2.269538	.2400759	-9.45	0.000	-2.740078	-1.798998
_cut13						
grad2	1.668862	.1944031	8.58	0.000	1.287839	2.049885
grad3	1.088168	.2214103	4.91	0.000	.6542123	1.522125
grad4	.8398807	.1800081	4.67	0.000	.4870713	1.19269
grad5	1.738926	.2082026	8.35	0.000	1.330857	2.146996
_cons	-1.658144	.1928516	-8.60	0.000	-2.036126	-1.280162
_cut14						
grad2	1.236492	.1748523	7.07	0.000	.8937878	1.579196
grad3	.7111903	.1962875	3.62	0.000	.3264739	1.095907
grad4	.8942416	.1603878	5.58	0.000	.5798873	1.208596
grad5	1.601494	.212614	7.53	0.000	1.184778	2.01821
_cons	-.8794976	.1560477	-5.64	0.000	-1.185345	-.5736497

```
_cut15       |
       grad2 |   1.054328   .1857254     5.68   0.000     .6903128    1.418343
       grad3 |   .4492818   .1922622     2.34   0.019     .0724549    .8261087
       grad4 |   .8514867   .1742863     4.89   0.000     .5098918    1.193082
       grad5 |   1.664347   .2603417     6.39   0.000     1.154087    2.174608
       _cons |  -.1793792   .1425682    -1.26   0.208    -.4588077    .1000493
-------------+----------------------------------------------------------------
_cut16       |
       grad2 |   .8328642   .2000749     4.16   0.000     .4407247    1.225004
       grad3 |   .3282975   .2088543     1.57   0.116    -.0810494    .7376443
       grad4 |   .7664854   .1991798     3.85   0.000        .3761    1.156871
       grad5 |   1.409387   .2812294     5.01   0.000     .8581876    1.960587
       _cons |   .4440523   .1446524     3.07   0.002     .1605388    .7275658
-------------+----------------------------------------------------------------
_cut17       |
       grad2 |   .4664753   .2244554     2.08   0.038     .0265507    .9063999
       grad3 |   .1412586   .2524303     0.56   0.576    -.3534957    .6360128
       grad4 |   .5492549    .239509     2.29   0.022     .0798259    1.018684
       grad5 |   1.233669   .3404648     3.62   0.000       .56637    1.900968
       _cons |   1.233926   .1691784     7.29   0.000     .9023423     1.56551
------------------------------------------------------------------------------

Variance at level 1
------------------------------------------------------------------------------

    equation for log standard deviation:

    grad2: -.20109602 (.16420623)
    grad3: .4523848 (.15515899)
    grad4: -.13880995 (.17086862)
    grad5: .03002867 (.16735767)

Variances and covariances of random effects
------------------------------------------------------------------------------


***level 2 (essay)

    var(1): 1.6764643 (.40806034)
------------------------------------------------------------------------------
```

Here the coefficient [_cut16]_cons is the estimate of $\kappa_{61} = \alpha_{61}$, the sixth threshold for grader 1, whereas [_cut16]grad2 is the estimate of α_{62}, the difference between the sixth thresholds for graders 2 and 1.

A likelihood-ratio test again suggests that the more elaborate model is preferred:

```
. estimates store model3

. lrtest model2 model3
(log likelihoods of null models cannot be compared)

Likelihood-ratio test                                 LR chi2(24)  =     41.63
(Assumption: model2 nested in model3)                 Prob > chi2  =    0.0142
```

However, it may not be necessary to relax the parallel-regressions assumption for all graders. For instance, we can consider the null hypothesis that all thresholds for grader 2 are simply translated by a constant relative to those of grader 1

$$H_0\colon \alpha_{12} = \alpha_{22} = \cdots = \alpha_{72}$$

This null hypothesis can be tested using the test command:

```
. test [_cut11=_cut12=_cut13=_cut14=_cut15=_cut16=_cut17]: grad2
 ( 1)  [_cut11]grad2 - [_cut12]grad2 = 0
 ( 2)  [_cut11]grad2 - [_cut13]grad2 = 0
 ( 3)  [_cut11]grad2 - [_cut14]grad2 = 0
 ( 4)  [_cut11]grad2 - [_cut15]grad2 = 0
 ( 5)  [_cut11]grad2 - [_cut16]grad2 = 0
 ( 6)  [_cut11]grad2 - [_cut17]grad2 = 0
           chi2(  6) =   25.60
         Prob > chi2 =    0.0003
```

The parallel-regressions assumption is clearly violated for grader 2.

For graders 3–5, we use

```
. test [_cut11=_cut12=_cut13=_cut14=_cut15=_cut16=_cut17]: grad3
 ( 1)  [_cut11]grad3 - [_cut12]grad3 = 0
 ( 2)  [_cut11]grad3 - [_cut13]grad3 = 0
 ( 3)  [_cut11]grad3 - [_cut14]grad3 = 0
 ( 4)  [_cut11]grad3 - [_cut15]grad3 = 0
 ( 5)  [_cut11]grad3 - [_cut16]grad3 = 0
 ( 6)  [_cut11]grad3 - [_cut17]grad3 = 0
           chi2(  6) =   11.68
         Prob > chi2 =    0.0696
. test [_cut11=_cut12=_cut13=_cut14=_cut15=_cut16=_cut17]: grad4
 ( 1)  [_cut11]grad4 - [_cut12]grad4 = 0
 ( 2)  [_cut11]grad4 - [_cut13]grad4 = 0
 ( 3)  [_cut11]grad4 - [_cut14]grad4 = 0
 ( 4)  [_cut11]grad4 - [_cut15]grad4 = 0
 ( 5)  [_cut11]grad4 - [_cut16]grad4 = 0
 ( 6)  [_cut11]grad4 - [_cut17]grad4 = 0
           chi2(  6) =    4.15
         Prob > chi2 =    0.6565
. test [_cut11=_cut12=_cut13=_cut14=_cut15=_cut16=_cut17]: grad5
 ( 1)  [_cut11]grad5 - [_cut12]grad5 = 0
 ( 2)  [_cut11]grad5 - [_cut13]grad5 = 0
 ( 3)  [_cut11]grad5 - [_cut14]grad5 = 0
 ( 4)  [_cut11]grad5 - [_cut15]grad5 = 0
 ( 5)  [_cut11]grad5 - [_cut16]grad5 = 0
 ( 6)  [_cut11]grad5 - [_cut17]grad5 = 0
           chi2(  6) =    6.84
         Prob > chi2 =    0.3358
```

For graders 4 and 5, the parallel-regressions assumption appears to be reasonable. We could therefore fit a more parsimonious model using

```
matrix a = e(b)
eq thr2: grad2 grad3
gllamm grade grad4 grad5, i(essay) link(soprobit) s(het) thresh(thr2) from(a)
   adapt skip
```

The `thresh()` option can be used more generally to relax the parallel-regressions assumption of constant covariate effects across categories s. As in section 7.2.3, this assumption corresponds to the proportional-odds assumption in ordinal logit models. To relax the parallel-regressions assumption for a covariate x_{2i}, we can simply move it to the threshold model (7.9) to estimate $S-1$ threshold parameters α_{s2} $(s=1,\ldots,S-1)$ instead of one regression parameter β_2. We cannot estimate both β_2 and the α_{s2} since such a model would not be identified.

Maximum likelihood estimates for the sequence of models developed in this section are given in table 7.2, using the notation for the final model. For "Model 1", the measurement error variances are set to 1, and the thresholds of all graders are set to be equal; i.e., in models (7.8) and (7.9) the following constraints are in place

$$\delta_i = 0 \quad \text{and} \quad \alpha_{si} = 0, \quad i = 2, 3, 4, 5$$

but the intercepts β_i for raters 2–5 are free parameters. "Model 2" is the same, except that the constraints for the δ_i are relaxed. In "Model 1" and "Model 2", $\alpha_{s1} \equiv \kappa_s$. Finally, in "Model 3" the constraints for both the δ_i and for the α_{si} are relaxed, but the intercepts are set to zero: $\beta_i = 0$.

To facilitate comparison of estimates between models, we also report estimates of the "reduced-form" thresholds:

$$\frac{\kappa_{si}}{\sqrt{\theta_i}} = \begin{cases} \frac{\alpha_{s1}-\beta_i}{\exp(\delta_i)} & \text{for } i = 1 \\ \frac{\alpha_{s1}+\alpha_{si}-\beta_i}{\exp(\delta_i)} & \text{for } i > 1 \end{cases}$$

(*Continued on next page*)

Table 7.2: Maximum likelihood estimates for essay grading data (for models 1 and 2, $\alpha_{s1} \equiv \kappa_s$)

	Model 1					Model 2					Model 3				
Grader i	1	2	3	4	5	1	2	3	4	5	1	2	3	4	5
β_i	0	−1.1	−0.6	−0.7	−1.3	0	−1.4	−0.7	−0.8	−1.6	0	0	0	0	0
α_{1i}	−2.8	0	0	0	0	−3.4	0	0	0	0	−3.2	1.7	1.1	0.6	1.3
α_{2i}	−1.8	0	0	0	0	−2.2	0	0	0	0	−2.3	1.7	1.2	1.0	1.5
α_{3i}	−1.3	0	0	0	0	−1.5	0	0	0	0	−1.7	1.7	1.1	0.8	1.7
α_{4i}	−0.7	0	0	0	0	−0.9	0	0	0	0	−0.9	1.2	0.7	0.9	1.6
α_{5i}	−0.2	0	0	0	0	−0.2	0	0	0	0	−0.2	1.1	0.4	0.9	1.7
α_{6i}	0.3	0	0	0	0	0.4	0	0	0	0	0.4	0.8	0.3	0.8	1.4
α_{7i}	0.9	0	0	0	0	1.1	0	0	0	0	1.2	0.5	0.1	0.5	1.2
δ_i	0	0	0	0	0	0	0.2	0.7	−0.1	0.0	0	−0.2	0.5	−0.1	0.0
$\frac{\alpha_{1i}-\beta_i}{\exp(\delta_i)}$	−2.8	−1.6	−2.1	−2.1	−1.5	−3.4	−1.6	−1.3	−2.9	−1.8	−3.2	−1.8	−1.3	−2.9	−1.8
$\frac{\alpha_{1i}+\alpha_{2i}-\beta_i}{\exp(\delta_i)}$	−1.8	−0.7	−1.2	−1.2	−0.5	−2.4	−0.7	−0.8	−1.6	−0.7	−2.3	−0.8	−0.7	−1.5	−0.7
$\frac{\alpha_{1i}+\alpha_{3i}-\beta_i}{\exp(\delta_i)}$	−1.3	−0.1	−0.6	−0.6	0.0	−1.5	−0.1	−0.4	−0.8	0.0	−1.7	0.0	−0.4	−0.9	0.0
$\frac{\alpha_{1i}+\alpha_{4i}-\beta_i}{\exp(\delta_i)}$	−0.7	0.4	−0.1	−0.1	0.6	−0.9	0.4	−0.1	−0.1	0.7	−0.9	0.4	−0.1	0.0	0.7
$\frac{\alpha_{1i}+\alpha_{5i}-\beta_i}{\exp(\delta_i)}$	−0.2	1.0	0.5	0.5	1.1	−0.2	1.0	0.3	0.7	1.3	−0.2	1.1	0.2	0.8	1.4
$\frac{\alpha_{1i}+\alpha_{6i}-\beta_i}{\exp(\delta_i)}$	0.3	1.4	0.9	1.0	1.6	0.4	1.4	0.5	1.4	1.9	0.4	1.6	0.5	1.4	1.8
$\frac{\alpha_{1i}+\alpha_{7i}-\beta_i}{\exp(\delta_i)}$	0.9	2.0	1.5	1.5	2.1	1.1	2.0	0.9	2.1	2.6	1.2	2.1	0.9	2.0	2.4
ψ			1.41					2.06					1.68		
Log likelihood			−1808.09					−1767.63					−1746.81		

7.13 Summary and further reading

We started by introducing cumulative models for ordinal responses, with special emphasis on the proportional odds model. Both the generalized linear and latent-response formulations of the model were outlined and the problem of identification in ordinal regression models was discussed. Finally, we discussed various extensions to cumulative models for ordinal responses. The most important is relaxing the parallel-regressions or proportional-odds assumption in cumulative models, as shown in (7.9) using the `thresh()` option in `gllamm` (see also exercise 5.2).

Cumulative models are sometimes applied to discrete-time duration or survival data. However, this is appropriate only if there is no left or right censoring or if the model is extended to handle censoring (see Rabe-Hesketh, Yang, and Pickles 2001b; Skrondal and Rabe-Hesketh 2004, sec. 12.3). More common approaches to modeling discrete-time survival data are discussed in chapter 8.

Other models that are sometimes used for ordinal responses include the continuation ratio logit model, the complementary log-log model, the adjacent-category logit model, and the stereotype model. The continuation ratio logit model, commonly used for discrete-time survival data and discussed in chapter 8, can be fitted using logistic regression if the data are expanded appropriately as shown in section 8.2.3. This model is appropriate for ordinal responses if the categories represent stages in some progression, such as levels of educational attainment. The ordinal complementary log-log model is analogous to the dichotomous version discussed in section 8.6 and can be fitted by using `gllamm` with the `link(ocll)` option. The adjacent-category and stereotype models are special cases of the multinomial logit model. See, for instance, Fahrmeir and Tutz (2001), Agresti (2002), Long and Freese (2006), or Greenland (1994) for introductions to these models (mostly without random effects). Such models with random effects can be fitted in `gllamm` using the `link(mlogit)` option. See Zheng and Rabe-Hesketh (2007) for implementing the adjacent-category logit model in `gllamm` for item-response models.

Generalized estimating equations (GEE) can be used for ordinal responses, but this is currently not implemented in Stata. It is unfortunately not possible to use conditional maximum likelihood estimation for ordinal models with fixed cluster-specific intercepts, in contrast to the dichotomous case.

We have not discussed multilevel modeling of nominal responses in this book. Both ordinal and nominal variables are categorical but the categories are unordered in the nominal case. For instance, the categories could be political parties in an election. Such variables are also called unordered categorical or polytomous variables and the response is sometimes called a discrete choice. A nominal response can also be a ranking of unordered categories, such as a preference ordering of different soft drinks. A comprehensive but demanding treatment of multilevel modeling of nominal responses is provided by Skrondal and Rabe-Hesketh (2003b) who used `gllamm` to fit the models.

Useful books discussing random-effects models for ordinal responses include Hedeker and Gibbons (2006) and Johnson and Albert (1999), the latter adopting a Bayesian approach. We also recommend the book chapter by Rabe-Hesketh and Skrondal (2008), the encyclopedia entry by Hedeker (2005), and the review article by Agresti and Natarajan (2001). Skrondal and Rabe-Hesketh (2004, chap. 10) analyze several datasets with ordinal responses using various approaches including growth-curve modeling.

7.14 Exercises

7.1 Respiratory-illness data

Koch et al. (1989) analyzed data from a clinical trial comparing two treatments for respiratory illness. In each of two centers, eligible patients were randomized to active treatment or placebo. The respiratory status was determined prior to randomization and during four visits after randomization. The dichotomized response was analyzed by Davis (1991) and Everitt and Pickles (1999), among others. Here we analyze the original ordinal response.

The dataset `respiratory.dta` has the following variables:

- `center`: center (1,2)
- `patient`: patient identifier
- `drug`: treatment group (1: active treatment; 0: placebo)
- `male`: dummy variable for patient being male
- `age`: age of patient in years
- `bl`: respiratory status at baseline (0: terrible; 1: poor; 2: fair; 3: good; 4: excellent)
- `v1` to `v4`: respiratory status at visits 1–4 (0: terrible; 1: poor; 2: fair; 3: good; 4: excellent)

1. Reshape the data by stacking the responses for visits 1–4 into a single variable.
2. Fit a proportional odds model with `drug`, `male`, `age`, and `bl` as covariates and a random intercept for patients.
3. Check whether there is a linear trend for the logits over time (after controlling for the patient-specific covariates) and whether the slope differs between treatment groups.
4. For your chosen model, plot the model-implied posterior cumulative probabilities over time for some of the patients.

7.2 Verbal-aggression data

Consider the data `aggression.dta` from Vansteelandt (2000) and De Boeck and Wilson (2004) described in exercise 6.4. Use 5-point adaptive quadrature to speed up estimation, which will be slow.

1. Fit the following "explanatory item-response model" for the original ordinal response y_{ij} (0: no; 1: perhaps; 2: yes)

$$\text{logit}\{\Pr(y_{ij} > s|\mathbf{x}_{ij}, \zeta_j)\} = \beta_2 x_{2ij} + \beta_3 x_{3ij} + \beta_4 x_{4ij} + \beta_5 x_{5ij} + \zeta_j - \kappa_s$$

   where $s = 0, 1$ and $\zeta_j|\mathbf{x}_{ij} \sim N(0, \psi)$ can be interpreted as the latent trait "verbal aggressiveness". Interpret the estimated coefficients.
2. Now extend the above model by including a latent regression, allowing verbal aggressiveness (now denoted η_j instead of ζ_j) to depend on the personal characteristics w_{1j} and w_{2j}:

$$\text{logit}\{\Pr(y_{ij} > s|\mathbf{x}_{ij}, \eta_j)\} = \beta_2 x_{2ij} + \beta_3 x_{3ij} + \beta_4 x_{4ij} + \beta_5 x_{5ij} + \eta_j - \kappa_s$$

$$\eta_j = \gamma_1 w_{1j} + \gamma_2 w_{2j} + \zeta_j$$

   Substitute the level-2 model for η_j into the level-1 model for the item responses and fit the resulting model.
3. ❖ Relax the proportional-odds assumption for x_{2ij} (*do* versus *want*). Interpret the estimates, and perform a likelihood-ratio test to assess whether there is any evidence that the proportional odds assumption is violated for this variable.

7.3 Smoking-intervention data

Gibbons and Hedeker (1994) and Hedeker and Gibbons (2006) analyzed data from a subset of the Television School and Family Smoking Prevention and Cessation Project (TVSFP) (see Flay et al. 1989).

Schools were randomized to one of four conditions given by different combinations of two factors: (1) TV, a media (television) intervention (1: present; 0: absent) and (2) CC, a social-resistance classroom curriculum (1: present; 0: absent). The outcome measure is the tobacco and health knowledge scale (THKS) score, defined as the number of correct answers to seven items on tobacco and health knowledge. This variable has been collapsed into four ordinal categories.

Students are nested in classes, which are nested in schools. We will consider two-level models in this exercise but revisit the data for three-level modeling in exercise 10.8.

The dataset `tvsfpors.dta` has the following variables:

- `school`: school identifier
- `class`: class identifier
- `thk`: ordinal THKS score postintervention (4 categories)
- `prethk`: ordinal THKS score preintervention (4 categories)
- `cc`: social-resistance classroom curriculum (dummy variable)
- `tv`: television intervention (dummy variable)

1. Investigate how tobacco and health knowledge is influenced by the interventions by fitting a two-level, random-intercept proportional odds model with students nested in schools. Include the covariates `cc`, `tv`, and their interaction, and control for `prethk`.
    a. Obtain estimated odds ratios for the interventions, and interpret these.
    b. Calculate the estimated residual intraclass correlation for the latent responses underlying the observed ordinal response.
2. Fit a two-level, random-intercept proportional odds model with students nested in classes (instead of schools).
    a. Obtain estimated odds ratios for the interventions, and interpret these.
    b. Calculate the estimated residual intraclass correlation for the latent responses underlying the observed ordinal response.
    c. Does class or school appear to induce more dependence among students?

See also exercise 10.8.

7.4 Essay-grading data

Here we consider the dataset from Johnson and Albert (1999) that was analyzed in section 7.9.

The dataset `essays.dta` has the following variables:

- `essay`: identifier for essays
- `grade`: grade on 10-point scale
- `grader`: identifier of expert grader
- `wordlength`: average word length
- `sqrtwords`: square root of the number of words in the essay
- `commas`: number of commas $\times$ 100 divided by number of words
- `errors`: percentage of spelling errors
- `prepos`: percentage of prepositions
- `sentlength`: average sentence length

1. For simplicity, collapse the variable `grade` into four categories {1,2}, {3,4}, {5,6}, and {7,8,9,10}. Fit an ordinal probit model without covariates and with a random intercept for essays. Obtain the estimated intraclass correlation for the latent responses.
2. Include dummy variables for graders 2–5 to allow some graders to be more generous in their grading than others. Does this model fit better?
3. Include the six essay characteristics as further covariates. Interpret the estimated coefficients.
4. Extend the model to investigate whether the graders differ in the importance they attach to the length of the essay (`sqrtwords`). Discuss your findings.

7.5 Attitudes-to-abortion data

Wiggins et al. (1991) analyzed data from the British Social Attitudes (BSA) survey. All adults aged 18 or over and living in private households in Britain were eligible to participate. A multistage sample was drawn, and in this dataset we have identifiers for districts that were drawn at stage 2. A subset of the respondents in the 1983 survey were followed up each year until 1986. Here we analyze the subset of respondents with complete data at all four waves that can be downloaded from the web page of the Centre for Multilevel Modelling at the University of Bristol.

The respondents were asked for each of seven circumstances whether abortion should be allowed by law. The circumstances included "The woman decides on her own that she does not wish to have the child" and "The woman became pregnant as a result of rape". The variables in the dataset `abortion.dta` are

- `district`: district identifier
- `person`: subject identifier
- `year`: year (1–4)
- `score`: number of items (circumstances) where respondents answered "yes" to the question if abortion should be allowed by law (0–7)
- `male`: dummy variable for being male
- `age`: age in years
- `religion`: religion (1: catholic; 2: protestant; 3: other; 4: none)
- `party`: party chosen (1: conservative; 2: labour; 3: liberal; 4: other; 5: none)
- `class`: self-assessed social class (1: middle class; 2: upper working class; 3: lower working class)

1. Recode the variable `score` to merge the relatively rare responses 0, 1, and 2 into one category.
2. Fit an ordinal logistic regression model with the recoded `score` variable as the response and with `male`, `age`, and dummy variables for `religion` and `year` as covariates. Use the `vce(cluster person)` option to obtain appropriate standard errors. You could either generate the dummy variables yourself or use `xi`; see [R] **xi**. Obtain odds ratios, and interpret them.
3. Now include a normally distributed random intercept for subjects in the above model. You can use the `nip(5)` option to speed up estimation and get relatively accurate estimates.
   a. Is there any evidence for between-subject residual variability in attitudes to abortion?
   b. Compare the estimated odds ratios for this model with the odds ratios for the model not including a random intercept. Explain why they differ the way they do.
   c. Obtain the estimated residual intraclass correlation for the latent responses.

d. The model does not take the clustering of subjects within districts into account. The `cluster()` option can be used to obtain standard errors based on the sandwich estimator that take clustering into account. Either include the `cluster(district)` option in the `gllamm` command when fitting the model, or fit the model without this option, and then issue the command

```
gllamm, cluster(district)
```

7.6 Marriage data

Kenny, Kashy, and Cook (2006) analyzed data from Acitelli (1997) on 148 married couples. Both husbands and wives were asked to rate their closeness to their spouse, their commitment to the marriage, and their satisfaction with the marriage. The length of the marriage at the time of data collection was also recorded.

The variables in the dataset `marriage.dta` are

- `couple`: couple identifier
- `husband`: dummy variable for being the husband
- `close`: closeness, rated from 1 to 4 (from less close to closer)
- `commit`: commitment, rated from 1 to 4
- `satis`: satisfaction, rated from 1 to 4
- `lmarr`: length of marriage in years

1. Fit an ordinal probit model for the variable `close` for both spouses including a random intercept for couples, and a dummy variable for the wife. Assume that husbands and wives have the same threshold parameters.
2. Obtain the estimated residual intraclass correlation for the latent responses.
3. Investigate whether length of marriage has an impact on closeness, allowing for different regression coefficients for husbands and wives.
   a. Write down the model using the latent-response formulation.
   b. Fit the model.
4. Interpret the estimated odds ratios.
5. State the following null hypotheses in terms of the model parameters and conduct hypothesis testing (with two-sided alternatives) using Stata:
   a. The coefficient of length of marriage is zero for husbands.
   b. The coefficient of length of marriage is zero for wives.
   c. The coefficients of length of marriage are zero both for husbands and wives (jointly).
   d. The coefficients of length of marriage are equal for husbands and wives.
6. ❖ In the model from step 3, relax the assumption that the residual variance in the latent-response formulation is the same for husbands and wives. Use a likelihood-ratio test to compare this model with the model from step 3.

7.7 Recovery after surgery

Davis (1991, 2002) analyzed data from a clinical trial comparing the effects of different dosages of anesthetic on postsurgical recovery. Sixty young children undergoing outpatient surgery were randomly assigned to four groups of size 15, receiving 15, 20, 25, and 30 milligrams of anesthetic per kilogram of body weight. Recovery scores were assigned upon admission to the recovery room (0 minutes) and 5, 15, and 30 minutes after admission to the recovery room. The score was an ordinal variable with categories 1 (least favorable) through 6 (most favorable).

The variables in `recovery.dta` are

- `dosage`: dosage of anesthetic in milligram/kilogram
- `id`: subject identifier, taking values 1–15 in each dosage group
- `age`: age of child in months
- `duration`: duration of surgery in minutes
- `score1`, `score2`, `score3`, `score4`: recovery scores 0, 5, 15, and 30 minutes after admission to the recovery room

1. Reshape the data to long form, stacking the recovery scores at the four occasions into a single variable, and generating an identifier `occ` for the four occasions. (You can specify several variables in the `i()` option of the `reshape` command if one variable does not uniquely identify the individuals.)
2. Construct a variable `time` taking the values 0, 5, 15, and 30 at the four occasions. Fit a random-intercept proportional-odds model with dummy variables for the dosage groups, `age`, `duration`, and `time` as covariates.
3. Compare the model from step 2 with a model including `dosage` as a continuous covariate instead of the dummy variables for dosage groups, using a likelihood-ratio test at the 5% significance level.
4. Extend the model chosen in step 3 to include an interaction between dosage and `time`. Test the interaction using a Wald test at the 5% level of significance.
5. Fit the model selected in step 4 and interpret the estimated odds ratios and random-intercept variance.

8 Discrete-time survival

8.1 Introduction

In this chapter, we consider survival data, also referred to as failure-time, time-to-event, event-history, or duration data. Examples include time between surgery and death, duration in first employment, time to failure of light bulbs, and time to dropout from school.

The response variable is the time from some event, such as the beginning of employment, to another event, such as unemployment. The subject is said to become at risk of (or eligible for) the event of interest after the initial event has occurred. Alternatively, we can view the response variable as the duration spent in one state (employment) until transition to another state (unemployment). A state is called absorbing if it is not possible to leave the state, a canonical example being death.

Special methods are needed for survival data mainly because of censoring, where the time of the event is not known exactly for some individuals. Other special features that must be handled by survival analysis are truncation and time-varying covariates.

In this chapter, we focus on events in discrete time, occurring at relatively few time points. The survival models are specified in terms of the discrete-time *hazard*, or in other words the conditional probability of the event occurring at a time point given that it has not already occurred. A convenient feature of these survival models is that they become models for dichotomous responses when the data have been expanded to so-called person–period data. Logit and probit models can then be used, as well as complementary log-log models that have not been introduced yet. It is probably a good idea to read chapter 6 before embarking on this chapter.

We also discuss multilevel discrete-time survival models where random effects, often called *frailties* in this context, are included to handle unobserved heterogeneity between clusters and within-cluster dependence. Discrete-time frailty models are applied to data on time from birth to death (or censoring) for different children nested in mothers.

8.1.1 Censoring and truncation

The left panel of figure 8.1 shows the event histories of several hypothetical individuals in terms of calendar time.

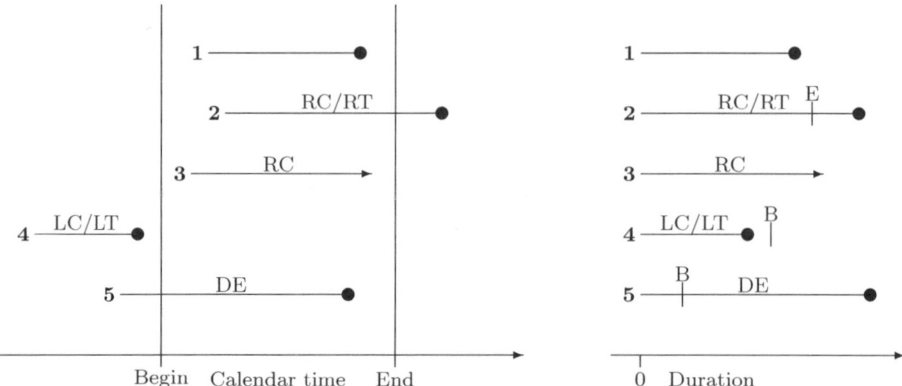

Figure 8.1: Illustration of different types of censoring and truncation (dots represent events and arrowheads represent censoring)

The lines start when the subjects become at risk of experiencing the event of interest. For most subjects, the line ends when the event occurs as indicated by a •. The vertical lines represent the beginning and end of the observation period, from the recruitment of subjects into the study until the end of the study. The right panel of figure 8.1 shows the same event histories as the left panel but now the time axis represents the time from the onset of risk so that the time when the event occurs is also the survival time or duration. Here B and E represent the beginning and end of the time spent in the study.

Subject 1, represented by the top line in the figure, is ideal in the sense that he becomes at risk within the observation period and experiences the event within the observation period, so that the survival time is known.

When *censoring* occurs, the survival time is not exactly known. In one form of right censoring (RC), often called generalized type I censoring, the event is not observed before the end of the observation period, and all we know is that the survival time exceeds the time between becoming at risk and the end of the observation period (situation 2 in the figure). Another form of right censoring occurs when the person stops being at risk of the event under investigation before the end of the observation period; for instance, he may experience a competing event such as dying from another disease than the one under investigation or he may drop out of the study (situation 3). For this type of censoring, the time the subject ceases being at risk is indicated by an arrowhead. It is usually assumed that censoring is noninformative in the sense that the survival times for the competing events are conditionally independent of the survival time of interest, given the covariates. If this assumption is true, we need not model the survival times for the competing events.

When all that is known is that the event occurred before observation began (situation 4), the survival time is called left censored (LC). For example, the event of interest may be onset of drug use, but all we may know at the start of observation is that a person has used drugs but not when he or she started this practice. Left censoring is less common than right censoring because in most studies observation either begins when people become at risk for the event, or those who have already experienced the event are not eligible for inclusion in the study.

In *current status data*, each individual is either left or right censored. For instance, surveys sometimes ask individuals whether they have had their sexual debut, producing a left-censored response at their current age if the answer is yes and a right-censored response if the answer is no.

When we only know that the event occurred within a time interval, but not precisely when it occurred, we say that the time is *interval censored*. For instance, time of HIV infection may be known only to lie between the dates of a negative and positive HIV test, employment status may be known only at each wave of a panel survey, and age at first marriage may be rounded to years. Interval censoring is also sometimes referred to as grouping, and the resulting data as discrete-time or grouped-time survival data.

Truncation occurs when a subject is not included in the study due to the timing of his or her event. *Left truncation* (LT) occurs when subjects are excluded from the study because the event of interest occurred before observation began (situation 4). This can happen only if individuals become at risk before observation begins (situations 4 and 5 in the figure). Situation 5, where the subject enters the study after having already been at risk for a period, is called *delayed entry* (DE). Delayed entry and left truncation are common in epidemiological cohort studies where only disease-free individuals are typically followed up, thus including individuals who have already been at risk and excluding individuals who developed the disease before observation began. We will discuss delayed entry and left truncation in more detail in section 8.2.5.

Right truncation (RT) occurs when a person is excluded from the study because their event happened after the end of the observation period (situation 2). This can happen in retrospective studies, for instance when investigating the incubation period of AIDS in patients who have developed the disease. Any person who has yet to experience the event is then not included in the study since he is not known to be at risk.

In this chapter, we discuss methods that are suitable when left truncation and right censoring occur, but we assume that there is no left censoring or right truncation. As discussed in section 8.1.3, interval censoring can also be handled and is indeed one of the reasons why the discrete-time methods discussed in this chapter become appropriate, rather than continuous-time methods.

8.1.2 Time-varying covariates and different time-scales

Another special feature of survival data are *time-varying covariates*—covariates that change between the time a person becomes at risk and experiences the event. For instance, if the event is reoffending after release from prison, employment status could be a time-varying covariate. A person's risk of reoffending could be lower during periods of employment than during periods of unemployment.

As discussed in section 5.4, several different time scales can be of interest. The survival time itself is an age-like time scale because it is the time since the subject-specific event of becoming at risk (which is birth in studies of longevity). Period (current calendar time) and cohort (calendar time when becoming at risk) may also be of interest. Unless all subjects became at risk at the same calendar time, one of these time scales can be included, and period would be a time-varying covariate. The subject-specific timing of other events may also be of interest. For instance, if the survival time is the time from surgery to death, age could be included as a time-varying covariate, or alternatively, age at the time of surgery could be included as a time-invariant covariate.

8.1.3 Discrete- versus continuous-time survival data

It is useful to distinguish between discrete-time survival data and continuous-time survival data. In *continuous-time survival data*, the exact survival and censoring times are recorded in relatively fine time units. Methods for continuous time can then be used, which typically assume that all survival times are unique and that there are no pairs of individuals with identical or tied survival times.

In this chapter, we will consider *discrete-time survival data* characterized by relatively few possible survival (or censoring) times with many people sharing the same survival time. Discrete-time data can result from interval censoring, where the event occurs in continuous time but we only know the time interval within which it occurred. In this case we assume that the intervals are the same for all individuals. Alternatively, the time scale is sometimes inherently discrete, examples being the number of menstrual cycles to conception and the number of elections to a change of government.

8.2 Single-level models for discrete-time survival data

8.2.1 Discrete-time hazard and discrete-time survival

We now consider data used by Long, Allison, and McGinnis (1993) and provided with the book by Allison (1995). Three hundred and one male biochemists who received their doctorates in 1956 or 1963 and were assistant professors at research universities in the United States some time in their careers were followed up for ten years from the beginning of their assistant professorships.

The event of interest is promotion to associate professor, which usually corresponds to receiving tenure (a permanent position). Censoring occurs when assistant professors leave their research university for a job outside academia or for a position at a college or university in which teaching is the primary mission. If some of these transitions occur because of concerns about not being able to pass tenure, censoring becomes informative, unless these concerns are captured by the covariates in the model. Here we assume that censoring is noninformative.

Initially, we will not include covariates and only use the following variables:

- `id`: person identifier
- `dur`: number of years from beginning of assistant professorship to promotion or censoring
- `event`: dummy variable for promotion to associate professor (1: promoted; 0: censored)

We will use the notation T for the time in years to promotion, which can take on integer values $t = 1, 2, \ldots, 10$. If the variable `event` equals 1, we know that T equals `dur`, and if `event` equals 0, we know that T is greater than `dur`.

Discrete-time survival models are specified in terms of the *discrete-time hazard*, defined as the conditional probability that the event occurs at time t, given that it has not yet occurred:

$$h_t \equiv \Pr(T = t|T > t - 1) = \Pr(T = t|T \geq t)$$

Stata's `ltable` command for life tables provides us with all the information we need to estimate these hazards:

```
. use http://www.stata-press.com/data/mlmus2/promotion
. ltable dur event, noadjust
                     Beg.                                  Std.
    Interval        Total   Deaths    Lost    Survival    Error      [95% Conf. Int.]
---------------------------------------------------------------------------------------
    1       2         301        1       1      0.9967    0.0033      0.9767    0.9995
    2       3         299        1       6      0.9933    0.0047      0.9737    0.9983
    3       4         292       17      12      0.9355    0.0143      0.9008    0.9584
    4       5         263       42      10      0.7861    0.0243      0.7337    0.8294
    5       6         211       53       9      0.5887    0.0297      0.5280    0.6442
    6       7         149       46       7      0.4069    0.0303      0.3473    0.4656
    7       8          96       31       6      0.2755    0.0282      0.2217    0.3319
    8       9          59       15       2      0.2055    0.0262      0.1567    0.2590
    9      10          42        7       6      0.1712    0.0248      0.1258    0.2227
   10      11          29        4      25      0.1476    0.0241      0.1043    0.1981
---------------------------------------------------------------------------------------
```

(See section 8.4 for an explanation of the `noadjust` option.)

We see that there are 301 individuals at the beginning of the first year, one of whom gets promoted (somewhat inappropriately for this particular application denoted as `Deaths` in the output) and one of whom is censored (`Lost`). The estimated hazard for this interval is $\widehat{h}_1 = 1/301 = 0.0033$. At the beginning of year 2, there are 299 individuals left in the sample who have not yet experienced the event (301 minus 1 promoted minus 1 censored) and are hence at risk. One of these individuals gets promoted in year 2, so $\widehat{h}_2 = 1/299 = 0.0033$. For the following years, we have $\widehat{h}_3 = 17/292 = 0.0582$, $\widehat{h}_4 = 42/263 = 0.1597$, $\widehat{h}_5 = 53/211 = 0.2512$, $\widehat{h}_6 = 46/149 = 0.3087$, $\widehat{h}_7 = 31/96 = 0.3229$,

$\widehat{h}_8 = 15/59 = 0.2542$, $\widehat{h}_9 = 7/42 = 0.1667$, and $\widehat{h}_{10} = 4/29 = 0.1379$. The individuals who are at risk at a given interval and contribute to the denominator for the estimated hazard are called the *risk set* for that interval.

We can obtain these estimated hazards using the `ltable` command with the `hazard` and `noadjust` options:

```
. ltable dur event, hazard noadjust
                    Beg.      Cum.      Std.                 Std.
  Interval         Total   Failure     Error     Hazard      Error    [95% Conf. Int.]
-----------------------------------------------------------------------------------------
   1     2           301    0.0033  0.0033     0.0033     0.0033     0.0001     0.0123
   2     3           299    0.0067  0.0047     0.0033     0.0033     0.0001     0.0123
   3     4           292    0.0645  0.0143     0.0582     0.0141     0.0339     0.0890
   4     5           263    0.2139  0.0243     0.1597     0.0246     0.1151     0.2115
   5     6           211    0.4113  0.0297     0.2512     0.0345     0.1882     0.3232
   6     7           149    0.5931  0.0303     0.3087     0.0455     0.2260     0.4041
   7     8            96    0.7245  0.0282     0.3229     0.0580     0.2194     0.4461
   8     9            59    0.7945  0.0262     0.2542     0.0656     0.1423     0.3981
   9    10            42    0.8288  0.0248     0.1667     0.0630     0.0670     0.3109
  10    11            29    0.8524  0.0241     0.1379     0.0690     0.0376     0.3023
-----------------------------------------------------------------------------------------
```

We see that the estimated hazard reaches a maximum in year 7 and then declines; those who have not been promoted by the end of year 9 have only a 14% chance of being promoted in year 10. Perhaps some of these assistant professors will never be promoted.

The discrete-time survival function is the probability of not experiencing the event by time t

$$S_t \equiv \Pr(T > t)$$

For $t = 1$, this is simply 1 minus the probability that the event occurs at time 1 (given that it has not yet occurred, which is impossible), $S_1 = (1 - h_1)$. For $t = 2$, S_2 is the probability that the event did not occur at time 1 and that it did not occur at time 2, given that it did not occur at time 1, and can be expressed as $S_2 = \Pr(T > 2) = \Pr(T > 2|T > 1)\Pr(T > 1) = (1 - h_2)(1 - h_1)$. In general, we have

$$S_t = \prod_{s=1}^{t} (1 - h_s) \tag{8.1}$$

Substituting the estimated hazards gives the estimated survival function given in the output of the `ltable` command without the `hazard` option shown on page 335.

8.2.2 Data expansion for discrete-time survival analysis

By expanding the data appropriately and defining a dichotomous variable `y` taking the values 0 and 1, we can obtain the estimated hazards as the proportion of 1s observed each year. As we will see later, this data expansion is necessary for conducting the most common type of discrete-time survival analysis.

Consider the first two individuals in the data:

```
. list id dur event if id<3

     +-----------------+
     | id   dur  event |
     |-----------------|
  1. |  1    10      0 |
  2. |  2     4      1 |
     +-----------------+
```

The person with id=1 was censored at $t = 10$. This person therefore represents one of the 301 individuals at risk in year 1 (i.e., is part of the risk set in year 1), one of the 299 individuals at risk in year 2, and so forth up to and including year 10. In the expanded dataset, this person should contribute one observation for each of the 10 years, with y equal to zero, so that the observation contributes to the denominator but not to the numerator of the estimated hazard. Similarly, the person with id=2 should be represented by four observations, for years 1–4. For the first three years, y should be 0, and for the fourth year y should be 1 since the event occurs that year and the observation should contribute to the numerator as well as the denominator of the estimated hazard.

In general, each person should be represented by a row of data for each year the person was at risk. We therefore expand the data to obtain dur rows per person and create a new variable year that labels the years

```
. expand dur
. by id, sort: generate year = _n
. list id dur year event if id<3, sepby(id)

     +------------------------+
     | id   dur   year  event |
     |------------------------|
  1. |  1    10      1      0 |
  2. |  1    10      2      0 |
  3. |  1    10      3      0 |
  4. |  1    10      4      0 |
  5. |  1    10      5      0 |
  6. |  1    10      6      0 |
  7. |  1    10      7      0 |
  8. |  1    10      8      0 |
  9. |  1    10      9      0 |
 10. |  1    10     10      0 |
     |------------------------|
 11. |  2     4      1      1 |
 12. |  2     4      2      1 |
 13. |  2     4      3      1 |
 14. |  2     4      4      1 |
     +------------------------+
```

The response variable y should be 1 if a promotion occurs for the person that year and 0 otherwise:

```
. generate y = 0
. replace y = event if year==dur
```

The variables id, year, and y for the first two individuals are shown in figure 8.2.

Original data

id	dur	event
1	10	0
2	4	1

⟶

Person–year data

id	year	y
1	1	0
1	2	0
1	3	0
1	4	0
1	5	0
1	6	0
1	7	0
1	8	0
1	9	0
1	10	0
2	1	0
2	2	0
2	3	0
2	4	1

Figure 8.2: Expansion of original data to person–year data for first two assistant professors (first one right censored in year 10, second one promoted in year 4)

The data are now in person–year or person–period form where each year of observation for each assistant professor has one record or observation for each period he or she is at risk of being promoted.

8.2.3 Estimation via regression models for dichotomous responses

We can obtain the sample hazards or estimated hazards $\widehat{h}_t$ for each year t by finding the proportion of 1s each year, for instance using the tabulate command:

```
. tabulate year y, row
```

Key
frequency *row percentage*

year	y 0	1	Total
1	300 99.67	1 0.33	301 100.00
2	298 99.67	1 0.33	299 100.00
3	275 94.18	17 5.82	292 100.00
4	221 84.03	42 15.97	263 100.00
5	158 74.88	53 25.12	211 100.00
6	103 69.13	46 30.87	149 100.00
7	65 67.71	31 32.29	96 100.00
8	44 74.58	15 25.42	59 100.00
9	35 83.33	7 16.67	42 100.00
10	25 86.21	4 13.79	29 100.00
Total	1,524 87.54	217 12.46	1,741 100.00

The penultimate column in the output above gives the estimated hazards as percentages, and these agree with the estimated hazards on page 336 using the `ltable` command.

Alternatively, we can obtain estimated hazards as predicted probabilities by using a logistic regression model where the covariates are dummy variables for each year

$$\text{logit}\{\Pr(y_{si}=1|\mathbf{d}_{si})\} \;=\; \beta_1+\alpha_2 d_{2si}+\cdots+\alpha_{10}d_{10,si}$$

Here y_{si} is an indicator for the event occurring at time s for person i, d_{2si}, ..., $d_{10,si}$ are dummy variables for years 2–10, and $\mathbf{d}_{si} = (d_{2si}, \ldots, d_{10,si})'$ is a vector containing all the dummy variables for professor i.

The Stata command to fit this model by maximum likelihood is

```
. xi: logit y i.year
i.year            _Iyear_1-10         (naturally coded; _Iyear_1 omitted)
Logistic regression                               Number of obs   =       1741
                                                  LR chi2(9)      =     251.17
                                                  Prob > chi2     =     0.0000
Log likelihood = -529.15641                       Pseudo R2       =     0.1918

------------------------------------------------------------------------------
           y |      Coef.   Std. Err.      z    P>|z|     [95% Conf. Interval]
-------------+----------------------------------------------------------------
    _Iyear_2 |    .006689   1.416576     0.00   0.996     -2.76975    2.783128
    _Iyear_3 |   2.920225   1.032373     2.83   0.005     .8968117    4.943638
    _Iyear_4 |   4.043289    1.01571     3.98   0.000     2.052534    6.034045
    _Iyear_5 |   4.611479   1.014165     4.55   0.000     2.623753    6.599206
    _Iyear_6 |   4.897695   1.017242     4.81   0.000     2.903937    6.891452
    _Iyear_7 |   4.963382   1.025171     4.84   0.000     2.954084    6.972681
    _Iyear_8 |   4.627643   1.045336     4.43   0.000     2.578822    6.676464
    _Iyear_9 |   4.094345   1.083864     3.78   0.000      1.97001    6.218679
   _Iyear_10 |   3.871201   1.137248     3.40   0.001     1.642236    6.100166
       _cons |  -5.703782   1.001665    -5.69   0.000     -7.66701   -3.740555
------------------------------------------------------------------------------
```

where the xi prefix command with "i." preceding the categorical covariate year produces dummy variables for year and includes them in the model, omitting the dummy variable for year 1. Predicted probabilities, which here correspond to estimated discrete-time hazards, can be obtained using the predict command with the pr option:

```
. predict haz, pr
```

Finally, we list the estimated hazards, the time-specific event indicator y_{is} and the variable event, for the years where each of the first two assistant professors in the dataset are at risk of promotion:

```
. list id year haz y event if id<3, sepby(id)

     +-------------------------------+
     | id   year        haz   y   event |
     |-------------------------------|
  1. |  1      1   .0033223   0       0 |
  2. |  1      2   .0033445   0       0 |
  3. |  1      3   .0582192   0       0 |
  4. |  1      4   .1596958   0       0 |
  5. |  1      5   .2511848   0       0 |
  6. |  1      6   .3087248   0       0 |
  7. |  1      7   .3229167   0       0 |
  8. |  1      8   .2542373   0       0 |
  9. |  1      9   .1666667   0       0 |
 10. |  1     10    .137931   0       0 |
     |-------------------------------|
 11. |  2      1   .0033223   0       1 |
 12. |  2      2   .0033445   0       1 |
 13. |  2      3   .0582192   0       1 |
 14. |  2      4   .1596958   1       1 |
     +-------------------------------+
```

The estimated hazards are identical to those obtained using the ltable command.

We could use any other link function, such as the probit link, or complementary log-log link introduced in section 8.6, to obtain the same predicted hazards.

We can plot the discrete-time hazards by selecting an assistant professor who has observations for all periods such as professor 1 (the same results would be obtained here if we picked another professor). The Stata command is

```
. twoway (line haz year if id==1, connect(stairstep)) legend(off)
> xtitle(Year) ytitle(Discrete-time hazard)
```

which produces the graph in figure 8.3.

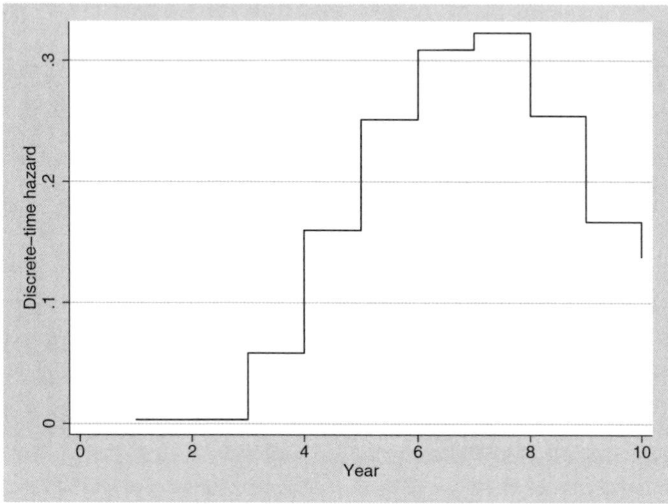

Figure 8.3: Discrete-time hazard (conditional probability of promotion given that promotion has not yet occurred)

Here we have used the `connect(stairstep)` option to plot a step function (but you can omit this option).

The likelihood from the binary regression models based on the expanded data is just the required likelihood for discrete-time survival data and the resulting predicted hazards are maximum likelihood estimates. To see this, consider the required likelihood contribution for a person who was censored at time t, which is just the corresponding probability $S_{ti} \equiv \Pr(T_i > t)$. From (8.1), this probability is given by

$$S_{ti} = \prod_{s=1}^{t}(1 - h_{si})$$

where h_{si} is the discrete-time hazard at time s for person i. In the expanded dataset, someone who is censored at time t is represented by a row of data for $s = 1, 2, \ldots, t$ with $\mathsf{y} = 0$. The corresponding likelihood contributions from binary regression are the model-implied probabilities of the observed responses, $\Pr(y_{si}=0)=(1-h_s)$, and these are simply multiplied together since the observations are taken as independent, giving the required likelihood contribution for discrete-time survival.

For an assistant professor who was promoted at time t, the likelihood contribution should be

$$\begin{aligned}\Pr(T_i=t) &= \Pr(T_i=t|T_i>t-1)\Pr(T_i>t-1)\\ &= h_{ti}\prod_{s=1}^{t-1}(1-h_{si}).\end{aligned}$$

Such a person is represented by a row of data for $s=1,2,\dots,t$ with `y` equal to zero for $s\leq t-1$ and `y` equal to 1 for $s=t$. The likelihood contribution from binary regression is therefore $\Pr(y_{1i}=0)\times\cdots\times\Pr(y_{t-1,i}=0)\times\Pr(y_{ti}=1)$ as required.

We can therefore interpret the logistic regression model as a linear model for the logit of the discrete-time hazard

$$\text{logit}\{\Pr(y_{si}=1|\mathbf{d}_{si})\} \;=\; \beta_1+\alpha_2 d_{2si}+\cdots+\alpha_{10}d_{10,si} \;=\; \text{logit}\{\underbrace{\Pr(T_i=s|T_i\geq s,\mathbf{d}_{si})}_{h_{si}}\}$$

No functional form is imposed on the relationship between discrete-time hazard and year because dummy variables are used for years 2–10 (the intercept represents year 1).

8.2.4 Including covariates

Although it is interesting to investigate how the population averaged or marginal hazard evolves over time, the main purpose of survival analysis is usually to estimate the effects of covariates on the hazard. Here regression models become useful.

Time-constant covariates

The following time-constant or person-specific covariates are available:

- `undgrad`: selectivity of undergraduate institution (scored from 1–7) (x_{2i})
- `phdmed`: dummy variable for having a Ph.D. from a medical school (1: yes; 0: no) (x_{3i})
- `phdprest`: a measure of prestige of the Ph.D. institution (ranges from 0.92 to 4.62) (x_{4i})

We specify a logistic regression with these three covariates added to the dummy variables for years 2–10

$$\text{logit}\{\Pr(y_{si}=1|\mathbf{d}_{si},\mathbf{x}_i)\} \;=\; \beta_1+\alpha_2 d_{2si}+\cdots+\alpha_{10}d_{10,si}+\beta_2 x_{2i}+\beta_3 x_{3i}+\beta_4 x_{4i}$$

where $\mathbf{x}_i = (x_{2i}, x_{3i}, x_{4i})'$ is the vector of covariates. The first part of the linear predictor, from β_1 to $\alpha_{10}d_{10,si}$ determines the so-called *baseline hazards*, the hazards when the covariates $\mathbf{x}_i$ are all zero. The model could be described as semiparametric because no assumptions are made regarding the functional form for the baseline hazards, whereas the effects of covariates are assumed to be linear and additive on the logit scale.

We can fit this model by using

```
. xi: logit y i.year undgrad phdmed phdprest
i.year            _Iyear_1-10          (naturally coded; _Iyear_1 omitted)
Logistic regression                               Number of obs   =       1741
                                                  LR chi2(12)     =     260.79
                                                  Prob > chi2     =     0.0000
Log likelihood = -524.34273                       Pseudo R2       =     0.1992

------------------------------------------------------------------------------
           y |      Coef.   Std. Err.      z    P>|z|     [95% Conf. Interval]
-------------+----------------------------------------------------------------
    _Iyear_2 |   .0063043   1.416703     0.00   0.996    -2.770383    2.782992
    _Iyear_3 |   2.923037   1.032552     2.83   0.005     .8992722    4.946802
    _Iyear_4 |    4.05151   1.015925     3.99   0.000     2.060334    6.042685
    _Iyear_5 |   4.619307    1.01441     4.55   0.000     2.631099    6.607515
    _Iyear_6 |   4.924626   1.017614     4.84   0.000     2.930139    6.919112
    _Iyear_7 |   4.992069   1.025703     4.87   0.000     2.981728    7.002409
    _Iyear_8 |   4.690531   1.046322     4.48   0.000     2.639776    6.741285
    _Iyear_9 |    4.16777   1.085192     3.84   0.000     2.040833    6.294707
   _Iyear_10 |   3.964635   1.139095     3.48   0.001      1.73205     6.19722
     undgrad |   .1576609     .06007     2.62   0.009      .039926    .2753959
      phdmed |  -.0950034   .1665181    -0.57   0.568    -.4213728    .2313661
    phdprest |   .0650372   .0854488     0.76   0.447    -.1024394    .2325138
       _cons |  -6.671891   1.081551    -6.17   0.000    -8.791692   -4.552089
------------------------------------------------------------------------------
```

This model assumes that the difference in log odds between individuals with different covariates is constant over time. For instance, the predicted log odds of individuals 1 and 4 can be obtained and plotted by using

```
. predict lo, xb
. twoway (line lo year if id==1, connect(stairstep) lpatt(solid))
>        (line lo year if id==4, connect(stairstep) lpatt(dash)),
>        legend(off) xtitle(Year) ytitle(Log odds)
```

giving the graph in figure 8.4.

(*Continued on next page*)

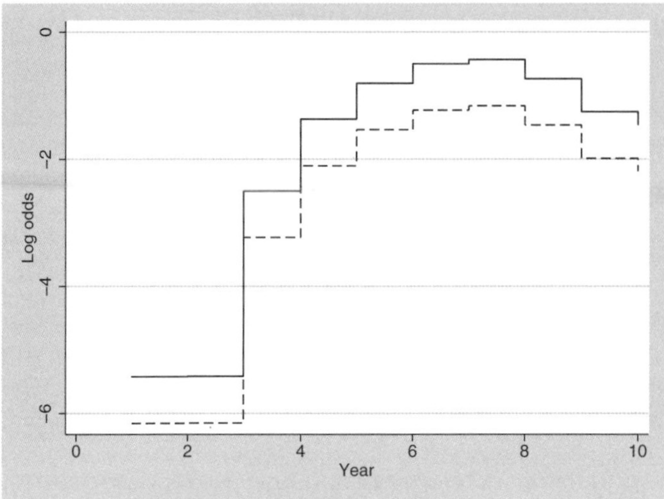

Figure 8.4: Predicted log odds of promotion given that promotion has not yet occurred for professors 1 (solid) and 4 (dashed)

A constant difference in the log odds corresponds to a constant ratio of the odds. For this reason, the model is often called a proportional odds model, not to be confused with the ordinal logistic regression model of the same name discussed in chapter 7. We will see in section 8.6 how time-varying covariates can be used to relax the proportionality assumption.

The model is also sometimes referred to as a *continuation-ratio logit* model or sequential logit model because the logit can be written as

$$\begin{aligned}
\text{logit}\{\Pr(T_i = s|T_i \geq s, \mathbf{d}_i, \mathbf{x}_i)\} &= \underbrace{\ln\left\{\frac{\Pr(T_i = s|T_i \geq s, \mathbf{d}_i, \mathbf{x}_i)}{\Pr(T_i > s|T_i \geq s, \mathbf{d}_i, \mathbf{x}_i)}\right\}}_{\text{Odds}(T_i=s|T_i\geq s, \mathbf{d}_i, \mathbf{x}_i)} \\
&= \ln\left\{\frac{\Pr(T_i = s, T_i \geq s, \mathbf{d}_i, \mathbf{x}_i)/\Pr(T_i \geq s, \mathbf{d}_i, \mathbf{x}_i)}{\Pr(T_i > s, T_i \geq s, \mathbf{d}_i, \mathbf{x}_i)/\Pr(T_i \geq s, \mathbf{d}_i, \mathbf{x}_i)}\right\} \\
&= \ln\left\{\frac{\Pr(T_i = s, T_i \geq s, \mathbf{d}_i, \mathbf{x}_i)}{\Pr(T_i > s, T_i \geq s, \mathbf{d}_i, \mathbf{x}_i)}\right\} \\
&= \ln\left\{\frac{\Pr(T_i = s|\mathbf{d}_i, \mathbf{x}_i)}{\Pr(T_i > s|\mathbf{d}_i, \mathbf{x}_i)}\right\}
\end{aligned}$$

which is the log of the odds (or ratio of probabilities) of stopping at time s versus continuing beyond time s. This becomes a model for the odds of continuing versus stopping if we replace 0 by 1 and 1 by 0 in the response variable (see exercise 8.5).

We can obtain the estimated odds ratios associated with the three covariates by using the `or` option with the `logit` command or by exponentiating the coefficients. For `undergrad`, the odds ratio is estimated as exp(0.157) = 1.17, implying that the odds of being promoted in any given year (given that promotion has not already occurred) increase 17% for every unit increase in the selectivity of the undergraduate institution.

Log-odds curves are not easy to interpret, and usually survival curves are presented instead. The estimated hazards $\widehat{h}_{si}$ can be obtained from the log odds `lo` using the `invlogit()` function. To get the cumulative products in (8.1) for each person, we make use of Stata's `sum()` function combined with the `by` prefix and apply this to the logarithms, using the relationship

$$\widehat{S}_{ti} = \prod_{s=1}^{t}(1-\widehat{h}_{si}) = \exp\left\{\sum_{s=1}^{t}\ln(1-\widehat{h}_{si})\right\}$$

```
. generate ln_one_m_haz = ln(1-invlogit(lo))
. by id (year), sort: generate ln_surv = sum(ln_one_m_haz)
. generate surv = exp(ln_surv)
. twoway (line surv year if id==1, connect(stairstep) lpatt(solid))
>        (line surv year if id==4, connect(stairstep) lpatt(dash)),
>        legend(off) xtitle(Year) ytitle(Survival)
```

The graph is given in figure 8.5.

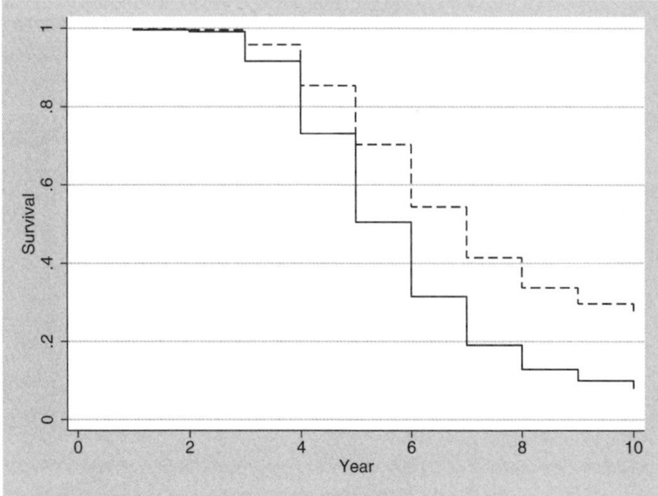

Figure 8.5: Predicted probability of remaining an assistant professor for professors 1 (solid) and 4 (dashed)

Time-varying covariates

The dataset contains four types of time-varying information; the cumulative number of papers published each year, the cumulative number of citations each year to all papers previously published, whether the employer is the same as at the start of the assistant professorship, and the prestige of the current employer. This information is coded in the following variables:

- `art1`–`art10`: the cumulative number of articles published in each of the ten years
- `cit1`–`cit10`: the cumulative number of citations to all previous articles in each of the ten years
- `jobtime`: number of years until change of employer from start of assistant professorship, coded as missing for those who did not change employer
- `prest1`: a measure of prestige of the assistant professor's first employing institution (ranges from 0.65 to 4.6)
- `prest2`: prestige of the assistant professor's second employing institution, coded as missing for those who did not change employers. (No one had more than two employers.)

Since the data are in person–year form with one observation for each year each person is at risk, it is straightforward to include time-varying information: we must create variables that take on the appropriate values for each person in each year. For the first two variables above, this is most easily achieved by reading the data in their original form and using the `reshape` command:

```
. use http://www.stata-press.com/data/mlmus2/promotion, clear
. reshape long art cit, i(id) j(year)
(note: j = 1 2 3 4 5 6 7 8 9 10)
Data                               wide   ->   long
-----------------------------------------------------------------------------
Number of obs.                      301   ->    3010
Number of variables                  29   ->      12
j variable (10 values)                    ->   year
xij variables:
                     art1 art2 ... art10  ->   art
                     cit1 cit2 ... cit10  ->   cit
-----------------------------------------------------------------------------
```

```
. list id year art cit dur event if id<3, sepby(id)
```

	id	year	art	cit	dur	event
1.	1	1	0	0	10	0
2.	1	2	0	0	10	0
3.	1	3	2	1	10	0
4.	1	4	2	1	10	0
5.	1	5	2	1	10	0
6.	1	6	2	1	10	0
7.	1	7	2	1	10	0
8.	1	8	2	1	10	0
9.	1	9	2	1	10	0
10.	1	10	2	1	10	0
11.	2	1	8	27	4	1
12.	2	2	10	44	4	1
13.	2	3	14	57	4	1
14.	2	4	18	63	4	1
15.	2	5	.	.	4	1
16.	2	6	.	.	4	1
17.	2	7	.	.	4	1
18.	2	8	.	.	4	1
19.	2	9	.	.	4	1
20.	2	10	.	.	4	1

We see that the first assistant professor stopped publishing after year 3, was cited only once in 10 years and did not get promoted within 10 years, whereas the second professor published 18 papers and had 63 citations by the fourth year when he or she got promoted.

The `reshape` command has produced 10 rows of data for each person although only `dur` rows are required. We therefore delete the excess rows of data and create the variables `year` and `y` as before:

```
. drop if year>dur
. generate y = 0
. replace y = event if year==dur
```

The time-varying covariates `art` and `cit` are now ready for inclusion in the model. Following Allison (1995), we will use the prestige of the current employer as the third time-varying covariate. Initially, prestige is `prest1` for everyone and then changes to `prest2` for those who changed employers in the year given in `jobtime`:

```
. generate prestige = prest1
. replace prestige = prest2 if year>=jobtime
```

`jobtime` is missing for those who never changed employers. Since Stata interprets missing values as large numbers, the logical expression `year>=jobtime` is never true for those who did not change employers, and therefore `prestige` remains equal to `prest1` for those individuals as required. The data for the first two assistant professors are given in table 8.1.

Table 8.1: Expanded data with time-constant and time-varying covariates for first two assistant professors

		Covariates						Response
		Time-constant			Time-varying			
id	year	undgrad	phdmed	phdprest	art	cit	prestige	y
i	s	x_{2i}	x_{3i}	x_{4i}	x_{5si}	x_{6si}	x_{7si}	y_{si}
1	1	7	0	2.21	0	0	2.36	0
1	2	7	0	2.21	0	0	2.36	0
1	3	7	0	2.21	2	1	2.36	0
1	4	7	0	2.21	2	1	2.36	0
1	5	7	0	2.21	2	1	2.36	0
1	6	7	0	2.21	2	1	2.36	0
1	7	7	0	2.21	2	1	2.36	0
1	8	7	0	2.21	2	1	2.36	0
1	9	7	0	2.21	2	1	2.36	0
1	10	7	0	2.21	2	1	2.36	0
2	1	6	0	2.21	8	27	1.84	0
2	2	6	0	2.21	10	44	1.84	0
2	3	6	0	2.21	14	57	2.88	0
2	4	6	0	2.21	18	63	2.88	1

As in figure 8.2, the expanded data has 10 records for the first assistant professor and 4 records for the second. For the first professor, the response y_{si} is equal to 0 throughout the 10 years ($s = 1, \ldots, 10$), and for the second professor, the response is 0 for the first 3 years ($s = 1, 2, 3$) and then 1 in year $s = 4$. There are three time-constant variables `undgrad`, `phdmed`, and `phdprest` (x_{2i}, x_{3i}, and x_{4i}) and three time-varying variables `art`, `cit`, and `prestige` (x_{5si}, x_{6si}, and x_{7si}).

Using expanded data of the above form, we can fit a logistic discrete-time hazards model including both the time-constant and time-varying covariates,

$$\begin{aligned}\text{logit}\{\Pr(y_{si} = 1|\mathbf{d}_{si}, \mathbf{x}_{si})\} &= \beta_1 + \alpha_2 d_{2si} + \cdots + \alpha_{10} d_{10,si} \\ &\quad + \beta_2 x_{2i} + \beta_3 x_{3i} + \beta_4 x_{4i} + \beta_5 x_{5si} + \beta_6 x_{6si} + \beta_7 x_{7si}\end{aligned}$$

where $\mathbf{x}_{si} = (x_{2i}, x_{3i}, x_{4i}, x_{5si}, x_{6si}, x_{7si})'$.

This model can be fitted using the command:

```
. xi: logit y i.year undgrad phdmed phdprest art cit prestige, or
i.year            _Iyear_1-10         (naturally coded; _Iyear_1 omitted)

Logistic regression                               Number of obs   =       1741
                                                  LR chi2(15)     =     303.80
                                                  Prob > chi2     =     0.0000
Log likelihood = -502.83976                       Pseudo R2       =     0.2320

------------------------------------------------------------------------------
           y | Odds Ratio   Std. Err.      z    P>|z|     [95% Conf. Interval]
-------------+----------------------------------------------------------------
    _Iyear_2 |   .9126305   1.293372    -0.06   0.949     .0567518    14.67609
    _Iyear_3 |   15.10098   15.61379     2.63   0.009     1.990206    114.5809
    _Iyear_4 |   44.34841   45.12662     3.73   0.000     6.035871    325.8488
    _Iyear_5 |   73.16559   74.37594     4.22   0.000     9.977559    536.5244
    _Iyear_6 |   93.20305   95.14621     4.44   0.000     12.60324    689.2521
    _Iyear_7 |   97.98715   100.9335     4.45   0.000     13.01286    737.8459
    _Iyear_8 |   69.18404   72.81962     4.03   0.000     8.791729    544.4244
    _Iyear_9 |   37.24828   40.90507     3.29   0.001     4.328521    320.5332
   _Iyear_10 |   32.10185   36.91246     3.02   0.003     3.371101    305.6949
     undgrad |    1.21355   .0767456     3.06   0.002      1.07208    1.373688
      phdmed |   .7916123   .1357166    -1.36   0.173     .5656917    1.107759
    phdprest |   1.022572   .0951069     0.24   0.810     .8521685    1.227049
         art |   1.076152   .0194367     4.06   0.000     1.038723    1.114929
         cit |   1.000092   .0013031     0.07   0.944      .997541    1.002649
    prestige |   .7816668   .0887699    -2.17   0.030     .6256841     .976536
------------------------------------------------------------------------------
```

The maximum likelihood estimates of the odds ratios for this model are also presented in table 8.2.

(*Continued on next page*)

Table 8.2: Maximum likelihood estimates for logistic discrete-time hazards model for promotions of assistant professors

	Odds ratio	(95% CI)
$\exp(\alpha_2)$ [_Iyear_2]	0.91	(0.1, 14.7)
$\exp(\alpha_3)$ [_Iyear_3]	15.10	(2.0, 114.6)
$\exp(\alpha_4)$ [_Iyear_4]	44.35	(6.0, 325.8)
$\exp(\alpha_5)$ [_Iyear_5]	73.17	(10.0, 536.5)
$\exp(\alpha_6)$ [_Iyear_6]	93.20	(12.6, 689.3)
$\exp(\alpha_7)$ [_Iyear_7]	97.99	(13.0, 737.8)
$\exp(\alpha_8)$ [_Iyear_8]	69.18	(8.8, 544.4)
$\exp(\alpha_9)$ [_Iyear_9]	37.25	(4.3, 320.5)
$\exp(\alpha_{10})$ [_Iyear_10]	32.10	(3.4, 305.7)
$\exp(\beta_2)$ [undgrad]	1.21	(1.1, 1.4)
$\exp(\beta_3)$ [phdmed]	0.79	(0.6, 1.1)
$\exp(\beta_4)$ [phdprest]	1.02	(0.9, 1.2)
$\exp(\beta_5)$ [art]	1.08	(1.0, 1.1)
$\exp(\beta_6)$ [cit]	1.00	(1.0, 1.0)
$\exp(\beta_7)$ [prestige]	0.78	(0.6, 1.0)
Log likelihood	−502.84	

At the 5% level, only the effects of `undergrad`, `art`, and `prestige` are significant. As might be expected, a more selective undergraduate institution and more published articles are associated with a greater odds of promotion (given that promotion has not yet occurred), whereas being employed at a more prestigious institution is associated with a decreased odds of promotion. Regarding the coefficients of the dummy variables, their exponentials represent the odds ratio of promotion in a given year (given that promotion did not occur earlier) compared with year 1. So, for instance, the odds of promotion are 98 times as great in year 7 as they are in year 1. The reason for these large odds ratios and wide confidence intervals is that the odds of promotion in year 1 are low and poorly estimated. It would perhaps be better to use a different year as reference category.

It is also interesting to consider a parametric form for the discrete-time baseline hazard as a function of time instead of using dummy variables. Allison (1995) specifies a polynomial relationship for these data (see also exercise 8.1).

8.2.5 Handling left-truncated data

In the promotion data, all individuals were sampled at the beginning of their assistant professorships, that is when they became at risk for promotion to associate professor. Such a sampling design, where the start of the observation period coincides with the time individuals become at risk, is sometimes called sampling the inflow (here the inflow to the state of assistant professor). This design is typical in experimental studies where for instance duration to death following surgery is investigated.

However, in observational studies we often have more complex sample designs where the times individuals become at risk do not necessarily coincide with the start of the observation period. Delayed entry means that individuals become at risk before entering the study, and individuals who have already experienced the event of interest are not included in the study. For instance, when investigating the duration of unemployment a *stock sample* includes individuals from the stock of unemployed at a given time and their times to employment (or censoring) are recorded. Those not unemployed at the start of observation are not eligible for the study, and the sample is in this sense truncated. If we analyze unemployment durations among those selected to the sample, without any correction, we are likely to underestimate the hazard of employment in the general population. The sampling is said to be *length biased* because the probability of selection at a given point in time depends on the time spent at risk.

In truncated samples, we require the conditional hazard of survival, given that the subject is selected to the study. Consider an individual who we know has already been at risk for a time t_b when he enters the study. This situation is illustrated by case 5 shown in the right panel of figure 8.1, where B denotes the beginning of the study. Since the individual would not have been included in the study if the event had occurred at a time $t \leq t_b$, we must condition on the fact that $t > t_b$. The likelihood contribution for this person if he is right censored at time $t > t_b$ therefore becomes

$$\Pr(T_i > t | T_i > t_b) = \frac{\Pr(T_i > t, T_i > t_b)}{\Pr(T_i > t_b)} = \frac{\prod_{s=1}^{t}(1-h_{si})}{\prod_{s=1}^{t_b}(1-h_{si})} = \prod_{s=t_b+1}^{t}(1-h_{si})$$

and the likelihood contribution if he experiences the event at time $t > t_b$ is

$$\Pr(T_i = t | T_i > t_b) = \frac{\Pr(T_i = t, T_i > t_b)}{\Pr(T_i > t_b)} = \frac{h_{ti}\prod_{s=1}^{t-1}(1-h_{si})}{\prod_{s=1}^{t_b}(1-h_{si})} = h_{ti}\prod_{s=t_b+1}^{t-1}(1-h_{si})$$

The correct likelihood contribution under delayed entry is thus simply obtained by letting the individual start contributing observations after entering the study and discarding the preceding periods $t = 1, \ldots, t_b$.

The above correction cannot be used if the time at risk before entering the study t_b is unknown. Fortunately, bias can be avoided here by simply discarding subjects with late entry; however, this can incur a substantial loss in efficiency.

Different kinds of sampling designs for survival analysis are discussed in, for instance, Lancaster (1990), Hamerle (1991), Guo (1993), Jenkins (1995), and Klein and Moeschberger (2003).

8.3 How does birth history affect child mortality?

Pebley and Stupp (1987) and Guo and Rodriguez (1992) analyzed data on child mortality in Guatemala. The data come from a retrospective survey conducted in 1974–76 by the Instituto Nutrición de Centroamérica Y Panamá (UNCAP) and RAND.

Most of the data come from a female life history survey that covered all women aged 15–49 who lived in six villages and towns. The survey questionnaire asked about the complete birth history, maternal education, and related subjects. The data include all children except multiple births (twins or triplets), stepchildren, and children born more than 15 years before the survey. Cases with missing socioeconomic data or inconsistent dates of events have been deleted.

The outcome of interest is the children's length of life. Since most of the mothers have several children, this is an example of multilevel or clustered survival data.

The variables we will use here are

- `kidid`: child identifier i
- `momid`: mother identifier j
- `time`: time in months from birth to death or censoring
- `death`: dummy variable for death (1: death; 0: censoring)
- `mage`: mother's age at time of birth of child
- `border`: birth order of child
- `p0014`, `p1523`, `p2435`, `p36up`: dummy variables for time between birth of index child and birth of previous child being 0–14, 15–23, 24–35, and 36 or more months, respectively (the reference category is no previous birth)
- `pdead`: dummy variable for previous child having died before conception of the index child
- `f0011` and `f1223`: dummy variables for next child being born within 0–11 and 12–23 months after the birth of the index child respectively, and before the death of the index child (the reference category is no subsequent birth within 23 months and before the death of the index child)

8.4 Data expansion

Pebley and Stupp (1987) treat these data as continuous-time survival data and use a piecewise exponential model (see chap. 9) with constant hazards in the intervals <1, 1–5, 6–11, 12–23, and >23 months. Here we will treat these same intervals as discrete-time intervals and use discrete-time survival models. When the exact timing is known, the piecewise exponential model described in section 9.12 may be more appropriate because it takes into account that individuals were at risk for only parts of the time interval within which they either died or were censored. In contrast, discrete-time survival analysis treats these individuals as being at risk for the entire interval. We will return to this issue shortly.

First, we use the `egen` function `cut()` to categorize the variable `time` and then the `table` command to make sure that we have done this correctly:

```
. use http://www.stata-press.com/data/mlmus2/mortality, clear
. egen discrete = cut(time), at(0 1 6 12 24 61) icodes
. table discrete, contents(min time max time)
```

discrete	min(time)	max(time)
0	.25	.25
1	1	5
2	6	11
3	12	23
4	24	60

The variable `discrete` takes on consecutive integer values from 0 to 4 because we used the `icodes` option in the `egen` command. However, for the data expansion to work, we require that `discrete` starts from 1, so we add 1

```
. replace discrete = discrete + 1
```

Before expanding the data, we summarize them in a life table using the `ltable` command so that we can check that our data expansion is correct:

```
. ltable discrete death, noadjust
```

Interval		Beg. Total	Deaths	Lost	Survival	Std. Error	[95% Conf. Int.]	
1	2	3120	109	9	0.9651	0.0033	0.9580	0.9710
2	3	3002	92	94	0.9355	0.0044	0.9263	0.9436
3	4	2816	66	126	0.9136	0.0051	0.9031	0.9230
4	5	2624	94	238	0.8808	0.0059	0.8687	0.8919
5	6	2292	42	2250	0.8647	0.0063	0.8518	0.8765

Returning to the issue of how to deal with individuals who died or were censored within an interval, `ltable` with the `noadjust` option treats these individuals as being at risk during the entire interval (assuming that the event and censoring take place at the end of the interval). This gives estimated hazards of $\widehat{h}_1 = 109/3120 = 0.0349$ and so on. The data expansion method described in section 8.2.2 yields these same estimates. If time is truly discrete, this is the correct method for estimating the hazards.

However, if time is treated as interval censored as here, we do not know when during the first interval the 109+9=118 individuals died or were censored. To allow for partial contributions to the risk set for those who were censored or died during the interval, the *actuarial adjustment* consists of excluding half of the 118 individuals from the denominator, giving $\widehat{h}_1 = 109/(3120 - 118/2) = 0.0356$. This can be motivated by assuming that the hazard of removal from the risk set follows a uniform distribution over the interval. The estimated hazards from the actuarial method are provided by `ltable` without the noadjust option. In this dataset, the adjustment makes a considerable difference in the last interval where 2250 individuals were censored. The data expansion method yields improved estimates of the hazards if smaller intervals are used.

We now expand the data and produce a table for comparison with the life table above:

```
. expand discrete
. by kidid, sort: generate interval = _n
. generate y = 0
. replace y = death if interval == discrete
. tabulate interval y, row
```

Key
frequency *row percentage*

interval	y 0	1	Total
1	3,011 96.51	109 3.49	3,120 100.00
2	2,910 96.94	92 3.06	3,002 100.00
3	2,750 97.66	66 2.34	2,816 100.00
4	2,530 96.42	94 3.58	2,624 100.00
5	2,250 98.17	42 1.83	2,292 100.00
Total	13,451 97.09	403 2.91	13,854 100.00

The row totals and numbers of 1s agree with the Beg. Total and Deaths in the life-table so we can trust our data expansion.

8.5 ❖ Proportional hazards and interval censoring

The most popular methods for modeling covariate effects for continuous time assume that the continuous-time hazards are proportional

$$h(z|\mathbf{x}_{ij}) = h_0(z)\exp(\beta_2 x_{2ij} + \cdots + \beta_p x_{pij})$$

where $h(z|\mathbf{x}_{ij})$ is the continuous-time hazard function for subject i in cluster j at time z, and $h_0(z)$ is the *baseline hazard function* (the hazard when $x_{2ij} = \cdots = x_{pij} = 0$). It follows that continuous-time hazards are proportional in the sense that the hazard ratio

$$\frac{h(z|\mathbf{x}_{ij})}{h(z|\mathbf{x}_{i'j'})} = \exp\{\beta_2(x_{2ij} - x_{2i'j'}) + \cdots + \beta_p(x_{pi} - x_{pi'j'})\}$$

does not depend on time. We see that an exponentiated coefficient, say $\exp(\beta_2)$, represents the *hazard ratio* for a one unit change in x_{2ij}, controlling for the other covariates.

It can be shown that the proportionality assumption translates to the following relationship for the survival function

$$S(z|\mathbf{x}_{ij}) \equiv \Pr(Z_{ij} > z|\mathbf{x}_{ij}) = S_0(z)^{\exp(\beta_2 x_{2ij} + \cdots + \beta_p x_{pij})}$$

where Z_{ij} is the continuous survival time for subject i in cluster j and $S_0(z)$ is the *baseline survival function*, the survival function when the covariates are all zero.

If the survival times are interval censored, so we only observe integer values $T_{ij} = t$ if $z_{t-1} < Z_{ij} \leq z_t$ $(t=1,2,\ldots)$, then the discrete-time hazard is given by

$$\begin{aligned} h_{tij} \equiv \Pr(T_{ij} = t|\mathbf{x}_{ij}, T_{ij} > t-1) &= \frac{\Pr(Z_{ij} > z_{t-1}|\mathbf{x}_{ij}) - \Pr(Z_{ij} > z_t|\mathbf{x}_{ij})}{\Pr(Z_{ij} > z_{t-1}|\mathbf{x}_{ij})} \\ &= 1 - \frac{S(z_t|\mathbf{x}_{ij})}{S(z_{t-1}|\mathbf{x}_{ij})} \end{aligned}$$

It follows that

$$1 - h_{tij} = \frac{S(z_t|\mathbf{x}_{ij})}{S(z_{t-1}|\mathbf{x}_{ij})} = \left\{\frac{S_0(z_t)}{S_0(z_{t-1})}\right\}^{\exp(\beta_2 x_{2ij} + \cdots + \beta_p x_{pij})}$$

and

$$\ln(1 - h_{tij}) = \exp(\beta_2 x_{2ij} + \cdots + \beta_p x_{pij})\{\ln S_0(z_t) - \ln S_0(z_{t-1})\}$$

The right-hand side is negative, so we reverse the signs and then take the logarithms and obtain

$$\ln\{-\ln(1 - h_{tij})\} = \beta_2 x_{2ij} + \cdots + \beta_p x_{pij} + \underbrace{\ln\{\ln S_0(z_{t-1}) - \ln S_0(z_t)\}}_{\alpha_t}$$

This model, which is linear in the covariates, contains the same regression parameters as the continuous-time proportional-hazards model, and time-specific constants α_t. The model is a generalized linear model with a complementary log-log link $g(h_{tij}) = \ln\{-\ln(1 - h_{tij})\}$.

By estimating the parameters α_t freely for each time-point t, we are making no assumption regarding the shape of the baseline survival function $S_0(z_t)$ or the corresponding baseline hazard function within the time intervals $z_{t-1} < Z_{ij} \leq z_t$. This is in contrast to the piecewise exponential model discussed in section 9.12, which assumes constant baseline hazards within time intervals.

8.6 Complementary log-log models

We will use the complementary log-log link here instead of the logit link because, as shown in section 8.5, this link function follows if a proportional hazards model holds in continuous time and the survival times are interval censored.

Hence, for child i of mother j we consider the complementary log-log discrete-time survival model

$$\text{cloglog}(h_{sij}) \equiv \ln\{-\ln(1-h_{sij})\} \;=\; \alpha_1 d_{1sij} + \cdots + \alpha_5 d_{5sij}$$

where we have included one dummy variable for each period but no constant (instead of a constant and dummy variables for all but the first period as in section 8.2). The model can alternatively be written as

$$h_{sij} = \text{cloglog}^{-1}(\alpha_1 d_{1sij} + \cdots + \alpha_5 d_{5sij}) \equiv 1 - \exp\{-\exp(\alpha_1 d_{1sij} + \cdots + \alpha_5 d_{5sij})\}$$

As mentioned in section 8.5, the exponentiated regression coefficients can be interpreted as hazard ratios in continuous time. Although probit and logit models are symmetric with

$$\begin{aligned} \text{probit}(h_{sij}) &= \Phi^{-1}(h_{sij}) = -\Phi^{-1}(1-h_{sij}) \\ \text{logit}(h_{sij}) &= -\text{logit}(1-h_{sij}) \end{aligned}$$

the complementary log-log link is not symmetric,

$$\text{cloglog}(h_{sij}) \;\neq\; -\text{cloglog}(1-h_{sij})$$

This means that switching the 0s and 1s in the response variable will not merely reverse the sign of the estimated coefficients as for logit and probit models. It is worth noting that the complementary log-log and logit models are similar when the time intervals are small.

By using dummy variables for the time intervals, we are not making any assumption regarding the shape of the discrete-time hazard function, and in the model without further covariates, we will get the same hazard estimates whatever link we use.

The `cloglog` command fits a model with a complementary log-log link:

```
. quietly tabulate interval, generate(int)
. cloglog y int1-int5, nocons
Complementary log-log regression                  Number of obs     =      13854
                                                  Zero outcomes     =      13451
                                                  Nonzero outcomes  =        403
                                                  Wald chi2(5)      =    4943.80
Log likelihood = -1811.6791                       Prob > chi2       =     0.0000

------------------------------------------------------------------------------
           y |      Coef.   Std. Err.      z    P>|z|     [95% Conf. Interval]
-------------+----------------------------------------------------------------
        int1 |  -3.336513   .0957877   -34.83   0.000    -3.524253   -3.148772
        int2 |  -3.469723   .1042614   -33.28   0.000    -3.674072   -3.265374
        int3 |  -3.741583   .1230944   -30.40   0.000    -3.982844   -3.500323
        int4 |  -3.310976   .1031478   -32.10   0.000    -3.513142   -3.108809
        int5 |  -3.990277   .1543055   -25.86   0.000    -4.292711   -3.687844
------------------------------------------------------------------------------
```

We can estimate the hazards for the children in each time-interval using

```
. predict haz, pr
```

and list the estimated hazards (which are the same for all children) for child 101:

```
. sort kidid interval
. list kidid interval haz if kidid==101

     +-----------------------------+
     | kidid   interval        haz |
     |-----------------------------|
  1. |   101          1   .0349359 |
  2. |   101          2  .03064624 |
  3. |   101          3   .0234375 |
  4. |   101          4  .03582317 |
  5. |   101          5  .01832461 |
     +-----------------------------+
```

The marginal (over the observed covariates) hazard function estimated above is of some interest, but we are more interested in the relationships between covariates and the death hazard. Understanding these relationships might provide insight into the etiology of child mortality and could potentially guide targeted interventions in order to reduce mortality.

Mother's age at birth is known to be a risk factor for child mortality. We thus allow for linear and quadratic functions of age by including the covariates `mage` (x_{2ij}) and `mage2` (x_{3ij}), the latter variable constructed as

```
. generate mage2 = mage^2
```

Since birth order is a known risk factor we also include `border` (x_{4ij}).

The other covariates relate to the timing of other births in the family. As discussed by Pebley and Stupp (1987), birth spacing may be important for several reasons, for instance several children of similar ages may compete for scarce resources (food, clothing, and parental time). We therefore include dummy variables for the time intervals between the births of the index child and the previous child p0014 (x_{5ij}), p1523 (x_{6ij}), p2435 (x_{7ij}), and p36up (x_{8ij}).

However, Pebley and Stupp point out that there may be a spurious relationship between the previous birth interval and the death of the index child if the previous birth interval is shortened by the previous child's death and if the risks of dying are correlated between children in the same family. For this reason, the survival status of the previous child pdead (x_{9ij}) is included as a covariate.

Also, the birth interval to the next child may be shortened by the index child's death. Therefore, the dummy variables f0011 and f1223 for the next birth interval being 0–11 months and 12–23 months, respectively, have been set to zero if the index child died within the corresponding time intervals. The birth of a subsequent child can affect the index child only after it has occurred, for a short birth interval (0–11 months) when the index child is at least 12 months old (interval 4 onward) and for a long birth interval (12–23 months) when the index child is at least 24 months old (interval 5). We therefore follow Guo and Rodriguez (1992) in defining the time-varying dummy variables comp12 ($x_{10,sij}$), comp24e ($x_{11,sij}$), and comp24l ($x_{12,sij}$) for competition with the subsequent child when the index child is 12–23 months old (interval 4) and over 24 months old (interval 5), differentiating in the latter interval between competition with children born earlier and later:

```
. generate comp12 = f0011*(interval==4)
. generate comp24e = f0011*(interval==5)
. generate comp24l = f1223*(interval==5)
```

Multiplying variables with dummy variables for time intervals (or other functions of time) is a device for allowing the effects of these variables to change over time, thus relaxing the proportional-hazards assumption.

We now include all the above covariates in the complementary log-log model:

$$\ln\{-\ln(1-h_{sij})\} \;=\; \alpha_1 d_{1sij} + \cdots + \alpha_5 d_{5sij} + \beta_2 x_{2ij} + \cdots + \beta_{12} x_{12,sij}$$

Estimates with robust standard errors for clustered data can be obtained using

```
. cloglog y int1-int5 mage mage2 border p0014 p1523 p2435 p36up pdead
> comp12 comp24e comp241, noconstant eform vce(cluster momid)
Complementary log-log regression                Number of obs     =      13854
                                                Zero outcomes     =      13451
                                                Nonzero outcomes  =        403

                                                Wald chi2(16)     =    4512.16
Log pseudolikelihood = -1784.1899               Prob > chi2       =     0.0000

                              (Std. Err. adjusted for 851 clusters in momid)

------------------------------------------------------------------------------
             |               Robust
           y |     exp(b)   Std. Err.      z    P>|z|     [95% Conf. Interval]
-------------+----------------------------------------------------------------
        int1 |   .2318333     .17652    -1.92   0.055      .052127    1.031071
        int2 |   .2042155   .1572742    -2.06   0.039     .0451383    .9239144
        int3 |   .1560266    .121621    -2.38   0.017     .0338607    .7189539
        int4 |   .2300307     .17649    -1.92   0.055     .0511335    1.034824
        int5 |   .1221787   .0922591    -2.78   0.005     .0278125    .5367244
        mage |   .8617077   .0499925    -2.57   0.010     .7690897    .9654794
       mage2 |   1.002562   .0010004     2.56   0.010     1.000603    1.004525
      border |   1.063391   .0364623     1.79   0.073     .9942743    1.137311
       p0014 |    1.73225   .3591742     2.65   0.008     1.153766    2.600779
       p1523 |   .8764676   .1549602    -0.75   0.456     .6197873     1.23945
       p2435 |   .7697006   .1391446    -1.45   0.148     .5400621    1.096983
       p36up |   .6681858   .1366279    -1.97   0.049     .4475556    .9975793
       pdead |   1.118051    .178947     0.70   0.486     .8170047    1.530026
      comp12 |   4.909878   2.279632     3.43   0.001     1.976342    12.19774
     comp24e |   4.436944   3.158988     2.09   0.036     1.099134    17.91089
     comp241 |   .8476116   .3038621    -0.46   0.645     .4198052    1.711378
------------------------------------------------------------------------------
```

We see that mother's age at birth (mage and mage2), being born within 14 months of the previous child (p0014), and getting a new sibling within the first 11 months of life (comp12 and comp24e) have statistically significant effects on the death hazard at the 5% level. For example, the estimated hazard of death increases 73% if the index child was born within 14 months of the previous child. Having a new sibling within the first 11 months of life increases the estimated hazard 4.91 fold at age 12 to 23 months and 4.44 fold at age 24–60 months. It appears that competition with children born shortly before, but even more so shortly after, the index child is an important risk factor. The estimates are also given in table 8.3. The exponentiated coefficients of the dummy variables do not represent hazard ratios here because the overall constant was omitted. The corresponding estimates are therefore not given in the table.

(Continued on next page)

Table 8.3: Maximum likelihood estimates for complementary log-log models with and without random intercept for Guatemalan child mortality data

	No random intercept		Random intercept	
	Hazard ratio	(95% CI)*	Hazard ratio	(95% CI)
Fixed part				
$\exp(\beta_2)$ [mage]	0.86	(0.77, 0.97)	0.86	(0.76, 0.96)
$\exp(\beta_3)$ [mage2]	1.00	(1.00, 1.00)	1.00	(1.00, 1.00)
$\exp(\beta_4)$ [border]	1.06	(0.99, 1.14)	1.06	(0.99, 1.14)
$\exp(\beta_5)$ [p0014]	1.73	(1.15, 2.60)	1.79	(1.17, 2.74)
$\exp(\beta_6)$ [p1523]	0.88	(0.62, 1.24)	0.90	(0.62, 1.30)
$\exp(\beta_7)$ [p2435]	0.77	(0.54, 1.10)	0.79	(0.54, 1.15)
$\exp(\beta_8)$ [p36up]	0.67	(0.45, 1.00)	0.68	(0.45, 1.03)
$\exp(\beta_9)$ [pdead]	1.12	(0.82, 1.53)	0.95	(0.67, 1.35)
$\exp(\beta_{10})$ [comp12]	4.91	(1.98, 12.20)	4.94	(1.98, 12.34)
$\exp(\beta_{11})$ [comp24e]	4.44	(1.10, 17.91)	4.53	(1.07, 19.17)
$\exp(\beta_{12})$ [comp241]	0.85	(0.42, 1.71)	0.85	(0.41, 1.73)
Random part				
$\sqrt{\psi}$			0.43	
Log likelihood	$-1,784.19$		$-1,782.70$	

*Using robust standard errors for clustered data

8.7 A random-intercept complementary log-log model

8.7.1 Model specification

To accommodate dependence among the survival times of different children of the same woman, after conditioning on the observed covariates, we include a random intercept ζ_j for mother j in the complementary log-log model

$$\ln\{-\ln(1-h_{sij})\} \;=\; \alpha_1 d_{1sij} + \cdots + \alpha_5 d_{5sij} + \beta_2 x_{2ij} + \cdots + \beta_{12} x_{12,sij} + \zeta_j$$

where

$$\zeta_j | \mathbf{d}_{ij}, \mathbf{x}_{ij} \sim N(0, \psi)$$

The model can also be written in terms of a continuous latent response

$$y^*_{sij} \;=\; \alpha_1 d_{1sij} + \cdots + \alpha_5 d_{5sij} + \beta_2 x_{2ij} + \cdots + \beta_{12} x_{12,sij} + \zeta_j + \epsilon_{sij}$$

where ϵ_{sij} has a standard extreme-value type-1 or Gumbel distribution, given the covariates and the random intercept ζ_j. The standard Gumbel distribution has a mean of about 0.577 (called Euler's constant) and a variance of $\pi^2/6$, and is asymmetric.

The exponentiated random intercept, $\exp(\zeta_j)$, is called a *shared frailty* because it is a mother-specific disposition or "frailty" that is shared among children nested in a mother. Notice that frailties are sometimes also included in single-level models, a practice that we do not generally recommend because such models are often not well identified.

8.7.2 Estimation using Stata

Two commands are available for fitting the random-intercept complementary log-log model, `xtcloglog` and `gllamm` with the `link(cll)` option. Whereas `xtcloglog` can be used only for a two-level random-intercept model, `gllamm` can also be used for random-coefficient and higher-level models. However, `xtcloglog` is considerably faster than `gllamm`.

The syntax for `xtcloglog` is analogous to that for `xtlogit`:

```
. xtcloglog y int1-int5 mage mage2 border p0014 p1523 p2435 p36up pdead
> comp12 comp24e comp24l, noconstant eform i(momid)

Random-effects complementary log-log model      Number of obs      =     13854
Group variable: momid                           Number of groups   =       851

Random effects u_i ~ Gaussian                   Obs per group: min =         1
                                                               avg =      16.3
                                                               max =        40

                                                Wald chi2(16)      =   2582.92
Log likelihood  = -1782.7021                    Prob > chi2        =    0.0000

------------------------------------------------------------------------------
           y |     exp(b)   Std. Err.      z    P>|z|     [95% Conf. Interval]
-------------+----------------------------------------------------------------
        int1 |   .2333078   .1806049    -1.88   0.060     .0511693    1.063774
        int2 |   .2071299   .1606492    -2.03   0.042     .0452954    .9471781
        int3 |   .1591151   .1239453    -2.36   0.018     .0345664    .7324346
        int4 |   .2358834   .1833431    -1.86   0.063     .0514151    1.082191
        int5 |   .1261628   .0998945    -2.61   0.009     .0267275    .5955317
        mage |   .8559014   .0513772    -2.59   0.010     .7609018    .9627618
       mage2 |    1.00268   .0010671     2.52   0.012     1.000591    1.004774
      border |   1.059005   .0375963     1.61   0.106     .9878226    1.135316
       p0014 |   1.790715   .3899374     2.68   0.007     1.168619    2.743974
       p1523 |   .8967051   .1696206    -0.58   0.564     .6189227    1.299161
       p2435 |   .7923456   .1489941    -1.24   0.216     .5480915     1.14545
       p36up |   .6810089   .1444591    -1.81   0.070      .449357    1.032082
       pdead |    .948696   .1702308    -0.29   0.769     .6674064     1.34854
      comp12 |   4.941595   2.307624     3.42   0.001     1.978671     12.3413
     comp24e |   4.534147   3.335828     2.05   0.040     1.072151    19.17499
     comp24l |   .8459522   .3100246    -0.46   0.648     .4124748    1.734979
-------------+----------------------------------------------------------------
    /lnsig2u |  -1.672034   .6326815                     -2.912067   -.4320015
-------------+----------------------------------------------------------------
     sigma_u |   .4334333   .1371126                      .2331592    .8057347
         rho |   .1025014   .0582035                      .0319916    .2829852
------------------------------------------------------------------------------
Likelihood-ratio test of rho=0: chibar2(01) =     2.98 Prob >= chibar2 = 0.042
```

The estimated residual correlation among the latent responses for two children of the same mother is only

$$\widehat{\rho} = \frac{\widehat{\psi}}{\widehat{\psi} + \pi^2/6} = \frac{0.433^2}{0.433^2 + \pi^2/6} = 0.10$$

(given as `rho` in the output) suggesting that there is not much dependence among the survival times for children of the same mother after controlling for the observed covariates.

Due to the small within-mother correlation the estimated hazard ratios are close to those from the model without a random intercept. Not surprisingly, we see from the output that the test of the null hypothesis of zero between-mother variance, H_0: $\psi = 0$, is barely significant at the 5% level. However, `pdead`, a dummy variable for the previous child having died before conception of the index child is included in the model, so some of the dependence among the siblings is captured by the fixed part of the model.

Another way of assessing the degree of unobserved heterogeneity is in terms of the median hazard ratio. Imagine randomly drawing children with the same covariate values but having different mothers j and j'. The hazard ratio comparing the child whose mother has the larger random intercept to the other child is given by $\exp(|\zeta_j - \zeta_{j'}|)$. Following the derivation in section 6.10.2, the median hazard ratio is given by

$$\text{HR}_{\text{median}} = \exp\left\{\sqrt{2\psi}\Phi^{-1}(3/4)\right\}$$

Plugging in the parameter estimates, we obtain $\widehat{\text{HR}}_{\text{median}}$:

```
. display exp(sqrt(2*.433433^2)*invnormal(3/4))
1.5120099
```

When two children with the same covariate values but different mothers are randomly sampled, the hazard ratio comparing the child with the larger hazard to the child with the smaller hazard will exceed 1.51 in 50% of the samples, which is of moderate magnitude compared to the estimated hazard ratios for some of the covariates.

8.8 ❖ Marginal and conditional survival probabilities

We now consider the model-implied conditional survival probabilities, given that $\zeta_j = 0$, as well as the corresponding marginal survival probabilities, integrating over the random-intercept distribution.

The predicted conditional or cluster-specific survival probabilities for $\zeta_j = 0$ are given by

$$S_{tij}^{C} = \prod_{s=1}^{t} \{1 - \text{cloglog}^{-1}(\widehat{\alpha}_1 d_{1sij} + \cdots + \widehat{\beta}_{12} x_{12,sij} + \underbrace{0}_{\zeta_j})\}$$

and the predicted marginal or population-averaged survival probabilities are given by

$$S^M_{tij} = \int \left[\prod_{s=1}^{t} \{1 - \text{cloglog}^{-1}(\widehat{\alpha}_1 d_{1sij} + \cdots + \widehat{\beta}_{12} x_{12,sij} + \zeta_j)\} \right] \phi(\zeta_j; 0, \widehat{\psi})\, d\zeta_j$$

The product inside the square brackets is the probability that a child with given covariates and a given value of the random intercept ζ_j survives past time t. Integrating this over the random-intercept distribution gives the corresponding survival probability, averaged over all children with the same covariate values. As explained in section 6.9, marginal probabilities are different from conditional probabilities evaluated at the population mean of the random intercept. Indeed, the marginal survival curve need not correspond to any individual survival curve since it represents the average over the selected population of individuals who are still at risk.

For a mother with one child in the data who was censored at time t, the contribution to the marginal likelihood has exactly the same form as S^M_{tij}. By constructing an appropriate prediction dataset, we can therefore obtain S^M_{tij} by calculating the marginal likelihood contributions using `gllapred` with the `ll` option, but first we must fit the model using `gllamm`:

```
. gllamm y int1-int5 mage mage2 border p0014 p1523 p2435 p36up pdead
> comp12 comp24e comp241, noconstant eform i(momid) link(cll)
> family(binom) adapt

number of level 1 units = 13854
number of level 2 units = 851

Condition Number = 27451.05

gllamm model

log likelihood = -1782.7021
```

y	exp(b)	Std. Err.	z	P>|z|	[95% Conf.	Interval]
int1	.2333055	.1806192	-1.88	0.060	.0511619	1.063907
int2	.2071278	.1606618	-2.03	0.042	.0452888	.9472962
int3	.1591135	.123955	-2.36	0.018	.0345614	.7325256
int4	.2358809	.1833574	-1.86	0.063	.0514076	1.082325
int5	.1261616	.0999021	-2.61	0.009	.0267237	.5956051
mage	.8559022	.051382	-2.59	0.010	.7608944	.962773
mage2	1.00268	.0010672	2.51	0.012	1.000591	1.004774
border	1.059005	.0375963	1.61	0.106	.9878228	1.135316
p0014	1.790712	.3899405	2.68	0.007	1.168612	2.74398
p1523	.8967037	.1696224	-0.58	0.564	.618919	1.299164
p2435	.7923441	.148996	-1.24	0.216	.5480876	1.145454
p36up	.6810076	.1444604	-1.81	0.070	.4493541	1.032084
pdead	.9487002	.1702352	-0.29	0.769	.6674043	1.348556
comp12	4.941589	2.307621	3.42	0.001	1.978668	12.34128
comp24e	4.534112	3.335808	2.05	0.040	1.07214	19.17489
comp241	.8459486	.3100238	-0.46	0.648	.4124726	1.734974

```
Variances and covariances of random effects
------------------------------------------------------------------------------

***level 2 (momid)

    var(1): .18785803 (.11886675)
------------------------------------------------------------------------------
```

Now we will produce a small prediction dataset from the data for child 101 because this child has a row of data for each interval (we will not save the data first, but sometimes this may be a good idea):

```
. keep if kidid==101
```

It is required that y be zero for each interval, which is already true. We can change the covariates to whatever values we want and then save the data:

```
. replace pdead = 0
. replace p2435 = 0
. save junk, replace
```

We now replicate the data five times for this child, creating unique identifiers 1–5 in kidid for the replications. To obtain S^M_{tij} for $t = 1, \ldots, 5$, we let child 1 have data for interval 1 only (corresponding to censoring at $t = 1$), child 2 for intervals 1 and 2, etc. The likelihood contributions for children 1–5 will then correspond to the marginal survival probabilities for times 1–5, respectively.

We first produce data for the five children by appending the data for child 101 to itself four times, changing kidid appropriately each time

```
. replace kidid = 1
. append using junk
. replace kidid = 2 if kidid==101
. append using junk
. replace kidid = 3 if kidid==101
. append using junk
. replace kidid = 4 if kidid==101
. append using junk
. replace kidid = 5 if kidid==101
```

and then delete the unnecessary rows of data:

```
drop if interval > kidid
```

The data now look like the following:

```
. sort kidid interval
. list kidid interval y, sepby(kidid)
```

	kidid	interval	y
1.	1	1	0
2.	2	1	0
3.	2	2	0
4.	3	1	0
5.	3	2	0
6.	3	3	0
7.	4	1	0
8.	4	2	0
9.	4	3	0
10.	4	4	0
11.	5	1	0
12.	5	2	0
13.	5	3	0
14.	5	4	0
15.	5	5	0

For the log-likelihood contributions to correspond to $\ln S^M_{tij}$, we must have one child per mother (otherwise we would get joint survival probabilities for all children):

```
. replace momid = kidid
```

The log-likelihood contributions are obtained using `gllapred` with the `ll` option, and with the `fsample` option to get predictions for the full sample (not just the estimation sample). The marginal survival probabilities S^M_{tij}, called `msurv`, are then obtained by exponentiation:

```
. gllapred loglik, ll fsample
. generate msurv = exp(loglik)
```

The conditional hazards can be obtained by defining a variable `zeta1` for the random-intercept values and then using the `mu` (and not `marginal`) option, together with the `us(zeta)` option:

```
. generate zeta1 = 0
. gllapred chaz, mu us(zeta) fsample
(mu will be stored in chaz)
```

To get the corresponding conditional survival probabilities S^C_{tij}, we use commands similar to those on page 345:

```
. generate ln1mchaz = ln(1-chaz)
. by kidid (interval), sort: generate sln1mchaz = sum(ln1mchaz)
. generate csurv = exp(sln1mchaz)
```

We can now plot the survival probabilities as a function of `interval`:

```
. twoway (line msurv interval if kidid==interval, sort lpatt(solid))
>        (line csurv interval if kidid==5, sort lpatt(dash)),
>        legend(order(1 "marginal" 2 "conditional"))
>        ytitle(Survival probability)
```

This command produces figure 8.6.

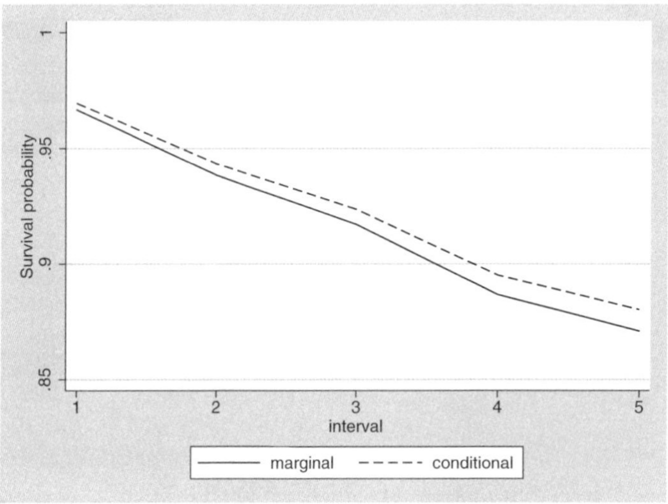

Figure 8.6: Predicted conditional and marginal survival functions

We see that the conditional and marginal survival curves are similar because of the relatively small random-intercept variance.

8.9 Summary and further reading

Special challenges in modeling survival data include censoring, truncation, and time-varying covariates. We have demonstrated how discrete-time survival analysis can proceed by using standard models for dichotomous responses once the data have been expanded appropriately. Using a logit link gives the continuation-ratio logit model in which the conditional odds of the event occurring given that it has not yet occurred are proportional. Using a complementary log-log link is consistent with proportional hazards in continuous time. In both models, proportionality can be relaxed by constructing appropriate time-varying covariates; see also exercises 8.4, 8.5, and 8.6.

Good introductory books on survival analysis in general include Collett (2003b) and Hosmer and Lemeshow (1999) for medicine and Allison (1984, 1995), Singer and Willett (2003), and Box-Steffensmeier and Jones (2004) for social science. Useful introductory articles on single-level discrete-time survival modeling include Allison (1982) and Singer and Willett (1993).

To accommodate clustered or multilevel survival data, we can include random effects in discrete-time survival models. Here we have included a random intercept, the exponential of which is usually called a *frailty* in the survival context, but random coefficients could of course also be included in the same manner as demonstrated in other chapters of the book. Multilevel discrete-time survival models are discussed for instance in the papers by Hedeker, Siddiqui, and Hu (2000), Barber et al. (2000), Rabe-Hesketh, Yang, and Pickles (2001b), and Grilli (2005). Models with discrete random effects, and many other issues in survival analysis, are discussed in Vermunt (1997).

We have considered multilevel survival data where subjects experience an event (death) that is *absorbing* (can only occur once), and subjects are nested in clusters. An alternative type of multilevel survival data, sometimes called multivariate survival data, arises if subjects can experience multiple events: either events of different types or the same event repeatedly. The latter type of data are called *recurrent-event* data. Several decisions must be made when analyzing such multiple-event data. For instance, is a subject at risk of the kth event before experiencing the $(k-1)$th event? Does the hazard of an event depend on whether other events have already occurred? The answers to these questions determine the definition of the time scale and risk sets, and other aspects of the analysis. A nice survey is provided by Kelly and Lim (2000) for continuous-time survival data, but the same issues are relevant for discrete-time survival data.

In this chapter, we have focused on modeling discrete-time survival data. Multilevel models for continuous-time survival data using Poisson regression are discussed in chapter 9.

8.10 Exercises

8.1 Assistant professor promotions data

1. Fit the logistic regression considered on page 349.
2. Modify the model from step 1 by using a low-order polynomial of time instead of dummy variables for time. Start with a linear relationship and include successively higher-order terms until the highest order term is not significant at the 5% level.
3. Produce graphs of the model-implied discrete-time hazards for professors 1 and 4 using the models from steps 1 and 2.
4. Fit the model from step 1, but with a complementary log-log link instead of a logit link.

5. Interpret the estimated exponentiated regression coefficients from step 4 for `undgrad` and `art`.
6. For individuals 1 and 4, plot the model-implied survival functions for the models with a logit link (from step 1) and a complementary log-log link (from step 4). Compare the curves for the two different models for the same individual.

8.2 Teacher turnover data

Here we consider data from Singer (1993) discussed in Singer and Willett (2003). The careers of 3,941 special educators newly hired in Michigan public schools between 1972 and 1978 were tracked for 13 years. The outcome of interest is the number of years the teachers worked at the school, from the date they were hired until they left the school.

The variables in the dataset `teachers.dta` are

- `id`: teacher identifier
- `t`: number of years teacher worked at the school, i.e. time to leaving school (no censoring) or end of study (censoring)
- `censor`: censoring indicator, equal to 1 if censored and 0 otherwise

1. Expand the data to person–years.
2. Obtain predicted hazards using a regression model of your choice with dummy variables for the periods.
3. Calculate and plot the predicted survival function over time.
4. Fit a logistic discrete-time survival model with a polynomial in time. What order of the polynomial seems to be required?
5. Produce a graph of the model-implied survival functions for the models in step 2 and 4.

8.3 First sexual intercourse data

We now analyze the age at first intercourse data from Capaldi, Crosby, and Stoolmiller (1996) that is supplied with the book by Singer and Willett (2003). One hundred eighty at-risk boys (who had not had intercourse before) were followed from grade 6–12 and the date of first intercourse was recorded.

The variables in `firstsex.dta` are

- `id`: boy identifier (j)
- `time`: grade (school-year) in which either intercourse occurred (no censoring) or the boy dropped out of the study (censoring). Children are about 6 years old in first grade, so their approximate age in a given grade is the grade plus 5 years
- `censor`: censoring indicator, equal to 1 if censored and 0 otherwise

- `pt`: a dummy variable for parental transition having occurred before seventh grade (0: lived with both parents; 1: experienced one or more parental transitions)
- `pas`: parental antisocial behavior when the boy was in fourth grade (based on arrests and drivers license suspensions, drug-use, subscales 2 and 9 of the Minnesota Multiphasic Personality Inventory, and for the mother, her age at birth)

1. Expand the data to person–years, remembering that boys in this study could not experience the event before grade 7 (an example of a stock sample; see section 8.2.5).
2. Fit a logistic regression model with a dummy variable for each grade and `pt` as the only covariate, and interpret the exponentiated estimated regression coefficient of `pt`.
3. Plot the predicted survival functions separately for boys who have and have not had parental transitions (variable `pt`).
4. Extend the model by including `pas` as a further covariate and interpret the exponentiated estimated coefficients of `pt` and `pas`.
5. Is there evidence for an interaction between `pas` and `pt`?

8.4 Child mortality data

Consider some further analyses of the child mortality data from Pebley and Stupp (1987) and Guo and Rodriguez (1992).

In addition to the variables listed in section 8.3, the following variables are included in `mortality.dta`:

- `sex`: sex of child (0: female; 1: male)
- `home`: dummy variable for child being born at home (1: home; 0: clinic or hospital)
- `edyears`: mother's years of education
- `income`: family annual income in quetzales in 1974–1975; the quetzal was pegged (equal to) the U.S. dollar. Missing responses are coded as −9

1. Repeat the analyses presented in section 8.7.2.
2. Add the variables given above to the analysis.
3. Interpret the estimated hazard ratios for these new variables.
4. Relax the proportional-hazards assumption for home birth. Specifically, include an interaction between the first time interval and home birth to test if home birth has a greater effect on the hazard of death shortly after, or during, birth than later on.

8.5 Brothers' school transition data

Mare (1994) analyzed data from the 1973 Occupational Changes in a Generation II survey (OCG) to investigate the relationship between the educational attainment of fathers and their sons. He selected male respondents who had at least one brother and considered the years of schooling of the respondent, his oldest brother and his father.

The variables in `brothers.dta` are

- `id`: family identifier
- `brother`: brother identifier (0: respondent; 1: respondent's brother)
- `time`: length of schooling (1: < 12 years; 2: 12 years; 3: ≥ 13 years)
- `feduc`: length of schooling for father (1: 0–8 years; 2: 9–11 years; 3: 12 years; 4: 13–15 years; 5: ≥ 16 years)
- `freq`: number of families

The data are in collapsed form with `freq` indicating the number of families having the given levels of education for the respondent, his brother and his father.

1. Expand the data so that there is one row of data for the transition to 12th grade for every subject and another row of data for the transition to 13th grade for those brothers who did make the transition to 12th grade. Define a variable `transition` equal to 1 and 2 for the two transitions and a dummy variable `event`, equal to 1 if the transition occurred and 0 otherwise. This is just the expansion to person–period data discussed in this chapter except that the response variable is now a dummy variable for continuing in education ("censoring"), not for leaving.
2. For the respondents (and not their brothers), fit a continuation-ratio logit model with dummy variables `tr1` and `tr2` for the two transitions and dummy variables `ed2` to `ed5` for father's education levels 2 to 5. Use `freq` as frequency weights (see `help weights`).
3. Now relax the proportional-odds assumption by allowing the effect of father's education to be different for the first and second transitions. Use a likelihood-ratio test (at the 5% level) to choose the more appropriate model.
4. Fit the model selected in step 2 for both brothers, assuming that the covariate effects are the same for both brothers, and including a random intercept for families. Use `gllamm`, specifying `freq` as level-2 weights, by first renaming `freq` to `wt2` and then using the `gllamm` option `weight(wt)`.
5. ❖ Interpret the parameter estimates for the random-intercept model.

8.6 Blindness data

Diabetic retinopathy is a complication associated with diabetes mellitus and is a major cause of visual loss. Huster, Brookmeyer, and Self (1989) describe data from the Diabetic Retinopathy Study (DRS) conducted in 1971 to investigate the effectiveness of laser photocoagulation in delaying the onset of blindness in patients with diabetic retinopathy. One eye of each patient was randomly selected for treatment, and patients were assessed at 4-month intervals. The endpoint used to assess the treatment effect was the occurrence of visual acuity less than 5/200 at two consecutive assessments. The data considered by Huster, Brookmeyer, and Self (1989) and provided by Ross and Moore (1999) are a 50% sample of the high-risk patients as defined by DRS criteria. Following Ross and Moore (1999), we treat the data as discrete-time data using the intervals [6,10), [10,14), ..., [50,54), [54,58), [58,66), [66,83).

The variables in `blindness.dta` are

- `id`: patient identifier
- `eye`: eye identifier
- `treat`: dummy variable for eye being treated by photocoagulation
- `time`: onset of blindness in months
- `discrete`: discrete-time intervals 1: [6,10); 2: [10,14) ... 12: [50,54); 13: [54,58); 14: [58,66); 15: [66,83)
- `censor`: dummy variable for blindness occurring
- `late`: dummy variable for late onset, or adult diabetes (onset at age 20 or later)

1. Expand the data to eye periods, with periods defined by the variable `discrete`.
2. Fit a logistic discrete-time survival model with dummy variables for time intervals, `treat`, `late` and their interaction as covariates, and a random intercept for patients.
3. Test the proportional-odds assumption for `treat` by including an interaction between `treat` and a new variable `midp` defined as the middle of the discrete-time intervals. (Hint: except for the last two intervals, `midp` = 4 + 4*`discrete`.)
4. ❖ For the selected model, plot the model-implied marginal survival functions for the two eyes for patients with late onset diabetes.

See also exercise 9.6.

8.7 Cigarette data

Hedeker, Siddiqui, and Hu (2000) analyzed data from a subset of the Television School and Family Smoking Prevention and Cessation Project (TVSFP; see Flay et al. 1989). The data are supplied with the MIXOR program (Hedeker and Gibbons 1996).

In 1986, schools in Los Angeles were randomized to one of four conditions given by different combinations of two factors: (1) TV: a media (television) intervention (1: present; 0: absent) and (2) CC: a social-resistance classroom curriculum (1: present; 0: absent). The intervention took place when the students were in seventh grade (when the children were about 12 years old). The students were assessed pre- and postintervention in 1986 and again a year later and two years later. At each of the four time points, the students were asked: "Have you ever smoked a cigarette?". To estimate the effect of the intervention only those who answered "no" to the question before the intervention were included in the study. The first time the students answered "yes" after the intervention will be the event of interest in this exercise.

In addition to the clustering of students in classes, the classes are clustered in schools. We will consider two-level models in this exercise but will revisit the data for three-level modeling in exercise 10.9.

The variables in `cigarette.dta` are

- `school`: school identifier
- `class`: class identifier
- `time`: postintervention time point (1: 7th grade; 2: 8th grade; 3: 9th grade)
- `event`: dummy variable for student responding "yes" to the question about smoking
- `cc`: social-resistance classroom curriculum (dummy variable)
- `tv`: television intervention (dummy variable)

1. Expand the data to person–period data.
2. Estimate the discrete-time model that assumes the continuous-time hazards to be proportional. Include `cc`, `tv`, and their interaction as explanatory variables and specify a random intercept for classes. Use dummy variables for periods.
3. Interpret the exponentials of the estimated regression coefficients.
4. Obtain the estimated residual intraclass correlation of the latent responses.

See also exercise 10.9.

9 Counts

9.1 Introduction

An important outcome in many investigations is a count of how many times some event has occurred. For instance, we could count the number of epileptic seizures in a week for a patient, the number of patents awarded to a company in a 5-year period, or the number of murders in a year in a city. Counts are nonnegative, integer-valued responses, taking on values 0, 1, 2, ...

In this chapter, we will discuss Poisson models for counts. After introducing single-level Poisson regression models, we will describe multilevel Poisson regression models, which include random effects to model dependence and unobserved heterogeneity. Some new issues that arise in modeling counts, such as using offsets and dealing with overdispersion, will be considered. We also briefly describe the negative binomial model for count data, as well as fixed-effects models and generalized estimating equations.

Three applications are considered: longitudinal modeling of number of doctor visits, survival analysis of the child mortality data discussed in chapter 8, and small area estimation or disease mapping of lip cancer incidence. On first reading, it might be a good idea to concentrate on the first application.

9.2 What are counts?

9.2.1 Counts versus proportions

In this chapter, we consider counts of events that could in principle occur any time during a time interval (or anywhere within a spatial region). For instance, von Bortkiewicz (1898) observed 14 corps of the Prussian army from 1875 to 1894 and counted the number of deaths from horse kicks (see the logo on the spine of this book). Other examples of such counts are the number of times a person visits a doctor during a year, the number of times a person blinks in an hour, and the number of violent fights occurring in a school during a week. The term *count* without further qualification usually implies this type of count. As we discuss in section 9.3, a Poisson distribution is often appropriate in this case.

Another type of count is a count of events or "successes" that can only occur at a predetermined number of "trials". The count y is then less than or equal to the number of trials n, and the response can be expressed as a sample proportion $p = y/n$. For instance, when counting the number of retired people in a city block at a given point in time, each of n inhabitants is either retired (success) or not (failure), and the count cannot exceed n. A meaningful summary then is the proportion of the city block's population that is retired. An appropriate distribution for the corresponding count is often the binomial distribution for n trials. If there are covariates, the same models as those for dichotomous responses are used, such as the logit or probit model. The only difference is that a binomial denominator n_i must be specified for each unit i. In `xtmelogit`, the `binomial()` option can be used to specify a variable containing the values n_i, and in `gllamm`, the `denom()` option. For low probabilities of success and large n_i, the binomial distribution is well-approximated by the Poisson distribution. We do not consider proportions further in this chapter.

9.2.2 Counts as aggregated event-history data

Counts can be thought of as aggregated versions or summaries of more detailed data on the occurrences of some kind of event. For instance, when considering the number of crimes, we usually aggregate over spatial regions and time intervals instead of retaining the original information on the locations and timings of the individual crimes. The size of spatial regions and time intervals determines the resolution at which we can investigate spatial and temporal variation. When counts are broken down by time intervals, the analysis can be referred to as survival analysis, a common model being the piecewise exponential model we will discuss in section 9.12.

Covariate information can be considered by aggregating within levels of the covariates, for example producing separate counts of crimes for each type of victim classified by race and gender. If spatial and temporal variation are also of interest, we could obtain counts by region, time interval, race, and gender. It is important to remember that the cells in such a contingency table, which become the cases or observations for analysis, are not units in the usual sense. However, some of the variables defining the cells may be viewed as defining units or clusters that could be characterized by unobserved covariates and hence modeled by including random effects. In the crime example, regions, or neighborhoods, might be considered as such clusters. As we will see in section 9.10, it is possible to include random intercepts, say, for regions, even if the data have been aggregated to one count per region. Although such aggregated data provide no information on the dependence among counts within regions (because there is only one count per region), they do provide information on the random-intercept variance via the phenomenon known as overdispersion, to be discussed later.

9.3 Single-level Poisson models for counts

It is often assumed that events occur independently of each other and at a constant *incidence rate* λ (pronounced lambda) defined as the instantaneous probability of new event per time interval. It follows that the number of events y occurring in a time interval of length t has a Poisson distribution

$$\Pr(y|\mu) \;=\; \frac{\exp(-\mu)\mu^y}{y!}$$

where μ is the expectation of y and is given by

$$\mu \;=\; \lambda t$$

The incidence rate λ is also often called the *intensity*, and the time interval t is sometimes referred to as the *exposure*.

If we observe two counts y_1 and y_2 during two successive intervals of length t, and if the incidence rate λ is the same for both intervals, we can either add the counts together, thus obtaining an interval of length $2t$, or consider the counts separately. A convenient property of the Poisson distribution is that both approaches yield the same likelihood, up to a multiplicative constant:

$$\begin{aligned}
\Pr(y_1|\mu=\lambda t)\Pr(y_2|\mu=\lambda t) &= \frac{\exp(-\lambda t)(\lambda t)^{y_1}}{y_1!}\frac{\exp(-\lambda t)(\lambda t)^{y_2}}{y_2!} \\
&= \frac{\exp(-\lambda 2t)(\lambda t)^{y_1+y_2}}{y_1!y_2!} \\
&= \frac{\exp(-\lambda 2t)(\lambda 2t)^{y_1+y_2}}{(y_1+y_2)!} \times \frac{(y_1+y_2)!}{2^{y_1+y_2}y_1!y_2!} \\
&\propto \Pr(y_1+y_2|\mu=\lambda 2t) \qquad (9.1)
\end{aligned}$$

The multiplicative constant does not affect parameter estimation because it does not depend on the parameters. Therefore, no information regarding λ is lost by aggregating the data. We must, however, keep track of the exposure by using offsets as discussed below.

When counts are observed for different units or subjects i characterized by covariates, the mean μ_i is usually modeled using a log-linear model. For one covariate x_i, a multiplicative regression model for the expected counts is specified as

$$\mu_i \equiv E(y_i|x_i) \;=\; \exp(\beta_1+\beta_2 x_i) \;=\; \exp(\beta_1)\times\exp(\beta_2 x_i) \qquad (9.2)$$

which can alternatively be written as an additive log-linear model

$$\ln(\mu_i) \;=\; \beta_1+\beta_2 x_i$$

If we think of this as a generalized linear model, the link function $g(\cdot)$ is just the natural logarithm and the inverse link function $h(\cdot)\equiv g^{-1}(\cdot)$ is the exponential.

A nice feature of the log link is that if the exposure t is the same for all subjects, the exponentiated coefficient $\exp(\beta_2)$ can be interpreted as the *incidence-rate ratio* for a unit increase in x_i. This can be seen by substituting $\mu_i=\lambda_i t$ in (9.2):

$$\lambda_i t \;=\; \exp(\beta_1+\beta_2 x_i) \;=\; \exp(\beta_1)\exp(\beta_2 x_i)$$

Then the ratio for two subjects i and i' becomes

$$\frac{\lambda_i}{\lambda_{i'}} = \frac{\lambda_i t}{\lambda_{i'} t} = \frac{\exp(\beta_1)\exp(\beta_2 x_i)}{\exp(\beta_1)\exp(\beta_2 x_{i'})} = \exp\{\beta_2(x_i - x_{i'})\} = \exp(\beta_2)^{(x_i - x_{i'})}$$

When the covariate changes by 1 unit, $x_i - x_{i'} = 1$, the ratio of the incidence rates $\lambda_i/\lambda_{i'}$, or *incidence-rate ratio* (IRR), is hence $\exp(\beta_2)$.

The log-linear model for the expectation is combined with the assumption that, conditional on the covariate x_i, the count y_i has a Poisson distribution with mean μ_i

$$\Pr(y_i|x_i) = \frac{\exp(-\mu_i)\mu_i^{y_i}}{y_i!}$$

If different subjects have different exposures t_i, the natural logarithm of the exposure must be included as an *offset*, a covariate with regression coefficient set to 1:

$$\lambda_i t_i = \exp\{\beta_1 + \beta_2 x_i + \underbrace{\ln(t_i)}_{\text{offset}}\} = \exp(\beta_1)\exp(\beta_2 x_i) t_i$$

Then t_i cancels out, and we get

$$\lambda_i = \exp(\beta_1)\exp(\beta_2 x_i)$$

Following the logic of (9.1), it is clear that we can sum the counts for all subjects sharing the same covariate value and let i denote the corresponding group of subjects. If we also sum the corresponding exposures, we obtain the same maximum likelihood estimates of the parameters β_1 and β_2 using the aggregated data as using the unit-level data.

For the Poisson distribution, the conditional variance of the counts, given the covariate, equals the conditional expectation:

$$\text{Var}(y_i|x_i) = \mu_i \tag{9.3}$$

In practice, the conditional sample variance is often larger or smaller than the conditional variance implied by the model; phenomena known as *overdispersion* or *underdispersion*, respectively. Overdispersion could be due to variability in the incidence rates λ_i that is not fully accounted for by the included covariates and is more common than underdispersion. We will return to this issue in section 9.9.

9.4 Did the German health-care reform reduce the number of doctor visits?

Government expenditures on health care surged in Germany in the 80s and 90s. To reduce the expenditure, a major health-care reform took place in 1997. The reform raised the copayments for prescription drugs by up to 200% and imposed upper limits on reimbursement of physicians by the state insurance. Given the large share of Gross Domestic Product (GDP) spent on health, it is of interest to investigate whether the reform was a success in the sense that the number of doctor visits decreased after the reform.

To address this research question, Winkelmann (2004) analyzed data from the German Socio-Economic Panel (SOEP Group 2001). We will consider a subset of his data, comprised of women working full time in the 1996 panel wave preceding the reform and the 1998 panel wave following the reform.

The dataset `drvisits.dta` has the following variables:

- `id`: person identifier (j)
- `numvisits`: self-reported number of visits to a doctor during the 3 months before the interview (y_{ij})
- `reform`: dummy variable for interview being during the year after the reform versus the year before the reform (x_{2i})
- `age`: age in years (x_{3ij})
- `educ`: education in years (x_{4ij})
- `married`: dummy variable for being married (x_{5ij})
- `badh`: dummy variable for self-reported current health being classified as "very poor" or "poor" (versus "very good", "good", or "fair") (x_{6ij})
- `loginc`: logarithm of household income (in 1995 German Marks, based on OECD weights for household members) (x_{7ij})

9.5 Longitudinal data structure

We start by exploring the participation patterns for the two panel waves using the `xtdescribe` command:

```
. use http://www.stata-press.com/data/mlmus2/drvisits
. xtdescribe if numvisit<., i(id) t(reform)
      id:  3, 4, ..., 9189                                 n =       1518
  reform:  0, 1, ..., 1                                    T =          2
           Delta(reform) = 1 unit
           Span(reform)  = 2 periods
           (id*reform uniquely identifies each observation)
Distribution of T_i:   min      5%      25%       50%       75%     95%     max
                         1       1        1         1         2       2       2
     Freq.  Percent    Cum.   Pattern
                             
       709    46.71   46.71   11
       418    27.54   74.24   .1
       391    25.76  100.00   1.
                             
      1518   100.00           XX
```

(The i() and t() options are not necessary if the xtset command has been used to declare the data to be panel data.) Fewer than half of the subjects provide responses for both occasions (having the pattern "11"). Some of the missing data is due to attrition (dropout) and nonresponse, and some is due to younger people entering and older people leaving the sample because the sample is restricted to those aged between 20 and 60 years at any point in time (a "rotating panel").

9.6 Single-level Poisson regression

9.6.1 Model specification

Before including random effects to model longitudinal dependence, we consider ordinary Poisson regression for the number of doctor visits.

The expected number of visits μ_{ij} at occasion i for subject j is specified as a log-linear model

$$\ln(\mu_{ij}) \;=\; \nu_{ij} = \beta_1 + \beta_2 x_{2i} + \cdots + \beta_7 x_{7ij}$$

or equivalently as an exponential model for the expected number of visits:

$$\mu_{ij} \;=\; \exp(\nu_{ij})$$

The number of doctor visits y_{ij} is assumed to have a Poisson distribution with expectation μ_{ij}, given the covariates. We do not have to include an offset because doctor visits were counted for the same interval, namely, 3 months, for all subjects at both occasions. The exponentiated regression coefficients can therefore be interpreted as rate ratios, or as ratios of expected counts for any length of interval we like to think about.

9.6.2 Estimation using Stata

We first fit the Poisson regression model with the poisson command, using the irr option (for incidence-rate ratio) to obtain exponentiated estimates, and store the estimates for later use:

```
. poisson numvisit reform age educ married badh loginc summer, irr
Poisson regression                                Number of obs   =       2227
                                                  LR chi2(7)      =    1429.00
                                                  Prob > chi2     =     0.0000
Log likelihood = -5942.6924                       Pseudo R2       =     0.1073

------------------------------------------------------------------------------
    numvisit |        IRR   Std. Err.      z    P>|z|     [95% Conf. Interval]
-------------+----------------------------------------------------------------
      reform |   .8689523   .0230968    -5.28   0.000     .8248423    .9154212
         age |   1.004371   .0013088     3.35   0.001     1.001809    1.006939
        educ |   .9894036   .0059465    -1.77   0.076      .977817    1.001127
     married |   1.042542    .029055     1.49   0.135     .9871229    1.101073
        badh |   3.105111   .0941052    37.39   0.000     2.926039    3.295142
      loginc |   1.160559   .0418632     4.13   0.000     1.081342     1.24558
      summer |   1.010269   .0408237     0.25   0.800     .9333421    1.093536
------------------------------------------------------------------------------

. estimates store ordinary
```

We can alternatively view the Poisson regression model as a generalized linear model and use the `glm` command with the `family(poisson)` and `link(log)` options. (The log link is the default link for the Poisson distribution, so the `link()` option could be omitted.) To facilitate interpretation, we use the `eform` option to get exponentiated regression coefficients:

```
. glm numvisit reform age educ married badh loginc summer,
> family(poisson) link(log) eform
Generalized linear models                          No. of obs      =      2227
Optimization     : ML                              Residual df     =      2219
                                                   Scale parameter =         1
Deviance         =  7419.853221                    (1/df) Deviance =  3.343782
Pearson          =  9688.740471                    (1/df) Pearson  =  4.366264

Variance function: V(u) = u                        [Poisson]
Link function    : g(u) = ln(u)                    [Log]

                                                   AIC             =  5.344133
Log likelihood   =  -5942.69244                    BIC             =  -9685.11

------------------------------------------------------------------------------
             |                 OIM
    numvisit |        IRR   Std. Err.      z    P>|z|     [95% Conf. Interval]
-------------+----------------------------------------------------------------
      reform |   .8689523   .0230968    -5.28   0.000     .8248423    .9154212
         age |   1.004371   .0013088     3.35   0.001     1.001809    1.006939
        educ |   .9894036   .0059465    -1.77   0.076      .977817    1.001127
     married |   1.042542    .029055     1.49   0.135     .9871229    1.101073
        badh |   3.105111   .0941052    37.39   0.000     2.926039    3.295142
      loginc |   1.160559   .0418632     4.13   0.000     1.081342     1.24558
      summer |   1.010269   .0408237     0.25   0.800     .9333421    1.093536
------------------------------------------------------------------------------
```

The estimates from `poisson` and `glm` are identical, as would be expected, and are displayed under the heading "Poisson" in table 9.1. We do not report the corresponding confidence intervals in the table since they assume that repeated counts for a person are independent (given the covariates) and are thus not trustworthy. The estimated incidence-rate ratio for `reform` is 0.87, implying a 13% reduction in the number of doctor visits per month (or per year) between 1996 and 1998 for given covariate values.

Table 9.1: Estimates for different kinds of Poisson regression: Ordinary, GEE, random-intercept (RI), and random-coefficient (RC)

	Marginal effects			Conditional effects			
	Poisson	GEE Poisson		RI Poisson		RC Poisson	
	Est	Est	(95% CI)†	Est	(95% CI)	Est	(95% CI)
Fixed part:							
Incidence-rate ratios (IRRs)							
$\exp(\beta_2)$ [reform]	0.87	0.88	(0.80, 0.98)	0.95	(0.90, 1.02)	0.90	(0.81, 1.00)
$\exp(\beta_3)$ [age]	1.00	1.01	(1.00, 1.01)	1.01	(1.00, 1.01)	1.00	(1.00, 1.01)
$\exp(\beta_4)$ [educ]	0.99	0.99	(0.97, 1.01)	1.01	(0.98, 1.03)	1.01	(0.98, 1.03)
$\exp(\beta_5)$ [married]	1.04	1.04	(0.90, 1.19)	1.08	(0.97, 1.20)	1.09	(0.97, 1.22)
$\exp(\beta_6)$ [badh]	3.11	3.02	(2.54, 3.58)	2.47	(2.19, 2.78)	3.03	(2.61, 3.52)
$\exp(\beta_7)$ [loginc]	1.16	1.15	(0.98, 1.34)	1.10	(0.96, 1.25)	1.14	(0.98, 1.32)
$\exp(\beta_8)$ [summer]	1.01	0.97	(0.82, 1.16)	0.87	(0.76, 0.98)	0.91	(0.78, 1.07)
Random part							
$\sqrt{\psi_{11}}$				0.90		0.95	
$\sqrt{\psi_{22}}$						0.93	
$\rho_{21} = \psi_{21}/(\sqrt{\psi_{11}}\sqrt{\psi_{22}})$						−0.49	
Log likelihood	−5,942.69			−4,643.36		−4,513.73	

†Based on the sandwich estimator

9.7 Random-intercept Poisson regression

The ordinary Poisson regression model makes the unrealistic assumption that the number of doctor visits before the reform y_{1j} is independent of the number of visits after the reform y_{2j} for the same person j, given the included covariates.

9.7.1 Model specification

One way to address the dependence within persons is to include a person-specific random intercept ζ_{1j} in the Poisson regression model

$$\begin{aligned}\mu_{ij} \equiv E(y_{ij}|\mathbf{x}_{ij}, \zeta_{1j}) &= \exp(\beta_1 + \beta_2 x_{2i} + \cdots + \beta_7 x_{7ij} + \zeta_{1j}) \\ &= \exp\{(\beta_1 + \zeta_{1j}) + \beta_2 x_{2i} + \cdots + \beta_7 x_{7ij}\} \\ &= \exp(\zeta_{1j}) \exp(\beta_1 + \beta_2 x_{2i} + \cdots + \beta_7 x_{7ij})\end{aligned}$$

where $\zeta_{1j}|\mathbf{x}_{ij} \sim \mathrm{N}(0, \psi_{11})$ and the ζ_{1j} are independent across persons j. The exponential of the random intercept, $\exp(\zeta_{1j})$, is sometimes called a *frailty*. The number of visits y_{1j}

and y_{2j} for a person j at the two occasions are specified as conditionally independent given the random intercept ζ_{1j} (and the covariates $\mathbf{x}_{ij}$).

As always in random-effects models, the regression coefficients have conditional or cluster-specific interpretations. In the present application, the clusters are persons, so the coefficients represent person-specific effects. Interestingly, we can also interpret the coefficients as marginal or population-averaged effects because the relationship between the marginal expectation of the count (given $\mathbf{x}_{ij}$ but averaged over ζ_{1j}) and the covariates is

$$\mu_{ij}^M = \exp\{(\beta_1+\psi_{11}/2) + \beta_2 x_{2i} + \cdots + \beta_7 x_{7ij}\}$$

The intercept is therefore the only parameter that is not the same in marginal and conditional log-linear Poisson regression models.

The marginal or population-averaged variance (given $\mathbf{x}_{ij}$ but averaged over ζ_{1j}) is

$$\text{Var}(y_{ij}|\mathbf{x}_{ij}) = \mu_{ij}^M + (\mu_{ij}^M)^2\{\exp(\psi_{11}) - 1\} \tag{9.4}$$

If $\psi_{11} > 0$, this variance is greater than the marginal mean. Therefore, the variance–mean relationship for Poisson models in (9.3) is violated and the variance is greater than that implied by a Poisson model, producing overdispersion.

The intraclass correlation between the counts y_{1j} and y_{2j} at the two occasions depends on the values of the observed covariates (and the exposure if this is included as an offset in the linear predictor) and therefore cannot be used as a simple measure of dependence. In contrast to logit, probit, or complementary log-log models for dichotomous and ordinal responses, Poisson models cannot be formulated in terms of latent responses underlying the observed responses. Thus we cannot define an intraclass correlation in terms of latent responses as shown in section 6.10.1 for dichotomous responses.

As a measure of heterogeneity, we can consider randomly drawing pairs of persons j and j' with the same covariate values and forming the incidence-rate ratio, or ratio of expected counts for the same exposure, comparing the person with the larger random intercept to the person with the smaller random intercept, given by $\exp(|\zeta_j - \zeta_{j'}|)$. Following the derivation in section 6.10.2, the median incidence-rate ratio is given by

$$\text{IRR}_{\text{median}} = \exp\left\{\sqrt{2\psi_{11}}\Phi^{-1}(3/4)\right\}$$

9.7.2 Estimation using Stata

Random-intercept Poisson models can be fitted using `xtpoisson`, `xtmepoisson`, and `gllamm`. All three programs use numerical integration as discussed in section 6.11.1. The details of implementation and speed considerations (discussed in section 6.11.2) are the same as those for logistic models, with `xtpoisson` corresponding to `xtlogit` and `xtmepoisson` to `xtmelogit`. The predictions available for each of the commands also correspond to their logistic counterparts.

Using xtpoisson

The syntax for `xtpoisson` is similar as that for `xtlogit`, but we must specify the `normal` option because `xtpoisson` otherwise assumes a gamma distribution for the frailty $\exp(\zeta_{1j})$ as briefly described in section 9.9.2. The command for fitting the random-intercept model therefore is

```
. xtpoisson numvisit reform age educ married badh loginc summer, i(id) normal
> irr

Random-effects Poisson regression               Number of obs      =      2227
Group variable: id                              Number of groups   =      1518

Random effects u_i ~ Gaussian                   Obs per group: min =         1
                                                               avg =       1.5
                                                               max =         2

                                                Wald chi2(7)       =    253.16
Log likelihood  = -4643.3608                    Prob > chi2        =    0.0000

------------------------------------------------------------------------------
    numvisit |        IRR   Std. Err.      z    P>|z|     [95% Conf. Interval]
-------------+----------------------------------------------------------------
      reform |   .9547597   .0310874    -1.42   0.155      .895733    1.017676
         age |   1.006003   .0028278     2.13   0.033     1.000475     1.01156
        educ |   1.008656   .0127767     0.68   0.496     .9839221    1.034011
     married |   1.077904   .0595749     1.36   0.175     .9672414    1.201228
        badh |   2.466573   .1519255    14.66   0.000     2.186076    2.783061
      loginc |   1.097455   .0747067     1.37   0.172     .9603805    1.254095
      summer |   .8672783   .0562749    -2.19   0.028      .763707    .9848956
-------------+----------------------------------------------------------------
    /lnsig2u |  -.2016963   .0611823    -3.30   0.001    -.3216115   -.0817812
-------------+----------------------------------------------------------------
     sigma_u |   .9040703   .0276566                      .8514575    .9599341
------------------------------------------------------------------------------
Likelihood-ratio test of sigma_u=0: chibar2(01) =  2598.66 Pr>=chibar2 = 0.000
```

We see that the estimated effect of `reform` is to reduce the incidence rate, and therefore the expected number of visits in a given period, by about 5% (controlling for the other variables), but this is not significant at the 5% level. Each extra year of age is associated with an estimated 0.6% increase in the incidence rate; the incidence-rate ratio for a 10-year increase in age is estimated as $1.006^{10} = 1.06$, corresponding to a 6% increase, for given values of the other covariates. The estimates also suggest that being in bad health more than doubles the incidence rate and that the incidence rate decreases by 13% in the summer, controlling for the other variables. The other coefficients are not significant at the 5% level. The estimates are also reported under the heading "RI Poisson" in table 9.1.

From the last line of output, we see that we can reject the null hypothesis that the random-intercept variance is zero, $\psi_{11} = 0$, with a likelihood-ratio statistic of 2,598.66 and $p < 0.001$.

The median incidence-rate ratio is estimated as

```
. display exp(sqrt(2)*.90)*invnormal(3/4)
2.4084741
```

Half the time, the ratio of the expected number of visits (in a given period) for two randomly chosen persons with the same covariate values, comparing the one with the larger expected value to the other one, will exceed 2.41. We could avoid having to say "comparing the one with the larger expected value to the other one" by saying the following: Half the time the ratio of the expected number of visits will lie in the range from 0.41 (= 1/2.41) to 2.41, and the other half of the time the ratio will lie outside that range.

Using xtmepoisson

The syntax for `xtmepoisson` is exactly as that for `xtmelogit` except that we use the `irr` option to get exponentiated regression coefficients instead of the `or` option. We fit the model and store the estimates using the following commands:

```
. xtmepoisson numvisit reform age educ married badh loginc summer || id:, irr
Mixed-effects Poisson regression                Number of obs      =      2227
Group variable: id                              Number of groups   =      1518

                                                Obs per group: min =         1
                                                               avg =       1.5
                                                               max =         2

Integration points =   7                        Wald chi2(7)       =    253.06
Log likelihood = -4643.2823                     Prob > chi2        =    0.0000

------------------------------------------------------------------------------
    numvisit |        IRR   Std. Err.      z    P>|z|     [95% Conf. Interval]
-------------+----------------------------------------------------------------
      reform |   .9547758   .0310889    -1.42   0.155     .8957463    1.017695
         age |   1.006003   .0028283     2.13   0.033     1.000475    1.011562
        educ |   1.008659   .0127792     0.68   0.496     .9839206    1.034019
     married |   1.077902    .059584     1.36   0.175     .9672229    1.201245
        badh |   2.466399   .1519349    14.65   0.000     2.185887    2.782909
      loginc |    1.09743   .0747148     1.37   0.172     .9603412    1.254088
      summer |   .8672584   .0562778    -2.19   0.028     .7636823    .9848822
------------------------------------------------------------------------------

------------------------------------------------------------------------------
  Random-effects Parameters  |   Estimate   Std. Err.     [95% Conf. Interval]
-----------------------------+------------------------------------------------
id: Identity                 |
                   sd(_cons) |   .9043832   .0276826      .8517217    .9603006
------------------------------------------------------------------------------
LR test vs. Poisson regression:   chibar2(01) =  2598.82 Prob>=chibar2 = 0.0000

. estimates store xtri
```

The estimates are similar to those produced by `xtpoisson`.

We could also fit the model more quickly by using one quadrature point giving the Laplace method. This method works well in this example, but we do not recommend it in general because the estimates can be poor.

Using gllamm

The random-intercept Poisson model can be fitted as follows using `gllamm`:

```
. generate cons = 1
. eq ri: cons
. gllamm numvisit reform age educ married badh loginc summer,
> family(poisson) link(log) i(id) eqs(ri) eform adapt
number of level 1 units = 2227
number of level 2 units = 1518

Condition Number = 723.77595

gllamm model

log likelihood = -4643.3427
```

numvisit	exp(b)	Std. Err.	z	P>|z|	[95% Conf.	Interval]
reform	.9547481	.0310831	-1.42	0.155	.8957293	1.017656
age	1.006002	.0028266	2.13	0.033	1.000477	1.011557
educ	1.008646	.0127702	0.68	0.497	.9839247	1.033988
married	1.077896	.059554	1.36	0.175	.9672696	1.201174
badh	2.466857	.15192	14.66	0.000	2.186367	2.78333
loginc	1.097486	.0746823	1.37	0.172	.9604527	1.25407
summer	.8673159	.0562616	-2.19	0.028	.7637672	.9849033

```
Variances and covariances of random effects
------------------------------------------------------------------------------

***level 2 (id)

    var(1): .81691979 (.04972777)
------------------------------------------------------------------------------
. estimates store glri
```

Here we specified an equation `ri` for the random intercept using the `eqs()` option. This is not necessary since `gllamm` would include a random intercept by default. However, defining the intercept explicitly from the start makes it easier to use estimates from this model as starting values for random-coefficient models because the parameter matrix will have the correct column labels.

The estimates are close to those using `xtlogit` and `xtmepoisson`, including the random-intercept standard deviation, estimated by `gllamm` as

```
. display sqrt(.81691979)
.90383615
```

We could also obtain robust standard errors using the command

```
gllamm, robust eform
```

There is no likelihood-ratio test for the random-intercept variance in the `gllamm` output, but we can perform this test ourselves by comparing the random-intercept model (with estimates stored under `glri`) with the ordinary Poisson model (with estimates stored under `ordinary`):

```
. lrtest ordinary glri, force
Likelihood-ratio test                                 LR chi2(1)  =   2598.70
(Assumption: ordinary nested in glri)                 Prob > chi2 =    0.0000
```

We used the `force` option to force Stata to perform the test, which is necessary because the two models being compared were fitted using different commands. We must divide the p-value by 2 as discussed in section 2.6.2, which makes no difference to the conclusion here that the random-intercept variance is highly significant.

9.8 Random-coefficient Poisson regression

9.8.1 Model specification

The random-intercept Poisson regression model accommodates dependence among the repeated counts. However, it assumes that the effect of the health-care reform is the same for all persons. In this section, we relax this assumption by introducing an additional person-level random coefficient ζ_{2j} for `reform` (x_{2i}):

$$\begin{aligned}\ln(\mu_{ij}) &= \beta_1 + \beta_2 x_{2i} + \cdots + \beta_7 x_{7ij} + \zeta_{1j} + \zeta_{2j} x_{2ij} \\ &= (\beta_1 + \zeta_{1j}) + (\beta_2 + \zeta_{2j}) x_{2i} + \cdots + \beta_7 x_{7ij} \qquad (9.5)\end{aligned}$$

As usual, we assume that given the covariates $\mathbf{x}_{ij}$, the random intercept and random coefficient have a bivariate normal distribution with zero mean and covariance matrix

$$\boldsymbol{\Psi} = \begin{bmatrix} \psi_{11} & \psi_{12} \\ \psi_{21} & \psi_{22} \end{bmatrix}, \qquad \psi_{21} = \psi_{12}$$

The correlation between the random intercept and random coefficient becomes

$$\rho_{21} = \frac{\psi_{21}}{\sqrt{\psi_{11}\psi_{22}}}$$

The above specification allows the effect of the health-care reform $\beta_2 + \zeta_{2j}$ to vary randomly over persons j.

The marginal expected count is given by

$$\begin{aligned}E(y_{ij}|\mathbf{x}_{ij}) &= \exp\{\beta_1 + \beta_2 x_{2ij} + \cdots + \beta_7 x_{7ij} + (\psi_{11} + 2\psi_{21} x_{2ij} + \psi_{22} x_{2ij}^2)/2\} \\ &= \exp\{\beta_1 + \psi_{11}/2 + \beta_2 x_{2ij} + \psi_{21} x_{2ij} + (\psi_{22}/2) x_{2ij}^2 + \cdots + \beta_7 x_{7ij}\} \\ &= \exp\{\beta_1 + \psi_{11}/2 + (\beta_2 + \psi_{21} + \psi_{22}/2) x_{2ij} + \cdots + \beta_7 x_{7ij}\}\end{aligned}$$

where the last equality holds only if x_{2i} is a dummy variable, since $x_{2i} = x_{2i}^2$ in this case.

Hence, the multiplicative effect of the reform on the *marginal* expected count now no longer equals $\exp(\beta_2)$ as in the random-intercept model, but $\exp(\beta_2 + \psi_{21} + \psi_{22}/2)$. In general, the marginal expected count is given by the exponential of the sum of the linear predictor and half the variance of the random part of the model. Unfortunately, this variance is usually not a linear function of the variable x_{2ij} having a random coefficient (except if x_{2i} is a dummy variable), so that the variable no longer has a simple multiplicative effect on the marginal expectation.

9.8.2 Estimation using Stata

Using xtmepoisson

The syntax is the same as that for random-coefficient models using `xtmixed` or `xtmelogit`, apart from the `irr` option that is used to obtain exponentiated regression coefficients:

```
. xtmepoisson numvisit reform age educ married badh loginc summer || id: reform,
> cov(unstructured) irr

Mixed-effects Poisson regression                Number of obs      =      2227
Group variable: id                              Number of groups   =      1518

                                                Obs per group: min =         1
                                                               avg =       1.5
                                                               max =         2

Integration points =   7                        Wald chi2(7)       =    241.11
Log likelihood = -4513.7299                     Prob > chi2        =    0.0000

------------------------------------------------------------------------------
    numvisit |        IRR   Std. Err.      z    P>|z|     [95% Conf. Interval]
-------------+----------------------------------------------------------------
      reform |   .9022792   .0483866    -1.92   0.055      .812257    1.002278
         age |   1.003456   .0028317     1.22   0.222      .997921    1.009021
        educ |   1.008894   .0128121     0.70   0.486     .9840931    1.034321
     married |   1.086874    .064118     1.41   0.158     .9681982    1.220097
        badh |   3.028323   .2322948    14.44   0.000     2.605606    3.519619
      loginc |   1.135636     .08665     1.67   0.096     .9778943    1.318824
      summer |   .9140246   .0741941    -1.11   0.268     .7795847    1.071649
------------------------------------------------------------------------------

------------------------------------------------------------------------------
  Random-effects Parameters  |   Estimate   Std. Err.     [95% Conf. Interval]
-----------------------------+------------------------------------------------
id: Unstructured             |
                  sd(reform) |   .9303106   .0561787      .8264688      1.0472
                   sd(_cons) |   .9541109   .0357016      .8866413    1.026715
           corr(reform,_cons) |  -.4908243   .0506047     -.5835351    -.385484
------------------------------------------------------------------------------
LR test vs. Poisson regression:      chi2(3) =  2857.93   Prob > chi2 = 0.0000
Note: LR test is conservative and provided only for reference.
```

We store the estimates

```
. estimates store xtrc
```

for use in the likelihood-ratio test below.

The random-intercept Poisson regression model is nested in the random-coefficient model (with $\psi_{22} = \psi_{21} = 0$). We can therefore perform a likelihood-ratio test using

```
. lrtest xtri xtrc
(log likelihoods of null models cannot be compared)
Likelihood-ratio test                                 LR chi2(2)  =    259.08
(Assumption: xtri nested in xtrc)                     Prob > chi2 =    0.0000
```

As discussed in section 4.6, this test is conservative and we should divide the p-value by 2. We conclude that the random-intercept Poisson model is rejected in favor of the random-coefficient Poisson model. The estimates for the latter model are reported under the heading "RC Poisson" in table 9.1.

Using gllamm

We now fit the random-coefficient Poisson regression model in `gllamm`, using the estimates from the random-intercept model (stored under `glri`) as starting values:

```
. estimates restore glri
. matrix a = e(b)
. eq rc: reform
. gllamm numvisit reform age educ married badh loginc summer,
> family(poisson) link(log) i(id) nrf(2) eqs(ri rc) from(a) eform adapt
number of level 1 units = 2227
number of level 2 units = 1518

Condition Number = 812.85837

gllamm model

log likelihood = -4513.8005
```

numvisit	exp(b)	Std. Err.	z	P>|z|	[95% Conf.	Interval]
reform	.9023139	.048376	-1.92	0.055	.8123103	1.00229
age	1.003457	.0028304	1.22	0.221	.9979246	1.00902
educ	1.008889	.0128058	0.70	0.486	.9841001	1.034303
married	1.086858	.0640872	1.41	0.158	.9682361	1.220013
badh	3.02813	.2322063	14.45	0.000	2.605564	3.519226
loginc	1.135641	.0866072	1.67	0.095	.9779708	1.31873
summer	.9140484	.0741615	-1.11	0.268	.7796627	1.071597

```
Variances and covariances of random effects
------------------------------------------------------------------------------

***level 2 (id)

    var(1): .90914641 (.06767415)
    cov(2,1): -.43462173 (.07121035) cor(2,1): -.49034779

    var(2): .86413301 (.10415938)
------------------------------------------------------------------------------
```

We store the estimates for later use

```
. estimates store glrc
```

It was not necessary to add two new elements to the matrix of starting values for the two new parameters being estimated. This is because the column labels of the matrix `a` let `gllamm` know which parameters are supplied, and `gllamm` sets the starting values of the other parameters to zero. By specifying the same equation `ri` for the random intercept in both models, we ensure that `gllamm` recognizes the previous estimate of the (square root of the) random-intercept variance as the starting value for that parameter in the random-coefficient model.

Robust standard errors can be obtained using

```
gllamm, robust eform
```

9.8.3 Interpretation of estimates

The estimated incidence-rate ratios have changed somewhat compared with the random-intercept Poisson model. Note in particular that the estimated incidence-rate ratio for `reform` now implies a 10% reduction in the expected number of visits per year *for a given person* and is nearly significant at the 5% level.

Instead of thinking of this model as a random-coefficient model, we could view it as a model with a random intercept ζ_{1j} for 1996 and a random intercept $\zeta_{1j}+\zeta_{2j}$ for 1998. In 1996, the random-intercept variance is

$$\text{Var}(\zeta_{1j}) \;=\; \psi_{11}$$

which is estimated as 0.909, and in 1998 the variance is

$$\text{Var}(\zeta_{1j}+\zeta_{2j}) \;=\; \psi_{11}+\psi_{22}+2\psi_{21}$$

and is estimated as 0.904. The covariance between the intercepts is

$$\text{Cov}(\zeta_{1j}, \zeta_{1j}+\zeta_{2j}) \;=\; \psi_{11}+\psi_{21}$$

which is estimated as 0.475.

In contrast, the conventional random-intercept model (9.4) with the same random intercept ζ_{1j} in 1996 and 1998 had a single parameter ψ_{11} representing both the random-intercept variance at the two occasions as well as the covariance across time. Assuming that this parameter "tries" to produce variances and covariance close to the estimated values using the more flexible model, it is not surprising that the estimate $\widehat{\psi}_{11}=0.817$ is between 0.909 and 0.475.

The reason we can identify the three parameters of the random part is that the variance parameters for 1996 and 1998 determine the relationship between marginal mean and marginal variance (given the covariates) at these two time points as shown in (9.4), and the covariance parameter affects the covariance between the counts. In the random-intercept model, all three properties were determined by one parameter, whereas the random-coefficient model uses a separate parameter for overdispersion at each time-point and for dependence across time.

As discussed in section 4.10, an analogous linear random-coefficient model with only two time points is not identified because it includes a fourth parameter, the level-1 variance θ, in the random part. The analogous random-coefficient logistic or probit model for dichotomous responses is also not identified because random intercepts have no effect on the marginal variance–mean relationship in these models (see sec. 6.9).

9.9 Overdispersion in single-level models

The assumption of the Poisson model that the variance of the count is equal to the expectation (given the covariates) is often violated. The most common violation is overdispersion or extra-Poisson variability, meaning that the variance is larger than the expectation. Overdispersion could be due to unobserved covariates that vary between the units of observation. For this to be meaningful, the "cases" or records of data must correspond to units such as subjects, occasions, houses, or schools, that could potentially be characterized by omitted covariates. If the data have been aggregated over units or disaggregated within units, the methods described here are not appropriate.

In this section, we consider methods for dealing with overdispersion in single-level models, as a preparation for the next section where we discuss overdispersion in two-level models. We therefore ignore clustering for simplicity, but note that this is inappropriate for the health-care reform data.

9.9.1 Normally distributed random intercept

As discussed in section 9.7.1, random intercepts induce overdispersion. Even when we do not have clustered data, random intercepts can be included to model overdispersion. The model and its implied marginal mean and variance are exactly as those for two-level models but the difference is that the random intercept varies between the level-1 units and hence does not produce any dependence among groups of observations.

The model can be written as

$$\ln(\mu_{ij}) \;=\; \beta_1 + \beta_2 x_{2i} + \cdots + \beta_7 x_{7ij} + \zeta_{ij}^{(1)} \qquad (9.6)$$

where $\zeta_{ij}^{(1)}|\mathbf{x}_{ij} \sim N(0, \psi^{(1)})$ and we have included a (1) superscript to denote that the random intercept varies at level 1 (and not at level 2 or higher as elsewhere in this book).

The marginal expectation becomes

$$\mu_{ij}^M \equiv E(y_{ij}|\mathbf{x}_{ij}) \;=\; \exp(\underbrace{\beta_1 + \psi^{(1)}/2}_{\beta_1^M} + \beta_2 x_{2i} + \cdots + \beta_7 x_{7ij}) \qquad (9.7)$$

and the marginal variance is

$$\text{Var}(y_{ij}|\mathbf{x}_{ij}) \;=\; \mu_{ij}^M + (\mu_{ij}^M)^2\{\exp(\psi^{(1)}) - 1\} \tag{9.8}$$

which is larger than the marginal expectation μ_{ij}^M if $\psi^{(1)} > 0$.

To fit the model with a normally distributed random intercept at level 1, we first generate an identifier obs for the level-1 observations

```
. generate obs = _n
```

and then specify obs as the "clustering" variable in the xtpoisson command:

```
. xtpoisson numvisit reform age educ married badh loginc summer, i(obs) normal
> irr

Random-effects Poisson regression               Number of obs      =      2227
Group variable: obs                             Number of groups   =      2227

Random effects u_i ~ Gaussian                   Obs per group: min =         1
                                                               avg =       1.0
                                                               max =         1

                                                Wald chi2(7)       =    272.60
Log likelihood  = -4546.8881                    Prob > chi2        =    0.0000

------------------------------------------------------------------------------
    numvisit |        IRR   Std. Err.      z    P>|z|     [95% Conf. Interval]
-------------+----------------------------------------------------------------
      reform |    .881623   .0466248    -2.38   0.017     .7948166      .97791
         age |   1.002419   .0026053     0.93   0.353     .9973256    1.007538
        educ |   1.005101   .0117969     0.43   0.665     .9822433     1.02849
     married |   1.084023   .0602281     1.45   0.146     .9721784    1.208735
        badh |   3.203538   .2429967    15.35   0.000     2.760985    3.717027
      loginc |   1.151836   .0840599     1.94   0.053     .9983224    1.328956
      summer |   .9576537   .0786351    -0.53   0.598     .8152943    1.124871
-------------+----------------------------------------------------------------
    /lnsig2u |  -.1143904   .0548408    -2.09   0.037    -.2218765   -.0069043
-------------+----------------------------------------------------------------
     sigma_u |   .9444097   .0258961                       .894994    .9965538
------------------------------------------------------------------------------
Likelihood-ratio test of sigma_u=0: chibar2(01) =  2791.61 Pr>=chibar2 = 0.000
```

We see from the estimated standard deviation of the level-1 random-intercept of 0.94 and the highly significant likelihood-ratio test that there is evidence for overdispersion.

Another consequence of including a random intercept, in addition to overdispersion, is that it produces a larger marginal probability of zeros than the ordinary Poisson model. Thus the problem of excess zeros often observed in count data is addressed to some extent. So-called zero-inflated Poisson (ZIP) models are tailor-made to address this problem, but these models will not be discussed here.

9.9.2 Negative binomial models

As previously mentioned, the random-intercept model with a normally distributed random intercept does not have a closed-form likelihood and is hence fitted using numerical integration. An appealing and computationally more efficient approach is to alter the model specification so that a closed-form likelihood is achieved. This can be accomplished by specifying a gamma distribution either for the exponentiated random intercept or for the cluster-specific mean, yielding two different kinds of negative binomial models.

Mean dispersion or NB2

The *mean dispersion* or NB2 version of the negative binomial model can be written as a random-intercept model as in (9.6), but instead of assuming a normal distribution for the random intercept $\zeta_{ij}^{(1)}$, we assume a gamma distribution for the frailty $\exp(\zeta_{ij}^{(1)})$ with mean 1 and variance α (pronounced alpha).

It follows from this specification that the marginal mean is

$$\mu_{ij}^{M} \equiv E(y_{ij}|\mathbf{x}_{ij}) \;=\; \exp(\beta_1 + \beta_2 x_{2ij} + \cdots + \beta_7 x_{7ij}) \tag{9.9}$$

and that the marginal variance has the same quadratic form as (9.8), with

$$\text{Var}(y_{ij}|\mathbf{x}_{ij}) \;=\; \mu_{ij}^{M} + (\mu_{ij}^{M})^2\alpha \tag{9.10}$$

so that α corresponds to $\exp(\psi^{(1)})-1$.

This model can be fitted using the `nbreg` command with the `dispersion(mean)` option:

```
. nbreg numvisit reform age educ married badh loginc summer, irr
> dispersion(mean)
Negative binomial regression                      Number of obs   =       2227
                                                  LR chi2(7)      =     303.15
Dispersion     = mean                             Prob > chi2     =     0.0000
Log likelihood = -4562.0459                       Pseudo R2       =     0.0322

------------------------------------------------------------------------------
    numvisit |        IRR   Std. Err.      z    P>|z|     [95% Conf. Interval]
-------------+----------------------------------------------------------------
      reform |   .8734045   .0447241    -2.64   0.008     .7900022    .9656119
         age |   1.004806   .0024754     1.95   0.052     .9999656    1.009669
        educ |   .9971352   .0115579    -0.25   0.805     .9747375    1.020047
     married |   1.081049    .057606     1.46   0.144     .9738393    1.200062
        badh |   3.118932   .2335543    15.19   0.000     2.693181    3.611987
      loginc |   1.142179    .081637     1.86   0.063     .9928749    1.313934
      summer |   .9424437    .074725    -0.75   0.455     .8067981    1.100895
-------------+----------------------------------------------------------------
    /lnalpha |   .0007291   .0475403                     -.0924481    .0939064
-------------+----------------------------------------------------------------
       alpha |   1.000729    .047575                      .9116965    1.098457
------------------------------------------------------------------------------
Likelihood-ratio test of alpha=0:  chibar2(01) = 2761.29 Prob>=chibar2 = 0.000
```

The same estimates can be obtained using `xtpoisson` as shown in section 9.9.1 but without the `normal` option. (When `xtpoisson` is used without the `normal` option and with the `i(id)` option, a gamma distribution with mean 1 and variance α is assumed for the level-2 random intercept, which yields the same variance function as the single-level model while also accounting for within-cluster dependence.)

Constant dispersion or NB1

The *constant dispersion* or NB1 version of the negative binomial model cannot be derived from a random-intercept model but is instead obtained by assuming that the subject-specific expected count μ_{ij} has a gamma distribution with expectation $\exp(\beta_1 + \beta_2 x_{2i} + \cdots + \beta_7 x_{7ij})$ and variance $\exp(\beta_1 + \beta_2 x_{2i} + \cdots + \beta_7 x_{7ij})\delta$.

From this specification it follows that the count y_{ij} has the same expectation as that for the negative binomial with mean dispersion given in (9.9) and variance

$$\text{Var}(y_{ij}|\mathbf{x}_{ij}) = \mu_{ij}^M(1+\delta)$$

Unlike the random-intercept models, the variance for the constant-dispersion negative binomial model is a constant multiple of the expectation.

This model can be fitted using the following command:

```
. nbreg numvisit reform age educ married badh loginc summer, irr
> dispersion(constant)
Negative binomial regression                      Number of obs   =       2227
                                                  LR chi2(7)      =     226.58
Dispersion     = constant                         Prob > chi2     =     0.0000
Log likelihood = -4600.3276                       Pseudo R2       =     0.0240

------------------------------------------------------------------------------
    numvisit |        IRR   Std. Err.      z    P>|z|     [95% Conf. Interval]
-------------+----------------------------------------------------------------
      reform |   .9002629   .0401958    -2.35   0.019     .8248293    .9825952
         age |   1.001107   .0022372     0.50   0.621     .9967317    1.005501
        educ |   1.007226   .0097883     0.74   0.459     .9882232    1.026595
     married |   1.038237   .0490102     0.79   0.427     .9464885    1.138879
        badh |   2.596117    .148113    16.72   0.000     2.321462    2.903265
      loginc |   1.131328   .0685746     2.04   0.042     1.004601    1.274041
      summer |    1.00864   .0693101     0.13   0.900     .8815455    1.154058
-------------+----------------------------------------------------------------
    /lndelta |   .9885984   .0532775                      .8841765     1.09302
-------------+----------------------------------------------------------------
       delta |   2.687465   .1431814                       2.42099    2.983271
------------------------------------------------------------------------------
Likelihood-ratio test of delta=0:  chibar2(01) = 2684.73 Prob>=chibar2 = 0.000
```

9.9.3 Quasilikelihood or robust standard errors

Generalized linear models are fitted using an algorithm that depends only on the expectation and variance of the response as a function of the covariates (although the log likelihood is used to monitor convergence). In the quasilikelihood approach, this same algorithm (or set of estimating equations) is used with a variance-expectation relationship of our choice, even when there is no statistical model that implies such a relationship. For the expectation, we retain the expression from the conventional Poisson regression model

$$\ln(\mu_{ij}) = \beta_1 + \beta_2 x_{2i} + \cdots + \beta_7 x_{7ij}$$

However, for the variance we relax the assumption inherent in the Poisson distribution that it is equal to the expectation by introducing the proportionality parameter ϕ (pronounced phi)

$$\text{Var}(y_{ij}|\mathbf{x}_{ij}) = \phi\mu_{ij}$$

This variance has the same form as the variance for the negative binomial model type NB1, whereas random intercept-models have variances that are quadratic functions of the mean.

The estimates of the regression parameters are the same as the maximum likelihood estimates for an ordinary Poisson model, and the proportionality parameter ϕ is estimated by a simple moment estimator, most commonly by the Pearson chi-squared statistic divided by the residual degrees of freedom. The scale parameter affects the standard errors only; they are multiplied by $\sqrt{\widehat{\phi}}$.

We can use the `glm` command with the `scale()` option to obtain maximum quasi-likelihood estimates. Here we use the `scale(x2)` option to estimate the proportionality parameter by the Pearson chi-squared statistic divided by the residual degrees of freedom:

```
. glm numvisit reform age educ married badh loginc summer,
> family(poisson) link(log) eform scale(x2)

Generalized linear models                          No. of obs      =       2227
Optimization     : ML                              Residual df     =       2219
                                                   Scale parameter =          1
Deviance         =  7419.853221                    (1/df) Deviance =   3.343782
Pearson          =  9688.740471                    (1/df) Pearson  =   4.366264

Variance function: V(u) = u                        [Poisson]
Link function    : g(u) = ln(u)                    [Log]

                                                   AIC             =   5.344133
Log likelihood   =  -5942.69244                    BIC             =   -9685.11

------------------------------------------------------------------------------
             |                 OIM
    numvisit |        IRR   Std. Err.      z    P>|z|     [95% Conf. Interval]
-------------+----------------------------------------------------------------
      reform |   .8689523   .0482622    -2.53   0.011     .7793268    .9688851
         age |   1.004371   .0027347     1.60   0.109     .9990249    1.009745
        educ |   .9894036   .0124256    -0.85   0.396     .9653472    1.014059
     married |   1.042542   .0607123     0.72   0.474     .9300882    1.168593
        badh |   3.105111   .1966385    17.89   0.000     2.742665    3.515454
      loginc |   1.160559   .0874758     1.98   0.048     1.001173     1.34532
      summer |   1.010269   .0853037     0.12   0.904     .8561784    1.192091
------------------------------------------------------------------------------
(Standard errors scaled using square root of Pearson X2-based dispersion)
```

We see from the output next to `(1/df) Pearson` that the proportionality parameter is estimated as $\widehat{\phi} = 4.366264$, compared with the corresponding parameter for the negative binomial model of type NB1, $1 + \widehat{\delta} = 3.687465$.

Since the quasilikelihood method changes only the standard errors, an alternative but similar approach is to use the sandwich estimator for the standard errors by specifying the `vce(robust)` option in the `glm` command.

9.10 Level-1 overdispersion in two-level models

As seen in (9.4), we would expect that including a random intercept ζ_{1j} at level 2 has, at least to some degree, addressed the problem of overdispersion. However, as discussed in section 9.8.3, the model uses a single parameter to induce both overdispersion for the level-1 units and dependence among level-1 units in the same cluster.

Sometimes there may be additional overdispersion at level 1 not accounted for by the random effect at level 2. For instance, in the health-care reform data, there may be unobserved heterogeneity between occasions within persons because medical problems (representing unobserved occasion-specific covariates) can lead to several extra doctor visits within the same 3-month period. After conditioning on the person-level random effect, the counts at the occasions are then overdispersed.

The most natural approach to handling level-1 overdispersion in a two-level model is by including a level-1 random intercept in addition to the level-2 random effect(s). The model then becomes a three-level model, with random effects at two nested levels, as discussed in chapter 10 (see exercise 10.10).

Hausman, Hall, and Griliches (1984) suggested another approach. Their model, which can be fitted using Stata's `xtnbreg` command, has a closed-form likelihood and is specified as follows. The conditional expectation μ_{ij} is assumed to have a gamma distribution with mean $\exp(\beta_1 + \beta_2 x_{2i} + \cdots + \beta_7 x_{7ij})\delta_j$ and variance $\exp(\beta_1 + \beta_2 x_{2i} + \cdots + \beta_7 x_{7ij})\delta_j^2$. For a cluster j with a given δ_j, the conditional expectation of the count becomes

$$\begin{aligned} E(y_{ij}|\mathbf{x}_{ij}, \delta_j) &= \exp(\beta_1 + \beta_2 x_{2i} + \cdots + \beta_7 x_{7ij})\delta_j \\ &= \exp\{(\beta_1 + \ln\delta_j) + \beta_2 x_{2i} + \cdots + \beta_7 x_{7ij}\} \end{aligned} \tag{9.11}$$

and the conditional variance becomes

$$\mathrm{Var}(y_{ij}|\mathbf{x}_{ij}, \delta_j) = E(y_{ij}|\mathbf{x}_{ij}, \delta_j)(1 + \delta_j) \tag{9.12}$$

The counts for the level-1 units in cluster j therefore have a cluster-specific intercept $\beta_1 + \ln\delta_j$ and are subject to multiplicative overdispersion (of the NB1 form) with a cluster-specific overdispersion factor $(1 + \delta_j)$. It is then assumed that $1/(1 + \delta_j)$ has a beta distribution. The syntax for fitting this model is

```
xtnbreg numvisit reform age educ married badh loginc summer, irr i(id)
```

A weakness of this model is that the subject-specific intercept and level-1 overdispersion factor are both determined by the same parameter δ_j. It is therefore not possible to have heterogeneity at level 2 without having overdispersion at level 1 or vice versa. A consequence is that, although the single-level Poisson model is a special case of the single-level negative binomial model (with $\delta_j = 0$), the two-level Poisson model is not a special case of the two-level negative binomial model. Another problem with the model is that the parameters of the beta distribution are difficult to interpret. We therefore do not recommend using this model.

Perhaps the simplest approach to handling overdispersion at level 1 in a two-level random-intercept Poisson model is to use the sandwich estimator for the standard errors. At the time of writing this book, the only command that can provide these standard errors is `gllamm` with the `robust` option. Restoring the random-intercept model estimates

```
. estimates restore glri
```

we can simply issue the command

```
. gllamm, robust eform

number of level 1 units = 2227
number of level 2 units = 1518

Condition Number = 723.77659

gllamm model

log likelihood = -4643.3427

Robust standard errors
```

numvisit	exp(b)	Std. Err.	z	P>|z|	[95% Conf.	Interval]
reform	.9547481	.0503036	-0.88	0.379	.8610748	1.058612
age	1.006002	.0031322	1.92	0.055	.9998817	1.01216
educ	1.008646	.0127823	0.68	0.497	.9839016	1.034012
married	1.077896	.0708484	1.14	0.254	.9476075	1.226097
badh	2.466857	.2880487	7.73	0.000	1.962236	3.101249
loginc	1.097486	.0956032	1.07	0.286	.9252303	1.301811
summer	.8673159	.0722128	-1.71	0.087	.7367263	1.021053

```
Variances and covariances of random effects
------------------------------------------------------------------------------

***level 2 (id)

    var(1): .81691979 (.0523264)
------------------------------------------------------------------------------
```

We see that the robust confidence intervals are somewhat wider than those using model-based standard errors.

9.11 Other approaches to two-level count data

9.11.1 Conditional Poisson regression

Instead of using random intercepts ζ_{1j} for persons, we could use fixed intercepts α_j by including a dummy variable for each person j (and omitting the overall intercept). This method would be analogous to the fixed-effects estimator of within-person effects discussed for linear models in section 3.7.2. In contrast to logistic regression, this approach yields consistent estimates of the within-person effects. However, it is impractical when there are many clusters.

Instead, we can eliminate the person-specific intercepts by constructing a likelihood that is conditional on the sum of the counts for each person, a sufficient statistic for the person-specific intercept. Conditional or fixed-effects Poisson regression can be performed using Stata's `xtpoisson` command with the `fe` option:

```
xtpoisson numvisit reform age educ married badh loginc summer, fe irr i(id)
```

An advantage of the fixed-effects approach is that it does not make any assumptions regarding the distribution of the person-specific intercept. A disadvantage is that effects of variables that do not vary within persons, such as gender, cannot be estimated. Furthermore, only persons having observations in both years contribute to the analysis; in the current application, this means that we lose the information from as many as 809 persons. We also lose the information from 95 persons who had zero doctor visits at both occasions. This dramatic depletion of the sample may partly explain why `xtpoisson` with the `fe` option did not converge for this problem.

9.11.2 Conditional negative binomial regression

We can also specify a conditional negative binomial model by treating the parameter δ_j in (9.11) and (9.12) as fixed. However, while the fixed-effects Poisson model does not allow for overdispersion, the fixed-effects negative binomial model does not allow for no overdispersion. Another oddity of the fixed effects negative binomial model is that effects of cluster-specific covariates can be estimated (in contrast to other fixed-effects models) because their inclusion affects the variance function. The syntax for fitting this model in Stata is

```
xtnbreg numvisit reform age educ married badh loginc summer, fe irr i(id)
```

9.11.3 Generalized estimating equations

Generalized estimating equations (GEE) can be used to estimate *marginal effects*, as in ordinary Poisson regression, but taking the dependence among units nested in clusters into account. The ideas are the same as discussed for dichotomous responses in section 6.14.2.

GEE estimates using the default exchangeable correlation structure but with robust sandwich-based standard errors are obtained using

```
. xtgee numvisit reform age educ married badh loginc summer,
> i(id) family(poisson) link(log) vce(robust) eform

GEE population-averaged model                   Number of obs      =      2227
Group variable:                         id      Number of groups   =      1518
Link:                                  log      Obs per group: min =         1
Family:                            Poisson                     avg =       1.5
Correlation:                  exchangeable                     max =         2
                                                Wald chi2(7)       =    235.37
Scale parameter:                         1      Prob > chi2        =    0.0000

                                   (Std. Err. adjusted for clustering on id)
```

numvisit	IRR	Semi-robust Std. Err.	z	P>|z|	[95% Conf.	Interval]
reform	.8849813	.0467574	-2.31	0.021	.7979238	.9815371
age	1.005253	.0033549	1.57	0.116	.9986986	1.01185
educ	.9906936	.0117057	-0.79	0.429	.9680144	1.013904
married	1.038283	.072958	0.53	0.593	.904698	1.191593
badh	3.020459	.264062	12.64	0.000	2.544821	3.584997
loginc	1.150186	.0914521	1.76	0.078	.9842114	1.34415
summer	.9741462	.0862448	-0.30	0.767	.8189626	1.158735

These estimates are reported under the heading “GEE Poisson” in table 9.1.

9.11.4 Marginal and conditional estimates when responses are missing at random

From the discussion in section 9.7.1 we would expect the estimated conditional effects using random-intercept Poisson models to be similar to the estimated marginal effects using ordinary Poisson regression or GEE (apart from the intercept). However, it is evident from table 9.1 that the estimates are different.

This discrepancy is probably due to the extremely unbalanced nature of the data. As discussed for linear models in section 5.9.1, if the responses are missing at random (missingness does not depend on unobserved responses, given the observed responses and covariates) and if the probability of a response being missing at an occasion depends on the observed response at the other occasion, maximum likelihood estimation of the correct model gives consistent parameter estimates. However, maximum likelihood estimation of an incorrect model, such as Poisson regression that ignores within-subject dependence, gives inconsistent estimates. GEE is also not consistent if the probability of a response being missing at an occasion depends on the observed response at another occasion after controlling for covariates. It is often claimed that GEE requires data to be missing completely at random (MCAR) for consistency, but missingness can actually depend on the covariates.

❖ Simulation

We now simulate complete data for the random-intercept Poisson model (with parameter values equal to those we have estimated for the data) and produce missing data that are missing at random (MAR). We then compare estimates based on the resulting incomplete data using generalized estimating equations (GEE) with an exchangeable correlation structure and using maximum likelihood estimation for ordinary and random-intercept Poisson models. This simulation is similar to the one described in section 5.9.1.

We can use the postestimation command `gllasim` for `gllamm` to simulate responses from the model just fitted in `gllamm`. To simulate complete data, we must first ensure that each person has two rows of data, one for 1996 and one for 1998. We create a variable `num` containing the number of rows of data per person using

```
. egen num = count(numvisit), by(id)
```

For those with one row only, we want to expand the data by 2, and for those with two rows, we do not want to expand the data, which is equivalent to expanding them by a factor of 1:

```
. generate mult = 3 - num
. expand mult
(809 observations created)
```

We now generate an occasion identifier `occ` equal to `reform` for those who already have complete data and equal to 0 and 1 for arbitrarily chosen rows otherwise

```
. by id (reform), sort: generate occ = _n - 1
```

To keep track of which responses were actually missing in the original data, we create the dummy variable `missing` and replace the corresponding responses by missing values:

```
. generate missing = (occ!=reform & num==1)
. replace numvisit = . if missing==1
(809 real changes made, 809 to missing)
```

For persons with missing data, the variable `reform` takes on the same value at both occasions. We rectify this by replacing the value of `reform` at the occasion when the response was missing by `occ`:

```
. replace reform = occ if missing==1
(809 real changes made)
```

We are now ready to simulate new responses `y`. To ensure that the results can be replicated, we first sort the data and then set the random-number seed to an arbitrary number:

```
. sort id reform
. set seed 1211
```

We then retrieve the `gllamm` estimates for the random-intercept Poisson model and use the `gllasim` command with the `fsample` option to simulate responses for the full sample, not just the estimation sample:

```
. estimates restore glri
. gllasim y, fsample
(simulated responses will be stored in y)
```

For those who visited the doctor at least twice in 1996, we now replace the response for 1998 by a missing value with a probability of 0.9:

```
. by id (reform), sort: generate drop = (y[1]>2 & uniform()<.9)
. replace y = . if drop==1 & reform==1
(466 real changes made, 466 to missing)
```

Since the responses are missing at random (MAR), maximum likelihood estimation of the true model that generated the data should yield consistent estimates:

```
. xtpoisson y reform age educ married badh loginc summer, i(id) normal irr
Random-effects Poisson regression               Number of obs      =      2570
Group variable: id                              Number of groups   =      1518
Random effects u_i ~ Gaussian                   Obs per group: min =         1
                                                               avg =       1.7
                                                               max =         2
                                                Wald chi2(7)       =    213.48
Log likelihood  = -4840.7573                    Prob > chi2        =    0.0000
```

y	IRR	Std. Err.	z	P>|z|	[95% Conf.	Interval]
reform	1.029429	.036059	0.83	0.408	.9611257	1.102586
age	1.006054	.0028652	2.12	0.034	1.000453	1.011685
educ	1.020096	.0126798	1.60	0.109	.9955447	1.045254
married	1.151819	.0669211	2.43	0.015	1.027849	1.290742
badh	2.660478	.2005024	12.98	0.000	2.295146	3.083962
loginc	1.04616	.0756244	0.62	0.532	.9079598	1.205395
summer	.8274459	.0645118	-2.43	0.015	.7101919	.9640588
/lnsig2u	-.1261919	.0571691	-2.21	0.027	-.2382412	-.0141425
sigma_u	.9388534	.0268367			.8877007	.9929537

```
Likelihood-ratio test of sigma_u=0: chibar2(01) =  2871.89 Pr>=chibar2 = 0.000
```

We see that the estimates above are close to the true parameter values, which are the estimates for the original data.

(*Continued on next page*)

However, using ordinary Poisson regression give different estimates:

```
. poisson y reform age educ married badh loginc summer, irr
Poisson regression                                Number of obs   =       2570
                                                  LR chi2(7)      =    1426.49
                                                  Prob > chi2     =     0.0000
Log likelihood = -6276.7036                       Pseudo R2       =     0.1020
```

y	IRR	Std. Err.	z	P>|z|	[95% Conf.	Interval]
reform	.6131784	.0177075	-16.94	0.000	.5794363	.6488855
age	1.004305	.0012989	3.32	0.001	1.001763	1.006854
educ	1.01852	.0057876	3.23	0.001	1.007239	1.029926
married	1.128861	.031214	4.38	0.000	1.069311	1.191727
badh	2.719263	.0854277	31.84	0.000	2.556879	2.891961
loginc	1.108184	.0392507	2.90	0.004	1.033863	1.187847
summer	.8433938	.0343386	-4.18	0.000	.7787066	.9134545

Specifically, the intervention now looks effective because those who visited the doctor frequently in 1998 were more likely to drop out, and due to the within-person dependence between responses, it is these same people who tended to visit the doctor frequently in 1996. Loosely speaking, the random-intercept model "knows" about the within-person dependence and hence makes the correct "adjustment" whereas the ordinary Poisson model does not.

Interestingly, generalized estimating equations (GEE) does not work well although we allow for within-person dependence by specifying an exchangeable correlation structure:

```
. xtgee y reform age educ married badh loginc summer, i(id) family(poisson)
> link(log) vce(robust) eform
GEE population-averaged model                   Number of obs      =      2570
Group variable:                         id      Number of groups   =      1518
Link:                                  log      Obs per group: min =         1
Family:                            Poisson                     avg =       1.7
Correlation:                  exchangeable                     max =         2
                                                Wald chi2(7)       =    242.22
Scale parameter:                         1      Prob > chi2        =    0.0000
                                   (Std. Err. adjusted for clustering on id)
```

y	IRR	Semi-robust Std. Err.	z	P>|z|	[95% Conf.	Interval]
reform	.7609233	.0251568	-8.26	0.000	.7131804	.8118624
age	1.004157	.0030258	1.38	0.169	.9982446	1.010105
educ	1.017779	.0134357	1.33	0.182	.9917829	1.044456
married	1.131598	.0699859	2.00	0.046	1.002416	1.277428
badh	2.718051	.2167291	12.54	0.000	2.324799	3.177824
loginc	1.119422	.0891531	1.42	0.157	.9576402	1.308535
summer	.8264793	.0788856	-2.00	0.046	.6854672	.9964999

9.12 How does birth history affect child mortality?

This section assumes some familiarity with survival analysis, and it is probably beneficial if you have already read chapter 8.

We revisit the Guatemalan child mortality data from Pebley and Stupp (1987) and Guo and Rodriguez (1992), which were described in detail in section 8.3. The data contain the survival times in months (variable `time`) of several children per mother (variable `momid`) together with a dummy variable, `death`, indicating whether the survival time ended in death or censoring. For instance, for mother 26, these variables are

```
. list kidid time death if momid==26, noobs
```

kidid	time	death
2601	60	0
2602	.25	1
2603	48	1
2604	.25	1
2605	60	0

We see that children 2601 and 2605 (children 1 and 5 of mother 26) were censored at 60 months, whereas children 2602 and 2604 died at .25 months and child 2603 died at 48 months.

9.12.1 Simple piecewise exponential survival model

Following Pebley and Stupp (1987), we could model how the incidence rate of deaths varies over the age bands less than a month, 1–5 months, 6–11 months, 12–23 months, and 24 or more months. Ignoring covariates, we can simply produce counts of the total number of deaths occurring in each of these age bands across all children. If we also keep track of the corresponding exposure, the total time of observation for each age band summed over all children, we can model the incidence rate as a function of age band using Poisson regression for the counts with the log exposure as an offset.

To calculate the total exposure for each age band, it is useful to begin by expanding the data to person–period data in essentially the same way as in chapter 8. Stata's `stsplit` command can do this; however, to use it we must first declare the data as survival data using the `stset` command:

(Continued on next page)

```
. stset time, failure(death) id(kidid)

                id:  kidid
     failure event:  death != 0 & death < .
obs. time interval:  (time[_n-1], time]
 exit on or before:  failure

------------------------------------------------------------------------------
     3120  total obs.
        0  exclusions
------------------------------------------------------------------------------
     3120  obs. remaining, representing
     3120  subjects
      403  failures in single failure-per-subject data
 131512.5  total analysis time at risk, at risk from t =         0
                             earliest observed entry t =         0
                                  last observed exit t =        60
```

We now expand the data using the `stsplit` command, specifying the limits of the age bands using the option `at(0 1 6 12 24 60)`:

```
. stsplit dur, at(0 1 6 12 24 60)
(10474 observations (episodes) created)
```

This command produced new variables `dur`, `_t0`, `_t`, and `_d` with the following values for the children of mother 26:

```
. list kidid dur _t0 _t _d if momid==26, sepby(kidid) noobs
```

kidid	dur	_t0	_t	_d
2601	0	0	1	0
2601	1	1	6	0
2601	6	6	12	0
2601	12	12	24	0
2601	24	24	60	0
2602	0	0	.25	1
2603	0	0	1	0
2603	1	1	6	0
2603	6	6	12	0
2603	12	12	24	0
2603	24	24	48	1
2604	0	0	.25	1
2605	0	0	1	0
2605	1	1	6	0
2605	6	6	12	0
2605	12	12	24	0
2605	24	24	60	0

Each child is represented by a row of data for each interval or age band it survived into and in which it was hence at risk. The age bands are labeled by their lower limits in the variable `dur`. Children 2601 and 2605 survived through all 5 age bands and are therefore represented by 5 rows of data. Children 2602 and 2604 died during the first age band and were therefore only at risk during that age band (labeled 0 in `dur`). Child 2603 died at 48 months and was therefore at risk into the fifth age band (labeled 24 in `dur`). The variables `_t0` and `_t` give the start and stop time each child was at risk in each age band, and `_d` indicates whether the child died in the age band. For example, child 2603 was at risk at 24 months, the beginning of the fifth age band and died at age 48 months, as indicated by `_t` being 48 and `_d` being 1 for the fifth age band.

We can obtain the total number of deaths in each age band by summing `_d` within age bands. To calculate the corresponding total exposures, we first create a variable `length` equal to the exposure for each child within each age band and then sum `length` within age bands:

```
. generate length = _t - _t0
. table dur, contents(sum length sum _d)
```

dur	sum(length)	sum(_d)
0	3031.5	157
1	14410	54
6	16212	75
12	28918	81
24	68941	36

The exposure for the first age band was 3031.5 child-months and 157 children died within that age band. Therefore an estimated constant incidence rate for this age band would be $157/3031.5 = 0.05178954$, and similarly for the other age bands.

We could collapse the data to the table shown above and perform Poisson regression to estimate these incidence rates. However, as shown in section 9.3, it does not matter whether we aggregate counts over units or time periods for which the incidence rate is constant, so we can obtain the same results with the original data.

We specify a Poisson regression model for age band s, child i, and mother j,

$$\ln(\mu_{sij}) \;=\; \alpha_1 d_{1sij} + \cdots + \alpha_5 d_{5sij} + \ln(t_{sij})$$

where $(d_{1sij}, \ldots, d_{5sij})'$ are dummy variables for the age bands and t_{sij} is the exposure or length of time child i of mother j spent in band s. We can alternatively express the model in terms of the piecewise-constant incidence rate λ_{sij}

$$\lambda_{sij} \equiv \ln\left(\frac{\mu_{sij}}{t_{sij}}\right) \;=\; \alpha_1 d_{1sij} + \cdots + \alpha_5 d_{5sij}$$

The model could be called a piecewise-constant Poisson model, where the word "piece" corresponds to the age bands. However, since a Poisson model for the number of events is equivalent to an exponential model for the time intervals between events, the model is usually referred to as a piecewise-constant exponential model or a piecewise exponential model. It can be shown that the incidence rate is just the continuous-time *hazard* defined as the instantaneous probability of the event *given that it has not yet occurred.* The conditioning is enforced by removing individuals from the data once they have experienced the event and by defining their exposure as the time until they experienced the event. This is essential because death is an absorbing event that cannot recur for the same child.

To fit the model in Stata, we first generate five dummy variables for the age bands:

```
. quietly tabulate dur, generate(ageband)
```

We include these dummies as covariates in the Poisson regression (and therefore omit the intercept using the `noconstant` option) and specify the exposure using the `exposure()` option:

```
. poisson _d ageband1-ageband5, exposure(length) noconstant irr
Poisson regression                              Number of obs   =      13594
                                                Wald chi2(5)    =   10083.68
Log likelihood = -2132.5591                     Prob > chi2     =     0.0000

------------------------------------------------------------------------------
          _d |        IRR   Std. Err.      z    P>|z|     [95% Conf. Interval]
-------------+----------------------------------------------------------------
    ageband1 |   .0517895   .0041333   -37.10   0.000     .0442903    .0605585
    ageband2 |   .0037474     .00051   -41.05   0.000     .0028701    .0048929
    ageband3 |   .0046262   .0005342   -46.56   0.000     .0036892    .0058011
    ageband4 |    .002801   .0003112   -52.90   0.000     .0022529    .0034825
    ageband5 |   .0005222    .000087   -45.34   0.000     .0003767    .0007239
      length | (exposure)
------------------------------------------------------------------------------
```

The estimated incidence rate for the first age band is the same as we calculated above.

9.12.2 Piecewise exponential survival model with covariates and frailty

The advantage of not aggregating the data is that we can incorporate covariate information in the model. We start by including the same observed covariates as in section 8.6:

$$\lambda_{sij} \;=\; \alpha_1 d_{1sij} + \cdots + \alpha_5 d_{5sij} + \beta_2 x_{2ij} + \cdots + \beta_{12} x_{12,sij}$$

This model can be fitted in Stata by using the following commands:

```
. generate mage2 = mage^2
. generate comp12 = f0011*(ageband4==1)
. generate comp24e = f0011*(ageband5==1)
. generate comp24l = f1223*(ageband5==1)
```

```
. poisson _d ageband1-ageband5 mage mage2 border pdead p0014 p1523 p2435 p36up
> comp12 comp24e comp241, exposure(length) nocons irr
Poisson regression                                Number of obs   =      13594
                                                  Wald chi2(16)   =    9905.61
Log likelihood = -2108.7006                       Prob > chi2     =     0.0000
```

_d	IRR	Std. Err.	z	P>|z|	[95% Conf.	Interval]
ageband1	.3375566	.2532262	-1.45	0.148	.0775885	1.468575
ageband2	.0245332	.0186095	-4.89	0.000	.0055473	.1084995
ageband3	.0303346	.0229222	-4.63	0.000	.0068981	.1333968
ageband4	.0180935	.0136752	-5.31	0.000	.0041132	.0795915
ageband5	.0032229	.0025034	-7.39	0.000	.0007032	.0147712
mage	.8612339	.0501274	-2.57	0.010	.7683828	.9653051
mage2	1.002581	.001035	2.50	0.013	1.000555	1.004612
border	1.063634	.0355569	1.85	0.065	.9961777	1.135658
pdead	1.103248	.1649263	0.66	0.511	.8230493	1.478839
p0014	1.71365	.3640741	2.54	0.011	1.130004	2.59875
p1523	.8845005	.1648386	-0.66	0.510	.6138541	1.274474
p2435	.7718101	.1424211	-1.40	0.160	.5375754	1.108107
p36up	.6764905	.141174	-1.87	0.061	.4493947	1.018346
comp12	2.246906	1.609712	1.13	0.258	.5517892	9.149482
comp24e	4.933551	3.633516	2.17	0.030	1.164816	20.89594
comp241	1.07556	.4053648	0.19	0.847	.5138396	2.251343
length	(exposure)					

The model assumes that the incidence rate is the same for all children with the same covariate values, regardless of whether they have the same mother. We can introduce between-mother heterogeneity by using a piecewise exponential survival model with a random intercept ζ_j for mother j

$$\lambda_{sij} \;=\; \alpha_1 d_{1sij} + \cdots + \alpha_5 d_{5sij} + \beta_2 x_{2ij} + \cdots + \beta_{12} x_{12,sij} + \zeta_j$$

We typically specify either that the random intercept ζ_j has a normal distribution with expectation 0 and variance ψ or that the frailty $\exp(\zeta_j)$ has a gamma distribution with expectation 1 and variance α, given the covariates.

(Continued on next page)

The model with a gamma-distributed frailty can be fitted in Stata using

```
. xtpoisson _d ageband1-ageband5 mage mage2 border pdead p0014 p1523 p2435 p36up
> comp12 comp24e comp241, i(momid) exposure(length) nocons irr

Random-effects Poisson regression               Number of obs      =     13594
Group variable: momid                           Number of groups   =       851

Random effects u_i ~ Gamma                      Obs per group: min =         1
                                                               avg =      16.0
                                                               max =        40

                                                Wald chi2(16)      =   8796.54
Log likelihood  = -2107.0551                    Prob > chi2        =    0.0000
```

_d	IRR	Std. Err.	z	P>|z|	[95% Conf.	Interval]
ageband1	.3710262	.2874212	-1.28	0.201	.0812846	1.693562
ageband2	.0271691	.0212782	-4.60	0.000	.0058538	.1261004
ageband3	.0337345	.0263341	-4.34	0.000	.0073047	.1557928
ageband4	.0202894	.0158516	-4.99	0.000	.0043878	.0938193
ageband5	.0036443	.0029231	-7.00	0.000	.0007566	.0175539
mage	.8559554	.0514063	-2.59	0.010	.7609048	.9628796
mage2	1.002689	.0010689	2.52	0.012	1.000596	1.004786
border	1.059148	.0376965	1.61	0.106	.9877825	1.13567
pdead	.9272842	.1655893	-0.42	0.672	.6534463	1.315878
p0014	1.7738	.3859226	2.63	0.008	1.158004	2.717059
p1523	.9075818	.1718846	-0.51	0.609	.6261508	1.315505
p2435	.7960679	.149776	-1.21	0.225	.5505554	1.151063
p36up	.6898139	.1463948	-1.75	0.080	.4550791	1.045628
comp12	2.210407	1.593665	1.10	0.271	.5379858	9.081837
comp24e	4.959532	3.677954	2.16	0.031	1.1593	21.21708
comp241	1.076996	.4067595	0.20	0.844	.5137269	2.257852
length	(exposure)					
/lnalpha	-1.540714	.632364			-2.780125	-.3013038
alpha	.214228	.1354701			.0620307	.739853

```
Likelihood-ratio test of alpha=0: chibar2(01) =     3.29 Prob>=chibar2 = 0.035
```

Alternatively, specifying the `normal` option would produce estimates for a model with a normal random intercept or log-normal frailty. The above estimates are close to those reported (and interpreted) for the random-intercept complementary log-log model in section 8.6, with the exception of the estimated effect of `comp12`, which is considerably smaller here.

We can also use the `streg` command with the `distribution(exponential)` option to fit the same models. For the model without a random intercept, the syntax is

```
streg ageband1-ageband5 mage mage2 border pdead p0014 p1523
    p2435 p36up comp12 comp24e comp241, dist(exponential) noconstant
```

To include a random intercept, use the `frailty(gamma)` and `shared()` options:

```
streg ageband1-ageband5 mage mage2 border pdead p0014 p1523
    p2435 p36up comp12 comp24e comp241, dist(exponential) frailty(gamma)
    shared(momid) noconstant
```

The piecewise exponential survival model takes into account the *precise time* that subjects leave the risk set (due to experiencing the event or censoring) in contrast to the discrete-time survival models discussed in chapter 8, which assume that subjects are at risk throughout the intervals. Although this advantage of the piecewise exponential model may appear important when modeling survival in continuous time, a cost of using the piecewise exponential model is that we have to assume *constant hazards* (given the covariates) in each interval, an assumption not made in discrete-time survival modeling.

9.13 Which Scottish counties have a high risk of lip cancer?

We now consider models for disease mapping or small-area estimation. Clayton and Kaldor (1987) presented and analyzed data on lip cancer for each of 56 Scottish counties over the period 1975–1980. These data have also been analyzed by Breslow and Clayton (1993) and Leyland (2001) among many others.

The dataset `lips.dta` has the following variables:

- `county`: county identifier (1 to 59)
- `o`: observed number of lip cancer cases
- `e`: expected number of lip cancer cases
- `x`: percentage of population working in agriculture, fishing, or forestry

The expected number of lip cancer cases is based on the age-specific lip cancer rates for the whole of Scotland and the age distribution of the counties.

9.14 Standardized mortality ratios

The standardized mortality ratio (SMR) for a county is defined as the ratio of the incidence rate to that expected if the age-specific incidence rates were equal to those of a reference population (e.g., Breslow and Day 1987), here the whole of Scotland. The crude estimate of the SMR for county j is obtained using

$$\widehat{\mathrm{SMR}}_j = \frac{o_j}{e_j}$$

where o_j is the observed number of cases and e_j is the expected number of cases.

We first read the data and then calculate the crude SMRs as percentages:

```
. use http://www.stata-press.com/data/mlmus2/lips, clear
. generate smr = 100*o/e
```

The number of observed lip cancer cases, the expected number of cases, and crude standardized mortality ratios are presented in table 9.2. The crude SMRs are also shown in the map in figure 9.1. When the SMR exceeds 100, the county has more cases than would be expected given the age distribution.

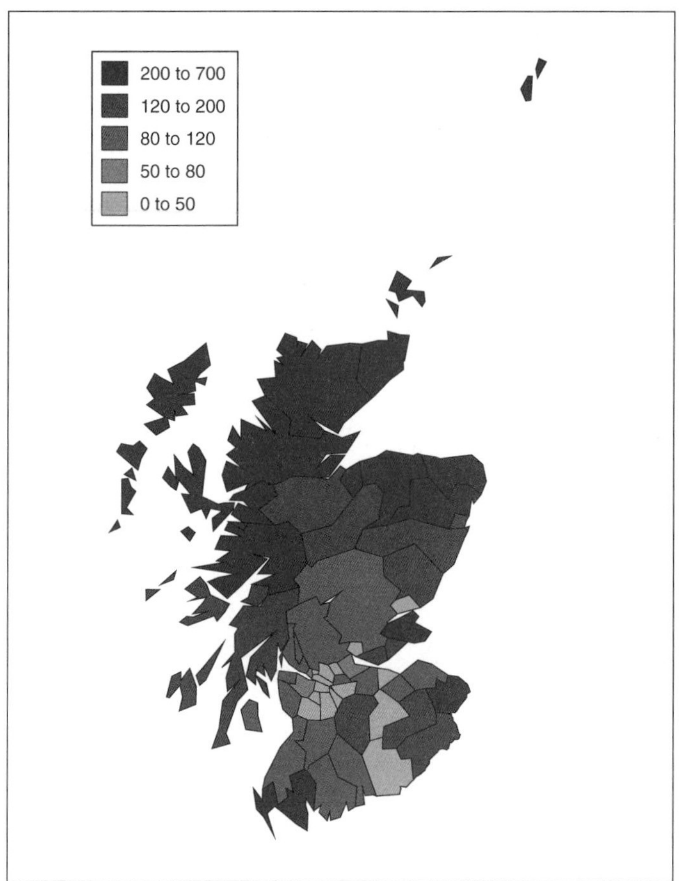

Figure 9.1: Map of crude SMR in percentages (*Source:* Skrondal and Rabe-Hesketh 2004)

Table 9.2: Observed and expected numbers of lip cancer cases and various SMR estimates (in percentages) for Scottish counties

		Obs	Exp	Crude	Predicted SMRs	
County	#	o_j	e_j	SMR	Norm.	NPMLE
Skye. Lochalsh	1	9	1.4	652.2	470.7	342.6
Banf. Buchan	2	39	8.7	450.3	421.8	362.4
Caithness	3	11	3.0	361.8	309.4	327.1
Berwickshire	4	9	2.5	355.7	295.2	321.6
Ross. Cromarty	5	15	4.3	352.1	308.5	327.6
Orkney	6	8	2.4	333.3	272.0	311.1
Moray	7	26	8.1	320.6	299.9	322.2
Shetland	8	7	2.3	304.3	247.8	292.5

– continued

County	#	Obs o_j	Exp e_j	Crude SMR	Predicted SMRs Norm.	NPMLE
Lochaber	9	6	2.0	303.0	239.0	280.1
Gordon	10	20	6.6	301.7	279.1	319.9
W. Isles	11	13	4.4	295.5	262.5	315.5
Sutherland	12	5	1.8	279.3	219.2	254.3
Nairn	13	3	1.1	277.8	198.4	222.7
Wigtown	14	8	3.3	241.7	210.9	249.6
NE. Fife	15	17	7.8	216.8	204.6	245.3
Kincardine	16	9	4.6	197.8	178.9	171.4
Badenoch	17	2	1.1	186.9	151.9	163.2
Ettrick	18	7	4.2	167.5	154.7	136.7
Inverness	19	9	5.5	162.7	154.2	128.4
Roxburgh	20	7	4.4	157.7	149.1	130.3
Angus	21	16	10.5	153.0	147.8	117.1
Aberdeen	22	31	22.7	136.7	135.0	116.4
Argyll. Bute	23	11	8.8	125.4	123.3	116.4
Clydesdale	24	7	5.6	124.6	122.9	116.7
Kirkcaldy	25	19	15.5	122.8	121.6	116.4
Dunfermline	26	15	12.5	120.1	119.1	116.3
Nithsdale	27	7	6.0	115.9	116.3	115.5
E. Lothian	28	10	9.0	111.6	111.4	115.9
Perth. Kinross	29	16	14.4	111.3	111.1	116.3
W. Lothian	30	11	10.2	107.8	108.5	115.9
Cumnock-Doon	31	5	4.8	105.3	107.2	111.5
Stewartry	32	3	2.9	104.2	109.1	109.4
Midlothian	33	7	7.0	99.6	102.7	113.1
Stirling	34	8	8.5	93.8	97.3	112.9
Kyle. Carrick	35	11	12.3	89.3	92.2	114.1
Inverclyde	36	9	10.1	89.1	92.6	112.4
Cunninghame	37	11	12.7	86.8	89.7	113.3
Monklands	38	8	9.4	85.6	89.4	109.4
Dumbarton	39	6	7.2	83.3	89.4	104.9
Clydebank	40	4	5.3	75.9	85.7	94.5
Renfrew	41	10	18.8	53.3	59.1	40.6
Falkirk	42	8	15.8	50.7	57.9	40.9
Clackmannan	43	2	4.3	46.3	68.8	65.5
Motherwell	44	6	14.6	41.0	50.7	37.5
Edinburgh	45	19	50.7	37.5	40.8	36.2
Kilmarnock	46	3	8.2	36.6	53.2	42.2
E. Kilbride	47	2	5.6	35.8	57.9	49.7
Hamilton	48	3	9.3	32.1	48.5	38.8
Glasgow	49	28	88.7	31.6	33.8	36.2

– continued

County	#	Obs o_j	Exp e_j	Crude SMR	Predicted SMRs Norm.	NPMLE
Dundee	50	6	19.6	30.6	39.8	36.2
Cumbernauld	51	1	3.4	29.1	63.1	57.8
Bearsden	52	1	3.6	27.6	61.1	55.4
Eastwood	53	1	5.7	17.4	46.4	40.6
Strathkelvin	54	1	7.0	14.2	40.8	37.8
Tweeddale	55	0	4.2	0.0	43.2	40.8
Annandale	56	0	1.8	0.0	64.9	60.6

Source: Clayton and Kaldor (1987)

9.15 Random-intercept Poisson regression

An important limitation of crude SMRs is that estimates for counties with small populations are imprecise. This problem can be addressed by using random-intercept Poisson models in conjunction with empirical Bayes prediction. The resulting SMRs are shrunken toward the overall SMR, thereby borrowing strength from other counties (see sec. 2.9.2).

9.15.1 Model specification

We consider a random-intercept Poisson regression model where the observed number of lip cancer cases in county j is assumed to have a Poisson distribution with mean μ_j

$$\ln(\mu_j) \;=\; \ln(e_j) + \beta_1 + \zeta_j$$

Here $\zeta_j \sim N(0, \psi)$ is a random intercept representing unobserved heterogeneity between counties and $\ln(e_j)$ is an offset, a covariate with regression coefficient set to 1. Here the purpose of the offset is to ensure that $\beta_1 + \zeta_j$ can be interpreted as a model-based county-specific log SMR. This interpretation becomes clear by subtracting the offset from both sides of the equation:

$$\ln(\mu_j) - \ln(e_j) \;=\; \ln(\underbrace{\mu_j/e_j}_{\text{SMR}_j}) \;=\; \beta_1 + \zeta_j$$

9.15.2 Estimation using gllamm

We will use `gllamm` for estimation because this is the only command that can provide the required posterior expectations of the standardized mortality ratios at the time of writing this book. We must first generate a variable for the offset

```
. generate lne = ln(e)
```

and pass this variable to `gllamm` using the `offset()` option. We also specify a Poisson distribution and log link using the `family()` and `link()` options, respectively:

```
. gllamm o, i(county) offset(lne) family(poisson) link(log) adapt
number of level 1 units = 56
number of level 2 units = 56

Condition Number = 1.2562952

gllamm model

log likelihood = -181.32392

------------------------------------------------------------------------------
           o |      Coef.   Std. Err.      z    P>|z|     [95% Conf. Interval]
-------------+----------------------------------------------------------------
       _cons |    .080211   .1167416     0.69   0.492    -.1485983    .3090204
         lne |   (offset)
------------------------------------------------------------------------------


Variances and covariances of random effects
------------------------------------------------------------------------------


***level 2 (county)

    var(1): .58432671 (.14721754)
------------------------------------------------------------------------------
```

The maximum likelihood estimates for the random-intercept model are given in table 9.3 under "Normal ML". We could also include the percentage of the population in county j working in agriculture, fishing, or forestry as a county-level covariate; see exercise 9.7.

Table 9.3: Estimates for random-intercept models for Scottish lip cancer data

	Normal ML		NPMLE ($C=4$)	
	Est	(SE)	Est	(SE)
β_1 [_cons]	0.08	(0.12)	0.08	(0.12)
ψ	0.58		0.63†	
Log likelihood	−181.32		−174.39	

† Derived from discrete distribution

9.15.3 Prediction of standardized mortality ratios

We will now consider predictions of the SMRs using empirical Bayes, as discussed for logistic regression models in section 6.13.2. Specifically, we would like to estimate the posterior expectation of the SMR,

$$\widetilde{\text{SMR}}_j = \int \exp(\widehat{\beta}_1 + \zeta_j) \text{ Posterior}(\zeta_j|o_j, e_j) \, d\zeta_j \tag{9.13}$$

It may be tempting to to obtain the predicted SMRs by simply plugging in the estimate $\widehat{\beta}_1$ and the predicted random intercept $\widetilde{\zeta}_j$ in the exponential function. However, this would be incorrect because the expectation of a nonlinear function of a random variable is not equal to the nonlinear function of the expectation of the random variable.

Empirical Bayes predictions of the SMRs can be obtained using `gllapred` with the `mu` option to get posterior means on the scale of the response and the `nooffset` option to omit the offset $\ln(e_j)$ from the linear predictor:

```
. gllapred mu, mu nooffset
(mu will be stored in mu)
. generate thet = 100*mu
. sort county
. list county thet in 1/10, clean noobs
    county       thet
         1   470.7203
         2   421.7975
         3   309.4238
         4   295.1885
         5   308.5293
         6   272.0485
         7   299.8875
         8   247.8383
         9   238.9533
        10   279.1405
```

These empirical Bayes predictions are given under "Norm." in table 9.2 and displayed as a map in figure 9.2. The maps in figures 9.1 and 9.2 were drawn using Stata, and an annotated do-file to create the maps can be obtained by typing

```
. copy http://www.stata-press.com/data/mlmus2/scotmaps.do scotmaps.do
```

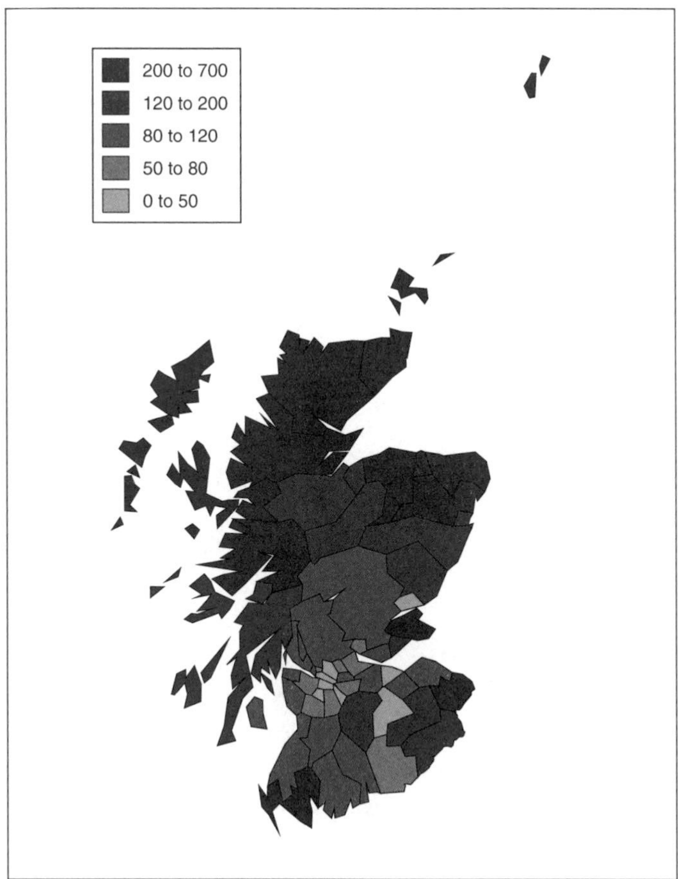

Figure 9.2: Map of SMRs assuming normally distributed random intercept (no covariate) (*Source:* Skrondal and Rabe-Hesketh 2004)

We can plot the empirical Bayes predictions against the crude SMRs by using

```
. twoway (scatter thet smr, msymbol(none) mlabpos(0) mlabel(county))
> (function y=x, range(0 600)), xline(108) yline(108)
> xtitle(Crude SMR) ytitle(Empirical Bayes SMR) legend(off)
```

The $y = x$ line has been superimposed, as well as lines $x = 108$ and $y = 108$, representing the SMR in percent $100 \times \exp(\widehat{\beta}_1)$ when ζ_j equals its mean 0. The resulting graph is given in figure 9.3. Shrinkage is apparent since counties with particularly high crude SMRs lie below the $y = x$ line (have predictions lower than the crude SMR) and counties with particularly low crude SMRs lie above the $y = x$ line.

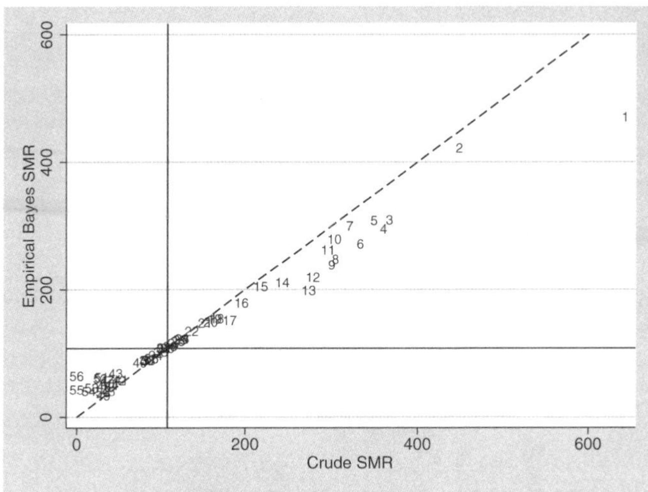

Figure 9.3: Empirical Bayes SMRs versus crude SMRs

9.16 ❖ Nonparametric maximum likelihood estimation

9.16.1 Specification

Instead of assuming that the random intercept ζ_j is normally distributed, we can relax this assumption using nonparametric maximum likelihood estimation (NPMLE).

It can been shown that the nonparametric estimator of the random-effects distribution is discrete with locations $\zeta_j = e_c$ and probabilities π_c $(c=1,\ldots,C)$, where the number of locations C with nonzero probabilities is determined to make the likelihood as large as possible.

Obviously, the probabilities must add to one

$$\pi_1 + \cdots + \pi_C \;=\; 1 \tag{9.14}$$

The discrete distribution of ζ_j can be parameterized so that it has zero mean,

$$\pi_1 e_1 + \cdots + \pi_C e_C \;=\; 0 \tag{9.15}$$

which has the advantage that we need not omit the intercept β_1 from the model

$$\ln(\mu_j) \;=\; \ln(e_j) + \beta_1 + \zeta_j$$

9.16.2 Estimation using gllamm

To obtain the nonparametric maximum likelihood estimates, we must use the maximum possible number of locations so that it is not possible to achieve a larger likelihood by introducing more locations.

This maximum number can be determined using the following iterative approach. First, we fit the model with a given number of locations C, for instance $C=2$. Keeping all the parameters equal to the resulting maximum likelihood estimates, we introduce another point with a small probability π_{C+1} and evaluate the likelihood for a large number of different locations for this point, e.g., by setting $e_{C+1} = -5 + l/100$, $(l = 1, \ldots, 1000)$. If the likelihood increases for any of these locations, we have evidence that another point is required. The model is then refitted with $C+1$ points, and we repeat the search for a location at which introducing a new point increases the likelihood. These cycles are repeated until the search does not identify any more locations.

For the Scottish lip cancer data, we will start with $C=2$ locations and maximize the likelihood with respect to β_1, e_1, and π_1, with e_2 and π_2 determined by the constraints in (9.14) and (9.15). This is accomplished in gllamm using the ip(f) and nip(2) options:

```
. gllamm o, i(county) offset(lne) family(poisson) link(log) ip(f) nip(2)
number of level 1 units = 56
number of level 2 units = 56

Condition Number = 9.5651503

gllamm model

log likelihood = -205.40139

------------------------------------------------------------------------------
           o |      Coef.   Std. Err.      z    P>|z|     [95% Conf. Interval]
-------------+----------------------------------------------------------------
       _cons |   .0314632    .1037764     0.30   0.762    -.1719348    .2348612
         lne |   (offset)
------------------------------------------------------------------------------


Probabilities and locations of random effects
------------------------------------------------------------------------------


***level 2 (county)

    loc1: -.85205, .54307
  var(1): .46272368
    prob: 0.3893, 0.6107
------------------------------------------------------------------------------
```

We see that the locations are estimated as $\widehat{e}_1 = -0.85$ and $\widehat{e}_2 = 0.54$ with estimated probabilities $\widehat{\pi}_1 = 0.39$ and $\widehat{\pi}_2 = 0.61$. The variance of this discrete distribution

$$\widehat{\mathrm{Var}}(\zeta_j) = \widehat{\pi}_1\widehat{e}_1^2 + \widehat{\pi}_2\widehat{e}_2^2$$

is given as 0.46 next to var(1).

We now gradually increase the number of locations until the likelihood cannot be increased any further. For $C=3$ locations, we use the nip(3) option and the previous estimates for $C = 2$ as starting values. We also use the gateaux() option (for the *Gâteaux derivative* or directional derivative) to specify the smallest and largest locations to be considered in the search, as well as the number of equally spaced steps to take within these limits. Finally, we must also supply the previous log likelihood using the `lf0()` option, which requires as its first argument the number of parameters in the previous model. It is convenient to obtain the parameter estimates, log likelihood, and number of parameters in the previous model as follows:

```
. matrix a = e(b)
. local ll = e(ll)
. local k = e(k)
. gllamm o, i(county) offset(lne) family(poisson) link(log) ip(f) nip(3)
> gateaux(-5 5 100) from(a) lf0(`k' `ll')
.............................................................................
> ....................
maximum gateaux derivative is 5.6798243

number of level 1 units = 56
number of level 2 units = 56

Condition Number = 9.2086563


gllamm model                                    Number of obs      =        56
                                                LR chi2(2)         =     61.92
Log likelihood = -174.44216                     Prob > chi2        =    0.0000

------------------------------------------------------------------------------
           o |      Coef.   Std. Err.      z    P>|z|     [95% Conf. Interval]
-------------+----------------------------------------------------------------
       _cons |   .0860002   .1189174     0.72   0.470    -.1470736    .3190741
         lne |   (offset)
------------------------------------------------------------------------------


Probabilities and locations of random effects
------------------------------------------------------------------------------

***level 2 (county)

    loc1: -1.1012, 1.1149, .07126
  var(1): .63591773
    prob: 0.2753, 0.241, 0.4836
------------------------------------------------------------------------------
```

To increase the number of locations to $C=4$, we use the same commands and options as before, except that we change the `nip()` option from `nip(3)` to `nip(4)`:

```
. matrix a = e(b)
. local ll = e(ll)
. local k = e(k)
. gllamm o, i(county) offset(lne) family(poisson) link(log) ip(f) nip(4)
> gateaux(-5 5 100) from(a) lf0(`k' `ll')
................................................................................
> ....................
maximum gateaux derivative is .00031851

number of level 1 units = 56
number of level 2 units = 56

Condition Number = 72.544698


gllamm model                                    Number of obs      =        56
                                                LR chi2(2)         =      0.11
Log likelihood = -174.38535                     Prob > chi2        =    0.9448

------------------------------------------------------------------------------
           o |      Coef.   Std. Err.      z    P>|z|     [95% Conf. Interval]
-------------+----------------------------------------------------------------
       _cons |   .0847742   .1188573     0.71   0.476    -.1481818    .3177303
         lne |   (offset)
------------------------------------------------------------------------------


Probabilities and locations of random effects
------------------------------------------------------------------------------


***level 2 (county)

    loc1: -1.1002, 1.0426, 1.2628, .06676
  var(1): .63397408
    prob: 0.2749, 0.1847, 0.0617, 0.4788
------------------------------------------------------------------------------
```

For $C=5$, we obtain

```
. matrix a = e(b)
. local ll = e(ll)
. local k = e(k)
. gllamm o, i(county) offset(lne) f(poiss) nip(5) ip(f)
> gateaux(-5 5 100) from(a) lf0(`k' `ll')
................................................................................
> ....................
maximum gateaux derivative is -.00001142
maximum gateaux derivative less than 0.00001
```

The output suggests that the likelihood cannot be increased by including a fifth point. However, we have only tried 100 locations. To make sure that there is no location at which the likelihood can be increased, we also used the option `gateaux(-5 5 1000)`, but this gave similar results.

We therefore conclude that nonparametric maximum likelihood estimator (NPMLE) has $C = 4$ locations. The nonparametric maximum likelihood estimates are shown in table 9.3 under "NPMLE". The estimates from NPMLE are similar to those obtained by assuming normality.

To visualize the discrete distribution, we can access the matrices of location and log-probability estimates stored by gllamm by using

```
. matrix locs = e(zlc2)'
. matrix lp = e(zps2)'
```

Here the apostrophe in e(zlc2)' transposes the row matrix to a column matrix that can be stored as a variable in the dataset using svmat:

```
. svmat locs
. svmat lp
```

We then transform the estimated log probabilities to probabilities and the log-incidence rates e_c to SMRs using

```
. generate p = exp(lp1)
. generate smrloc = 100*exp(_b[_cons] + locs1)
```

and finally produce a graph of the discrete distribution:

```
. twoway (dropline p smrloc), xtitle(Location) ytitle(Probability)
```

The graph is shown in figure 9.4.

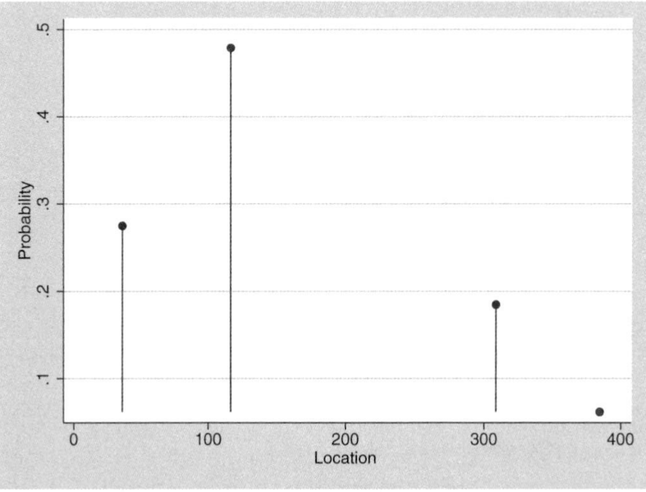

Figure 9.4: Lip cancer in Scotland: SMRs ("locations") and probabilities for nonparametric maximum likelihood estimate of random-intercept distribution

9.16.3 Prediction

Empirical Bayes predictions of the standardized mortality ratios (SMRs) can be obtained as before:

```
. gllapred mu2, mu nooffset
. generate thet2 = 100*mu2
. sort county
. list county thet thet2 in 1/10, clean noobs
    county       thet      thet2
         1   470.7203   342.6493
         2   421.7975   362.4183
         3   309.4238   327.1122
         4   295.1885   321.6138
         5   308.5293    327.567
         6   272.0485    311.061
         7   299.8875   322.2489
         8   247.8383   292.5374
         9   238.9533   280.1219
        10   279.1406   319.8951
```

Predicted SMRs based on nonparametric maximum likelihood estimation (NPMLE) are tabulated under "NPMLE" in table 9.2.

9.17 Summary and further reading

In this chapter, we have discussed Poisson regression modeling of clustered count data. Multilevel Poisson regression models with random effects were applied to longitudinal data on number of doctor visits, continuous-time survival analysis of child mortality, and small area estimation or disease mapping of standardized mortality ratios for lip cancer.

We started by outlining the motivation and some of the basic properties of the Poisson distribution, such as the equality of the expectation and variance. We then introduced random effects at the cluster level to accommodate within-cluster dependence, which must be tackled, for instance, when modeling longitudinal or panel data of counts. Both random-intercept and random-coefficient models were considered. Subsequently, we considered how different approaches to Poisson regression, such as quasilikelihood and introduction of a level-1 random intercept, can be used when there is overdispersion, or in other words when the variance of a count is larger than the expectation. We also briefly discussed other approaches to modeling clustered count data, such as conditional maximum likelihood, GEE, and NPMLE where the distributional assumption for the random intercept is relaxed (see also Rabe-Hesketh, Pickles, and Skrondal 2003).

Most of the issues discussed for prediction of random effects for dichotomous responses persist for counts, such as the choice between using the mean or the mode of the posterior distribution and the rather limited utility of diagnostics based on such predictions. We also experience shrinkage in empirical Bayes prediction as before. In small area estimation or disease mapping, inclusion of random effects is beneficial for "borrowing strength" from other counties in predicting standardized mortality ratios. Empirical Bayes predictions for disease mapping can be improved by modeling spatial dependence (see Skrondal and Rabe-Hesketh 2004, sec. 11.4), which results in an even smoother disease map.

Extensive treatments of methods for count data are given by Cameron and Trivedi (1998) and Winkelmann (2003) who also discuss some models for clustered data. A brief introduction to single-level models for counts is provided by Long (1997), and Hilbe (2007) is devoted to the negative binomial model. Use of Poisson regression for continuous-time survival analysis is discussed in Breslow and Day (1987), and models for disease mapping are described in Lawson, Browne, and Vidal-Rodeiro (2003) and Lawson et al. (1999).

9.18 Exercises

9.1 Epileptic-fit data

We now consider the famous longitudinal epilepsy data from Thall and Vail (1990), also analyzed by Breslow and Clayton (1993). The data come from a randomized controlled trial comparing the drug progabide for the treatment of epilepsy with a placebo. The outcomes are counts of epileptic seizures during the 2 weeks before each of four clinic visits.

The dataset `epilep.dta` has the following variables:

- `subj`: subject identifier (j)
- `y`: count of epileptic seizures over 2-week period (y_{ij})
- `visit`: visit time (z_{ij}) coded −0.3, −0.1, 0.1, 0.3
- `treat`: treatment group (x_{2j}) (1: progabide; 0: placebo)
- `lbas`: logarithm of quarter of number of seizures in 8 weeks preceding entry to trial (x_{3j})
- `lbas_trt`: interaction between `lbas` and `treat` (x_{4j})
- `lage`: logarithm of age (x_{5j})
- `v4`: dummy variable for fourth visit (x_{6ij})

The covariates `lbas`, `lbas_trt`, `lage`, and `v4` have all been mean centered, which will affect only the constant β_1 in the models below.

1. Model II in Breslow and Clayton is a log-linear (Poisson regression) model with covariates `lbas`, `treat`, `lbas_trt`, `lage`, and `v4` and a normally distributed random intercept for subjects. Fit this model using `gllamm`.
2. Breslow and Clayton also considered a random-coefficient model (Model IV) using the variable `visit` instead of `v4`. The effect of `visit`, denoted z_{ij}, varies randomly between subjects. The model can be written as

$$\log(\mu_{ij}) = \beta_1 + \beta_2 x_{2j} + \cdots + \beta_5 x_{5j} + \beta_6 z_{ij} + \zeta_{1j} + \zeta_{2j} z_{ij}$$

where the subject-specific random intercept ζ_{1j} and slope ζ_{2j} have a bivariate normal distribution, given the covariates. Fit this model using `gllamm`.

3. Plot the posterior mean counts versus time for twelve patients in each treatment group.

9.2 Headache data

Here we consider data from McKnight and van den Eeden (1993) that were previously analyzed by Hedeker (1999). A multiple-period, two-treatment, crossover trial was conducted to investigate if aspartame, an artificial sweetener, causes headaches. Twenty-seven patients were randomized to different sequences of aspartame and placebo. Hedeker found no evidence for sequence or period effects, and we will therefore ignore this aspect here.

The dataset `headache.dta` has the following variables:

- `id`: subject identifier
- `y`: number of headaches in week i for subject j, for some subjects counted over fewer than 7 days
- `days`: number of days for which headaches were counted (t_{ij})
- `aspartame`: dummy variable for aspartame versus placebo (x_{ij})

1. Fit the random-intercept model given below using `gllamm`

$$\begin{aligned}\log(\mu_{ij}) &= \underbrace{\ln(t_{ij})}_{\text{Offset}} + \beta_1 + \beta_2 x_{ij} + \zeta_{1j} \\ \mu_{ij} &= t_{ij} \underbrace{\exp(\beta_1 + \beta_2 x_{ij} + \zeta_{1j})}_{\text{Rate}}\end{aligned}$$

where $\ln(t_{ij})$, the logarithm of `days`, is an offset and $\zeta_{1j}|x_{ij} \sim N(0, \psi_{11})$. Interpret the coefficients.

2. Now also include a random slope ζ_{2j} for `aspartame`

$$\log(\mu_{ij}) = \ln(t_{ij}) + \beta_1 + \beta_2 x_{ij} + \zeta_{1j} + \zeta_{2j} x_{ij}$$

and assume that the random intercept and slope have a bivariate normal distribution, given the covariate. Is there any evidence that the random slope is required?

3. Produce a graph showing the empirical Bayes predictions of the headache rate per day for each patient (except patient 26, who never took aspartame) versus treatment. To plot the graph, it is helpful to pick out one observation for each combination of `id` and `aspartame` (using the `egen` function `tag()`) and to use the `connect(ascending)` option in the `twoway` command.

9.3 Police stops data

Gelman and Hill (2007) analyzed data on the number of times the police stopped individuals for questioning or searching on the streets of New York City. In particular, Gelman and Hill wanted to investigate whether ethnic minorities were stopped disproportionately often compared with whites. They used data collected by the police over a 15-month period in 1998–1999 in which the stops were classified by ethnicity, type of suspected crime, and the precinct (area) in which it occurred. Only the three largest ethnic groups (black, Hispanic, and white) are included here.

It could be argued that it would be reasonable if the police stopped individuals from the different ethnic groups in proportion to their population size, or in proportion to the number of crimes they have committed in the past. As a proxy for the latter, Gelman and Hill (2007) use the number of arrests in 1997.

The rows in the data in `police.dta` correspond to each possible combination of ethnicity, type of crime and precinct. The dataset contains the following variables:

- `eth`: ethnicity (1: black; 2: Hispanic; 3: white)
- `crime`: type of crime (1: violent crimes; 2: weapons offenses; 3: property crimes; 4: drug crimes)
- `precinct`: New York city precinct (1–75)
- `stops`: number of police stops over 15-month period in 1998–1999
- `arrests`: number of arrests in New York City in 1997
- `pop`: population size of each ethnic group in the precinct
- `prblack`: proportion of precinct population that is black
- `prhisp`: proportion of precinct population that is Hispanic

These data, provided with Gelman and Hill (2007), have had some noise added to protect confidentiality. In this exercise, we will analyze the data on violent crimes only, so you can the delete data on the other types of crimes.

1. For stops due to suspected violent crimes, fit a Poisson model with the number of police stops as the response variable, dummies for the ethnic minority groups (blacks and Hispanics) as explanatory variables, and with
   a. `pop` as an exposure
   b. `arrests` as an exposure

   Use robust standard errors that take the clustering within precincts into account. Interpret the estimated incidence-rate ratios for the two types of exposure and comment on the difference.
2. Fit the model from step 1 with `arrests` as the exposure but also including `prblack` and `prhisp` as further covariates.
3. Compare the estimated incidence-rate ratios for the minority groups between the model considered in step 2 and a model that uses fixed effects for precincts instead of the covariates `prblack` and `prhisp`. Does there appear to be much precinct level confounding in step 2?

4. Fit the model from step 2 but also include a normally distributed random intercept for precincts. Use `xtmepoisson`.
5. For the model in step 4, calculate the estimated median incidence-rate ratio, comparing the precinct with the larger random intercept with the precinct with the smaller random intercept for two randomly chosen precincts having the same covariate values. Interpret this estimate.
6. ❖ Extend the model from step 4 further by including random coefficients for the ethnic minority dummy variables, specifying a trivariate normal distribution for the random intercept and slopes with a freely estimated covariance matrix.
   a. Write down the model using a two-stage formulation as discussed in section 4.9.
   b. Fit the model using `xtmepoisson` (this will take a long time).
   c. Obtain empirical Bayes modal predictions of the intercepts and slopes and plot them using a scatterplot matrix.

9.4 Patent data

Hall, Griliches, and Hausman (1986) analyzed panel data on the number of patents awarded to 346 firms between 1975 and 1979. In particular, they considered the effect of research and development (R&D) expenditures in the current and previous years. The dataset has been analyzed many times. Good discussions can be found in Cameron and Trivedi (1998; 2005) who made the data available on the web page for their 1998 book.

The variables in the dataset `patents.dta` that we will use here are

- `cusip`: Compustat's (Committee on Uniform Security Identification Procedures) identifying number for firm j
- `year`: year i (1–5)
- `pat`: number of patents applied for during the year that were eventually granted (y_{ij})
- `scisect`: dummy variable for firm being in the scientific sector
- `logk`: logarithm of the book value of capital in 1972
- `logr`: logarithm of R&D spending during the year (in 1972 dollars) [$\ln(r_{ij})$]
- `logr1`, `logr2`, ..., `logr5`: Lagged `logr` variable 1 year, 2 years, ..., 5 years ago [$\ln(r_{i-1,j})$, $\ln(r_{i-2,j})$, ..., $\ln(r_{i-5,j})$]

Consider the following model, which allows the expected number of patents successfully applied for by firm j in year i to depend on the logarithms of R&D expenditure for the current and previous five years:

$$\ln(\mu_{ij}) = \alpha + \beta_0 \ln(r_{ij}) + \beta_1 \ln(r_{i-1,j}) + \cdots + \beta_5 \ln(r_{i-5,j}) \qquad (9.16)$$

We can consider the effect of multiplying the expenditures for a given firm by a constant a:

$$\ln(\mu_{ij}^a) = \alpha + \beta_0 \ln(ar_{ij}) + \beta_1 \ln(ar_{i-1,j}) + \cdots + \beta_5 \ln(ar_{i-5,j})$$

Taking derivatives with respect to a (using the chain rule), we find that

$$\frac{\partial \ln(\mu_{ij}^a)}{\partial a} = \frac{1}{\mu_{ij}^a}\frac{\partial \mu_{ij}^a}{\partial a} = \beta_0 \frac{r_{ij}}{ar_{ij}} + \beta_1 \frac{r_{i-1,j}}{ar_{i-1,j}} + \cdots + \beta_5 \frac{r_{i-5,j}}{ar_{i-5,j}}$$
$$= \frac{1}{a}(\beta_0 + \beta_1 + \cdots + \beta_5)$$

so that the *elasticity* is

$$\left(\frac{\partial \mu_{ij}^a}{\partial a}\right)\left(\frac{a}{\mu_{ij}^a}\right) = \frac{\partial \mu_{ij}^a}{\mu_{ij}^a} \Big/ \frac{\partial a}{a} = \beta_0 + \beta_1 + \cdots + \beta_5$$

The sum of the coefficients therefore represents the relative or percentage change in the expected number of patents per percentage change in total expenditures over the last 5 years (a given percentage change in a amounts to the same percentage change in ax).

1. Fit a Poisson model with the mean modeled as shown in (9.16) but also include a normally distributed random intercept for firm j and four dummy variables for years 2 to 5. Use `gllamm` with the `robust` option to obtain reasonable inferences even if there is overdispersion at level 1.
2. Use `lincom` to obtain the sum of the estimated coefficients of the contemporaneous and lagged log expenditures (the elasticity) and its 95% confidence interval. Also use the `test` or `lincom` command to test the null hypothesis that the elasticity is 1. Comment on your finding.
3. Repeat the above analysis (steps 1 and 2) also controlling for `logk`, a measure of the size of the firm, and `scisect`, a dummy variable for the firm being in the scientific sector.

9.5 School-absenteeism data

These data come from a sociological study by Quine (1975) and have previously been analyzed by Aitkin (1978). The sample included Australian aboriginal and white children from four age groups (final year in primary school and first three years in secondary school) who were classified as slow or average learners. The number of days absent from school during the school year was recorded for each child (children who had suffered a serious illness during the year were excluded).

The dataset `absenteeism.dta` has the following variables:

- `id`: child identifier
- `days`: number of days absent from school in one year
- `aborig`: dummy variable for child being aboriginal (versus white)
- `girl`: dummy variable for child being a girl
- `age`: age group (1: last year in primary school; 2: first year in secondary school; 3: second year in secondary school; 4: fourth year in secondary school)
- `slow`: dummy variable for child being a slow learner (1: slow; 0: average)

1. Use the `glm` command to fit an ordinary Poisson model with `aborig`, `girl`, dummy variables for age groups 2 to 4, and `slow` as covariates.
2. Repeat the analysis, but this time use quasilikelihood estimation to include an overdispersion parameter ϕ. How have the estimates and standard errors changed?
3. Fit a negative binomial model of type NB1 and compare the estimated overdispersion factor with that estimated in step 2.
4. Use `xtpoisson` to include a normally distributed random intercept for children.
5. Write down the model of step 4, and interpret the estimates.

9.6 Blindness data

Here will use the piecewise exponential survival model to analyze the data from exercise 8.6. The study was a randomized trial of a treatment to delay the onset of blindness in patients suffering from diabetic retinopathy. One eye of each patient was randomly selected for treatment, and patients were assessed at 4-month intervals. The endpoint used to assess the treatment effect was the occurrence of visual acuity less than 5/200 at two consecutive assessments.

The variables in the data `blindness.dta` are described in exercise 8.6.

1. Expand the data for analysis using the piecewise exponential model with time intervals [6,10), [10,14), ..., [50,54), [54,58), [58,66), [66,83).
2. Use Poisson regression to fit a piecewise exponential model with dummy variables for the intervals, `treat`, `late`, and the `treat` by `late` interaction as covariates.
3. Fit the same model as in step 2 but with a normally distributed random intercept.
   a. Is there evidence for within-subject dependence?
   b. Interpret the estimated incidence-rate ratios for `treat`, `late`, and their interaction.
4. For the model in step 3, test the proportional-hazards assumption for `treat` by including an interaction between `treat` and a variable containing the start times of the intervals. (Hint: you can make use of `_t0` or the variable created using the `stsplit` command in step 1.)

9.7 Lip-cancer data

1. Extend the model considered in section 9.15 by including `x`, the percentage of the population working in agriculture, fishing, or forestry, as a covariate.
   a. Write down the model and state all assumptions.
   b. Fit the model using `gllamm`.
   c. Interpret the exponentiated estimated coefficient of `x`.
2. Obtain the predicted standardized mortality ratios (SMRs) for this model.

9.8 Skin-cancer data

Langford and Lewis (1998) analyzed data from the Atlas of Cancer Mortality in the European Economic Community (Smans, Muir, and Boyle 1992). Malignant-melanoma (skin cancer) mortalities were defined as deaths recorded and certified by a medical practitioner as ICD-8 172. These data were collected between 1971 and 1980, although for the United Kingdom, Ireland, Germany, Italy, and The Netherlands, data were only available from 1975–1976 onwards and are aggregated over the period of data collection. As incidence was generally rising during this period, it is important to include nation as a variable in the analysis.

The geographical resolution of the data, referred to here as counties, was European Economic Community (EEC), now the European Union, levels II or III (identified by EEC statistical services). For example, these units are counties in England and Wales, départements in France and Regierungsbezirke in (West) Germany. Regions are the EEC level-I areas.

The main research question is how malignant melanoma (skin cancer) is associated with ultraviolet (UV) radiation exposure.

Seven variables are included in the data `skincancer.dta`:

- `nation`: nation identifier (labeled)
  1. Belgium
  2. W. Germany
  3. Denmark
  4. France
  5. UK
  6. Italy
  7. Ireland
  8. Luxembourg
  9. Netherlands
- `region`: region identifier (EEC level-I areas)
- `county`: county identifier (EEC level-II and level-III areas)
- `deaths`: number of male deaths due to malignant melanoma (skin cancer) during 1971–1980
- `expected`: expected number of male deaths due to malignant melanoma during 1971–1980, calculated from crude rates for all countries combined
- `uv`: epidemiological index of UV dose reaching the earth's surface in each county (see Langford and Lewis 1998), mean centered

1. Fit a Poisson model for the number of male deaths from skin cancer using the log expected number of male deaths as an offset, dummy variables for the nations as covariates, and a random intercept for counties.
2. Write down the model, and interpret the estimates.
3. Refit the model including `uv` and squared `uv` as further covariates.

4. Obtain predictions of the fixed part of the linear predictor without the offset (using the `xb` and `nooffset` options), and plot these against `uv` by nation. Interpret the graph.
5. For the model in step 3, obtain empirical Bayes predictions of the SMR for each county (as a percentage). Summarize the SMRs by nation.

See also exercise 10.11 for a four-level model.

Part IV

Models with nested and crossed random effects

10 Higher-level models with nested random effects

10.1 Introduction

We have until now considered two-level models where units are nested in groups or clusters.

In three-level models, the clusters themselves are nested in superclusters, forming a hierarchical structure. For example, we may have repeated measurement occasions (units) for patients (clusters) who are clustered in hospitals (superclusters). This three-level design is displayed in figure 10.1.

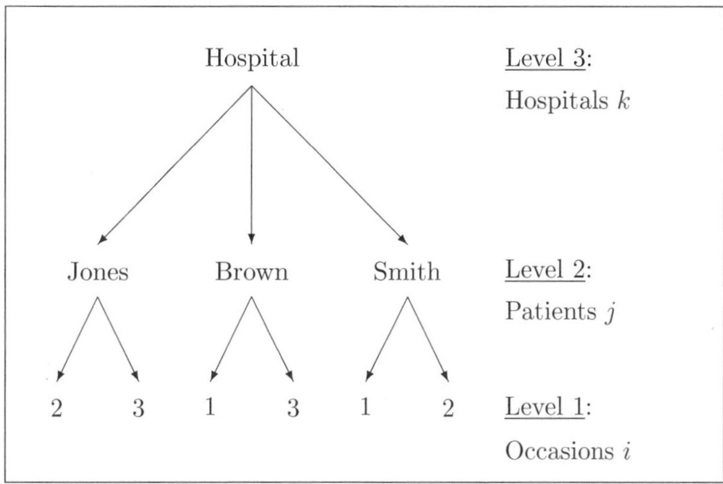

Figure 10.1: Illustration of three-level design

It seems reasonable to assume that measurements within the same hospital are correlated and that measurements within the same patient are even more correlated.

Other examples of three-level designs include repeated measurements on students nested in schools (exercise 10.1), or on states in regions (exercise 10.3), or on patients in families (exercise 10.6). Alternatively, there could be single measurements on children nested in mothers in communities (section 10.5), children nested in physicians in hospitals (exercise 10.7), or cows nested in herds in regions (exercise 10.4).

In this chapter, we consider three-level models for continuous and dichotomous responses. The ideas extend directly to models with four or more levels and other response types (see also exercises).

10.2 Do peak-expiratory-flow measurements vary between methods?

We first return to the peak-expiratory-flow measurements discussed in chapter 2. Peak expiratory flow was measured using two methods, the standard Wright peak flow meter and the Mini Wright peak flow meter, each on two occasions. We previously considered test–retest data using only the Mini Wright meter. Here we will analyze the full dataset to consider variability due to measurement method.

We first read the data and perform the necessary data manipulation, as discussed in section 2.5:

```
. use http://www.stata-press.com/data/mlmus2/pefr

. reshape long wp wm, i(id) j(occasion)
(note: j = 1 2)
Data                               wide   ->   long
-----------------------------------------------------------------------------
Number of obs.                       17   ->     34
Number of variables                   7   ->      6
j variable (2 values)                     ->   occasion
xij variables:
                                wp1 wp2   ->   wp
                                wm1 wm2   ->   wm
-----------------------------------------------------------------------------
```

We then stack the responses `wm` for the Mini Wright peak flow meter and `wp` for the Wright peak meter into a single response `w`, producing a string variable `meth` equal to the suffixes `m` and `p`. To do this using Stata's `reshape long` command with the `string` option, we must first define a new variable `i` for the `i()` option that takes on a different value for each observation in the wide dataset:

```
. generate i = _n

. reshape long w, i(i) j(meth) string
(note: j = m p)
Data                               wide   ->   long
-----------------------------------------------------------------------------
Number of obs.                       34   ->     68
Number of variables                   5   ->      5
j variable (2 values)                     ->   meth
xij variables:
                                  wm wp   ->   w
-----------------------------------------------------------------------------

. sort id meth occasion
```

```
. list id meth occasion w in 1/8, clean noobs

    id   meth   occasion     w
     1      m          1   512
     1      m          2   525
     1      p          1   494
     1      p          2   490
     2      m          1   430
     2      m          2   415
     2      p          1   395
     2      p          2   397
```

We also create a dummy variable for method m (the Mini Wright meter) using the encode command, which generates a numeric variable with successive integer values corresponding to the strings sorted in alphabetical order (here m becomes 1 and p becomes 2):

```
. encode meth, generate(method)
. recode method 2=0
```

10.3 Two-level variance-components models

10.3.1 Model specification

We start by simply fitting a two-level variance-components model to all four measurements, with a random intercept for subjects. This model, which ignores the fact that different methods were used, can be written as

$$y_{ijk} = \beta_1 + \zeta_k^{(3)} + \epsilon_{ijk} \tag{10.1}$$

where the index i is for occasions, j for methods, and k for subjects. $\zeta_k^{(3)} \sim N(0, \psi^{(3)})$ is a random intercept for subjects k, independently distributed from the residual error term $\epsilon_{ijk}|\zeta_k^{(3)} \sim N(0, \theta)$.

We can also allow for a systematic difference (or bias) between the methods j by including a dummy variable x_j (method) for the Mini Wright meter in the fixed part of the model:

$$y_{ijk} = \beta_1 + \beta_2 x_j + \zeta_k^{(3)} + \epsilon_{ijk} \tag{10.2}$$

with $\zeta_k^{(3)}|x_j \sim N(0, \psi^{(3)})$ and $\epsilon_{ijk}|x_j, \zeta_k^{(3)} \sim N(0, \theta)$.

10.3.2 Estimation using xtmixed

Maximum likelihood estimates for the two-level variance-components model (10.1) can be obtained using xtmixed:

```
. xtmixed w || id:, mle
Mixed-effects ML regression                     Number of obs      =        68
Group variable: id                              Number of groups   =        17

                                                Obs per group: min =         4
                                                               avg =       4.0
                                                               max =         4

                                                Wald chi2(0)       =         .
Log likelihood = -349.88896                     Prob > chi2        =         .

------------------------------------------------------------------------------
           w |      Coef.   Std. Err.      z    P>|z|     [95% Conf. Interval]
-------------+----------------------------------------------------------------
       _cons |   450.8971   26.63839    16.93   0.000     398.6868    503.1073
------------------------------------------------------------------------------

------------------------------------------------------------------------------
  Random-effects Parameters  |   Estimate   Std. Err.     [95% Conf. Interval]
-----------------------------+------------------------------------------------
id: Identity                 |
                   sd(_cons) |   109.1843   18.94853      77.70288    153.4203
-----------------------------+------------------------------------------------
                sd(Residual) |   23.83799   2.360313      19.63307    28.94351
------------------------------------------------------------------------------
LR test vs. linear regression: chibar2(01) =   134.62 Prob >= chibar2 = 0.0000
. estimates store model1
```

The estimates for model (10.1) are also presented under "Model 1" in table 10.1 on page 441. The between-subject standard deviation $\sqrt{\psi^{(3)}}$ is estimated as 109.2 and the within-subject standard deviation $\sqrt{\theta}$ as 23.8, giving an estimated intraclass correlation of $\widehat{\rho} = 109.18^2/(109.18^2 + 23.84^2) = 0.95$.

Maximum likelihood estimates for model (10.2), which also includes a fixed effect for the Mini Wright meter, are obtained using the command

```
. xtmixed w method || id:, mle
Mixed-effects ML regression                     Number of obs      =        68
Group variable: id                              Number of groups   =        17
                                                Obs per group: min =         4
                                                               avg =       4.0
                                                               max =         4
                                                Wald chi2(1)       =      1.11
Log likelihood = -349.33929                     Prob > chi2        =    0.2918

------------------------------------------------------------------------------
           w |      Coef.   Std. Err.      z    P>|z|     [95% Conf. Interval]
-------------+----------------------------------------------------------------
      method |   6.029412   5.719584     1.05   0.292    -5.180768    17.23959
       _cons |   447.8824   26.79146    16.72   0.000     395.3721    500.3926
------------------------------------------------------------------------------

------------------------------------------------------------------------------
  Random-effects Parameters  |   Estimate   Std. Err.     [95% Conf. Interval]
-----------------------------+------------------------------------------------
id: Identity                 |
                   sd(_cons) |   109.1981    18.9461      77.71949    153.4265
-----------------------------+------------------------------------------------
                sd(Residual) |   23.58245    2.33501       19.4226    28.63324
------------------------------------------------------------------------------
LR test vs. linear regression: chibar2(01) =   135.67 Prob >= chibar2 = 0.0000
. estimates store model2
```

These estimates are presented under "Model 2" in table 10.1. Including a fixed effect for `method` did not decrease the estimated within-subject standard deviation appreciably (from 23.84 to 23.58). Furthermore, the coefficient of `method` is not significant at the 5% level.

Both two-level variance-components models assume that the four measurements, using the two methods, are all mutually independent, conditional on the random intercept $\zeta_k^{(3)}$ (and after allowing for a main effect of method in the second model). To see if this appears reasonable, we can plot all four measurements against the subject identifier using different symbols for the methods

```
. twoway (scatter w id if method==0, msymbol(circle))
>        (scatter w id if method==1, msymbol(circle_hollow)),
> xtitle(Subject id) ytitle(Peak-expiratory-flow measurements)
> legend( order(1 "Wright" 2 "Mini Wright")) xlabel(1/17)
```

giving the graph in figure 10.2.

(Continued on next page)

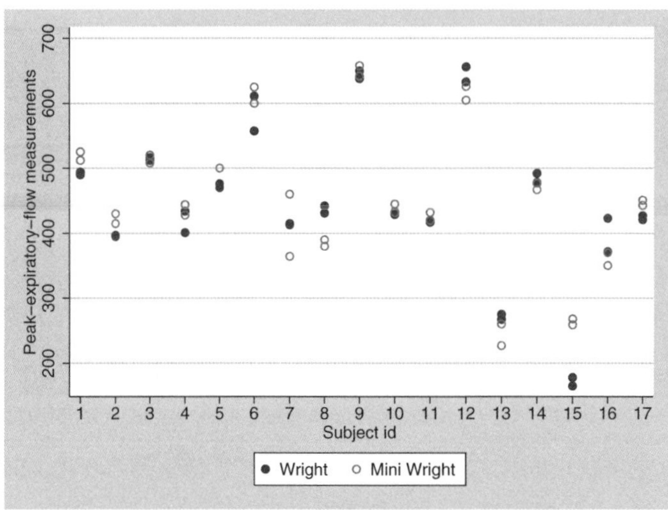

Figure 10.2: Scatterplot of peak expiratory flow measured by two methods versus subject

As would be expected, measurements on the same subjects are more similar than measurements on different subjects. This between-subject heterogeneity is modeled by the subject-level random intercept $\zeta_k^{(3)}$. However, the figure also suggests that for a given subject, the measurements using the same method tend to resemble each other more than measurements using the other method, violating the conditional independence assumption. This is not due to some constant shift of the measurements using one method relative to the other, but due to shifts that vary between subjects. For some subjects (e.g., the first two), the Wright peak flow meter measurements are lower, whereas for other subjects (e.g., subjects 8 and 12) the Mini Wright meter measurements are lower. Furthermore, the difference between methods is large for some subjects (e.g., subject 15) and small for others (e.g., subjects 3, 9, 10, and 11).

10.4 Three-level variance-components models

10.4.1 Model specification

We can accommodate the between-method within-subject heterogeneity apparent in figure 10.2 by including another random intercept $\zeta_{jk}^{(2)}$ for each combination of method and subject.

The three-level model can be written as

$$y_{ijk} \;=\; \beta_1 + \zeta_{jk}^{(2)} + \zeta_k^{(3)} + \epsilon_{ijk} \tag{10.3}$$

where $\zeta_{jk}^{(2)}$ is the random intercept for method j and subject k and $\zeta_k^{(3)}$ is the random intercept for subject k. The random effect for method is nested within subjects in the sense that it does not take on the same value for a given method across all subjects, but takes on a different value for each combination of method and subject.

We can also allow for systematic differences between the methods by including a dummy variable x_j for the Mini Wright meter:

$$y_{ijk} \;=\; \beta_1 + \beta_2 x_j + \zeta_{jk}^{(2)} + \zeta_k^{(3)} + \epsilon_{ijk} \qquad (10.4)$$

Using ANOVA terminology, there are two "factors" (or categorical explanatory variables): method and subject, with occasions treated as replicates as in chapter 2. Method is a fixed factor, whereas subject is a random factor, giving a two-way mixed-effects ANOVA model. The model in (10.4) includes a main effect of method β_2, a main effect of subject $\zeta_k^{(3)}$, and a method by subject interaction $\zeta_{jk}^{(2)}$. This random interaction takes on a different value for each subject and method combination and is therefore nested within subjects. It can be interpreted as a subject-specific bias of the methods (see Dunn 1992).

We make the following assumptions regarding the random intercepts and the residual error term:

$$\zeta_k^{(3)}|x_j \sim N(0,\psi^{(3)}), \quad \zeta_{jk}^{(2)}|x_j,\zeta_k^{(3)} \sim N(0,\psi^{(2)}), \quad \text{and} \quad \epsilon_{ijk}|x_j,\zeta_{jk}^{(2)},\zeta_k^{(3)} \sim N(0,\theta)$$

The random intercepts and residual error term are also assumed to be mutually independent and independent across replications.

The random part of the model (everything apart from $\beta_1+\beta_2 x_j$) is represented by a path diagram in figure 10.3. The rectangles represent observed variables, here the responses y_{11k}, y_{21k}, y_{12k}, and y_{22k} for subject k. The k subscript is implied by the label "subject k" inside the frame surrounding the diagram. The circles represent the random effects $\zeta_{1k}^{(2)}$ for method 1 and subject k, $\zeta_{2k}^{(2)}$ for method 2 and subject k, and $\zeta_k^{(3)}$ for subject k, again with k subscripts not shown. The long arrows represent regressions, here with regression coefficients set to 1, and the short arrows from below represent additive error terms ϵ_{ijk}. For instance, y_{12k} (with $i=1$ and $j=2$) is regressed on $\zeta_{2k}^{(2)}$ and $\zeta_k^{(3)}$ with additive error term ϵ_{12k}

$$y_{12k} \;=\; \zeta_{2k}^{(2)} + \zeta_k^{(3)} + \epsilon_{12k} + \cdots$$

where the "$\cdots$" indicates that the fixed part of the model (here $\beta_1+\beta_2 x_j$) is not shown.

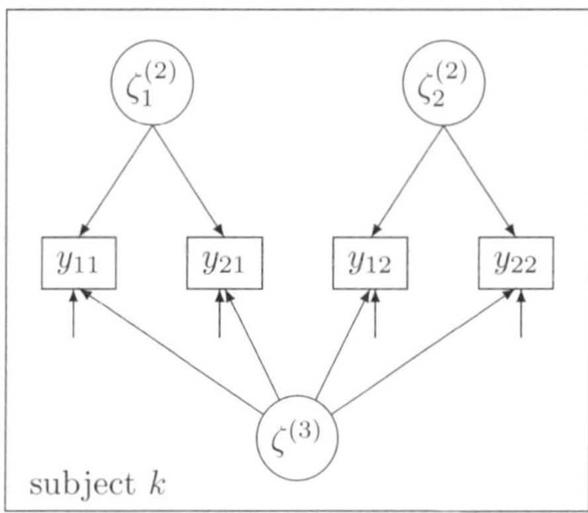

Figure 10.3: Path diagram of random part of reduced-form three-level model

The model is a simple example of a three-level model. Measurements (level 1) are nested in methods (level 2), which in turn are nested in subjects (level 3).

10.4.2 Different types of intraclass correlation

For three-level models, we can consider several types of intraclass correlations for pairs of responses.

For the same subject k, but different methods j and j', we obtain

$$\rho(\text{subject}) \equiv \text{Cor}(y_{ijk}, y_{i'j'k} \mid x_j, x_{j'}) \quad = \quad \frac{\psi^{(3)}}{\psi^{(2)} + \psi^{(3)} + \theta}$$

whereas for the same method j and subject k, we get

$$\rho(\text{method}, \text{subject}) \equiv \text{Cor}(y_{ijk}, y_{i'jk} \mid x_j) \quad = \quad \frac{\psi^{(2)} + \psi^{(3)}}{\psi^{(2)} + \psi^{(3)} + \theta}$$

This intraclass correlation can be thought of as the test–retest reliability for each method. The models considered here assume that both methods are equally reliable, an assumption we relax in exercise 10.12.

In both intraclass correlations, the numerator is equal to the variance shared by both measurements, and the denominator is just the total variance. In a proper three-level model, the variances of the random intercepts are positive, $\psi^{(2)} > 0$ and $\psi^{(3)} > 0$, and it follows that $\rho(\text{method}, \text{subject}) > \rho(\text{subject})$. This relationship makes sense since, as we saw in figure 10.2, measurements for the same subject are more correlated if they use the same method than if they use different methods.

We also see in figure 10.3 that responses for the same method (e.g., y_{11k} and y_{21k}) are connected via two paths whereas responses for different methods (e.g., y_{21k} and y_{12k}) are connected via only one path, making them less correlated.

10.4.3 Three-stage formulation

Raudenbush and Bryk (2002) use a three-stage formulation for three-level models. The level-1 model can be written as

$$y_{ijk} = \eta_{1jk} + \epsilon_{ijk}$$

where the intercept η_{1jk} varies between methods j and subjects k. This intercept is in turn modeled using the level-2 model

$$\eta_{1jk} = \pi_{11k} + \pi_{12}x_j + \zeta_{jk}^{(2)}$$

The intercept π_{11k} in this model has three subscripts, the last of which denotes that it varies over subjects k. The first two subscripts, say a and b (here 1 and 1) denote that the coefficient is the bth coefficient in the level-2 model for the ath coefficient η_{ajk} in the level-1 model.

A level-3 model is specified for the intercept π_{11k}:

$$\pi_{11k} = \gamma_{111} + \zeta_k^{(3)}$$

We see that level-3 models contain coefficients with three subscripts; γ_{abc} represents the cth coefficient in the level-3 model for the bth coefficient in the level-2 model for the ath coefficient in the level-1 model.

Substituting the expression for π_{11k} into the level-2 model and subsequently for η_{1jk} into the level-1 model, we obtain

$$y_{ijk} = \underbrace{\gamma_{111}}_{\beta_1} + \underbrace{\pi_{12}}_{\beta_2} x_j + \zeta_{jk}^{(2)} + \zeta_k^{(3)} + \epsilon_{ijk}$$

In the present application, there is only one covariate. See section 10.6.4 for a three-stage formulation involving covariates.

10.4.4 Estimation using xtmixed

When there are several nested levels, we simply specify an equation for the random part at each level, starting with the highest level and working our way down. In the current three-level application, we retain the specification of the random part for subjects, `|| id:`, which is now at level 3. Next, we specify the random part at level 2 as `|| method:`, which gives us

```
. xtmixed w method || id: || method:, mle
Computing standard errors:
Mixed-effects ML regression                     Number of obs      =        68

                ┌─────────────────────────────────────────────
                │     No. of       Observations per Group
 Group Variable │     Groups    Minimum    Average    Maximum
────────────────┼─────────────────────────────────────────────
             id │         17          4        4.0          4
         method │         34          2        2.0          2
────────────────┴─────────────────────────────────────────────

                                                Wald chi2(1)       =      0.60
Log likelihood = -344.99736                     Prob > chi2        =    0.4403

─────────────────────────────────────────────────────────────────────────────
            w │      Coef.   Std. Err.      z    P>|z|     [95% Conf. Interval]
──────────────┼──────────────────────────────────────────────────────────────
       method │   6.029412   7.812736     0.77   0.440    -9.283269    21.34209
        _cons │   447.8824   26.92328    16.64   0.000     395.1137     500.651
─────────────────────────────────────────────────────────────────────────────

─────────────────────────────────────────────────────────────────────────────
  Random-effects Parameters  │   Estimate   Std. Err.     [95% Conf. Interval]
─────────────────────────────┼───────────────────────────────────────────────
id: Identity                 │
                   sd(_cons) │   108.6455   19.04644      77.05293   153.1914
─────────────────────────────┼───────────────────────────────────────────────
method: Identity             │
                   sd(_cons) │   19.00385   4.789038      11.59673   31.14207
─────────────────────────────┼───────────────────────────────────────────────
                sd(Residual) │    17.7586   2.153546      14.00184    22.5233
─────────────────────────────────────────────────────────────────────────────
LR test vs. linear regression:       chi2(2) =   144.35   Prob > chi2 = 0.0000
Note: LR test is conservative and provided only for reference
. estimates store model3
```

The estimates are presented under "Model 3" in table 10.1. Both the syntax and the output start with the highest level random effects, here the subject-level random intercept at level 3. The between-subject standard deviation $\sqrt{\psi^{(3)}}$ is estimated as 108.65, and the between-methods within-subjects standard deviation $\sqrt{\psi^{(2)}}$ is estimated as 19.00. Finally, the standard deviation $\sqrt{\theta}$ between occasions, within methods and subjects, is estimated as 17.76. We could obtain the corresponding variances using the `variance` option.

Table 10.1: Maximum likelihood estimates for two-level and three-level models for peak-expiratory-flow data

	Two-level models		Three-level models	
	Model 1	Model 2	Model 3	Model 4
	Est (SE)	Est (SE)	Est (SE)	Est (SE)
Fixed part				
β_1	450.9 (26.6)	447.9 (26.8)	447.9 (26.9)	450.90 (26.6)
β_2		6.0 (5.7)	6.0 (7.8)	
Random part				
$\sqrt{\psi^{(2)}}$			19.0	19.5
$\sqrt{\psi^{(3)}}$	109.2	109.2	108.6	108.6
$\sqrt{\theta}$	23.8	23.6	17.8	17.8
Log likelihood	−349.89	−349.34	−345.00	−345.29

We now test the null hypothesis that the variance component for methods is zero, $\psi^{(2)}=0$, using a likelihood-ratio test:

```
. lrtest model2 model3
Likelihood-ratio test                                 LR chibar2(01)  =      8.68
(Assumption: model2 nested in model3)                 Prob > chibar2  =    0.0016
```

The p-value is based on the correct sampling distribution, taking into account that the null hypothesis is on the boundary of the parameter space, as indicated by the label `chibar2(01)`. The test suggests that the random effect for methods is required. However, the estimated fixed effect for method $\widehat{\beta}_2$ is small and not significant at the 5% level and is therefore removed:

```
. xtmixed w || id: || method:, mle
Mixed-effects ML regression                     Number of obs      =        68

-----------------------------------------------------------
                |   No. of       Observations per Group
 Group Variable |   Groups    Minimum    Average    Maximum
----------------+------------------------------------------
             id |       17          4        4.0          4
         method |       34          2        2.0          2
-----------------------------------------------------------

                                                Wald chi2(0)       =         .
Log likelihood = -345.29005                     Prob > chi2        =         .

------------------------------------------------------------------------------
           w |      Coef.   Std. Err.      z    P>|z|     [95% Conf. Interval]
-------------+----------------------------------------------------------------
       _cons |   450.8971   26.63838    16.93   0.000     398.6868    503.1073
------------------------------------------------------------------------------
```

```
------------------------------------------------------------------------------
  Random-effects Parameters  |   Estimate   Std. Err.     [95% Conf. Interval]
-----------------------------+------------------------------------------------
id: Identity                 |
                   sd(_cons) |   108.6037   19.05409      77.00244    153.1738
-----------------------------+------------------------------------------------
method: Identity             |
                   sd(_cons) |   19.47622   4.829488      11.97936    31.66473
-----------------------------+------------------------------------------------
                sd(Residual) |   17.75859   2.153546      14.00184     22.5233
------------------------------------------------------------------------------
LR test vs. linear regression:       chi2(2) =   143.81   Prob > chi2 = 0.0000
Note: LR test is conservative and provided only for reference
```

Estimates for this model are presented under "Model 4" in table 10.1.

The estimated intraclass correlation between measurements on the same subject using the same method is

$$\widehat{\rho}(\text{method}, \text{subject}) = \frac{19.48^2 + 108.60^2}{19.48^2 + 108.60^2 + 17.76^2} = 0.97$$

and the corresponding estimated intraclass correlation using different methods is

$$\widehat{\rho}(\text{subject}) = \frac{108.60^2}{19.48^2 + 108.60^2 + 17.76^2} = 0.94$$

In summary, there is no evidence of systematic bias between the methods, but there is some evidence for a subject by method interaction or subject-specific bias. The methods appear to have good test–retest reliability.

We will not repeat the analysis using `gllamm` because we generally do not recommend using `gllamm` for linear models and will demonstrate the use of `gllamm` in the next application (starting in section 10.5).

10.4.5 Empirical Bayes prediction using xtmixed

Empirical Bayes prediction is straightforward after estimation with `xtmixed`. We simply use the `predict` command with the `reffects` option, specifying as many variables as there are random effects (here 2). Keep in mind the ordering of random effects from the highest to the lowest levels:

```
. predict subj instr, reffects
. sort id method
. label define m 0 "Wright" 1 "Mini Wright"
. label values method m
```

```
. list id method subj instr if id<8 & occasion==1, clean noobs

 id        method        subj       instr
  1        Wright    53.14315   -8.504792
  1   Mini Wright    53.14315    10.21389
  2        Wright   -40.72008   -10.01413
  2   Mini Wright   -40.72008    8.704557
  3        Wright     61.6984    .9921207
  3   Mini Wright     61.6984    .9921207
  4        Wright   -23.60959   -6.913527
  4   Mini Wright   -23.60959    6.154234
  5        Wright    34.81049   -8.976175
  5   Mini Wright    34.81049    10.09569
  6        Wright    144.0732   -7.748988
  6   Mini Wright    144.0732    12.38243
  7        Wright   -37.05355    .1105382
  7   Mini Wright   -37.05355   -1.302193
```

We see that, just as suggested by figure 10.2, for subjects 1, 2, 4, 5, and 6 the Mini Wright meter appears to be positively biased compared with the Wright peak flow meter, whereas this bias is reversed for subject 7. For subject 3, the empirical Bayes predictions for both methods are small because all four measurements nearly coincide.

10.5 Did the Guatemalan immunization campaign work?

Pebley, Goldman, and Rodriguez (1996) and Rodriguez and Goldman (2001) analyzed data on Guatemalan families' decisions whether to immunize their children. Data are available from the National Survey of Maternal and Child Health (ENSMI) conducted in Guatemala in 1987. A nationally representative sample of 5,160 women aged between 15 and 44 was interviewed. The questionnaire included questions determining the immunization status of children who were born in the previous 5 years and alive at the time of the interview.

Beginning in 1986, the Guatemalan government undertook a series of campaigns to immunize the population against major childhood diseases. The immunization campaign visited most of the country and often located children in their own households. The full set of recommended immunizations included three doses of DPT vaccine (against diphtheria, pertussis or whooping cough, and tetanus), three doses of polio vaccine, one dose of BCG (antituberculosis), and one dose of measles vaccine.

The data considered here comprise 2,159 children aged 1–4 years for which we have community data on health services and who received at least one immunization during the campaign (87.8% of these children). The response variable of interest is whether the children received the full set of immunizations. An important explanatory variable is whether the child was at least 2 years old at the time of the interview, in which case the child was eligible to receive all immunizations during the campaign. If this variable is associated with immunization status, there is some indication that the campaign worked.

The dataset `guatemala.dta` comprises children i nested in mothers j nested in communities k. It contains the following subset of variables from Pebley, Goldman, and Rodriguez (1996):

- Level-1 (child):
  - `immun`: indicator variable for child receiving full set of immunizations (y_{ijk})
  - `kid2p`: dummy variable for child being at least 2 years old at time of the interview and hence eligible for full set of immunizations (x_{2ijk})
- Level-2 (mother):
  - `mom`: identifier for mothers (j)
  - Ethnicity (dummy variables with Latino as reference category)
    * `indNoSpa`: mother is indigenous, not Spanish speaking (x_{3jk})
    * `indSpa`: mother is indigenous, Spanish speaking (x_{4jk})
  - Mother's education (dummy variables with no education as reference category)
    * `momEdPri`: mother has primary education (x_{5jk})
    * `momEdSec`: mother has secondary education (x_{6jk})
  - Husband's education (dummy variables with no education as reference category)
    * `husEdPri`: husband has primary education (x_{7jk})
    * `husEdSec`: husband has secondary education (x_{8jk})
    * `husEdDK`: husband's education is not known (x_{9jk})
- Level-3 (community):
  - `cluster`: identifier for communities (k)
  - `rural`: dummy variable for community being rural ($x_{10,k}$)
  - `pcInd81`: percentage of population that was indigenous in 1981 ($x_{11,k}$)

10.6 A three-level logistic random-intercept model

10.6.1 Model specification

We specify a three-level random-intercept logit model for childhood immunization with children i nested in mothers j who are nested in communities k:

$$\begin{aligned}\text{logit}\{\Pr(y_{ijk}=1|\mathbf{x}_{ijk},\zeta_{jk}^{(2)},\zeta_k^{(3)})\} &= \beta_1+\beta_2x_{2ijk}+\cdots+\beta_{11}x_{11,k}+\zeta_{jk}^{(2)}+\zeta_k^{(3)} \\ &= (\beta_1+\zeta_{jk}^{(2)}+\zeta_k^{(3)})+\beta_2x_{2ijk}+\cdots+\beta_{11}x_{11,k} \quad (10.5)\end{aligned}$$

Here $\mathbf{x}_{ijk}=(x_{2ijk},\ldots,x_{11,k})'$ is a vector containing all covariates, $\zeta_{jk}^{(2)}|\mathbf{x}_{ijk},\zeta_k^{(3)} \sim N(0,\psi^{(2)})$ is a random intercept varying over mothers (level 2), and

$\zeta_k^{(3)}|\mathbf{x}_{ijk} \sim N(0, \psi^{(3)})$ is a random intercept varying over communities (level 3). As usual, the random effects $\zeta_{jk}^{(2)}$ and $\zeta_k^{(3)}$ are assumed independent of each other and across clusters, and $\zeta_{jk}^{(2)}$ is assumed independent across units as well.

The model can alternatively be written as a latent-response model (see sec. 6.2.2)

$$y_{ijk}^* = \beta_1 + \beta_2 x_{2ijk} + \cdots + \beta_{11} x_{11,k} + \zeta_{jk}^{(2)} + \zeta_k^{(3)} + \epsilon_{ijk}$$

where $\epsilon_{ijk}|\mathbf{x}_{ijk}, \zeta_{jk}^{(2)}, \zeta_k^{(3)}$ has a logistic distribution with variance $\pi^2/3$. Analogous to single-level logistic regression, the observed dichotomous responses are then presumed to be generated from the threshold model

$$y_{ijk} = \begin{cases} 1 & \text{if } y_{ijk}^* > 0 \\ 0 & \text{otherwise} \end{cases}$$

10.6.2 Different types of intraclass correlations for the latent responses

We can again consider different types of intraclass correlations for the latent responses of two children y_{ijk}^* and $y_{i'jk}^*$.

For the same community k but different mothers j and j', we obtain

$$\rho(\text{comm.}) \equiv \text{Cor}(y_{ijk}^*, y_{i'j'k}^* \mid \mathbf{x}_{ijk}, \mathbf{x}_{i'j'k}) = \frac{\psi^{(3)}}{\psi^{(2)} + \psi^{(3)} + \pi^2/3}$$

whereas for the same mother j (and then obviously the same community k), we get

$$\rho(\text{mother}, \text{comm.}) \equiv \text{Cor}(y_{ijk}^*, y_{i'jk}^* \mid \mathbf{x}_{ijk}, \mathbf{x}_{i'jk}) = \frac{\psi^{(2)} + \psi^{(3)}}{\psi^{(2)} + \psi^{(3)} + \pi^2/3}$$

In a three-level model, $\psi^{(2)} > 0$ and $\psi^{(3)} > 0$, and it follows that $\rho(\text{mother}, \text{comm.}) > \rho(\text{comm.})$. This makes sense since children of a given mother are more similar than children from the same community but with different mothers.

10.6.3 Different kinds of median odds ratios

As discussed for two-level models in section 6.10.2, we can quantify the unobserved heterogeneity by considering the median odds ratio for pairs of randomly sampled units having the same covariate values, where the unit with the larger random intercept is compared with the unit of the smaller random intercept.

In the three-level model for the Guatemalan data, comparing children of different mothers in the same community, gives the median odds ratio

$$\text{OR}(\text{comm.})_{\text{median}} = \exp\left\{\sqrt{2\psi^{(2)}}\Phi^{-1}(3/4)\right\}$$

and when comparing children of different mothers from different communities we get

$$\text{OR}_{\text{median}} = \exp\left\{\sqrt{2(\psi^{(2)} + \psi^{(3)})}\Phi^{-1}(3/4)\right\}$$

10.6.4 Three-stage formulation

Retaining the distributional assumptions for the random intercepts, we can specify the same model as in (10.5) using a three-stage formulation. Remember that the covariate `kid2p` at the child level was denoted x_{2ijk}. The child-level (level-1) model can then be written as

$$\text{logit}\{\Pr(y_{ijk} = 1|\eta_{1jk}, x_{2ijk})\} = \eta_{1jk} + \beta_2 x_{2ijk}$$

where the intercept η_{1ij} varies between mothers j and communities k. Denoting the seven covariates at the mother level as w_{2jk} to w_{8jk}, the mother-level (level-2) model for the intercept becomes

$$\eta_{1jk} = \pi_{11k} + \pi_{12}w_{2jk} + \cdots + \pi_{18}w_{8jk} + \zeta_{jk}^{(2)}$$

Here only the intercept π_{11k} has a k subscript and therefore requires a community-level (level-3) model

$$\pi_{11k} = \gamma_{111} + \gamma_{112}v_{2k} + \gamma_{113}v_{3k} + \zeta_k^{(3)}$$

where v_{2k} is `rural` and v_{3k} is `pcInd81`, the covariates at level 3.

Substituting the model for π_{11k} into the level-2 model and subsequently for η_{1jk} into the level-1 model, we obtain

$$\begin{aligned}\text{logit}\{\Pr(y_{ijk} = 1|\mathbf{x}_{ijk}, \zeta_{jk}^{(2)}, \zeta_k^{(3)})\} &= \underbrace{\gamma_{111}}_{\beta_1} + \beta_2 x_{2ijk} + \underbrace{\pi_{12}}_{\beta_3}\underbrace{w_{2jk}}_{x_{3jk}} + \cdots + \underbrace{\pi_{18}}_{\beta_9}\underbrace{w_{8jk}}_{x_{9jk}} \\ &\quad + \underbrace{\gamma_{112}}_{\beta_{10}}\underbrace{v_{2k}}_{x_{10,jk}} + \underbrace{\gamma_{113}}_{\beta_{11}}\underbrace{v_{3k}}_{x_{11,jk}} + \zeta_{jk}^{(2)} + \zeta_k^{(3)}\end{aligned}$$

See also exercise 10.1 for another example of the three-stage formulation.

10.7 Estimation of three-level logistic random-intercept models using Stata

Two Stata commands can be used to fit three-level (and higher-level) logistic regression models, `xtmelogit` and `gllamm`. For the applications considered here, `gllamm` is faster, so we start with `gllamm`.

10.7.1 Using gllamm

In `gllamm`, the clustering variables for the different levels in the model are given in the `i()` option, starting from level 2 and then going up the levels. This ordering is the reverse of that used in `xtmixed` (and `xtmelogit`).

When all the random effects are random intercepts, as in the present example, gllamm does not require the eqs() option. The gllamm command for the three-level random-intercept logit model (10.5) is therefore simply

```
. use http://www.stata-press.com/data/mlmus2/guatemala, clear
. gllamm immun kid2p indNoSpa indSpa momEdPri momEdSec husEdPri husEdSec
> husEdDK rural pcInd81, family(binomial) link(logit) i(mom cluster) nip(5)

number of level 1 units = 2159
number of level 2 units = 1595
number of level 3 units = 161

Condition Number = 10.125574

gllamm model

log likelihood = -1328.0727
```

immun	Coef.	Std. Err.	z	P>|z|	[95% Conf.	Interval]
kid2p	1.712282	.2139083	8.00	0.000	1.293029	2.131535
indNoSpa	-.2992919	.4837166	-0.62	0.536	-1.247359	.6487752
indSpa	-.2178983	.361165	-0.60	0.546	-.9257687	.4899722
momEdPri	.3789442	.2154968	1.76	0.079	-.0434219	.8013102
momEdSec	.3836724	.4605474	0.83	0.405	-.5189838	1.286329
husEdPri	.4934885	.2244022	2.20	0.028	.0536682	.9333087
husEdSec	.4466856	.4008267	1.11	0.265	-.3389202	1.232291
husEdDK	-.0079424	.3485074	-0.02	0.982	-.6910043	.6751195
rural	-.8642705	.300585	-2.88	0.004	-1.453406	-.2751347
pcInd81	-1.17417	.4953427	-2.37	0.018	-2.145023	-.2033158
_cons	-1.054729	.4085557	-2.58	0.010	-1.855484	-.2539746

```
Variances and covariances of random effects
------------------------------------------------------------------------------

***level 2 (mom)

    var(1): 5.427267 (1.318504)

***level 3 (cluster)

    var(1): 1.1338842 (.37262627)
------------------------------------------------------------------------------
```

Here we have used only 5 quadrature points because estimation would otherwise be slow for this sample. With as few as 5 points, the version of adaptive quadrature implemented in gllamm is sometimes unstable, so we have used ordinary quadrature by omitting the adapt option.

We now increase the number of quadrature points to the default of 8 per dimension and use adaptive quadrature to get more accurate results. Using the previous estimates as starting values, we get

```
. matrix a = e(b)

. gllamm immun kid2p indNoSpa indSpa  momEdPri momEdSec husEdPri husEdSec
> husEdDK rural pcInd81, family(binomial) link(logit) i(mom cluster)
> from(a) adapt

number of level 1 units = 2159
number of level 2 units = 1595
number of level 3 units = 161

Condition Number = 9.6723284

gllamm model

log likelihood = -1328.496
```

immun	Coef.	Std. Err.	z	P>|z|	[95% Conf.	Interval]
kid2p	1.713088	.2154478	7.95	0.000	1.290818	2.135357
indNoSpa	-.299316	.4778503	-0.63	0.531	-1.235885	.6372535
indSpa	-.1576852	.3568031	-0.44	0.659	-.8570065	.5416361
momEdPri	.3840802	.2170399	1.77	0.077	-.0413101	.8094705
momEdSec	.3616156	.4738778	0.76	0.445	-.5671679	1.290399
husEdPri	.4988694	.2274981	2.19	0.028	.0529813	.9447575
husEdSec	.4377393	.4042692	1.08	0.279	-.3546138	1.230092
husEdDK	-.0091557	.3519029	-0.03	0.979	-.6988728	.6805614
rural	-.8928454	.2990444	-2.99	0.003	-1.478962	-.3067291
pcInd81	-1.154086	.4935178	-2.34	0.019	-2.121363	-.1868085
_cons	-1.027388	.4061945	-2.53	0.011	-1.823514	-.2312612

```
Variances and covariances of random effects
------------------------------------------------------------------------------

***level 2 (mom)

    var(1): 5.1872364 (1.1927247)

***level 3 (cluster)

    var(1): 1.0274006 (.31690328)
------------------------------------------------------------------------------
```

The maximum likelihood estimates have changed somewhat but not drastically. Fitting the model with 12 points per dimension (not shown) gives nearly the same results as for 8 points, indicating that the latter estimates, reported under "Log odds" in table 10.2, are reliable.

To obtain estimated odds ratios with 95% confidence intervals, we can simply "replay" the results with the `eform` option:

```
. gllamm, eform

number of level 1 units = 2159
number of level 2 units = 1595
number of level 3 units = 161

Condition Number = 9.6723282

gllamm model

log likelihood = -1328.496
```

immun	exp(b)	Std. Err.	z	P>|z|	[95% Conf.	Interval]
kid2p	5.546059	1.194886	7.95	0.000	3.635758	8.46007
indNoSpa	.7413251	.3542425	-0.63	0.531	.2905774	1.891279
indSpa	.8541186	.3047522	-0.44	0.659	.4244307	1.718817
momEdPri	1.468263	.3186716	1.77	0.077	.9595316	2.246718
momEdSec	1.435647	.6803213	0.76	0.445	.5671293	3.634237
husEdPri	1.646858	.3746571	2.19	0.028	1.05441	2.57219
husEdSec	1.549201	.6262943	1.08	0.279	.7014443	3.421546
husEdDK	.9908861	.3486957	-0.03	0.979	.4971454	1.974986
rural	.4094889	.1224554	-2.99	0.003	.2278742	.7358499
pcInd81	.3153457	.1556287	-2.34	0.019	.1198682	.8296026

```
Variances and covariances of random effects
------------------------------------------------------------------------------

***level 2 (mom)

    var(1): 5.1872364 (1.1927247)

***level 3 (cluster)

    var(1): 1.0274006 (.31690327)
------------------------------------------------------------------------------
```

We store the estimates for later use

```
. estimates store glri8
```

The estimated odds ratios with confidence intervals are also reported under "Odds ratios" in table 10.2. The explanatory variable of main interest `kid2p` has a large estimated odds ratio of 5.55, adjusted for the other observed covariates and given the random intercepts at the mother and community levels. There is thus some evidence of an effect of the government campaign on child immunization, although it should be emphasized that the study is observational and thus vulnerable to confounding.

Table 10.2: Maximum likelihood estimates for three-level logistic random-intercept model (using 8-point adaptive quadrature in gllamm)

	Log odds $=\beta$		Odds ratios $=\exp(\beta)$	
	Est	(SE)	OR	(95% CI)
Fixed part				
β_1 [_cons]	−1.03	(0.41)		
β_2 [kid2p]	1.71	(0.22)	5.55	(3.64, 8.46)
β_3 [indNoSpa]	−0.30	(0.48)	0.74	(0.29, 1.89)
β_4 [indSpa]	−0.16	(0.36)	0.85	(0.42, 1.72)
β_5 [momEdPri]	0.38	(0.22)	1.47	(0.96, 2.25)
β_6 [momEdSec]	0.36	(0.47)	1.44	(0.57, 3.63)
β_7 [husEdPri]	0.50	(0.23)	1.65	(1.05, 2.57)
β_8 [husEdSec]	0.44	(0.40)	1.55	(0.70, 3.42)
β_9 [husEdDK]	−0.01	(0.35)	0.99	(0.50, 1.97)
β_{10} [rural]	−0.90	(0.30)	0.41	(0.23, 0.74)
β_{11} [pcInd81]	−1.15	(0.49)	0.32	(0.12, 0.83)
Random part				
$\psi^{(2)}$	5.19			
$\psi^{(3)}$	1.03			
Log likelihood		−1328.50		

To interpret the random part of the model, it is instructive to consider the estimated residual intraclass correlations for the latent responses. For children from the same community, we obtain $\widehat{\rho}(\text{comm.}) = 0.11$ if they have different mothers and $\widehat{\rho}(\text{mother, comm.}) = 0.65$ if they have the same mother.

Comparing children of different mothers from different communities, we obtain the estimated median odds ratio

```
. display exp(sqrt(2*(1.03+5.19))*invnormal(3/4))
10.793577
```

and comparing children of different mothers from the same community, we obtain the estimated median odds ratio

```
. display exp(sqrt(2*5.19)*invnormal(3/4))
8.7852271
```

10.7.2 Using xtmelogit

In xtmelogit, the clustering variables for the different levels in the model are given starting from the top level and then going down the levels, just as in xtmixed. In the current three-level application, we first specify random intercepts for communities using || cluster: and then random intercepts for mothers using || mom:.

Since estimation of this model is time consuming, we start by using the computationally efficient Laplace method, which is obtained in `xtmelogit` by specifying the `intpoints(1)` or `laplace` option (even this command will take a long time):

```
. xtmelogit immun kid2p indNoSpa indSpa momEdPri momEdSec husEdPri husEdSec
> husEdDK rural pcInd81 || cluster: || mom:, intpoints(1)
Mixed-effects logistic regression               Number of obs      =      2159

--------------------------------------------------------------------------
                |   No. of       Observations per Group       Integration
 Group Variable |   Groups    Minimum    Average    Maximum      Points
----------------+---------------------------------------------------------
        cluster |      161          1       13.4         55           1
            mom |     1595          1        1.4          3           1
--------------------------------------------------------------------------

                                                Wald chi2(10)      =     93.95
Log likelihood = -1359.709                      Prob > chi2        =    0.0000

------------------------------------------------------------------------------
       immun |      Coef.   Std. Err.      z    P>|z|     [95% Conf. Interval]
-------------+----------------------------------------------------------------
       kid2p |    1.28234    .158357     8.10   0.000     .9719658    1.592714
    indNoSpa |  -.2017538   .3312773    -0.61   0.543    -.8510454    .4475377
      indSpa |  -.0893714   .2481248    -0.36   0.719    -.5756871    .3969443
    momEdPri |    .262748   .1495222     1.76   0.079    -.0303101    .5558061
    momEdSec |   .2554452   .3288252     0.78   0.437    -.3890404    .8999307
    husEdPri |   .3687133   .1566719     2.35   0.019     .0616421    .6757846
    husEdSec |   .3216275   .2795014     1.15   0.250    -.2261852    .8694402
     husEdDK |   .0096085   .2430354     0.04   0.968    -.4667321    .4859491
       rural |  -.6601202   .2075705    -3.18   0.001    -1.066951   -.2532896
     pcInd81 |  -.8580104   .3431572    -2.50   0.012    -1.530586   -.1854347
       _cons |  -.7624847   .2861215    -2.66   0.008    -1.323272   -.2016969
------------------------------------------------------------------------------

------------------------------------------------------------------------------
  Random-effects Parameters  |   Estimate   Std. Err.     [95% Conf. Interval]
-----------------------------+------------------------------------------------
cluster: Identity            |
                   sd(_cons) |   .7207785   .1005431      .5483599    .9474101
-----------------------------+------------------------------------------------
mom: Identity                |
                   sd(_cons) |   1.111227   .1579589      .8410191    1.468248
------------------------------------------------------------------------------
LR test vs. logistic regression:     chi2(2) =    88.72   Prob > chi2 = 0.0000
Note: LR test is conservative and provided only for reference.
Note: log-likelihood calculations are based on the Laplacian approximation.
```

We now perform maximum likelihood estimation using adaptive quadrature with 5 integration points in `xtmelogit` by specifying the `intpoints(5)` option. We can save time by using the Laplace estimates as starting values instead of letting `xtmelogit` compute its own starting values. This is accomplished by using the `from()` option to pass the starting values to `xtmelogit` and the `refineopts(iterate(0))` option to specify 0 iterations for "refining" the starting values

```
. matrix a = e(b)

. xtmelogit immun kid2p indNoSpa indSpa momEdPri momEdSec husEdPri husEdSec
> husEdDK rural pcInd81 || cluster: || mom:, intpoints(5) from(a)
> refineopts(iterate(0))
Mixed-effects logistic regression               Number of obs      =      2159

--------------------------------------------------------------------------
                |   No. of       Observations per Group       Integration
 Group Variable |   Groups    Minimum    Average    Maximum      Points
----------------+---------------------------------------------------------
        cluster |      161          1       13.4         55           5
            mom |     1595          1        1.4          3           5
--------------------------------------------------------------------------

                                                Wald chi2(10)      =     91.18
Log likelihood = -1329.993                      Prob > chi2        =    0.0000

------------------------------------------------------------------------------
       immun |      Coef.   Std. Err.      z    P>|z|     [95% Conf. Interval]
-------------+----------------------------------------------------------------
       kid2p |   1.640996   .2003109     8.19   0.000     1.248394    2.033598
    indNoSpa |  -.2930592   .4533645    -0.65   0.518    -1.181637    .5955188
      indSpa |  -.1521804   .3383809    -0.45   0.653    -.8153947    .5110339
    momEdPri |   .3615499   .2047625     1.77   0.077    -.0397773     .762877
    momEdSec |   .3411993   .4484828     0.76   0.447    -.5378109     1.22021
    husEdPri |   .4748191   .2145763     2.21   0.027     .0542574    .8953809
    husEdSec |   .4164453   .3824737     1.09   0.276    -.3331894     1.16608
     husEdDK |  -.0073808   .3328711    -0.02   0.982    -.6597963    .6450346
       rural |  -.8526129   .2837777    -3.00   0.003    -1.408807   -.2964188
     pcInd81 |  -1.098836   .4679618    -2.35   0.019    -2.016024   -.1816473
       _cons |  -.9821477    .385403    -2.55   0.011    -1.737524   -.2267718
------------------------------------------------------------------------------

------------------------------------------------------------------------------
  Random-effects Parameters  |   Estimate   Std. Err.     [95% Conf. Interval]
-----------------------------+------------------------------------------------
cluster: Identity            |
                   sd(_cons) |   .9768667   .1458175      .7290813    1.308864
-----------------------------+------------------------------------------------
mom: Identity                |
                   sd(_cons) |   2.093691   .2102443      1.719636     2.54911
------------------------------------------------------------------------------
LR test vs. logistic regression:     chi2(2) =   148.15   Prob > chi2 = 0.0000

Note: LR test is conservative and provided only for reference.
```

The estimated standard deviation $\sqrt{\widehat{\psi}^{(2)}}$ of the mother-level random intercept and the estimated effects of the covariates are substantially larger than the estimates obtained from the Laplace method. It is evident that the rather crude Laplace method should be used with considerable caution.

However, the 5-point estimates using `xtmelogit` differ from the 8-point solution using `gllamm`, so it appears that more integration points are required. We therefore use 8-point adaptive quadrature with the 5-point estimates as starting values:

```
. matrix a = e(b)
. xtmelogit immun kid2p indNoSpa indSpa momEdPri momEdSec husEdPri husEdSec
> husEdDK rural pcInd81 || cluster: || mom:, intpoints(8) from(a)
> refineopts(iterate(0))
Mixed-effects logistic regression               Number of obs      =      2159

--------------------------------------------------------------------------
                |   No. of       Observations per Group       Integration
 Group Variable |   Groups    Minimum    Average    Maximum      Points
----------------+---------------------------------------------------------
        cluster |      161          1       13.4         55           8
            mom |     1595          1        1.4          3           8
--------------------------------------------------------------------------

                                                Wald chi2(10)      =     82.76
Log likelihood =  -1328.437                     Prob > chi2        =    0.0000

------------------------------------------------------------------------------
       immun |      Coef.   Std. Err.      z    P>|z|     [95% Conf. Interval]
-------------+----------------------------------------------------------------
       kid2p |   1.715209   .2156215     7.95   0.000     1.292599     2.13782
    indNoSpa |  -.3004943   .4779276    -0.63   0.530    -1.237215    .6362266
      indSpa |  -.1583792   .3573175    -0.44   0.658    -.8587087    .5419502
    momEdPri |   .3851238    .217334     1.77   0.076    -.0408429    .8110905
    momEdSec |   .3623701   .4742333     0.76   0.445    -.5671101     1.29185
    husEdPri |   .4999468     .22774     2.20   0.028     .0535845    .9463091
    husEdSec |   .4393951    .404856     1.09   0.278    -.3541081    1.232898
     husEdDK |  -.0091831   .3521293    -0.03   0.979    -.6993439    .6809777
       rural |  -.8959985    .300119    -2.99   0.003    -1.484221   -.3077761
     pcInd81 |   -1.15788   .4947361    -2.34   0.019    -2.127545   -.1882149
       _cons |  -1.027295   .4065235    -2.53   0.012    -1.824066   -.2305237
------------------------------------------------------------------------------

------------------------------------------------------------------------------
  Random-effects Parameters  |   Estimate   Std. Err.     [95% Conf. Interval]
-----------------------------+------------------------------------------------
cluster: Identity            |
                   sd(_cons) |   1.015413   .1566953      .7503913    1.374034
-----------------------------+------------------------------------------------
mom: Identity                |
                   sd(_cons) |   2.281915   .2604568      1.824499    2.854008
------------------------------------------------------------------------------
LR test vs. logistic regression:     chi2(2) =   151.27   Prob > chi2 = 0.0000
Note: LR test is conservative and provided only for reference.
```

These estimates are close to the `gllamm` estimates.

10.8 A three-level logistic random-coefficient model

We have already seen that the communities vary in their overall levels of immunization, both due to fixed effects and random effects. It would be interesting to investigate whether the effect of the campaign also varies randomly between communities. This can be achieved by including a random slope $\zeta_{2k}^{(3)}$ of `kid2p` (x_{2ijk}) at level 3. To keep the model simple we retain only `kid2p` and the community-level covariates:

$$\begin{aligned}\text{logit}\{\Pr(y_{ijk}=1|\mathbf{x}_{ijk},\zeta_{jk}^{(2)},\zeta_{1k}^{(3)},\zeta_{2k}^{(3)})\} &= \beta_1+\beta_2 x_{2ijk}+\beta_{10}x_{10,ijk}+\beta_{11}x_{11,ijk}\\ &\quad +\zeta_{jk}^{(2)}+\zeta_{1k}^{(3)}+\zeta_{2k}^{(3)}x_{2ijk}\\ &= (\beta_1+\zeta_{jk}^{(2)}+\zeta_{1k}^{(3)})+(\beta_2+\zeta_{2k}^{(3)})x_{2ijk}\\ &\quad +\beta_{10}x_{10,ijk}+\beta_{11}x_{11,ijk} \qquad (10.6)\end{aligned}$$

where $\zeta_{1k}^{(3)},\zeta_{2k}^{(3)}|\mathbf{x}_{ijk}$ have a bivariate normal distribution with zero means and covariance matrix

$$\boldsymbol{\Psi}^{(3)} = \begin{bmatrix}\psi_{11}^{(3)} & \psi_{12}^{(3)}\\ \psi_{21}^{(3)} & \psi_{22}^{(3)}\end{bmatrix}, \qquad \psi_{21}^{(3)}=\psi_{12}^{(3)} \qquad (10.7)$$

It is unusual to include a random slope for a given variable at a higher level and not at the lower level(s). Here we have done so because the "treatment" was applied at the community level, and we believe that its effect will vary more between communities than between mothers within communities. Furthermore, individual mothers do not provide much information on the mother-specific effect of `kid2p` because they do not have many children of each type.

10.9 Estimation of three-level logistic random-coefficient models using Stata

10.9.1 Using gllamm

We first fit the random-intercept model

$$\text{logit}\{\Pr(y_{ijk}=1|\mathbf{x}_{ijk},\zeta_{jk}^{(2)},\zeta_{k}^{(3)})\} = \beta_1+\beta_2 x_{2ijk}+\beta_{10}x_{10,k}+\beta_{11}x_{11,k}+\zeta_{jk}^{(2)}+\zeta_{k}^{(3)}$$

using the previous estimates for the three-level random-intercept model with the full set of covariates as starting values. These estimates are retrieved using

```
. estimates restore glri8
```

We then copy the starting values into the matrix `a` and pass that to `gllamm` using the `from()` option. We must also specify the `skip` option, because `a` contains extra parameters (the regression coefficients of the covariates we have dropped):

```
. matrix a = e(b)

. gllamm immun kid2p rural pcInd81, family(binomial) link(logit)
> i(mom cluster) from(a) skip adapt eform

number of level 1 units = 2159
number of level 2 units = 1595
number of level 3 units = 161

Condition Number = 5.4620658

gllamm model

log likelihood = -1335.0426

------------------------------------------------------------------------------
       immun |     exp(b)   Std. Err.      z    P>|z|     [95% Conf. Interval]
-------------+----------------------------------------------------------------
       kid2p |   5.369864   1.151166     7.84   0.000     3.527661    8.174097
       rural |   .3450857   .0979057    -3.75   0.000     .1978923    .6017623
     pcInd81 |   .1905651   .0678821    -4.65   0.000     .0948053    .3830485
------------------------------------------------------------------------------

Variances and covariances of random effects
------------------------------------------------------------------------------

***level 2 (mom)

    var(1): 5.213685 (1.2063283)

***level 3 (cluster)

    var(1): 1.0333637 (.31347077)
------------------------------------------------------------------------------
```

We store the estimates for later use

```
. estimates store glri0
```

The maximum likelihood estimates are shown under "Random intercept" in table 10.3. The estimate of the odds ratio for `kid2p` has not changed considerably compared with the estimate for the full model in table 10.2, suggesting that discarding the level-2 covariates does not seriously distort the estimate.

We then turn to estimation of the model with a random coefficient for `kid2p` at the community level, as well as random intercepts at both the community and mother levels. As in chapter 4, we specify equations for the intercept(s) and slope:

```
. generate cons = 1

. eq inter: cons

. eq slope: kid2p
```

The new model has two extra parameters: $\psi_{22}^{(3)}$ and $\psi_{21}^{(3)}$. We can therefore use the matrix containing the previous estimates as starting values if we add two more values

```
. matrix a = e(b)
. matrix a = (a,.2,0)
```

As in two-level random-coefficient models, we must use the `nrf()` option to specify the number of random effects. However, in the three-level case, we must specify two numbers of random effects, `nrf(1 2)`, for levels 2 and 3. In the `eqs()` option, we must then specify one equation for level 2 and two equations for level 3. To speed up estimation, we reduce the number of quadrature points per dimension for the two random effects at level 3 from the default of 8 to 4 using the `nip()` option:

```
. gllamm immun kid2p rural pcInd81, family(binomial) link(logit) i(mom cluster)
> nrf(1 2) eqs(inter inter slope) nip(8 4 4) from(a) copy adapt eform

number of level 1 units = 2159
number of level 2 units = 1595
number of level 3 units = 161

Condition Number = 7.1155372

gllamm model

log likelihood = -1330.8285
```

immun	exp(b)	Std. Err.	z	P>|z|	[95% Conf.	Interval]
kid2p	6.729718	1.975344	6.50	0.000	3.785711	11.96317
rural	.3296138	.0989118	-3.70	0.000	.1830516	.5935224
pcInd81	.1769139	.0667759	-4.59	0.000	.0844261	.3707207

```
Variances and covariances of random effects
------------------------------------------------------------------------------

***level 2 (mom)

    var(1): 5.8320415 (1.4186368)

***level 3 (cluster)

    var(1): 2.4200309 (1.0925729)
    cov(2,1): -1.5234084 (.94663252) cor(2,1): -.72984351

    var(2): 1.8003307 (.98477222)
------------------------------------------------------------------------------
. estimates store glrc
```

The estimated fixed-effects and variance at level 2 have not changed considerably compared with the three-level random-intercept model. At level 3, the estimated elements of the covariance matrix of the random intercept and slope are given under `***level 3 (cluster)`. The random-intercept variance, now interpretable as the residual between-community variance for children who were too young to be immunized at the time of the campaign (`kid2p`=0), has increased considerably to $\widehat{\psi}_{11}^{(3)} = 2.42$. The variance of the slope of `kid2p` can be interpreted as the residual variability in the effectiveness of the campaign across communities and is estimated as $\widehat{\psi}_{22}^{(3)} = 1.80$. The estimated correlation between the random intercepts and slopes is

$$\frac{\widehat{\psi}_{21}^{(3)}}{\sqrt{\widehat{\psi}_{11}^{(3)}\widehat{\psi}_{22}^{(3)}}} = -0.73$$

which suggests that, for given covariate values, the immunization campaign was less effective in communities where immunization rates are high for children who were too young to be immunized during the campaign (`kid2p`=0).

The random-intercept model, which is nested in the random-coefficient model, is rejected at the 5% significance level (using a conservative likelihood-ratio test):

```
. lrtest glrc glri0
Likelihood-ratio test                                 LR chi2(2)  =      8.43
(Assumption: glri0 nested in glrc)                    Prob > chi2 =    0.0148
```

The estimates are shown in table 10.3 under "Random coefficient". Practically the same estimates (not shown) are obtained with 8 quadrature points per dimension.

As a next step, we could include cross-level interactions between `kid2p` and the community-level covariates to try to explain the variability in the effect of `kid2p` between communities but will not pursue this here.

Table 10.3: Maximum likelihood estimates from `gllamm` for three-level logistic random-intercept and random-coefficient models

	Random intercept		Random coefficient	
	Est	(95% CI)	Est	(95% CI)
Fixed part: odds ratios				
$\exp(\beta_2)$ [`kid2p`]	5.37	(3.53, 8.17)	6.73	(3.79, 11.96)
$\exp(\beta_{10})$ [`rural`]	0.35	(0.20, 0.60)	0.33	(0.18, 0.59)
$\exp(\beta_{11})$ [`pcInd81`]	0.19	(0.09, 0.38)	0.18	(0.08, 0.37)
Random part				
$\psi^{(2)}$	5.21		5.83	
$\psi_{11}^{(3)}$	1.03		2.42	
$\psi_{22}^{(3)}$			1.80	
$\psi_{21}^{(3)}$			−1.52	
Log likelihood	−1,335.04		−1,330.83	

10.9.2 Using xtmelogit

We can fit the random-coefficient model using the following `xtmelogit` command:

```
. xtmelogit immun kid2p rural pcInd81 || cluster: kid2p, cov(unstructured)
> || mom:, intpoints(4 8) or
Mixed-effects logistic regression               Number of obs      =      2159

----------------------------------------------------------------------------
                |   No. of       Observations per Group       Integration
 Group Variable |   Groups    Minimum    Average    Maximum      Points
----------------+-----------------------------------------------------------
        cluster |      161          1       13.4         55          4
            mom |     1595          1        1.4          3          8
----------------------------------------------------------------------------

                                                Wald chi2(3)       =     60.78
Log likelihood = -1330.7479                     Prob > chi2        =    0.0000

------------------------------------------------------------------------------
       immun | Odds Ratio   Std. Err.      z    P>|z|     [95% Conf. Interval]
-------------+----------------------------------------------------------------
       kid2p |   6.758305     1.98633     6.50   0.000     3.798932    12.02303
       rural |   .3276506    .0988741    -3.70   0.000     .1813621    .5919369
     pcInd81 |   .1751773    .0666158    -4.58   0.000     .0831354    .3691217
------------------------------------------------------------------------------

------------------------------------------------------------------------------
  Random-effects Parameters  |   Estimate   Std. Err.     [95% Conf. Interval]
-----------------------------+------------------------------------------------
cluster: Unstructured        |
                   sd(kid2p) |   1.342272   .3678773       .784425    2.296835
                   sd(_cons) |   1.555729   .3501361       1.00083    2.418284
           corr(kid2p,_cons) |  -.7284218   .1377932     -.9052757   -.3363829
-----------------------------+------------------------------------------------
mom: Identity                |
                   sd(_cons) |   2.424341   .2929182      1.913147    3.072127
------------------------------------------------------------------------------
LR test vs. logistic regression:     chi2(4) =   163.69   Prob > chi2 = 0.0000
Note: LR test is conservative and provided only for reference.
```

We store the estimates

```
. estimates store xtrc
```

In the random part for the community level, specified after the first `||`, the syntax `cluster: kid2p` is used to include a random slope for `kid2p`, and a random intercept is included by default. The `cov(unstructured)` option is used to estimate the covariance matrix in (10.7) without any restrictions. Without this option, the covariance is set to zero, which is rarely recommended in practice (see sec. 4.4.2). The random part for the mother level is specified after the second `||`, where `mom:` means that there should be only a random intercept. We specified the `intpoints(4 8)` option to use 4 integration points for each random effect at level 3 and 8 integration points at level 2.

10.10 Prediction of random effects

10.10.1 Empirical Bayes prediction

At the time of writing this book, `gllapred` is the only command that can produce empirical Bayes predictions of the random effects for multilevel logistic models. After retrieving the `gllamm` estimates for the three-level random-coefficient model, we use `gllapred` with the `u` option:

```
. estimates restore glrc
. gllapred zeta, u
(means and standard deviations will be stored in zetam1 zetas1 zetam2 zetas2
> zetam3 zetas3)
. sort cluster mom
. list cluster mom zetam1 zetam2 zetam3 in 1/9, clean noobs
   cluster   mom      zetam1      zetam2      zetam3
         1     2   1.0097228    .15487673    .04783502
        36   185  -2.0246433   -.42102637   -.01032988
        36   186  -2.0246433   -.42102637   -.01032988
        36   187  -2.0246433   -.42102637   -.01032988
        36   188  -.69007018   -.42102637   -.01032988
        36   188  -.69007018   -.42102637   -.01032988
        36   189   1.1806865   -.42102637   -.01032988
        36   190   2.2574648   -.42102637   -.01032988
        36   190   2.2574648   -.42102637   -.01032988
```

Here `zetam1` is the predicted random intercept $\widetilde{\zeta}_{jk}^{(2)}$ for mothers, `zetam2` is the predicted random intercept $\widetilde{\zeta}_{1k}^{(3)}$ for communities, and `zetam3` is the predicted random slope $\widetilde{\zeta}_{2k}^{(3)}$ for communities. The order of these random effects is the same as in the `eqs()` option and always from the lowest to the highest level.

The random intercept for mothers can vary between mothers but is constant across multiple children of the same mother (mothers 188 and 190). Mothers 185, 186, and 187 all have the same predictions because they have the same covariate values and responses. We can plot the predicted random intercepts $\widetilde{\zeta}_{1k}^{(3)}$ and slopes $\widetilde{\zeta}_{2k}^{(3)}$ for the communities using

```
. egen pick_com = tag(cluster)
. twoway scatter zetam3 zetam2 if pick_com==1, xtitle(Intercept) ytitle(Slope)
```

The resulting graph is shown in figure 10.4.

(*Continued on next page*)

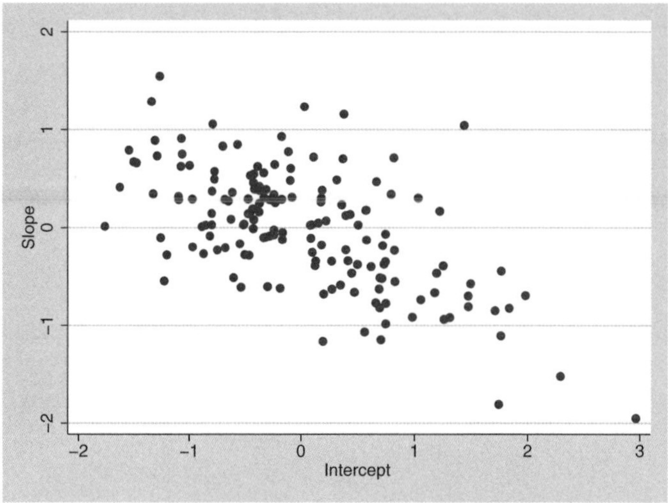

Figure 10.4: Predicted community-level random slopes versus random intercepts

10.10.2 Empirical Bayes modal prediction

The `predict` command for `xtmelogit` provides the modes of the posterior distributions instead of the means:

```
. estimates restore xtrc
. predict comms commi mother, reffects
```

Here we have chosen variable names `comms` and `commi` for the slopes and intercepts at the community level, respectively, and `mother` for the mother-level random intercepts, to remind us which random effects have been stored in which variable (the `predict` command also produces informative variable labels). `xtmelogit` starts from the highest level, and the random intercept always comes last, exactly as in the output.

We list the empirical Bayes modal predictions using

```
. sort cluster mom
. list cluster mom mother commi comms in 1/9, clean noobs

  cluster   mom      mother       commi       comms
        1     2     .634367    .0970523    .0302855
       36   185   -1.786311   -.4229506   -.0808257
       36   186   -1.786311   -.4229506   -.0808257
       36   187   -1.786311   -.4229506   -.0808257
       36   188   -.7136106   -.4229506   -.0808257
       36   188   -.7136106   -.4229506   -.0808257
       36   189    .8373249   -.4229506   -.0808257
       36   190    1.886907   -.4229506   -.0808257
       36   190    1.886907   -.4229506   -.0808257
```

These predictions are similar but not identical to the empirical Bayes counterparts listed (in the same order) earlier.

10.11 Different kinds of predicted probabilities

10.11.1 Predicted marginal probabilities

At the time of writing this book, marginal or population-averaged probabilities can be predicted for multilevel logistic regression models only by using `gllapred` after estimation using `gllamm`. Here we have to integrate out all three random effects (the random intercept at level 2 and that random intercept and slope at level 3). The command is the same as for two-level models discussed in section 6.13.1:

```
. estimates restore glrc
. gllapred margp, mu marginal
(mu will be stored in margp)
```

10.11.2 Predicted median or conditional probabilities

We can obtain median probabilities by setting the random effects equal to their median values of zero. These probabilities are just conditional probabilities, given that the random effects are zero:

```
. generate z1 = 0
. generate z2 = 0
. generate z3 = 0
. gllapred condp, mu us(z)
(mu will be stored in condp)
```

Plotting both these conditional probabilities and the marginal probabilities from the previous section using

```
. label define r 0 "Urban" 1 "Rural"
. label values rural r
. twoway (line condp pcInd81 if kid2p==0, lpatt(solid) sort)
>        (line condp pcInd81 if kid2p==1, lpatt(solid) sort)
>        (line margp pcInd81 if kid2p==0, lpatt(dash) sort)
>        (line margp pcInd81 if kid2p==1, lpatt(dash) sort),
>         by(rural) legend(order(1 "Conditional" 3 "Marginal"))
> xtitle(Percentage Indigenous) ytitle(Probability)
```

we obtain the graph in figure 10.5. It is clear that the marginal effects of both percentage indigenous (slopes of dashed curves) and the child being eligible for vaccination during the campaign (vertical distances between dashed curves) are attenuated compared with the conditional effects (solid curves) as usual.

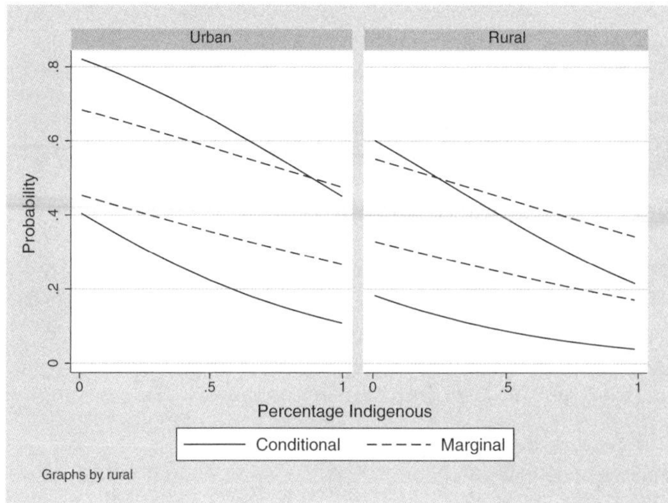

Figure 10.5: Predicted median or conditional probabilities with random effects set to zero (solid curves) and marginal probabilities (dashed curves). Curves higher up in each graph correspond to `kid2p` = 1 and curves lower down to `kid2p` = 0.

10.11.3 Predicted posterior mean probabilities

To obtain predicted probabilities for a given child based on the covariates and responses, we integrate the conditional probability, conditional on the random effects, over the posterior distribution of the random effects, as shown for a two-level model in (9.13). The `gllapred` command is also the same as for two-level models:

```
. gllapred postp, mu
(mu will be stored in postp)
```

We now list these predictions together with the empirical Bayes predictions of the random effects:

```
. sort cluster mom

. list cluster mom kid2p zetam1 zetam2 zetam3 postp in 1/9, clean noobs

  cluster   mom   kid2p      zetam1       zetam2       zetam3       postp
        1     2       1   1.0097228    .15487673    .04783502   .82746602
       36   185       1  -2.0246433   -.42102637   -.01032988   .34772187
       36   186       1  -2.0246433   -.42102637   -.01032988   .34772187
       36   187       1  -2.0246433   -.42102637   -.01032988   .34772187
       36   188       1  -.69007018   -.42102637   -.01032988   .55892132
       36   188       1  -.69007018   -.42102637   -.01032988   .55892132
       36   189       1   1.1806865   -.42102637   -.01032988   .79839256
       36   190       0   2.2574648   -.42102637   -.01032988   .70573537
       36   190       1   2.2574648   -.42102637   -.01032988   .90750043
```

For community 36, all covariates are constant except that `kid2p` changes from 0 to 1 for mother 190 (the other covariates are community specific). Mothers 185, 186, and 187 have a low predicted random intercept of −2.02 and correspondingly a low predicted probability of 0.35, whereas mother 190 has a large predicted random intercept of 2.26 and a correspondingly large predicted probability of 0.91 for the child with `kid2p` = 1. For the other child with `kid2p` = 0, the predicted probability of 0.71 is somewhat lower as expected.

10.12 Summary and further reading

We have introduced the idea of nested random effects for three-level datasets where there are two nested levels of clustering. The models allow the total variance to be attributed not just to within-cluster and between-cluster variability but also to between-cluster variability at the different levels. We discussed several intraclass correlations for this situation. All these ideas extend naturally to higher-level models (see exercises 10.11 and 10.4 for four-level examples). In this chapter, we considered applications with continuous and dichotomous responses. We refer to the exercises for examples with ordinal responses, counts, and discrete-time survival.

There are some differences between the terminology and notation used in this book and the Stata documentation and programs. Consider what we call a three-level dataset with students nested in classes nested in schools. Using the usual multilevel or hierarchical modeling conventions, we use the indices i for students, j for classes, and k for schools and call the corresponding levels 1, 2, and 3. In contrast, the `xtmixed`, `xtmelogit`, and `xtmepoisson` documentation uses i for schools, j for classes, presumably k for students and calls the first two levels 1 and 2. Contrary to common terminology, where multilevel models for such data are called three-level models, the Stata documentation calls these models two-level models because the elementary units are not considered a level.

Good books on linear multilevel models with several levels of nested random effects include Raudenbush and Bryk (2002), Goldstein (2003), and Snijders and Bosker (1999). A review of linear two- and three-level models with an application to political science is given by Steenbergen and Jones (2002). Bryk and Raudenbush (1988) discuss two- and three-level linear models in educational research.

10.13 Exercises

10.1 Math-achievement data

Raudenbush and Bryk (2002) and Raudenbush et al. (2004) discuss data from a longitudinal study of children's academic growth in the six primary school years. The data have a three-level structure with repeated observations on 1,721 students from 60 urban public primary schools.

The dataset `achievement.dta` has the following variables:

- Level 1 (occasion)
  - `math`: math-test score derived from an item response model
  - `year`: year of study minus 3.5 (values −2.5, −1.5, −0.5, 0.5, 1.5, 2.5) (a_{1ijk})
  - `retained`: indicator for child being retained in grade (1: retained; 0: not retained)
- Level 2 (child)
  - `child`: child identifier
  - `female`: dummy variable for being female
  - `black`: dummy variable for being African American (X_{1jk})
  - `hispanic`: dummy variable for being Hispanic (X_{2jk})
- Level 3 (school)
  - `school`: school identifier
  - `size`: number of students enrolled in the school
  - `lowinc`: percentage of students from low-income families (W_{1k})
  - `mobility`: percentage of students moving during the course of a school year

Raudenbush et al. (2004) specify a three-level model in three stages. The level-1 model is

$$Y_{ijk} \;=\; \pi_{0jk} + \pi_{1jk}a_{1ijk} + e_{ijk}, \quad e_{ijk} \sim N(0,\sigma^2)$$

where Y_{ijk} is the jth child's math achievement at occasion i in school k and a_{1ijk} is `year` at that occasion. The level-2 models are

$$\pi_{pjk} \;=\; \beta_{p0k} + \beta_{p1}X_{1jk} + \beta_{p2}X_{2jk} + r_{pjk}, \quad p = 0,1 \quad (r_{0jk}, r_{1jk})' \sim N(\mathbf{0}, \mathbf{T}_\pi)$$

where X_{1jk} is `black`, X_{2jk} is `hispanic`, and r_{pjk} is a random effect (intercept if $p = 0$, slope if $p = 1$) at level 2. The covariance matrix of the level-2 random effects is defined as

$$\mathbf{T}_\pi = \begin{bmatrix} \tau_{\pi 00} & \tau_{\pi 01} \\ \tau_{\pi 10} & \tau_{\pi 11} \end{bmatrix}, \quad \tau_{\pi 10} = \tau_{\pi 01}$$

Finally, the level-3 model is

$$\beta_{p0k} \;=\; \gamma_{p00} + \gamma_{p01}W_{1k} + u_{p0k}, \quad p = 0,1 \quad u_{p0k} \sim N(0, T_\beta)$$

where W_{1k} is `lowinc` and u_{p0k} is a random intercept at level 3.

1. Substitute the level-3 models into the level-2 models and then the resulting level-2 models into the level-1 model. Rewrite the final reduced-form model using the notation of this book.
2. Fit the model using `xtmixed`, and interpret the estimates.
3. Include some of the other covariates in the model, and interpret the estimates.

10.2 Instructional improvement data

West, Welch, and Galecki (2007) analyzed a dataset on first- and third-grade students from the Study of Instructional Improvement (SII) by Hill, Rowan, and Ball (2005).

The question of interest is how teachers' experience, mathematics preparation, and mathematics content knowledge affect students' gain in mathematics achievement scores from kindergarten to first grade.

The variables in the dataset `instruction.dta` used here are

- Level 1 (students)
  - `childid`: student identifier
  - `mathgain`: gain in math achievement score from spring of kindergarten to spring of first grade
  - `mathkind`: match achievement score in the spring of the kindergarten year
  - `girl`: dummy variable for being a girl
  - `minority`: dummy variable for being a minority student
  - `ses`: socioeconomic status
- Level 2 (class)
  - `classid`: class identifier
  - `yearstea`: first-grade teacher's years of teaching experience
  - `mathprep`: first-grade teacher's mathematics preparation (score based on number of mathematics content and methods courses)
  - `mathknow`: first-grade teacher's mathematics content knowledge, based on a 30-item scale with higher values indicating better knowledge
- Level 3 (school)
  - `schoolid`: school identifier
  - `housepov`: percentage of households in the neighborhood of the school below the poverty level

1. Treating `mathgain` as the response variable, write down the "unconditional" (without covariates) three-level random-intercept model for children nested in classes nested in schools, using
   a. A three-stage formulation
   b. The reduced form
2. Fit the model in Stata, obtain the estimated intraclass correlations and interpret them.
3. For the three-stage formulation in step 1, write down an extended level-1 model that includes the four student-level variables, `mathkind`–`ses`, as covariates. Similarly, write down an extended level-2 model by including the three class-level covariates, and an extended level-3 model by including the school-level covariate.

4. Fit the model from step 3. Interpret the estimated coefficients of the class-level covariates.
5. Fit an extended model that also includes the school means of `ses`, `minority`, and `mathkind`. Is there evidence that the within-school effects of these variables differ from the between-school effects?
6. Obtain empirical Bayes predictions of the random effects and produce graphs to assess their normality. Do the normality assumptions appear to be justified?

10.3 U.S. production data

A Cobb–Douglas production function expresses production P as a log-linear model of input, such as capital K and labor L

$$P_i = AK_i^{\beta_2}L_i^{\beta_3}e^{\epsilon_i}$$

so that

$$\ln(P_i) = \ln(A) + \beta_2\ln(K_i) + \beta_3\ln(L_i) + \epsilon_i$$

Thus after taking logarithms of the output and all input variables, the production function can be estimated using linear regression.

Baltagi, Song, and Jung (2001) analyzed data from Munnell (1990) on state productivity for 48 U.S. states from 9 regions over the period 1970–1986. They estimate a Cobb–Douglas production function with error components for region and state.

The variables in `productivity.dta` are

- `state`: state identifier
- `region`: region identifier
- `year`: year 1970–1986
- `private`: logarithm of private capital stock
- `hwy`: logarithm of highway component of public stock
- `water`: logarithm of water component of public stock
- `other`: logarithm of building and other components of public stock
- `unemp`: state unemployment rate

1. Fit a three-level model for the logarithm of private capital stock, `private`, with covariates `hwy`, `water`, `other`, and `unemp` and with random intercepts for `state` and `region`. Use `xtmixed` both with the `mle` and `reml` options.
2. Which components of public capital have a positive effect on private output?
3. Interpret the sizes of the estimated variance components. Also comment on any differences between the maximum likelihood and restricted maximum likelihood estimates.

See also exercise 11.3 in chapter 11.

10.4 Dairy-cow data I

Dohoo et al. (2001) and Dohoo, Martin, and Stryhn (2003) analyzed data on dairy cows nested in herds and regions of Reunion Island. One outcome considered was the time interval between calving (giving birth to a calf) and first service (attempt to inseminate the cow again). This outcome was available for several lactations (calvings) per cow.

The variables in the dataset `dairy.dta` used here and in the next exercise are

- `cow`: cow identifier
- `herd`: herd identifier
- `region`: geographic region
- `lncfs`: log of calving to first service interval (in log days)
- `fscr`: first service conception risk (dummy variable for cow becoming pregnant)
- `ai`: dummy variable for artificial insemination being used (versus natural) at first service
- `heifer`: dummy variable for being a young cow that has given birth only once

1. Fit a four-level random-intercept model with `lncfs` as the response variable and with random intercepts for cows, herds, and geographic regions. Do not include any covariates. Use restricted maximum likelihood (REML) estimation. There are only five geographic regions so that it is arguable that region should be treated as fixed.
2. Obtain the estimated intraclass correlations for (1) two observations for the same cow, (2) observations for two different cows from the same herd, and (3) observations for two different cows from different herds in the same region.
3. Fit a three-level model for lactations nested in cows nested in herds, including dummy variables for the five geographic regions using REML and omitting the constant. Compare the estimates for this model with the estimates using a four-level model.

10.5 Dairy-cow data II

Consider again the data of exercise 10.4 (see also exercise 6.6).

1. Fit a two-level random-intercept logistic regression model for the response variable `fscr`, an indicator for conception at the first insemination attempt (first service). Include a random intercept for cow and the covariates `lncfs`, `ai`, and `heifer`.
2. Now extend the model by including a random intercept for herds, as well. Use `xtmelogit` or `gllamm` with 5 integration points to speed up estimation. Is there any evidence for unobserved heterogeneity in fertility between herds?

10.6 Tower-of-London data

Rabe-Hesketh, Touloupulou, and Murray (2001a) analyzed data on patients with schizophrenia, their relatives, and controls. Cognitive performance was assessed by the Tower of London (a computerized task), which was repeated at three levels of difficulty. The data have a three-level structure with measurements at occasion i for person j in family k. We will consider the dichotomous response y_{ijk}, equal to 1 if the tower was completed in the minimum number of moves and 0 otherwise.

The dataset `towerl.dta` contains the following variables:

- `famnum`: family identifier
- `id`: subject identifier
- `dtlm`: indicator for completing the task in the minimum number of moves
- `level`: level of difficulty of the Tower of London (x_{ijk})
- `group`: group (1: controls; 2: relatives; 3: schizophrenics)

1. Fit the two-level random-intercept model (random intercept for subjects):

$$\text{logit}\{\Pr(y_{ijk}=1 \mid \mathbf{x}_{ijk}, \zeta_{jk}^{(2)})\} = \beta_0 + \beta_1 x_{ijk} + \beta_2 g_{2ijk} + \beta_3 g_{3ijk} + \zeta_{jk}^{(2)}$$

   where $\zeta_{jk}^{(2)}|\mathbf{x}_{ij} \sim N(0, \psi^{(2)})$ and g_{2ijk} and g_{3ijk} are dummy variables for groups 2 and 3, respectively.

2. Fit the three-level random-intercept model (random intercepts for subjects and families):

$$\text{logit}\{\Pr(y_{ijk}=1 \mid \mathbf{x}_{ijk}, \zeta_{jk}^{(2)}, \zeta_k^{(3)})\} = \beta_0+\beta_1 x_{ijk}+\beta_2 g_{2ijk}+\beta_3 g_{3ijk}+\zeta_{jk}^{(2)}+\zeta_k^{(3)}$$

   where $\zeta_{jk}^{(2)}|\mathbf{x}_{ijk}, \zeta_k^{(3)} \sim N(0, \psi^{(2)})$, independent from $\zeta_k^{(3)}|\mathbf{x}_{ijk} \sim N(0, \psi^{(3)})$.

10.7 Antibiotics data

Acute respiratory tract infection (ARI) is a common disease among children, pneumonia being a leading cause of death in young children in developing countries. In China, the standard medication for ARI is antibiotics, which has led to concerns about antibiotics misuse and resultant drug resistance. As a response, the World Health Organization (WHO) introduced a program of case management for ARI in children under 5-years old in China in the 1990s.

Here we consider data on physicians' prescribing behavior of antibiotics in two Chinese counties, only one of which was in the WHO program. These data have previously been analyzed by Yang (2001); see also Skrondal and Rabe-Hesketh (2003a).

Medical records were examined for medicine prescribed and a correct diagnosis determined from symptoms and clinical signs. The antibiotic prescription was defined as abuse if there were no clinical indications.

The dataset `antibiotics.dta` has the following variables:

- Level 1 (child):
  - `abuse`: classification of prescription (1: correct use; 2: abuse of one antibiotic; 3: abuse of several antibiotics)
  - `age`: age in years (0–4)
  - `temp`: body temperature, centered at 36 degrees Centigrade
  - `Paymed`: dummy variable for patient paying for his or her own medication
  - `Selfmed`: dummy variable for self-medication before seeing doctor
  - `Wrdiag`: dummy variable for diagnosis classified as wrong
- Level 2 (doctor):
  - `doc`: doctor identifier
  - `DRed`: doctor's education (ordinal with six categories from self-taught to medical school)
- Level 3 (hospital):
  - `hosp`: hospital identifier
  - `WHO`: dummy variable for hospital's being in the WHO program

1. Recode `abuse` into a dichotomous variable equal to 1 if there was abuse of one or more antibiotics and 0 otherwise.
2. Fit a three-level logistic regression model for the dichotomized variable `abuse` with random intercepts for doctors and hospitals and no covariates. Use adaptive quadrature with 5 quadrature points.
3. Add the dummy variable for hospital being in the WHO program. Obtain and interpret the estimated odds ratio for the WHO program, as well as its 95% confidence interval. Does this analysis suggest that the program has been effective?
4. Include all the other predictors in the model and reassess the apparent effect of the WHO program. Comment on the change in the estimated random-intercept variances between the model without covariates (step 2) and the model with all the covariates (step 4).

10.8 Smoking-intervention data

1. Use the data `tvsfpors.dta` described in exercise 7.3.
2. Investigate how tobacco and health knowledge is influenced by the interventions by fitting a three-level proportional odds model with students nested in classes nested in schools. Include `cc`, `tv`, and their interaction and `prethk` as covariates. Use 5-point adaptive quadrature.
3. Does this model fit better than the two-level models with a random intercept either at the class or at school levels?

10.9 Cigarette data

1. Use the data `cigarette.dta` described in exercise 8.7.
2. Expand the data to person-period data.
3. Fit the discrete-time survival model that assumes the continuous-time hazards to be proportional. Use dummy variables for periods; include `cc`, `tv`, their interaction, and `male` as explanatory variables; and specify random intercepts for classes and schools. (Five quadrature points at each level should suffice.)
4. Interpret the exponentials of the estimated regression coefficients.
5. Perform likelihood-ratio tests for each of the variance components using a 5% level of significance.

10.10 Health-care reform data

1. For the health-care reform data described in section 9.4, fit the random-intercept Poisson model specified on page 380.
2. Write down and fit the three-level model that also includes an occasion and person-specific random intercept. Use 5-point adaptive quadrature.
3. Does the three-level model appear to fit better than the two-level model?
4. ❖ Consider the total random part in the three-level model, and compare the variances for 1996 and 1998 and the covariance with the corresponding estimates for the random-coefficient model fitted in section 9.8 (see also section 9.8.3). Comment on the difference between the random-coefficient and three-level models. Is one more general than the other?

10.11 Skin-cancer data

Here we consider the data from Langford, Bentham, and McDonald (1998) that is described in exercise 9.8.

The variables in the dataset `skincancer.dta` are described in exercise 9.8.

1. Use `gllamm` to fit a Poisson model for the number of male deaths using the log expected number of male deaths as an offset (see page 410) and `uv` and squared `uv` as covariates. Include random intercepts for county, region, and nation. This is a four-level model requiring the option `i(county region nation)`. To speed up estimation in `gllamm`, use 8 quadrature points at level 2 and 5 points at levels 3 and 4 (the `nip(8 5 5)` option).
2. Write down the model, and interpret the estimates.
3. Obtain empirical Bayes predictions for the nations.
4. Instead of using a random intercept for nation, include dummy variables for all the nations and exclude the constant (use `xtmelogit`). How do the estimated regression coefficients and standard errors of the nation dummies compare with the empirical Bayes predictions and posterior standard deviations for the nations from the previous model?

10.12 ❖ Peak-expiratory-flow data I

The three-stage formulation of the three-level model for the peak-expiratory-flow data discussed in section 10.4.3 (excluding x_j) can be represented by the path diagram in the left panel of figure 10.6. Here we have already substituted the level-3 model into the level-2 model for η_{1jk}

$$\eta_{1jk} = \gamma_{111} + \zeta_k^{(3)} + \zeta_{jk}^{(2)}$$

and do not show the fixed part of the model γ_{111}.

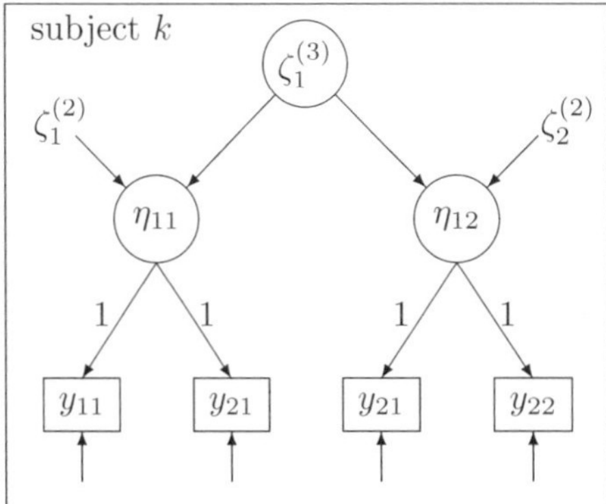

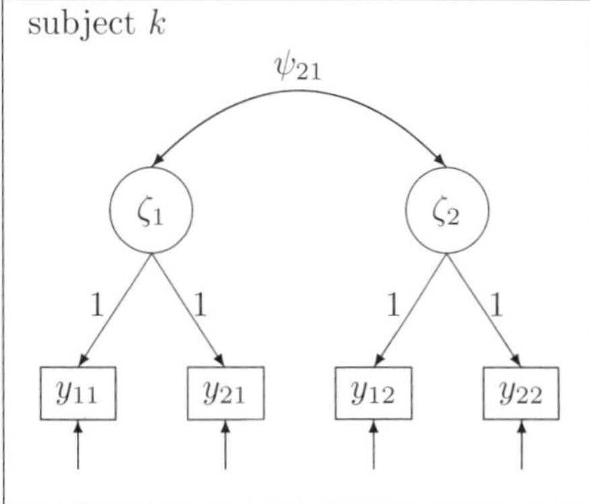

Figure 10.6: Path diagrams of equivalent models (left panel: three-stage formulation of three-level model; right panel: correlated random effects)

1. Express the correlation between η_{11k} and η_{12k} in the three-stage formulation represented in the left panel of figure 10.6 in terms of the variances $\psi^{(2)}$ at level 2 and $\psi^{(3)}$ at level 3.
2. The right panel in figure 10.6 represents a model with two method-specific correlated random intercepts ζ_{1k} for method 1 and ζ_{2k} for method 2. Defining dummy variables d_{1j} for method 1 and d_{2j} for method 2, the model can be written as a two-level random-coefficient model:

$$y_{ijk} = \beta_1 + \zeta_{1k}d_{1j} + \zeta_{2k}d_{2j} + \epsilon_{ijk}$$

   If the variances of the two random coefficients are constrained equal and if the covariance is positive, this model is equivalent to the previous three-level model. Verify this by fitting the above model by maximum likelihood using `xtmixed`. You can use the option `covariance(exchangeable)` to constrain the variances of the random coefficients to be equal.
3. Relax the assumption of equal variances of the random coefficients by using the option `covariance(unstructured)`. Does this model fit better than the model with variances constrained equal?

4. For the model from step 3, obtain the estimated method-specific intraclass correlations $\widehat{\rho}(\text{subject}|\text{method}{=}1)$ and $\widehat{\rho}(\text{subject}|\text{method}{=}2)$.
5. Fit the above model using `gllamm`.
6. Extend the model further (still using `gllamm`) by relaxing the assumption of equal measurement error variances θ using the `s()` option (see sec. 5.14). Is there any evidence that the measurement error variances differ between the methods? Again obtain the estimated method-specific intraclass correlations.

10.13 ❖ Peak-expiratory-flow data II

This exercise is a useful transition to the next chapter where we sometimes treat random intercepts for a factor as random coefficients of dummy variables for the levels of the factor.

For the peak-expiratory-flow data, we require a random intercept for method nested in subject. This can be achieved by having uncorrelated random coefficients at the subject level for a dummy variable d_{1j} for method 1 and d_{2j} for method 2. We also retain the random intercept at the subject level (now called ζ_{3k}):

$$y_{ijk} = \beta_1 + \zeta_{1k}d_{1j} + \zeta_{2k}d_{2j} + \zeta_{3k} + \epsilon_{ijk}$$

1. All three random effects should be mutually uncorrelated, but only the first two should have the same variance. This can be achieved by using two specifications (or equations) for the same level: `|| id: d1 d2, cov(identity) nocons || id:`, where the `covariance(identity)` option specifies equal variances and no covariance for the random coefficients of `d1` and `d2` (the dummy variables) and the `noconstant` option suppresses the random intercept since this is already specified by the next equation, `|| id:`. Random effects in different random part specifications (separated by `||`) are assumed to be uncorrelated. Also note that the `noconstant` option was not necessary because `xtmixed` puts the constant only in the last equation for a given level by default.

   Fit this model using `xtmixed`, and show that it equivalent to the three-level random-intercept model specified using `|| id: || method:`.

11 Crossed random effects

11.1 Introduction

In the previous chapter, we discussed higher-level hierarchical models where units are classified by some factor (for instance school) into top-level clusters at level L. The units in each top-level cluster are then (sub)classified by a further factor (for instance class) into clusters at level $L-1$, etc. The factors defining the classifications are nested in the sense that a lower-level cluster can only belong to one higher-level cluster (for instance, a class can only belong to one school).

We now discuss nonhierarchical models where units are *cross-classified* by two or more factors, with each unit potentially belonging to any combination of levels of the different factors.

A prominent example is panel data where the factor "individual" (or country or firm, etc.) is crossed with another factor "time" or occasion (see section 11.2 for such an example). So far, we have treated occasions as nested within individuals. However, if all individuals are affected similarly by some events or characteristics associated with the occasions, such as weather conditions, strikes, new legislation, etc., it seems reasonable to treat occasions as crossed with individuals, or to consider a "main effect" of time. If the factors individual and time are both treated as random, econometricians call the model a *two-way error-components model.*

Models with crossed random effects also arise in *generalizability theory* of measurement. Here a simple design is a two-way cross-classification of subjects and raters; see also exercise 11.4. In a similar context, the distinction between nested and crossed factors was illustrated in figure 2.8.

Factors are not always completely crossed. For instance, the high schools and elementary schools attended by students are not nested, but there are many combinations of high school and elementary school that do not occur in practice, perhaps because the schools are in different geographical regions. Such an example is considered in section 11.4.

Most of this chapter is devoted to linear mixed models for continuous responses because estimation of generalized linear mixed models with crossed effects is computationally demanding. However, we also discuss estimation of a crossed random-effects model for dichotomous data on the success of salamander mating.

11.2 How does investment depend on expected profit and capital stock?

Grunfeld (1958) and Boot and DeWit (1960), among others, analyzed data on 10 large American corporations collected annually from 1935 to 1954 to investigate how investment depends on the expected profit and capital stock of firms.

The variables in `grunfeld.dta` are

- `fn`: firm identifier i
- `firmname`: firm name
- `yr`: year j
- `I`: annual gross investment (in \$1,000,000) defined as amount spent on plant and equipment plus maintenance and repairs (y_{ij})
- `F`: market value of firm (in \$1,000,000) defined as value of all shares plus book value of all debts outstanding at the beginning of the year (x_{2ij})
- `C`: real value of capital stock (in \$1,000,000), defined as the deviation of stock of plant and equipment from stock in 1933 (x_{3ij})

Grunfeld proposed an investment theory, which Boot and DeWit (1960) summarize as follows (you can skip the rest of this section without loss of continuity if you like): Observed profits are rejected as an explanation of investment, and expected profits, measured as the market value of the firm $\mathtt{F}(t)$ at time t, are used instead. The desired capital stock $C^*(t)$ is assumed to be a linear function of the market value:

$$C^*(t) = c_1 + c_2\mathtt{F}(t)$$

The desired net investment is the difference between desired capital stock $C^*(t)$ and the existing capital stock $\mathtt{C}(t)$, $C^*(t) - \mathtt{C}(t)$. Assuming that a constant fraction q_1 of the desired net investment is made between t and $t+1$, the net investment for that year becomes

$$q_1\{C^*(t) - \mathtt{C}(t)\} = q_1 c_1 + q_1 c_2 \mathtt{F}(t) - q_1 \mathtt{C}(t)$$

Assuming that replacement investment plus maintenance and repairs equals a constant fraction q_2 of the existing capital stock $\mathtt{C}(t)$, the gross investment in the following year $\mathtt{I}(t+1)$ becomes

$$\begin{aligned}\mathtt{I}(t+1) &= q_1\{C^*(t) - \mathtt{C}(t)\} + q_2\mathtt{C}(t) \\ &= q_1 c_1 + q_1 c_2\mathtt{F}(t) + (q_2 - q_1)\mathtt{C}(t)\end{aligned}$$

Denoting the gross investment `I` for firm i in year j as y_{ij}, the market value `F` as x_{2ij} and the value of capital stock `C` as x_{3ij}, the investment equation can finally be written as

$$y_{ij} = \beta_1 + \beta_2 x_{2ij} + \beta_3 x_{3ij}$$

where $\beta_1 \equiv q_1 c_1$, $\beta_2 \equiv q_1 c_2$ and $\beta_3 \equiv q_2 - q_1$. Both β_2 and β_3 are expected to be positive. However, the intercept β_1 has limited meaning since capital stock has been measured as a deviation from the stock in 1933.

11.3 A two-way error-components model

11.3.1 Model specification

Since the investment behavior of corporations is surely not deterministic, statistical models including error terms have invariably been specified. Baltagi (2005) allows the effects of both firms and years on gross investment y_{ij} to vary by specifying the following two-way error-components model:

$$y_{ij} = \beta_1 + \beta_2 x_{2ij} + \beta_3 x_{3ij} + \zeta_{1i} + \zeta_{2j} + \epsilon_{ij} \tag{11.1}$$

Here x_{2ij} and x_{3ij} represent the market value and capital stock of firm i in year j. ζ_{1i} and ζ_{2j} are random intercepts for firms i and years j, respectively, and ϵ_{ij} is a residual error term. Given the covariates, the random intercepts have zero means, are independent of each other and across firms and years, and have variances ψ_1 and ψ_2. The random intercepts are also uncorrelated with ϵ_{ij}, which has zero mean and variance θ, and are independent across firms and years. We assume that ζ_{1j}, ζ_{2j}, and ϵ_{ij} are normally distributed given the covariates.

This model differs from the models considered so far because the two random intercepts represent the factors firm and year that are crossed instead of nested. The random intercept for firm ζ_{1i} is shared across all years for a given firm i, whereas the random intercept for year ζ_{2j} is shared by all firms in a given year j. The residual error ϵ_{ij} comprises both the interaction between year and firm and any other effect specific to firm i in year j. An interaction between firm and year could be due to some events occurring in some years being more beneficial (or detrimental) to some firms than others.

The `xtmixed` command is primarily designed for multilevel models with nested random effects. To fit models with crossed effects, we therefore use the following trick described by Goldstein (1987):

- Consider the entire dataset as an artificial level-3 unit a within which both firms and years are nested.
- Treat either years or firms as level-2 units j, and specify a random intercept $u^{(2)}_{ja}$ for them. It is best to choose the factor with more levels, i.e., years.
- For the other factor, here firm, specify a level-3 random intercept for each firm, $u^{(3)}_{pa}$, $(p = 1, \ldots, 10)$. This can be constructed by treating $u^{(3)}_{pa}$ as the random coefficient of the dummy variable d_{pij} for firm p, where

  $$d_{pij} = \begin{cases} 1 & \text{if } p = i \\ 0 & \text{otherwise} \end{cases}$$

  The ten random coefficients are then specified as having equal variance ψ_1 and being uncorrelated.

Here we have used the notation u for the random effects to avoid confusion between the different formulations. Model (11.1) can then be written as

$$\begin{aligned} y_{ija} &= \beta_1 + \beta_2 x_{2ij} + \beta_3 x_{3ij} + u_{ja}^{(2)} + \sum_p u_{pa}^{(3)} d_{pij} + \epsilon_{ija} \\ &= \beta_1 + \beta_2 x_{2ij} + \beta_3 x_{3ij} + \underbrace{u_{ja}^{(2)}}_{\zeta_{2j}} + \underbrace{u_{ia}^{(3)}}_{\zeta_{1i}} + \underbrace{\epsilon_{ija}}_{\epsilon_{ij}} \end{aligned}$$

11.3.2 Residual intraclass correlations

We can define two different residual intraclass correlations: one for correlations of observations on the same firm over time

$$\rho(\text{firm}) \equiv \text{Cor}(y_{ij}, y_{ij'}|x_{2ij}, x_{3ij}, x_{2ij'}, x_{3ij'}) = \frac{\psi_1}{\psi_1 + \psi_2 + \theta}$$

and one for the correlation for the same year across firms

$$\rho(\text{year}) \equiv \text{Cor}(y_{ij}, y_{i'j}|x_{2ij}, x_{3ij}, x_{2i'j}, x_{3i'j}) = \frac{\psi_2}{\psi_1 + \psi_2 + \theta}$$

11.3.3 Estimation

Fortunately, `xtmixed` makes it easy to specify a random intercept $u_{pa}^{(3)}$ for each level of a factor (here each firm) or, as specified above, random coefficients for the corresponding dummy variables. The syntax `R.fn` in the random part accomplishes this and also automatically sets all variances equal and all correlations to zero as required. This covariance structure is called `identity` in `xtmixed` because the covariance matrix is proportional to the 10×10 identity matrix (a matrix with ones on the diagonal and zeros elsewhere). We also do not need to create an artificial level-3 identifier because `xtmixed` accepts the cluster name `_all` for this purpose.

The command for fitting the crossed error-components model in `xtmixed` by maximum likelihood is

```
. use http://www.stata-press.com/data/mlmus2/grunfeld
. xtmixed I F C || _all: R.fn || yr:, mle
Mixed-effects ML regression                     Number of obs      =       200

-----------------------------------------------------------
                |   No. of       Observations per Group
 Group Variable |   Groups    Minimum    Average    Maximum
----------------+------------------------------------------
           _all |        1        200      200.0        200
             yr |       20         10       10.0         10
-----------------------------------------------------------

                                                Wald chi2(2)       =    661.07
Log likelihood = -1095.2485                     Prob > chi2        =    0.0000

------------------------------------------------------------------------------
           I |      Coef.   Std. Err.      z    P>|z|     [95% Conf. Interval]
-------------+----------------------------------------------------------------
           F |   .1099009   .0103779    10.59   0.000     .0895606    .1302413
           C |   .3092262   .0172179    17.96   0.000     .2754798    .3429726
       _cons |  -58.27126   27.76275    -2.10   0.036    -112.6853   -3.857263
------------------------------------------------------------------------------

------------------------------------------------------------------------------
  Random-effects Parameters  |   Estimate   Std. Err.     [95% Conf. Interval]
-----------------------------+------------------------------------------------
_all: Identity               |
                   sd(R.fn)  |   80.41164   18.42471       51.3196    125.9954
-----------------------------+------------------------------------------------
yr: Identity                 |
                  sd(_cons)  |   3.860619   15.29479      .0016384    9097.042
-----------------------------+------------------------------------------------
               sd(Residual)  |   52.34756   2.904361       46.9537    58.36104
------------------------------------------------------------------------------
LR test vs. linear regression:       chi2(2) =   193.11   Prob > chi2 = 0.0000
Note: LR test is conservative and provided only for reference
```

The market value ("expected profit") and stock value ("capital stock") of a firm both have positive effects on investment. The estimated residual standard deviation between firms is 80.41, and the estimated residual standard deviation between years is only 3.86. The remaining residual standard deviation, not due to additive effects of firms and years, is estimated as 52.35. The estimates are also given in table 11.1.

(*Continued on next page*)

Table 11.1: Maximum likelihood estimates of two-way error-components model for Grunfeld data

	Est	(SE)
Fixed part		
β_1	-58.27	(27.76)
β_2 [F]	0.11	(0.01)
β_3 [C]	0.31	(0.02)
Random part		
$\sqrt{\psi_1}$ [Firm]	80.41	
$\sqrt{\psi_2}$ [Year]	3.86	
$\sqrt{\theta}$	52.35	
Log likelihood	$-1,095.25$	

The residual intraclass correlation for firms is estimated as

$$\widehat{\rho}(\text{firm}) = \frac{80.41164^2}{80.41164^2 + 3.860619^2 + 52.34756^2} = 0.70$$

and the residual intraclass correlation for years is estimated as

$$\widehat{\rho}(\text{year}) = \frac{3.860619^2}{80.41164^2 + 3.860619^2 + 52.34756^2} = 0.002$$

Hence, there is a high correlation over years within firms and a negligible correlation over firms within years, given the covariates.

The reason for choosing year to be at level 3 and to have a random intercept for each firm at level 2 is to minimize the computational burden. With this formulation, the model has ten random effects at level 2 (one for each firm) and one random effect at level 3. If we instead had chosen firm to be at level 2 and to have a separate random intercept for each year at level 3, we would have required 20 random effects at level 3 (one for each year) and one random effect at level 2. The syntax would be

```
xtmixed I F C || _all: R.yr || fn:, mle
```

Although more computationally demanding, this setup produces identical estimates to the one used above.

11.3.4 Prediction

Having fitted the model, it is easy to obtain various predictions. For instance, we can obtain empirical Bayes (or BLUP) predictions of the random effects of firm and year using `predict` with the `reffects` option:

```
. predict firm year, reffects
. sort fn yr
. list fn firmname yr firm year if yr<1938 & fn<5, sepby(fn) noobs
```

fn	firmname	yr	firm	year
1	General Motors	1935	-10.2973	1.753753
1	General Motors	1936	-10.2973	.9433348
1	General Motors	1937	-10.2973	.0166125
2	US Steel	1935	157.526	1.753753
2	US Steel	1936	157.526	.9433348
2	US Steel	1937	157.526	.0166125
3	General Electric	1935	-172.8691	1.753753
3	General Electric	1936	-172.8691	.9433348
3	General Electric	1937	-172.8691	.0166125
4	Chrysler	1935	30.08079	1.753753
4	Chrysler	1936	30.08079	.9433348
4	Chrysler	1937	30.08079	.0166125

The predictions for the firms have all conveniently been placed into one variable `firm`, not into ten variables as might have been expected. We see that the prediction for firm 1 (General Motors) is −10.30 and the prediction for 1935 is 1.75. After controlling for market and stock value, 1935 was a year with higher investment than the average over the 19-year interval, and General Motors was a firm with lower investment than the average across firms.

We can visualize these effects by plotting the sum of these predicted random effects $\widetilde{\zeta}_{1i} + \widetilde{\zeta}_{2j}$ versus time with a separate line for each firm:

```
. generate reffpart = firm + year
. twoway (line reffpart yr, connect(ascending))
> (scatter reffpart yr if yr==1954, msymbol(none) mlabel(firmnam) mlabpos(3)),
> xtitle(Year) ytitle(Predicted random effects of firm and year)
> xscale(range(1935 1958)) legend(off)
```

It is clear from the resulting figure 11.1 that the between-firm variability is much more considerable than the between-year variability.

(Continued on next page)

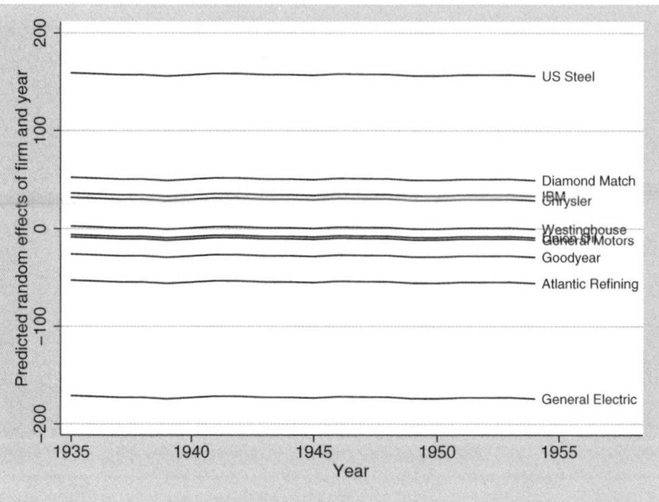

Figure 11.1: Predicted random part $\widetilde{\zeta}_{1i} + \widetilde{\zeta}_{2j}$ versus time for ten firms

The effect of year is hardly visible in figure 11.1, and we therefore produce another graph just showing the effect of year on its own

```
. twoway (line year yr if fn==1), xtitle(Year)
> ytitle(Predicted random effect for year)
```

which is given in figure 11.2. (Here we plotted the data for the first firm but would have obtained an identical graph if we had picked another firm.)

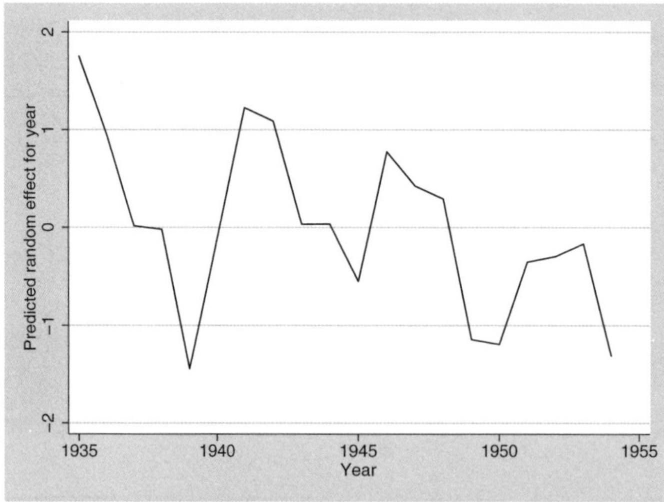

Figure 11.2: Predicted random effect of year $\widetilde{\zeta}_{2j}$

11.4 How much do primary and secondary schools affect attainment at age 16?

We now consider data described by Paterson (1991) on pupils (students) from Fife, Scotland, who are cross-classified by 148 primary schools (elementary schools) and 19 secondary schools (middle/high schools). The data are distributed with the MLwiN program (Rasbash et al. 2005) and have previously been analyzed by Goldstein (2003) and Rasbash (2005).

The dataset `fife.dta` has the following variables:

- `attain`: attainment score at age 16 [summary of passes in the Scottish Certificate of Education (SCE), a school exit examination] (y_{ijk})
- `pid`: identifier for primary school (up to age 12) (k)
- `sid`: identifier for secondary school (from age 12) (j)
- `vrq`: verbal-reasoning score from test taken in the last year of primary school
- `sex`: gender (1: female; 0: male)

The structure of these data is somewhat different than in the Grunfeld panel dataset considered earlier. First, not every combination of primary and secondary school exists. Second, many combinations of primary and secondary school occur multiple times.

We first explore this crossed structure in more detail. For this, it is useful to define a dummy variable taking the value 1 for exactly one observation for each combination of the primary-school and secondary-school identifiers. The `egen` function `tag()` is designed for creating such dummy variables:

```
. use http://www.stata-press.com/data/mlmus2/fife, clear
. egen pick_comb = tag(pid sid)
```

(*Continued on next page*)

Now we can count the number of unique values of sid in each primary school using the egen function total():

```
. egen numsid = total(pick_comb), by(pid)
. sort pid sid
. list pid sid numsid if pick_comb & pid<10, sepby(pid) noobs
```

pid	sid	numsid
1	1	3
1	9	3
1	18	3
2	7	1
3	5	1
4	6	2
4	9	2
5	1	1
6	1	4
6	3	4
6	5	4
6	11	4
7	4	4
7	6	4
7	18	4
7	19	4
8	1	3
8	9	3
8	19	3
9	1	6
9	3	6
9	6	6
9	17	6
9	18	6
9	19	6

We see that, for instance, children in this sample who attended primary school 1 ended up in three secondary schools, 1, 9, and 18. To obtain the frequency distribution of the number of secondary schools per primary school, numsid, we must first define a dummy variable equal to 1 for one child per primary school

```
. egen pick_pid = tag(pid)
```

and we can subsequently use the tabulate command:

```
. tabulate numsid if pick_pid

     numsid |      Freq.     Percent        Cum.
------------+-----------------------------------
          1 |         57       38.51       38.51
          2 |         50       33.78       72.30
          3 |         26       17.57       89.86
          4 |         10        6.76       96.62
          5 |          2        1.35       97.97
          6 |          3        2.03      100.00
------------+-----------------------------------
      Total |        148      100.00
```

There are at most 6 secondary schools per primary school, and for 90% of the primary schools there are at most 3 secondary schools per primary school.

Repeating the same commands as above but with **sid** and **pid** interchanged, we obtain the frequency table of the number of primary schools per secondary school:

```
. egen numpid = total(pick_comb), by(sid)
. egen pick_sid = tag(sid)
. tabulate numpid if pick_sid

     numpid |      Freq.     Percent        Cum.
------------+-----------------------------------
          7 |          1        5.26        5.26
         10 |          2       10.53       15.79
         12 |          1        5.26       21.05
         13 |          2       10.53       31.58
         14 |          4       21.05       52.63
         15 |          1        5.26       57.89
         16 |          1        5.26       63.16
         17 |          2       10.53       73.68
         18 |          2       10.53       84.21
         23 |          1        5.26       89.47
         26 |          1        5.26       94.74
         32 |          1        5.26      100.00
------------+-----------------------------------
      Total |         19      100.00
```

There are between 7 and 32 primary schools per secondary school, the median being between 13 and 14.

11.5 An additive crossed random-effects model

11.5.1 Specification

We first consider the following model for the attainment score y_{ijk} at age 16 for student i who went to secondary school j and primary school k:

$$y_{ijk} \;=\; \beta_1 + \zeta_{1j} + \zeta_{2k} + \epsilon_{ijk} \tag{11.2}$$

The random part of this model has exactly the same structure as in the previous application with additive (and independent) random effects $\zeta_{1j} \sim N(0, \psi_1)$ and $\zeta_{2k} \sim N(0, \psi_2)$ of the two cross-classified factors plus a residual error term $\epsilon_{ijk} \sim N(0, \theta)$.

11.5.2 Estimation using xtmixed

Since there are only 19 secondary schools compared with 148 primary schools, we treat primary schools as level-2 units and use 19 random effects for the secondary schools at level 3.

The `xtmixed` command for fitting model (11.2) by maximum likelihood is

```
. xtmixed attain || _all: R.sid || pid:, mle
Mixed-effects ML regression                     Number of obs      =      3435

--------------------------------------------------------------------
                |   No. of       Observations per Group
 Group Variable |   Groups    Minimum    Average    Maximum
----------------+---------------------------------------------------
           _all |        1       3435     3435.0       3435
            pid |      148          1       23.2         72
--------------------------------------------------------------------

                                                Wald chi2(0)       =         .
Log likelihood = -8574.5655                     Prob > chi2        =         .

------------------------------------------------------------------------------
      attain |      Coef.   Std. Err.      z    P>|z|     [95% Conf. Interval]
-------------+----------------------------------------------------------------
       _cons |    5.50401   .1749317    31.46   0.000      5.16115     5.84687
------------------------------------------------------------------------------

------------------------------------------------------------------------------
  Random-effects Parameters  |   Estimate   Std. Err.     [95% Conf. Interval]
-----------------------------+------------------------------------------------
_all: Identity               |
                   sd(R.sid) |   .5900527   .1371146      .3741892    .9304441
-----------------------------+------------------------------------------------
pid: Identity                |
                   sd(_cons) |   1.060357   .0971077      .8861331    1.268835
-----------------------------+------------------------------------------------
                sd(Residual) |   2.848066   .0351956      2.779912     2.91789
------------------------------------------------------------------------------
LR test vs. linear regression:       chi2(2) =   278.13   Prob > chi2 = 0.0000
Note: LR test is conservative and provided only for reference
. estimates store model1
```

The estimates are presented under "Additive" in table 11.2. We see that the estimated standard deviation of the primary school random effect is 1.06, which is considerably larger than the estimated standard deviation of the high school random effect, given by 0.59. Therefore, elementary schools appear to have a greater effect or to be more variable in their effects than secondary schools. However, neither of these estimated standard deviations is precise.

The standard deviation of ϵ_{ijk} is estimated as 2.85. This number reflects any interactions between primary and secondary schools (deviations of the means for the combinations of primary and secondary schools from the means implied by the additive effects) and variability within the groups of children belonging to the same combination of primary and secondary school.

11.6 Including a random interaction

11.6.1 Model specification

For many combinations of primary and secondary school, we have several observations because more than one child attended that combination of schools.

We can therefore include a random interaction term $\zeta_{3jk} \sim N(0, \psi_3)$ between secondary and primary schools in the model:

$$y_{ijk} = \beta_1 + \zeta_{1j} + \zeta_{2k} + \zeta_{3jk} + \epsilon_{ijk} \tag{11.3}$$

The interaction term takes on a different value for each combination of secondary and primary school to allow the assumption of additive (random) effects to be relaxed. For instance, some secondary schools may be more beneficial for children who attended particular elementary schools, perhaps because of similar instructional practices. ζ_{3jk} is independent of the other random terms and across combinations of primary and secondary school. The residual ϵ_{ijk} represents the deviation of an individual child's response from the mean for secondary school j and primary school k.

We could not include an interaction in the previous panel-data application because we had no replicates for any of the firm and year combinations so that the interaction would be completely confounded with the level-1 residual.

11.6.2 Intraclass correlations

We can consider several intraclass correlations. For children from the same primary school but different secondary schools,

$$\rho(\text{primary}) \equiv \text{Cor}(y_{ijk}, y_{i'j'k}) = \frac{\psi_2}{\psi_1 + \psi_2 + \psi_3 + \theta}$$

where $\psi_3 = 0$ if there is no interaction. For children from the same secondary school but different primary schools

$$\rho(\text{secondary}) \equiv \text{Cor}(y_{ijk}, y_{i'jk'}) = \frac{\psi_1}{\psi_1 + \psi_2 + \psi_3 + \theta}$$

Finally, for children from both the same secondary and the same primary schools

$$\rho(\text{secondary}, \text{primary}) \equiv \text{Cor}(y_{ijk}, y_{i'jk}) = \frac{\psi_1 + \psi_2 + \psi_3}{\psi_1 + \psi_2 + \psi_3 + \theta}$$

We could also condition on secondary school and consider the intraclass correlation due to being in the same primary school

$$\rho(\text{primary}|\text{secondary}) \equiv \text{Cor}(y_{ijk}, y_{i'jk}|j) = \frac{\psi_2 + \psi_3}{\psi_2 + \psi_3 + \theta}$$

where the between-secondary school variance ψ_1 vanishes since secondary school is held constant. The analogous expression for the intraclass correlation due to secondary school for a given primary school is

$$\rho(\text{secondary}|\text{primary}) \equiv \text{Cor}(y_{ijk}, y_{i'jk}|k) = \frac{\psi_1 + \psi_3}{\psi_1 + \psi_3 + \theta}$$

11.6.3 Estimation using xtmixed

The crossed random-effects model with interaction (11.3) can be fitted in `xtmixed` by augmenting the random-part specification of the previous command for the additive model (11.2) given by

```
xtmixed attain || _all: R.sid || pid:, mle
```

We require a random intercept taking on distinct values for each combination of primary and secondary school. We could achieve this by defining an identifier variable for the combinations using `egen` with the `group()` function

```
. egen comb = group(sid pid)
```

and specifying the last equation in the random part as `|| comb:`. This random intercept would be treated as nested within `pid`, the cluster identifier for the previous equation. A more convenient setup, not requiring the variable `comb`, is to specify the last equation as `|| sid:`. Since `sid` is now treated as nested in `pid`, the random intercept will take on a different value for each unique secondary school `sid` within primary school `pid`, i.e., for each combination of primary and secondary school:

```
. xtmixed attain || _all: R.sid || pid: || sid:, mle
Performing EM optimization:
Performing gradient-based optimization:
Computing standard errors:
Mixed-effects ML regression                     Number of obs      =      3435
```

Group Variable	No. of Groups	Observations per Group Minimum	Average	Maximum
_all	1	3435	3435.0	3435
pid	148	1	23.2	72
sid	303	1	11.3	72

```
                                                Wald chi2(0)       =         .
Log likelihood = -8573.9826                     Prob > chi2        =         .
```

attain	Coef.	Std. Err.	z	P>|z|	[95% Conf. Interval]	
_cons	5.501305	.1690353	32.55	0.000	5.170002	5.832608

```
Random-effects Parameters  |   Estimate   Std. Err.     [95% Conf. Interval]
---------------------------+------------------------------------------------
_all: Identity             |
                 sd(R.sid) |   .5595461   .1431865      .3388562    .9239667
---------------------------+------------------------------------------------
pid: Identity              |
                 sd(_cons) |   .9501046   .1547825      .6904009    1.307499
---------------------------+------------------------------------------------
sid: Identity              |
                 sd(_cons) |   .4907404   .2534124      .1783617    1.350212
---------------------------+------------------------------------------------
              sd(Residual) |   2.843847   .0353695      2.775362    2.914022
----------------------------------------------------------------------------
LR test vs. linear regression:       chi2(3) =   279.29   Prob > chi2 = 0.0000
Note: LR test is conservative and provided only for reference
. estimates store model2
```

The estimates are presented under "Interaction" in table 11.2:

Table 11.2: Maximum likelihood estimates for crossed random-effects models for Fife data

	Additive		Interaction	
	Est	(SE)	Est	(SE)
Fixed part				
β_1	5.50	(0.17)	5.50	(0.17)
Random part				
$\sqrt{\psi_1}$ [Secondary]	0.59		0.56	
$\sqrt{\psi_2}$ [Primary]	1.06		0.95	
$\sqrt{\psi_3}$ [Primary×Secondary]			0.49	
$\sqrt{\theta}$	2.85		2.84	
Log likelihood	$-8,574.57$		$-8,573.98$	

The estimated intraclass correlations for the models with and without the random interaction are given in table 11.3:

Table 11.3: Estimated intraclass correlations for Fife data

	Additive	Interaction
ρ(primary)	0.12	0.09
ρ(secondary)	0.04	0.03
ρ(secondary, primary)	0.15	0.15
ρ(primary|secondary)	0.12	0.12
ρ(secondary|primary)	0.04	0.06

We can perform a likelihood-ratio test to compare the model (11.3) with the random interaction term ζ_{3jk} and the model (11.2) without the interaction using the `lrtest` command:

```
. lrtest model1 model2
Likelihood-ratio test                               LR chibar2(01)  =      1.17
(Assumption: model1 nested in model2)              Prob > chibar2  =    0.1401
```

The p-value is automatically adjusted for testing on the boundary of parameter space as indicated by the label `chibar2(01)`. The interaction is not significant at the 5% level, and we return to the model with additive random effects:

```
. estimates restore model1
(results model1 are active now)
```

11.6.4 Some diagnostics

We can obtain empirical Bayes predictions of both the secondary and primary school random effects. If the model is correct, these predictions should have a normal distribution. (This is true only for linear models.)

Returning to the crossed random-effects model without an interaction (11.2), we now obtain these predictions and assess their normality using a normal Q–Q plot:

```
. predict secondary primary, reffects
. qnorm secondary if pick_sid, xtitle(Quantiles of normal distribution)
> ytitle(Quantiles of empirical Bayes predictions)
. qnorm primary if pick_pid, xtitle(Quantiles of normal distribution)
> ytitle(Quantiles of empirical Bayes predictions)
```

Here we used the previously defined dummy variable `pick_sid` to choose one observation per secondary school, and similarly for primary school. If this were not done, the `qnorm` command would compute the quantiles using the empirical distribution of the predictions for all students, which could be different from the quantiles required if the number of students per school varies. The graph for secondary schools is given in figure 11.3, and the graph for primary schools is given in figure 11.4. Both figures suggest that there are no serious violations of the normality assumption.

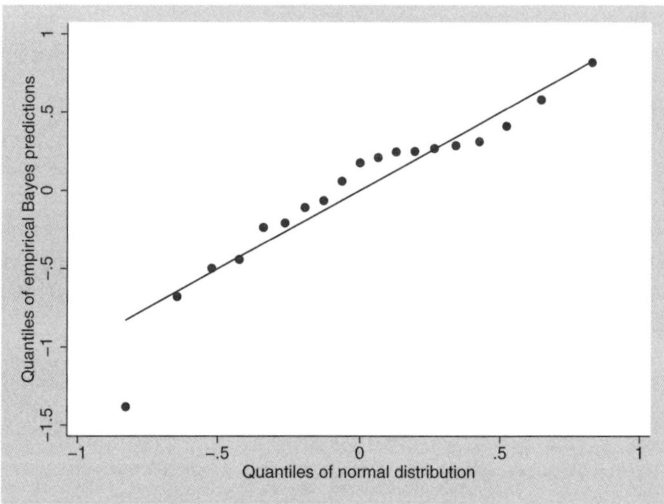

Figure 11.3: Normal Q–Q plot for secondary school predictions $\widetilde{\zeta}_{1j}$

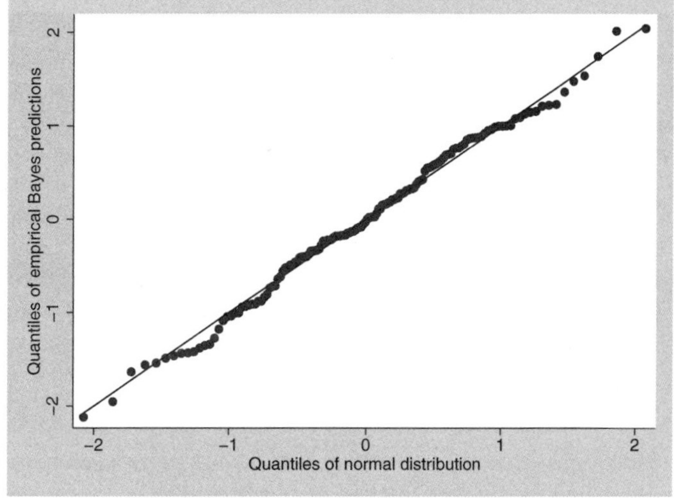

Figure 11.4: Normal Q–Q plot for primary school predictions $\widetilde{\zeta}_{2k}$

11.7 ❖ A trick requiring fewer random effects

We may have used more random effects than necessary for the primary and secondary school example. Imagine that both primary and secondary schools could be nested within regions. This would be the case if children attend both primary and secondary schools within the region in which they live and never move to a different region.

Suppose that no region has more than three secondary schools. In this case, we could arbitrarily number the secondary schools in each region from 1 to at most 3 in a variable `sec` and specify three corresponding random intercepts at the region level (with identifier `region`). The `xtmixed` syntax would be `|| region: R.sec`. Importantly, schools in different regions that happen to have the same value in `sec` would not have the same value of the corresponding random intercept because the intercept varies between regions. The random part for primary schools could then be specified as before, giving the `xtmixed` command (for the additive model)

```
xtmixed attain || region: R.sec || pid:, mle
```

This specification would require only three random intercepts at level 3 instead of 19.

In practice, there are no regions with insurpassable boundaries, but we could produce an identifier for a virtual level 3 within which both primary and secondary schools happen to be nested in the data. This approach becomes important if neither of the cross-classified factors has only a small number of levels and in generalized linear mixed models where computation takes a long time if there are many random effects.

The setup is shown in figure 11.5, where the students, shown as short vertical lines, can be viewed as nested in the primary schools, represented by vertical lines to their left. The students are also connected to the secondary schools to which they belong, shown as vertical lines to the right. The lines connecting students to secondary school cross each other because primary and secondary schools are crossed. To the left of primary schools, we see the virtual level 3 units to which they belong, shown as even longer vertical lines. Both primary and secondary schools are nested within these virtual units. The random effect for primary school can be modeled by one random intercept, nested within the virtual level 3. For secondary school, the model requires 3 random coefficients of dummy variables d_{i1}, d_{i2}, and d_{i3} for the (at most) 3 secondary schools (numbered 1, 2, 3 in the figure) per virtual level-3 unit. The blobs show where these dummy variables take the value 1.

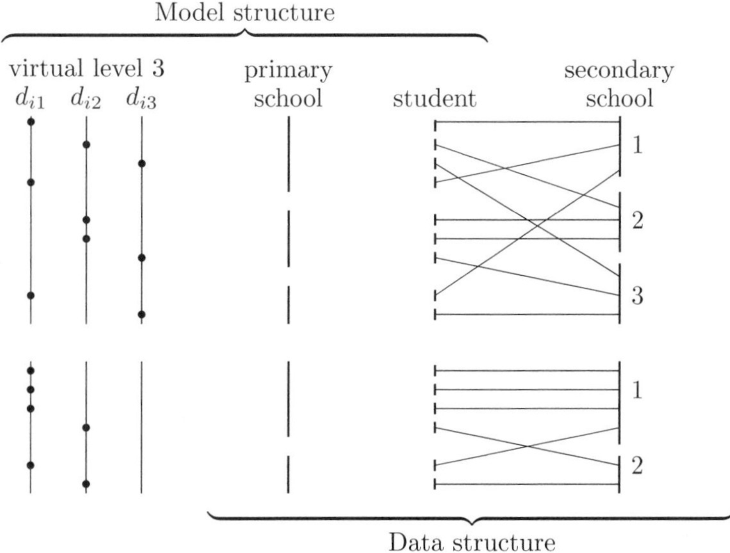

Figure 11.5: Model structure and data structure for students in primary schools crossed with secondary schools (*Source:* Skrondal and Rabe-Hesketh 2004)

The virtual level-3 clustering variable can be produced using the command `supclust` developed by Ben Jann and available from the Statistical Software Component (SSC) archive maintained by Kit Baum. The command can be downloaded using

```
. ssc install supclust
```

The syntax for generating a new variable `region` to serve as level-3 identifier is

```
. supclust pid sid, gen(region)
1 clusters in 3435 observations
```

Here one supercluster is found and thus the problem cannot be simplified. It is not possible to subdivide the sample into clusters within which both primary and secondary schools are nested.

To illustrate the trick, we will therefore delete primary and secondary school combinations that occur fewer than 3 times, which is also done in the MLwiN manual (Rasbash et al. 2005):

```
. egen num = count(attain), by(pid sid)
. drop if num<3
(168 observations deleted)
```

Now we will again try creating a virtual level-3 identifier:

```
. drop region
. supclust pid sid, gen(region)
6 clusters in 3267 observations
```

The command has identified six regions within which both sid and pid are nested. We create a new identifier for secondary schools, taking the values 1 to n_k within each region k:

```
. by region sid, sort: generate f = _n==1
. by region: generate sec = sum(f)
. table sec
```

sec	Freq.
1	1,020
2	769
3	267
4	292
5	467
6	99
7	249
8	104

There are at most 8 secondary schools per region, reducing the number of random effects required to 8 for secondary schools (compared with 19 previously) and 1 for primary schools.

In the xtmixed command, we simply replace _all by region and sid by sec:

```
. xtmixed attain || region: R.sec || pid:, mle
Mixed-effects ML regression                     Number of obs      =      3267
```

Group Variable	No. of Groups	Observations per Group Minimum	Average	Maximum
region	6	78	544.5	1330
pid	135	3	24.2	72

```
                                                Wald chi2(0)       =         .
Log likelihood = -8153.6587                     Prob > chi2        =         .
```

attain	Coef.	Std. Err.	z	P>|z|	[95% Conf.	Interval]
_cons	5.58194	.1812368	30.80	0.000	5.226722	5.937157

Random-effects Parameters	Estimate	Std. Err.	[95% Conf.	Interval]
region: Identity				
sd(R.sec)	.61966	.1546552	.3799356	1.010641
pid: Identity				
sd(_cons)	1.050122	.098281	.8741293	1.261548
sd(Residual)	2.84682	.0360606	2.777012	2.918382

```
LR test vs. linear regression:       chi2(2) =   263.22   Prob > chi2 = 0.0000
Note: LR test is conservative and provided only for reference
```

The estimates are not identical to the previous estimates because we dropped a small proportion of the observations. We would not generally recommend dropping any observations but did so here to illustrate the trick.

11.8 Do salamanders from different populations mate successfully?

We now consider the famous salamander mating data from three experiments conducted by S. Arnold and P. Verrell. The data were first introduced into the statistical literature by McCullagh and Nelder (1989) and then analyzed by Karim and Zeger (1992), Breslow and Clayton (1993), and many others. The purpose of the experiment was to investigate whether salamanders from two different populations, roughbutts and whitesides, which had been geographically isolated from each other, would cross-breed.

The first experiment in the summer of 1986 used two groups of 20 salamanders. Each group comprised 5 roughbutt males (RBM), 5 whitesides males (WSM), 5 roughbutt females (RBF), and 5 whitesides females (WSF). Within each group, 60 male–female pairs were formed so that each salamander had 3 partners from the same population and 3 partners from the other population. This design is shown under "Experiment 1" in table 11.4, where the pairs are indicated by a 1 (successful mating) or 0 (unsuccessful mating). Two further similar experiments were conducted in the fall of 1987 as shown under "Experiment 2" and "Experiment 3" in table 11.4. Experiment 2 actually used the same salamanders as experiment 1, with salamander 21 corresponding to salamander 1, salamander 31 to salamander 11, etc. Experiment 3 used a new set of salamanders.

(Continued on next page)

Table 11.4: Salamander mating data

Experiment 1

Group 1

		RBM					WSM				
		01	02	03	04	05	06	07	08	09	10
RBF	01	1			1	1	0			1	1
	02	1		1		1		1	1		1
	03	1	1	1			0		1	1	
	04		1		1	1		1		1	0
	05		1	1	1		1	1	1		
WSF	06	0	0		0			0	1		0
	07	1	0			1	1			1	1
	08			0	0	0	1	1			1
	09	1		1	0		1		1	1	
	10		0	1		0		1	1	0	

Group 2

		RBM					WSM				
		11	12	13	14	15	16	17	18	19	20
RBF	11		0	1	1		1			1	1
	12	0		0	1			0	0	0	
	13	1			1	0	0	0			1
	14	1	0			1			1	1	0
	15		0	0		0	1	0	1		
WSF	16		1		0	0			1	1	1
	17	0	0		0		0	1			0
	18	0	0	0			1	0		0	
	19			1	0	0			1	1	1
	20	0		0		0	0	0	1		

Experiment 2

Group 3

		RBM					WSM				
		21	22	23	24	25	26	27	28	29	30
RBF	21	1			1	1	0			0	0
	22	0		1		1		1	0		1
	23	1	0	1			1		1	1	
	24		0		1	0		0		0	0
	25		0	1	1		0	0	0		
WSF	26	1	0		1			1	1		1
	27	0	0			0	0			1	1
	28			0	0	0	0	0			0
	29	1		1	1		1		1	1	
	30		0	1		0		1	1	1	

Group 4

		RBM					WSM				
		31	32	33	34	35	36	37	38	39	40
RBF	31		0	0	1		0			1	1
	32	1		1	1			0	1	0	
	33	0			0	0	1	0			1
	34	1	1			1			1	0	1
	35		0	1		0	1	0	1		
WSF	36		0		0	0			1	1	0
	37	0	0		0		1	0			1
	38	0	0	0			1	1		1	
	39			1	0	0			1	1	0
	40	0		0		0	0	0	0		

Experiment 3

Group 5

		RBM					WSM				
		41	42	43	44	45	46	47	48	49	50
RBF	41	1			1	1	0			1	1
	42	0		0		0		0	1		0
	43	1	1	1			1		1	0	
	44		1		1	0		0		0	1
	45		1	0	1		0	0	1		
WSF	46	0	0		0			0	1		0
	47	0	0			0	0			1	1
	48			0	0	0	0	1			1
	49	0		0	0		0		1	1	
	50		0	0		0		0	1	1	

Group 6

		RBM					WSM				
		51	52	53	54	55	56	57	58	59	60
RBF	51		1	1	1		1			0	1
	52	0		1	0			0	0	0	
	53	1			0	1	1	1			1
	54	0	0			1			1	1	0
	55		1	1		1	1	0	1		
WSF	56		0		0	0			1	0	1
	57	0	0		0		1	1			1
	58	1	0	1			1	0		0	
	59			1	0	1			0	0	1
	60	0		0		1	1	1	1		

The dataset `salamander.dta` has the following variables:

- `y`: indicator for successful mating (1: successful; 0: unsuccessful)
- `male`: male identifier as in table 11.4 (1–60)
- `female`: female identifier as in table 11.4 (1–60)
- `experiment`: experiment number (1–3)
- `group`: identifiers of 6 groups of 20 salamanders as in table 11.4 with groups {1,2}, {3,4}, and {5,6} used in experiments 1, 2, and 3, respectively
- `rbm`, `wsm`, `rbf`, `wsf`: dummy variables for salamander being of the given type: roughbutt males, whitesides males, roughbutt females, and whitesides females, respectively

Since the design is complicated, we should first make sure that we understand the data correctly by, for instance, producing the cross-tabulation for the first groups for experiment 1:

```
. use http://www.stata-press.com/data/mlmus2/salamander, clear
. table female male if group==1, contents(mean y)
```

	male									
female	1	2	3	4	5	6	7	8	9	10
1	1			1	1	0			1	1
2	1		1		1		1	1		1
3	1	1	1			0		1	1	
4		1		1	1		1		1	0
5		1	1	1		1	1	1		
6	0	0		0			0	1		0
7	1	0			1	1			1	1
8			0	0	0	1	1			1
9	1		1	0		1		1	1	
10		0	1		0		1	1	0	

11.9 Crossed random-effects logistic regression

We now consider Model A from Karim and Zeger (1992) that treated the salamanders from experiments 1 and 2 as independent. A logistic random-effects regression model was specified that included the covariates `wsm` (x_{2i}), `wsf` (x_{3j}), and their interaction, as well as random intercepts ζ_{1i} for males i and ζ_{2j} for females j

$$\text{logit}(\Pr(y_{ij} = 1|x_{2i}, x_{3j}, \zeta_{1i}, \zeta_{2j}) \ = \ \beta_1 + \beta_2 x_{2i} + \beta_3 x_{3j} + \beta_4 x_{2i} x_{3j} + \zeta_{1i} + \zeta_{2j}$$

where as usual the random intercepts are assumed to be independently distributed as $\zeta_{1i}|x_{2i}, x_{3j} \sim N(0, \psi_1)$ and $\zeta_{2j}|x_{2i}, x_{3j} \sim N(0, \psi_2)$.

We can choose one of the sexes as a level and use random coefficients for the other. This choice is arbitrary because we have the same number of males and females. With random coefficients for males, the random part of the model can be specified as `|| _all: R.male || female:` but this would require 60 random coefficients.

We can instead exploit the fact that salamanders are nested within groups with no matings occurring across groups. This allows us to use the trick described in section 11.7 with `group` as the highest level in the model. We then only need 10 random coefficients for males; the first random coefficient multiplies a dummy variable for the salamander being one of the male salamanders 1, 11, 21, 31, 41, and 51, the second is for male salamanders 2, 12, 22, 32, 42, and 52, and so forth. These groups of 6 salamanders do not share the same value of the random coefficient because they are in different groups and the random coefficient is nested in groups.

Unfortunately, numerical integration is computationally demanding for crossed random-effects models. We will therefore initially use one integration point giving the Laplace approximation. We generally do not recommend relying on this approach alone without conducting some simulations to make sure that the estimates are reasonable.

We first relabel the males so that the groups of salamanders sharing the same random coefficient have the same label. (We do the same for females although this is not necessary.)

```
. generate m = male - (group-1)*10
. generate f = female - (group-1)*10
```

Next we generate the interaction term

```
. generate ww = wsf*wsm
```

and fit the model using `xtmelogit` with the `laplace` option:

```
. xtmelogit y wsm wsf ww || group: R.m || f:, laplace
Mixed-effects logistic regression               Number of obs      =       360

--------------------------------------------------------------------------
                |   No. of       Observations per Group       Integration
 Group Variable |   Groups    Minimum    Average    Maximum      Points
----------------+---------------------------------------------------------
          group |        6         60       60.0         60           1
              f |       60          6        6.0          6           1
--------------------------------------------------------------------------

                                                Wald chi2(3)       =     37.57
Log likelihood = -209.27659                     Prob > chi2        =    0.0000

------------------------------------------------------------------------------
           y |      Coef.   Std. Err.      z    P>|z|     [95% Conf. Interval]
-------------+----------------------------------------------------------------
         wsm |  -.7020417   .4614756    -1.52   0.128    -1.606517    .2024339
         wsf |   -2.90421   .5608211    -5.18   0.000    -4.003399   -1.805021
          ww |   3.588444   .6390801     5.62   0.000      2.33587    4.841018
       _cons |   1.008221   .3937705     2.56   0.010     .2364454    1.779997
------------------------------------------------------------------------------
```

```
------------------------------------------------------------------------------
  Random-effects Parameters  |   Estimate   Std. Err.     [95% Conf. Interval]
-----------------------------+------------------------------------------------
group: Identity              |
                     sd(R.m) |   1.020296    .2483604      .6331794    1.64409
-----------------------------+------------------------------------------------
f: Identity                  |
                  sd(_cons)  |   1.083683     .253092      .6856555   1.712769
------------------------------------------------------------------------------
LR test vs. logistic regression:     chi2(2) =    27.02   Prob > chi2 = 0.0000
Note: LR test is conservative and provided only for reference.
Note: log-likelihood calculations are based on the Laplacian approximation.
```

Having reduced the dimensionality of integration to 10 at level 3 and 1 at level 2 (compared with 60 at level 3 and 1 at level 2 without the trick), it becomes feasible to use adaptive quadrature in `xtmelogit` with 2 integration points. To speed things up, we use the estimates from the Laplace approximation as starting values. We use the `refineopts(iterate(0))` option to prevent `xtmelogit` from computing its own starting values.

```
. matrix a = e(b)
. xtmelogit y wsm wsf ww || group: R.m || f:, intpoints(2) from(a)
> refineopts(iterate(0))
Mixed-effects logistic regression               Number of obs      =       360

--------------------------------------------------------------------------
                |   No. of       Observations per Group       Integration
 Group Variable |   Groups    Minimum    Average    Maximum      Points
----------------+---------------------------------------------------------
          group |        6         60       60.0         60           2
              f |       60          6        6.0          6           2
--------------------------------------------------------------------------

                                                Wald chi2(3)       =     37.40
Log likelihood = -208.39862                     Prob > chi2        =    0.0000

------------------------------------------------------------------------------
           y |      Coef.   Std. Err.      z    P>|z|     [95% Conf. Interval]
-------------+----------------------------------------------------------------
         wsm |  -.6946282   .4668329    -1.49   0.137    -1.609604    .2203475
         wsf |   -2.91089   .5661133    -5.14   0.000    -4.020451   -1.801328
          ww |   3.587045   .6378699     5.62   0.000     2.336843    4.837247
       _cons |   1.003062   .3996581     2.51   0.012     .2197464    1.786378
------------------------------------------------------------------------------

------------------------------------------------------------------------------
  Random-effects Parameters  |   Estimate   Std. Err.     [95% Conf. Interval]
-----------------------------+------------------------------------------------
group: Identity              |
                     sd(R.m) |   1.058889    .2481504      .6689158   1.676213
-----------------------------+------------------------------------------------
f: Identity                  |
                  sd(_cons)  |   1.114577    .2522591      .7152534   1.736841
------------------------------------------------------------------------------
LR test vs. logistic regression:     chi2(2) =    28.78   Prob > chi2 = 0.0000
Note: LR test is conservative and provided only for reference.
```

This took only some minutes on our computer, so we increase the number of integration points to 3:

```
. matrix a=e(b)
. xtmelogit y wsm wsf ww || group: R.m || f:, intpoints(3) from(a)
> refineopts(iterate(0))
Mixed-effects logistic regression               Number of obs      =       360
```

Group Variable	No. of Groups	Observations per Group Minimum	Average	Maximum	Integration Points
group	6	60	60.0	60	3
f	60	6	6.0	6	3

```
                                                Wald chi2(3)       =     37.07
Log likelihood = -207.71151                     Prob > chi2        =    0.0000
```

y	Coef.	Std. Err.	z	P>|z|	[95% Conf. Interval]	
wsm	-.6962913	.4750504	-1.47	0.143	-1.627373	.2347903
wsf	-2.946427	.5785634	-5.09	0.000	-4.080391	-1.812464
ww	3.621843	.6442771	5.62	0.000	2.359083	4.884603
_cons	1.013944	.4097109	2.47	0.013	.2109251	1.816962

Random-effects Parameters	Estimate	Std. Err.	[95% Conf. Interval]	
group: Identity				
sd(R.m)	1.100838	.2558109	.6981082	1.735898
f: Identity				
sd(_cons)	1.161528	.2610426	.7477043	1.804385

```
LR test vs. logistic regression:     chi2(2) =    30.15   Prob > chi2 = 0.0000
Note: LR test is conservative and provided only for reference.
```

This command took a night to run.

The estimates using the Laplace approximation and adaptive quadrature with 3 integration points (AQ-3pt) are given in table 11.5 together with estimates using Monte Carlo EM (MCEM) from Vaida and Meng (2005), Markov chain Monte Carlo (MCMC) with noninformative priors from Karim and Zeger (1992), and penalized quasilikelihood (PQL) from Breslow and Clayton (1993). Treating the estimates from MCEM as the gold standard (because they are approximate maximum likelihood estimates employing Monte Carlo integration), the Laplace estimates are considerably better than the PQL estimates, and the adaptive quadrature estimates with just 3 quadrature points are nearly identical to the MCEM estimates.

Table 11.5: Different estimates for the salamander mating data

	Laplace	AQ-3pt	MCEM	MCMC	PQL
	Est (SE)	Est (SE)	Est	Est (SE)*	Est (SE)
Fixed part					
β_1 [_cons]	1.00 (0.39)	1.01 (0.41)	1.02	1.03 (0.43)	0.79 (0.32)
β_2 [wsm]	−0.70 (0.46)	−0.70 (0.48)	−0.70	−0.69 (0.50)	−0.54 (0.39)
β_3 [wsf]	−2.90 (0.56)	−2.95 (0.58)	−2.96	−3.01 (0.60)	−2.29 (0.43)
β_4 [wsm×wsf]	3.59 (0.64)	3.62 (0.64)	3.63	3.74 (0.68)	2.82 (0.50)
Random part					
$\sqrt{\psi_1}$ [Males]	1.02	1.10	1.11	1.16	0.79
$\sqrt{\psi_2}$ [Females]	1.08	1.16	1.18	1.22	0.84

* Range of 90% CI divided by 3.3

When the random intercepts are zero, $\zeta_{1i}=\zeta_{2j}=0$, the odds of success for combination RBM–RBF are estimated as

```
. lincom _cons, or
 ( 1)  [eq1]_cons = 0
```

y	Odds Ratio	Std. Err.	z	P>|z|	[95% Conf.	Interval]
(1)	2.75645	1.129348	2.47	0.013	1.23482	6.153139

For RBM–WSF, we get

```
. lincom _cons + wsf, or
 ( 1)  [eq1]wsf + [eq1]_cons = 0
```

y	Odds Ratio	Std. Err.	z	P>|z|	[95% Conf.	Interval]
(1)	.1447882	.0675077	-4.14	0.000	.0580576	.3610831

for WSM–RBF

```
. lincom _cons + wsm, or
 ( 1)  [eq1]wsm + [eq1]_cons = 0
```

y	Odds Ratio	Std. Err.	z	P>|z|	[95% Conf.	Interval]
(1)	1.373899	.5370913	0.81	0.416	.6385557	2.956042

and for WSM–WSF

```
. lincom _cons + wsm + wsf + ww, or
 ( 1)  [eq1]wsm + [eq1]wsf + [eq1]ww + [eq1]_cons = 0
```

y	Odds Ratio	Std. Err.	z	P>|z|	[95% Conf. Interval]	
(1)	2.699504	1.09926	2.44	0.015	1.215256	5.99653

Within the same population, the odds are high for both whitesides and roughbutts. Across populations the odds are considerably lower, particularly if the male is a roughbutt and the female a whiteside. However, for a given type of pairing, the median odds ratio comparing the more fertile of two randomly chosen pairs with the less fertile pair is large

```
. display exp(sqrt(2*(1.10^2+1.16^2))*invnormal(3/4))
4.5946103
```

indicating that there is considerable unexplained heterogeneity.

11.10 Summary and further reading

In this chapter, we discussed crossed-effects models for data with two crossed clustering variables, or cross-classified random factors. We described models with additive random effects of the factors, as well as a model including a random interaction between factors. The latter is identified only if there are several observations for at least some of the combinations of the categories of the factors.

We considered three typical examples requiring crossed random effects: panel data with random time effects, observational data on individuals cross-classified in two ways, and data from an experimental designs with crossed blocking factors. Another typical example where several raters (random factor 1) rate each of several objects (random factor 2) is discussed in exercises 11.4 and 11.7. Slightly more elaborate versions of models with crossed random effects are often applied to social-network data where each individual rates for instance how much they like every other individual. The crossed factors are the senders and the receivers of the judgments where each person is both a sender and a receiver. A nice social-network application is discussed in Snijders and Bosker (1999).

In all three examples considered in this chapter, there were only two random factors. In exercise 11.3, we consider a problem with three factors: occasions, states, and regions. Although states are nested in regions, both states and regions are cross-classified with occasions. In exercise 11.8, state of birth and state of residence are crossed, and mothers are nested within the cross-classifications.

Models with crossed random effects must be distinguished from *multiple membership models*. Both types of models share the common feature that a unit is not simply nested in a given cluster as in multilevel models. In multiple membership models, a unit is a member of several clusters with known weights designating the degree of membership. For instance, a student may attend several secondary schools with the time spent in each school representing the weight.

Goldstein (2003), Snijders and Bosker (1999), Raudenbush and Bryk (2002), and Leyland and Goldstein (2001) have chapters on models with crossed random effects; the latter also covers multiple membership models. We also recommend an excellent encyclopedia entry by Rasbash (2005) for crossed random effects models and multiple membership models. Linear two-way error-components models are extensively discussed in Hsiao (2002) and Baltagi (2005).

11.11 Exercises

11.1 Fife school data

Here we revisit the Fife school data analyzed in this chapter and described on page 481. In addition to the variables listed there, the dataset `fife.dta` contains a verbal reasoning score `vrq`, which was considered a measure of ability by Paterson (1991).

1. Fit the model with the same random part as in (11.2) but with covariates `sex` and `vrq`. Use maximum likelihood estimation.
2. Interpret the estimates, and discuss the change in the estimated variance components.

11.2 Airline cost data

Greene (2008) provides data on the annual total costs and output, fuel price and load factor for six U.S. airlines over 15 years from 1970 to 1984. The data were provided to Greene by Professor Moshe Kim and have also been analyzed by Greene (1997). The variables in `airlines.dta` are

- `airline`: airline identifier
- `year`: year number (1–15)
- `cost`: total annual cost in $1000
- `output`: annual output, in revenue passenger miles, index number
- `fuelprice`: fuel price
- `loadf`: load factor, the average capacity utilization of the fleet

Here the fuel price differs between airlines in a given year because different airlines use different mixes of types of planes and because there are regional differences in supply characteristics.

1. Write down a regression model for the log cost regressed on the log output, log fuel price, and load factor with random effects for airlines and years.
2. Fit the model using `xtmixed` with the `mle` option.

3. Perform likelihood-ratio tests for the null hypotheses that there is no residual variability between airlines and between years. If there are convergence problems when fitting the restricted models, use `xtreg` and specify the `force` option when performing the likelihood-ratio test.
4. Obtain empirical Bayes predictions of the random effects.
5. Plot the empirical Bayes predictions for the years against time.
6. Fit the model using fixed effects for years instead of random effects. Explain how and why the estimated coefficient of log fuel price changes compared with the model that treats year as random.

11.3 U.S. production data

In exercise 10.3, we considered a three-level model for state productivity.

1. Fit the three-level model described in exercise 10.3 using `xtmixed` with the `mle` option.
2. Add a random effect for year to this model. Choose the most efficient model specification in `xtmixed` that minimizes the number of random effects at a level. Fit the model by maximum likelihood.
3. Compare the models using a likelihood-ratio test.
4. Interpret the variance-component estimates for the model in step 2.

11.4 Video-ratings data

The Vancouver Sedative Recovery Scale (VSRS) was developed to measure recovery from sedation following pediatric open heart surgery. MacNab et al. (1994) report a study where 16 ICU staff were trained by videotape instruction to use the VSRS. To determine whether videotaped instruction produced adequate skill, an interobserver reliability study was carried out.

In a balanced incomplete design, 16 staff each rated a different subset of 16 video taped case examples using the VSRS so that each case was rated by 6 raters and each rater rated 6 cases. Two experts also rated all cases. The data, shown in table 11.6, were also analyzed by Dunn (2004).

Table 11.6: Rating data for 16 cases in incomplete block design

	Cases or Videos															
Raters	1	2	3	4	5	6	7	8	9	10	11	12	13	14	15	16
1	.	.	.	6	.	.	.	0	18	.	12	.	14	.	.	19
2	.	.	.	5	.	4	.	.	13	13	.	0	.	.	16	.
3	.	.	20	.	13	1	.	0	15	.	.	.	.	15	.	.
4	9	.	.	.	.	.	22	.	19	18	.	.	14	14	.	.
5	11	0	21	.	.	.	.	.	19	.	10	.	.	.	19	.
6	.	1	.	.	13	.	22	.	16	.	.	0	.	.	.	22
7	12	0	.	0	.	1	22	0	.	.	.	.	.	.	.	.
8	.	1	19	5	14	.	.	.	.	15	.	.	14	.	.	.
9	.	.	.	9	16	.	15	.	.	.	10	.	.	12	19	.
10	10	.	17	7	.	.	.	.	.	.	.	0	.	13	.	20
11	.	0	.	.	.	.	.	0	.	.	.	0	13	16	22	.
12	.	.	20	.	.	.	20	0	.	12	10	0	.	.	.	.
13	6	.	.	.	15	.	.	0	.	17	.	.	.	.	19	22
14	8	.	.	.	16	1	.	.	.	.	11	0	15	.	.	.
15	.	.	21	.	.	1	18	.	.	.	.	.	16	.	21	22
16	.	0	.	.	.	1	.	.	.	18	15	.	.	10	.	21
17	10	0	21	8	18	4	16	0	18	16	10	0	16	12	19	20
18	10	0	21	5	19	7	18	0	16	15	12	0	14	14	20	21

Source: MacNab et al. (1994)

The long version of these data in `videos.dta` contains the following variables:

- `rater`: rater identifier j
- `video`: video (case example) identifier i
- `y`: VSRS score
- `novice`: dummy variable for rater being a novice (not an expert)

1. In generalizability theory, the design of this study can be viewed as a one-facet crossed design; see Shavelson and Webb (1991). Here the *objects of the measurement* are the cases shown in the videos, and these are crossed with raters, who can be thought of as *conditions* of a *facet* of the measurement.

   We wish to generalize from this particular set of raters to a *universe* of raters, treating raters as a random sample of all *admissible* raters. Denoting cases or videos as i and raters as j, the model can be written as

   $$y_{ij} \;=\; \beta + \zeta_{1i} + \zeta_{2j} + \epsilon_{ij}, \quad \psi_1 \equiv \text{Var}(\zeta_{1i}), \psi_2 \equiv \text{Var}(\zeta_{2j}), \theta \equiv \text{Var}(\epsilon_{ij})$$

   Here ϵ_{ij} represents the sum of the case by rater interaction and any other sources of error. Fit this model for the nonexperts using restricted maximum likelihood.

2. Interchange `rater` and `video` in your command for step 1, and compare the estimates.
3. If the intention is to use the VSRS score from one rater for an *absolute decision*, not just for the purpose of ranking cases rated by the same rater, the generalizability coefficient is defined as

$$\phi = \frac{\psi_1}{\psi_1 + \psi_2 + \theta}$$

where $\psi_2+\theta$ represents the measurement error variance for absolute decisions. Obtain an estimate of this coefficient by plugging in the estimated variance components from steps 1 or 2.
4. If the intention is to use the mean score from N_r raters for an absolute decision, the measurement error variance can be divided by N_r and the generalizability coefficient is defined as

$$\phi = \frac{\psi_1}{\psi_1 + (\psi_2 + \theta)/N_r}$$

Estimate this generalizability coefficient for $N_r = 6$.

11.5 Neighborhood-effects data

We now consider the data from Garner and Raudenbush (1991), Raudenbush and Bryk (2002), and Raudenbush et al. (2004) previously analyzed in exercise 3.2. In that exercise, we ignored the fact that the students are not only nested in neighborhoods but also in schools, which are crossed with neighborhoods.

The variables in `neighborhood.dta` are

- Student level
  - `attain`: a measure of educational attainment
  - `p7vrq`: primary 7 verbal reasoning quotient
  - `p7read`: primary 7 reading test scores
  - `dadocc`: father's occupation scaled on the Hope–Goldthorpe scale in conjunction with the Registrar General's social-class index (Willms 1986)
  - `dadunemp`: dummy variable for father being unemployed (1: unemployed; 0: not unemployed)
  - `daded`: dummy variable for father's schooling being past the age of 15
  - `momed`: dummy variable for mother's schooling being past the age of 15
  - `male`: dummy variable for student being male
- Neighborhood level
  - `neighid`: neighborhood identifier
  - `deprive`: social deprivation score, derived from poverty concentration, health, and housing stock of local community

- School level
  - `schid`: school identifier

1. Fit a model for student educational attainment without covariates but with random intercepts of neighborhood and school using maximum likelihood.
2. Include a random interaction between neighborhood and school, and use a likelihood-ratio test to compare this model with the previous model.
3. Include the neighborhood-level covariate `deprive`, and discuss both the estimated coefficient of `deprive` and the changes in the standard deviation estimates for the random effects due to including this covariate.
4. Remove the neighborhood-by-school random interaction (which is no longer significant at the 5% level) and include all student-level covariates. Interpret the estimated coefficients and the change in the standard deviation estimates.
5. For the final model, estimate residual intraclass correlations due to being in the same neighborhood but not the same school, the same school but not the same neighborhood, and both in the same neighborhood and in the same school.
6. ❖ Use the `supclust` command to see if estimation can be simplified by defining a virtual level-3 identifier.

11.6 Nitrogen data

Littell et al. (2006) describe an experiment to evaluate the yield of two varieties of crop at five levels of nitrogen fertilization. The five levels of nitrogen were applied to 15 relatively large whole plots that formed a 5 × 3 grid. Because of substantial north-south and east-west gradients, the levels of nitrogen were applied in an incomplete Latin-square design, with each nitrogen level occurring in each row, but not in each column:

	Col 1	Col 2	Col 3	Col 4	Col 5
Row 1	Nit 1	Nit 2	Nit 5	Nit 4	Nit 3
Row 2	Nit 2	Nit 1	Nit 3	Nit 5	Nit 4
Row 3	Nit 3	Nit 4	Nit 1	Nit 2	Nit 5

Each whole plot was split into two subplots to which the two crops were randomly assigned.

Letting i denote the subplots, j the rows, and k the columns in which the whole plots are arranged, a crossed random-effects model without a fixed part can be written as

$$y_{ijk} \;=\; \beta_1 + \zeta_{1j} + \zeta_{2k} + \zeta_{3jk} + \epsilon_{ijk}$$

where $\zeta_{1j} \sim N(0, \psi_1)$ is the random effect of row, $\zeta_{2k} \sim N(0, \psi_2)$ is the random effect of column, $\zeta_{3jk} \sim N(0, \psi_3)$ is the random interaction between row and column, and $\epsilon_{ijk} \sim N(0, \theta)$ is a residual error term.

The variables in `nitrogen.dta` are

- `row`: row identifier j
- `col`: column identifier k
- `N`: nitrogen level (1, 2, 3, 4, 5)
- `G`: genotype or crop variety (1, 2)
- `y`: yield of the crop y_{ijk}

1. Fit the model given above and interpret the estimated variance components.
2. Fit a model with the same random part as in step 1 but also including fixed effects of nitrogen level (treated as unordered), genotype, and their interaction.
3. Perform a Wald test for the nitrogen by genotype interaction, and omit the interaction terms if this test is not significant at the 5% level.
4. Interpret the estimated regression coefficients.

11.7 Olympic skating data

In the 1932 Lake Placid Winter Olympics, seven figure skating pairs were judged by seven judges using two different criteria (program and performance). The ratings are provided by Gelman and Hill (2007) and are shown in table 11.7, where the countries of origin of judges and pairs are also given as abbreviations for France, the United States, Hungary, Canada, Norway, Austria, Finland, and the United Kingdom.

Table 11.7: Ratings of 7 skating pairs by 7 judges using 2 criteria (program and performance) in the 1932 Winter Olympics

	Judge													
	1		2		3		4		5		6		7	
Pair	(Hun)		(Nor)		(Aus)		(Fin)		(Fra)		(UK)		(US)	
1 (Fra)	5.6	5.6	5.5	5.5	5.8	5.8	5.3	4.7	5.6	5.7	5.2	5.3	5.7	5.4
2 (US)	5.5	5.5	5.2	5.7	5.8	5.6	5.8	5.4	5.6	5.5	5.1	5.3	5.8	5.7
3 (Hun)	6.0	6.0	5.3	5.5	5.8	5.7	5.0	4.9	5.4	5.5	5.1	5.2	5.3	5.7
4 (Hun)	5.6	5.6	5.3	5.3	5.8	5.8	4.4	4.8	4.5	4.5	5.0	5.0	5.1	5.5
5 (Can)	5.4	4.8	4.5	4.8	5.8	5.5	4.0	4.4	5.5	4.6	4.8	4.8	5.5	5.2
6 (Can)	5.2	4.8	5.1	5.6	5.3	5.0	5.4	4.7	4.5	4.0	4.5	4.6	5.0	5.2
7 (US)	4.8	4.3	4.0	4.6	4.7	4.5	4.0	4.0	3.7	3.6	4.0	4.0	4.8	4.8

The data in `olympics.dta` contain one row for each combination of skating pair and judge with separate variables for the two criteria. The variables are

- `pair`: skating pair
- `judge`: judge
- `pcountry`: country of origin of pair

- `jcountry`: country of origin of judge
- `program`: rating for program
- `performance`: rating for performance

1. Write down a linear model for the program rating of judge j for pair k that includes a fixed overall intercept and random intercepts for judges and pairs. Interpret the random intercepts in the context of this example.
2. Fit the model from step 1 using restricted maximum likelihood estimation.
3. Test if both random intercepts are needed by comparing the model from step 1 with the two nested models having only one random intercept.
4. Extend the model by constructing a dummy variable for judge and pair coming from the same country and including this as a covariate. Interpret the estimated regression coefficient and comment on its magnitude.
5. Reshape the data to long form, stacking the ratings using the two criteria into one response variable, and produce a dummy variable for the criterion being performance.
6. ❖ Fit the same model as in step 5 to the responses using both criteria. The model makes these strong assumptions (among others):
   - The mean rating is the same for the two criteria after controlling for whether the pair and judge are from the same country
   - The ratings for the same skating pair from the same judge using two different criteria are conditionally independent given the covariate and the random intercepts for judges and pairs

   Fit an extended model that relaxes these assumptions.

11.8 Smoking and birthweight data

We now consider the data from Abrevaya (2006), discussed in chapter 3. There we ignored the fact that mothers are nested in U.S. states of residence crossed with the states in which they were born. Here we consider the effects of the mother's state of residence and state of birth.

The variables in `smoking.dta` that we will use here are

- `momid`: mother identifier
- `birwt`: birthweight (in grams)
- `stateres`: mother's state of residence
- `mplbir`: mother's place (state) of birth
- `male`: dummy variable for child being male

1. Fit a model for birthweight with random effects for state of residence and state of birth and a fixed effect of `male`.
2. Fit the same model as in step 1 but also include a random interaction between state of residence and state of birth. Compare this model with the model from step 1 using a likelihood-ratio test.

3. Fit the model from step 2 but with an additional random intercept for mothers. Mothers are nested within the state of residence and state of birth pairs (mothers who moved between births could not be matched and were not included in the data). Compare this model with the model from step 2 using a likelihood-ratio test.
4. Interpret the parameter estimates from step 3.

A Syntax for gllamm, eq, and gllapred: The bare essentials

Here we describe the features of `gllamm` and `gllapred` that are most important for multilevel and longitudinal modeling. We also describe the `eq` command that is required for fitting random-coefficient models in `gllamm`. The full set of options for `gllamm`, `gllapred`, and `gllasim` is given in appendices B, C, and D, respectively. See Stata's *Longitudinal/Panel-Data Reference Manual* for [XT] **xtmixed**.

Title

gllamm — Generalized linear latent and mixed models

Syntax

`gllamm` *depvar* [*indepvars*] [*if*] [*in*], `i(`*varlist*`)` [*options*]

options	description
Model	
*`i(`*varlist*`)`	cluster identifiers from level 2 and up
`nrf(`#,...,#`)`	number of random effects at each level
`eqs(`*eqnames*`)`	equation for each random effect
`family(`*familyname*`)`	distribution of *depvar*, given the random effects; default is `family(gaussian)`
`link(`*linkname*`)`	link function; default is canonical link for `family()` specified
Model 2	
`noconstant`	suppress constant or intercept
`offset(`*varname*`)`	include *varname* in model with coefficient constrained to 1
Integration method	
`adapt`	use adaptive quadrature; default is ordinary quadrature
`nip(`#`)`	number of integration points
Reporting	
`eform`	report exponentiated coefficients

*`i(`*varlist*`)` is required.

familyname	description
`gaussian`	Gaussian (normal)
`poisson`	Poisson
`binomial`	Bernoulli

linkname	description
`identity`	identity
`log`	log
`logit`	logit
`probit`	probit
`ologit`	ordinal logit, proportional odds
`oprobit`	ordinal probit

Description

`gllamm` fits a wide range of models. Here we consider multilevel generalized linear models.

In the two-level case, units i (e.g., students) are nested in clusters j (e.g., schools). In this case, the `i()` option specifies a single variable, the identifier for the clusters. The conditional expectation or mean μ_{ij} of the response y_{ij} (*depvar*) given the covariates and random effects is linked to the linear predictor ν_{ij} via a *link function* $g(\cdot)$, specified using the `link()` option,

$$g(E[y_{ij}|\nu_{ij}]) \equiv g(\mu_{ij}) \;=\; \nu_{ij}$$

The distribution of y_{ij} given its mean is a member of the *exponential family*, specified using the `family()` option. In a simple random-coefficient model, the *linear predictor* has the form

$$\nu_{ij} \;=\; \beta_1 + \beta_2 x_{ij} + \zeta_{1j}^{(2)} + \zeta_{2j}^{(2)} x_{ij}$$

where x_{ij} is a covariate, β_1 is a *constant* or intercept, and β_2 is the slope or regression coefficient of x_{ij}. This *fixed part* of the model is specified using *indepvars* (the constant is not explicitly specified but included by default). $\zeta_{1j}^{(2)}$ is a *random intercept*, whereas $\zeta_{2j}^{(2)}$ is a *random slope* or *random coefficient*. These two *random effects* at level 2 represent deviations for cluster j from the mean intercept β_1 and slope β_2, respectively. The fact that there are two random effects is declared using the `nrf()` option. The `eqs()` option is used to specify two corresponding equations previously defined using the `eq` command. Each equation consists of a single variable, containing the values multiplying the random effect, 1 and x_{ij}, respectively.

In a three-level model for units i nested in clusters j, nested in superclusters k (e.g., students in classes in schools), an example of a linear predictor would be

$$\nu_{ijk} = \beta_1 + \beta_2 x_{ijk} + \zeta_{1jk}^{(2)} + \zeta_{2jk}^{(2)} x_{ijk} + \zeta_k^{(3)}$$

where $\zeta_k^{(3)}$ is the deviation for supercluster k from the mean intercept. The three-level structure can be communicated to `gllamm` by specifying the identifiers for levels 2 and 3 in the `i()` option. The `nrf()` option now requires two numbers of random effects, one for level 2 and one for level 3, here `nrf(2 1)`, and the `eqs()` option expects three equation names, first two equations for level 2 and then another equation for level 3. Here the first and last equation names could be the same.

Options

Model

`i(`*varlist*`)` gives the variables that define the hierarchical, nested clusters, from the lowest level (finest clusters) to the highest level, e.g., `i(student class school)`.

`nrf(`*#*`,...,`*#*`)` specifies the number of random effects at each level, i.e., for each variable in `i(`*varlist*`)`. The default is `nrf(1,...,1)`.

`eqs(`*eqnames*`)` specifies the equation names (defined before running `gllamm` using `eq`; see page 512). In multilevel models, these equations simply define the variables multiplying the random effects. The number of equations per level is specified in the `nrf()` option.

`family(`*family*`)` specifies the family to be used for the distribution of the response given the mean. The default is `family(gaussian)`.

`link(`*link*`)` specifies the link function to be used, linking the conditional expectation of the response to the linear predictor.

Model 2

`noconstant`, `offset(`*varname*`)`; see [R] **estimation options**.

Integration method

`adapt` specifies adaptive quadrature. The default is ordinary quadrature.

`nip(`*#*`)` specifies the number of integration points.

Reporting

`eform` displays the exponentiated coefficients and corresponding standard errors and confidence intervals. For `family(binomial) link(logit)` (i.e., logistic regression), exponentiated coefficients are odds ratios; for `family(poisson) link(log)` (i.e., Poisson regression), exponentiated coefficients are rate ratios.

Title

eq — Equations

Syntax

`eq` [`define`] *eqname*: *varlist*

Description

`eq` [`define`] can be used to define an equation having the name *eqname* for use with estimation commands such as `gllamm`. The equation is just a linear combination of the variables in *varlist*. For instance,

```
eq slope: x
```

implies the equation λx, whereas

```
eq longer: x1 x2 x3 x4
```

implies the equation $\lambda_1 x_1 + \lambda_2 x_2 + \lambda_3 x_3 + \lambda_4 x_4$. If used to specify a component of a model in an estimation command, the λ parameters will typically be estimated. In `gllamm`, the `eqs()` option sets the coefficient of the first variable to 1. The `thresh()` and `peqs()` options add an intercept or constant to the linear combination, giving $\lambda_0 + \lambda_1 x_1 + \lambda_2 x_2 + \lambda_3 x_3 + \lambda_4 x_4$ in the previous example.

Title

gllapred — predict command for gllamm

Syntax

`gllapred` *varname* [*if*] [*in*] [, *statistic option*]

statistic	description
`u`	empirical Bayes predictions of random effects
`ustd`	standardized empirical Bayes predictions
`xb`	fixed part of linear predictor
`linpred`	linear predictor with empirical Bayes predictions of random effects plugged in
`mu`	mean of response; by default posterior mean
`pearson`	Pearson residual; by default posterior mean

option	description
`marginal`	combined with `mu`, gives marginal or population-averaged mean

Description

`gllapred` is the prediction command for `gllamm`.

For numerical integration, `gllapred` uses the same number of integration points as the preceding `gllamm` command and the same method (adaptive or nonadaptive quadrature).

For predictions of quantities that are linear in the random effects (`u`, `ustd`, and `linpred`), posterior means or empirical Bayes predictions are substituted for the random effects. These are means of the posterior distributions of the random effects given the observed responses, with parameter estimates plugged in. With the `u` option, standard deviations of the posterior distributions are also returned. The `ustd` option returns the empirical Bayes predictions divided by their approximate sampling standard deviations. The sampling standard deviation is approximated by the square root of the difference between the estimated prior variance $\widehat{\psi}$ and the posterior variance. For linear models, this approximation is exact (except that the parameter estimates are plugged in).

For predictions of quantities that are nonlinear functions of the random effects, the expectation of the nonlinear function is returned, by default, with respect to the posterior distribution of the random effects. The `marginal` option specifies that the expectation should be taken with respect to the prior distribution of the random effects (not conditioning on the observed responses).

Only one of the statistics may be requested at a time. With the `u` and `ustd` options, *varname* is the prefix used for the variables that will contain the predictions.

Options

`u` returns posterior means (empirical Bayes predictions) and posterior standard deviations of the random effects in *varname*`m1`, *varname*`m2`, etc., and *varname*`s1`, *varname*`s2`, etc., respectively, where the order of the random effects is the same as in the call to `gllamm`.

`ustd` returns standardized empirical Bayes predictions of the random effects in *varname*`m1`, *varname*`m2`, etc. Each empirical Bayes prediction is divided by the square root of the difference between the estimated prior and posterior variances, which equals the sampling standard deviation in linear models but only approximates it in other models.

`xb` returns the fixed-effects part of the linear predictor in *varname*.

`linpred` returns the linear predictor, including the fixed- and random-effects part, where empirical Bayes predictions are substituted for the random effects.

`mu` returns the expectation of the response, for example, the predicted probability in the case of dichotomous responses. By default, the expectation is with respect to the posterior distribution of the random effects; also see the `marginal` option.

`pearson` returns Pearson residuals. By default, the posterior expectation with respect to the random effects is returned.

`marginal` together with the `mu` option gives the expectation of the response with respect to the prior distribution of the random effects. This is useful for looking at the marginal or population-averaged effects of covariates.

B Syntax for gllamm

Here we give the full syntax for `gllamm` with all available options.

Title

gllamm — Generalized linear latent and mixed models

Syntax

`gllamm` *depvar* [*indepvars*] [*if*] [*in*], `i(`*varlist*`)` [*options*]

options	description
Linear predictor	
*`i(`*varlist*`)`	cluster identifiers from level 2 and up
`nrf(#,...,#)`	number of random effects at each level
`eqs(`*eqnames*`)`	equation for each random effect
†`frload(#,...,#)`	free first factor loading
Linear predictor 2	
`noconstant`	suppress constant or intercept
`offset(`*varname*`)`	include *varname* in model with coefficient constrained to 1
Response model	
`family(`*familynames*`)`	distributions of *depvar*, given the random effects; default is `family(gaussian)`
†`fv(`*varname*`)`	variable assigning distributions to units or observations
`link(`*linknames*`)`	link functions; default is canonical link for `family()` specified (if only one *familyname* is given)
†`lv(`*varname*`)`	variable assigning links to units or observations
†`denom(`*varname*`)`	variable containing binomial denominator if `link(binomial)` is used; default denominator is 1
`s(`*eqname*`)`	equation for log of residual standard deviation
`thresh(`*eqnames*`)`	equations for threshold models
†`ethresh(`*eqnames*`)`	equations for alternative threshold models
†`expanded(`*varname*...`)`	combined with `link(mlogit)` specifies that data are in expanded form
†`basecategory(#)`	reference category for multinomial logit models
†`composite(`*varname*...`)`	composite link

*`i(`*varlist*`)` is required.

options (cont'd)	description
Random-effects distribution	
†nocorrel	uncorrelated random effects
ip(*string*)	continuous or discrete random effects
Structural model	
geqs(*eqnames*)	equations for regressions of random effects on covariates
†bmatrix(*matrix*)	matrix for regressions among random effects
†peqs(*eqname*)	equation for multinomial logit model for probabilities of discrete distribution
Weights	
weight(*varname*)	frequency weights at different levels
pweight(*varname*)	inverse probability sampling weights at different levels
Constraints	
†constraints(*clist*)	linear parameter constraints
Integration method	
adapt	use adaptive quadrature; default is ordinary quadrature
ip(m)	use spherical integration rules; default is cartesian product
nip(#)	number of integration points
SE/Robust	
robust	standard errors based on sandwich estimator
cluster(*varname*)	adjust standard errors for intragroup correlation
Reporting	
eform	report exponentiated coefficients
†level(#)	confidence level for confidence intervals; default is 95%
†trace	report model details and more detailed iteration log
†nolog	suppress iteration log
†nodisplay	do not display estimates
†allc	display all parameters as they are estimated (sometimes transformation of parameters usually reported)
lf0(# #)	specify number of parameters and log likelihood of nested model for likelihood-ratio test
†eval	evaluate log likelihood for parameters given in from(*matrix*)
†init	fit the model where random part is set to 0
†noest	do not fit the model
†dots	display dots each time the log likelihood is evaluated

options (cont'd)	description
Starting values	
`from(`*matrix*`)`	row matrix of starting values
`copy`	ignore equation and column names of matrix specified in `from(`*matrix*`)`, relying on values being in correct order
`skip`	matrix of starting values has extra parameters
† `long`	matrix of starting values is for model without constraints (with `constraints()` option)
† `search(#)`	number of values to try in search for starting values for random part of model
Max options	
† `iterate(#)`	maximum number of Newton–Raphson iterations
† `adoonly`	use ado-version of `gllamm`
Gâteaux derivative	
`gateaux(# # #)`	range and number of steps for Gâteaux derivative search

† Options not used elsewhere in this book.

familyname	description
`gaussian`	Gaussian (normal)
`poisson`	Poisson
`binomial`	Bernoulli or binomial (with `denom(`*varname*`)`)
† `gamma`	gamma

† Options not used elsewhere in this book.

linkname	description
`identity`	identity
`log`	log
† `reciprocal`	reciprocal (power −1)
`logit`	logit
`probit`	probit
† `sprobit`	scaled probit
`cll`	complementary log-log
`ologit`	ordinal logit, proportional odds
`oprobit`	ordinal probit
`soprobit`	scaled ordinal probit
† `ocll`	ordinal complementary log-log
† `mlogit`	multinomial logit

† Options not used elsewhere in this book.

Description

`gllamm` fits generalized linear latent and mixed models (GLLAMMs). The models include random effects or latent variables varying at different levels. Here we use *random effect* instead of *latent variable* because we regard the terms more or less as synonyms and because this book is essentially about random-effects models. GLLAMMs consist of a linear predictor, response model, random-effects distribution, and structural model.

Linear predictor

For an L-level model, the most general form of the linear predictor is

$$\begin{aligned}
\nu &= \beta_1 x_1 + \cdots + \beta_p x_p \\
&\quad + \eta_1^{(2)}(\lambda_{11}^{(2)} z_{11}^{(2)} + \cdots + \lambda_{1q_{12}}^{(2)} z_{1q_{12}}^{(2)}) + \cdots \\
&\quad + \eta_{M_2}^{(2)}(\lambda_{M_2,1}^{(2)} z_{M_2,1}^{(2)} + \cdots + \lambda_{M_2,1}^{(2)} z_{M_2,q_{M_2,2}}^{(2)}) + \cdots \\
&\quad + \eta_1^{(L)}(\lambda_{11}^{(L)} z_{11}^{(L)} + \cdots + \lambda_{1q_{1L}}^{(L)} z_{1q_{1L}}^{(L)}) + \cdots \\
&\quad + \eta_{M_L}^{(L)}(\lambda_{M_L,1}^{(L)} z_{M_L,1}^{(L)} + \cdots + \lambda_{M_L,1}^{(L)} z_{M_L,q_{M_L,L}}^{(L)}), \\
&\quad \text{where } \lambda_{m1}^{(l)} = 1, \text{ for } l = 2, \ldots, L, \ m = 1, \ldots, M_l
\end{aligned}$$

where we have omitted subscripts for the units at the different levels $(ijk\ldots)$ to simplify notation. Here the fixed part in the first line has the usual form. In the random part, $\eta_m^{(l)}$ is a random effect at level l, where l goes from 2 to L and at each level, m goes from 1 to M_l. Each random effect is multiplied by a linear combination of covariates $z_{mk}^{(l)}$ with coefficients $\lambda_{mk}^{(l)}$, $k = 1, \ldots, q_{ml}$. These coefficients sometimes function as factor loadings in measurement models. In ordinary random-coefficient models, each term in parentheses collapses to a single variable $z_m^{(l)}$.

Using vector notation and summation signs, we can write the linear predictor more compactly as

$$\nu = \mathbf{x}'\boldsymbol{\beta} + \sum_{l=2}^{L} \sum_{m=1}^{M_l} \eta_m^{(l)} \mathbf{z}_m^{(l)\prime} \boldsymbol{\lambda}_m^{(l)}, \quad \lambda_{m1}^{(l)} = 1$$

In `gllamm`, the `i()` option is used to define the levels 2 to L of the model, and the `nrf()` option is used to specify the numbers of random effects M_l at each level. The `eqs()` option is used to specify equations defining the linear combinations $\mathbf{z}_m^{(l)\prime}\boldsymbol{\lambda}_m^{(l)}$ multiplying the random effects.

Response model

The response model is a generalized linear model with unit-specific link function and distribution from the exponential family, specified using the `link()`, `lv()`, `family()`, and `fv()` options. In addition, the variance of the response, given the linear predictor, can depend on covariates as can the thresholds in ordinal response models (via the `s()` and `thresh()` options, respectively).

Random-effects distribution

The random effects can be either continuous or discrete. In the continuous case, the following applies to the disturbances if there is a structural model (see below). The random effects are multivariate normal with zero means. Random effects at the same level may be correlated but there are no correlations across levels.

In the discrete case, the M_l random effects at level l take on discrete values that can be thought of as locations in M_l dimensions. The locations, together with the associated probabilities are parameters of the model. Random effects at different levels are again assumed to be mutually independent. Discrete random effects can be used for latent-class or finite-mixture models or for *nonparametric maximum likelihood* estimation. The `ip()` option is used to specify whether the random effects are continuous or discrete.

Structural model

Continuous case

In the structural model, each random effect can be regressed on covariates w

$$\eta_m^{(l)} = \gamma_{m1}^{(l)} w_{m1}^{(l)} + \cdots + \gamma_{m,r_{ml}}^{(l)} w_{m,r_{ml}}^{(l)} + \zeta_m^{(l)}$$

where $\zeta_m^{(l)}$ is a disturbance or residual. These linear combinations of covariates are defined as equations and passed to `gllamm` using the `geqs()` option.

The structural model also allows random effects to be regressed on other random effects. These relations, together with the regressions on observed covariates given above, are easiest to express using a column vector $\boldsymbol{\eta}$ for all random effects from levels 2 to L and $\boldsymbol{\zeta}$ for the corresponding disturbances

$$\boldsymbol{\eta} = \mathbf{B}\boldsymbol{\eta} + \boldsymbol{\Gamma}\mathbf{w} + \boldsymbol{\zeta}$$

where $\mathbf{B}$ (specified using the `bmatrix()` option) is an upper triangular matrix of regression coefficients allowing random effects to be regressed on other random effects. The term $\boldsymbol{\Gamma}\mathbf{w}$ represents the regressions of random effects on covariates written more explicitly above.

Discrete case

In the discrete case, the `peqs()` option can be used to specify multinomial logit models for the probabilities of the different locations for the random effects.

Options

Linear predictor

`i(`*varlist*`)` gives the variables that define the hierarchical, nested clusters, from the lowest level (finest clusters) to the highest level, e.g., `i(student class school)`.

`nrf(`#`,...,`#`)` specifies the number of random effects M_l at each level l, i.e., for each variable in `i(`*varlist*`)`. The default is `nrf(1,...,1)`.

`eqs(`*eqnames*`)` specifies the equation names (defined before running `gllamm`) for the linear combinations $\mathbf{z}_m^{(l)\prime}\boldsymbol{\lambda}_m^{(l)}$ multiplying the random effects. The equations for the level-2 random effects are listed first, followed by those for the level-3 random effects, etc., the number of equations per level being specified in the `nrf()` option. If required, constants should be explicitly included in the equation definitions using variables equal to 1. If the option is not used, the random effects are assumed to be random intercepts, and only one random effect is allowed per level. The first coefficient $\lambda_{m1}^{(l)}$ is set to one, unless the `frload()` option is specified. The other coefficients are estimated together with the (co)variances of the random effects.

`frload(`#`,...,`#`)` lists the random effects for which the first coefficient $\lambda_{m1}^{(l)}$ should be freely estimated instead of set to 1. It is up to the user to define appropriate constraints to identify the model. Here the random effects are referred to as 1 2 3, etc., in the order in which they are defined by the `eqs()` option.

Linear predictor 2

`noconstant`, `offset(`*varname*`)`; see [R] **estimation options**.

Response model

`family(`*familynames*`)` specifies the family (or families) to be used for the response probabilities (or densities) given the random effects. The default is `family(gaussian)`. Several families may be given, in which case, the variable allocating families to units or observations must be given using `fv(`*varname*`)`.

`fv(`*varname*`)` is required if several families are specified in the `family()` option. The variable indicates which family applies to which unit or observation. A value of one refers to the first family specified in `family()`, etc.

`denom(`*varname*`)` gives the variable containing the binomial denominator for the responses whose family was specified as `binomial`. The default denominator is 1.

`s(`*eqname*`)` specifies that the log of the standard deviation (or of the coefficient of variation) at level 1 for normally (or gamma) distributed responses (or the scale for the `sprobit` or `soprobit` links) is given by the linear combination of covariates defined by `eqname`. This allows for heteroskedasticity at level 1. For example, if dummy variables for groups are used in the definition of *eqname*, different residual variances are estimated for different groups.

`link(`*linknames*`)` specifies the link functions linking the conditional means to the linear predictors. If a single family is specified, the default link is the canonical link. Several links may be given, in which case, the variable assigning links to units or observations must be given using `lv(`*varname*`)`.

`lv(`*varname*`)` is the variable whose values indicate which link applies to which unit or observation. See `fv(`*varname*`)` for details.

`thresh(`*eqnames*`)` specifies equations for the thresholds for ordinal responses. One equation is specified for each ordinal response, and constants are automatically added. This option allows the effects of some covariates to differ between the categories of the ordinal outcome rather than assuming a constant effect—the parallel regression assumption, or with the `ologit` link, the proportional odds assumption. Variables used in the model for the thresholds generally should not appear in the fixed part of the linear predictor.

`ethresh(`*eqnames*`)` is the same as `thresh(`*eqnames*`)`, except that a different parameterization is used for the threshold model. To ensure that $\kappa_{s-1} \leq \kappa_s$, the model is $\kappa_s = \kappa_{s-1} + \exp(\mathbf{x}'\boldsymbol{\alpha})$, for $s = 2, \ldots, S-1$, where S is the number of response categories.

`expanded(`*varname varname string*`)` is used with the `mlogit` link and specifies that the data have been expanded as illustrated below

```
A                                  B
choice                             response altern selected
   1                                   1       1       1
   2                                   1       2       0
                                       1       3       0
                                       2       1       0
                                       2       2       1
                                       2       3       0
```

where the variable `choice` is the multinomial response (possible values 1, 2, 3), `response` labels the original lines of data, `altern` gives the possible responses or alternatives, and `selected` indicates the response that was given. The syntax would be `expanded(response selected m)`, and `altern` would be used as the dependent variable. This expanded form allows the user to have alternative-specific covariates, apply different random effects to different alternatives, and have different alternative sets for different individuals. The third argument is `o` if one set of coefficients should be estimated for the explanatory variables and `m` if one set of coefficients is to be estimated for each category of the response except the reference category.

`basecategory(`*#*`)` specifies the value of the response to be used as the reference category when the `mlogit` link is used. This option is ignored if the `expanded()` option is used with the third argument equal to `m`.

`composite(`*varname varname* `...)` specifies that a composite link be used. The first variable is a cluster identifier (`cluster` below) so that linear predictors within the cluster can be combined into a single composite link. The second variable (`ind` below) indicates to which response the composite links defined by the subsequent weight variables belong. Observations with `ind`=0 have a missing link. The remaining variables (`c1` and `c2` below) specify weights for the composite links. The composite link based on the first weight variable will go to where `ind`=1, etc.

```
Example:

Data setup with form of inverse link          Interpretation of
h_i determined by link() and lv():            composite(cluster ind c1 c2)

cluster ind  c1  c2  inverse link               cluster  composite link
  1       1   1  0    h_1                          1        h_1 - h_2
  1       2  -1  1    h_2                          1        h_2 + h_3
  1       0   0  1    h_3             ==>          1        missing
  2       1   1  0    h_4                          2        h_4 + h_5
  2       2   1  1    h_5                          2        h_5 + 2*h_6
  2       0   0  2    h_6                          2        missing
```

Random-effects distribution

`nocorrel` may be used to constrain all correlations to zero if there are several random effects at any of the levels and if these are modeled as multivariate normal.

`ip(`*string*`)` if *string* is `g`, the random effects are multivariate normal (Gaussian), and if *string* is `f`, the random effects are discrete with freely estimated mass points. The default is Gaussian quadrature. With the `ip(f)` option, only `nip-1` mass-point locations are estimated, the last being determined by setting the mean of the mass-point distribution to 0. The `ip(fn)` option can be specified to estimate all `nip` masses freely—the user must then make sure that the mean is not modeled in the linear predictor, e.g., by specifying the `noconstant` option. See integration method for a description of `ip(m)`.

Structural model

`geqs(`*eqnames*`)` specifies equations for regressions of random effects on explanatory variables. The second character of the equation name indicates which random effect is regressed on the variables used in the equation definition; e.g., `f1:a b` means that the first random effect is regressed on `a` and `b` (without a constant).

`bmatrix(`*matrix*`)` specifies a square matrix **B** of regression coefficients for the dependence of the random effects on other random effects. The matrix must be upper diagonal and have number of rows and columns equal to the total number of random effects. Elements of *matrix* are 1 where coefficients in **B** should be estimated and 0 where they should be set to 0.

`peqs(`*eqname*`)` can be used with the `ip(f)` or `ip(fn)` options to allow the (prior) probabilities of the discrete distribution to depend on covariates via a multinomial logit model. A constant is automatically included in addition to the covariates specified in the `eq` command.

Weights

`weight(`*wt*`)` specifies that variables *wt*1, *wt*2, etc., contain frequency weights. The suffixes (1,2,...) in the variable names should not be included in the `weight()` option. They determine at what level each weight applies. If only some of the weight variables exist, e.g., only level 2 weights, the other weights are assumed to be equal to 1. For example, if the level-1 units are occasions (or panel waves) in longitudinal data and the level-2 units are individuals, and the only variable used in the analysis is a binary variable `result`, we can collapse dataset A into dataset B by defining level 1 weights as follows:

A

ind	occ	result
1	1	0
1	2	0
2	3	0
2	4	1
3	5	0
3	6	1

B

ind	occpat	result	wt1
1	1	0	2
2	2	0	1
2	3	1	1
3	4	0	1
3	5	1	1

The two occasions for individual 1 in dataset A have the same result. The first row in B therefore represents two occasions (occasions 1 and 2), as indicated by `wt1`. The variable `occpat` labels the unique patterns of responses at level 1.

The two individuals 2 and 3 in dataset B have the same pattern of results over the measurement occasions (both have two occasions with values 0 and 1). We can therefore collapse the data into dataset C by using level-2 weights:

B

ind	occpat	result	wt1
1	1	0	2
2	2	0	1
2	3	1	1
3	4	0	1
3	5	1	1

C

indpat	occpat	result	wt1	wt2
1	1	0	2	1
2	2	0	1	2
2	3	1	1	2

The variable `indpat` labels the unique patterns of responses at level 2, and `wt2` indicates that `indpat` 1 in dataset C represents one individual and `indpat` 2 represents two individuals; i.e., all the data for individual 2 are replicated once. Collapsing the data in this way can make `gllamm` run faster.

`pweight(`*varname*`)` specifies that variables *varname*1, *varname*2, etc., contain inverse probability sampling weights for levels 1, 2, etc. As far as the estimates and log likelihood are concerned, the effect of specifying these weights is the same as for frequency weights, but the standard errors will be different. Robust standard errors will automatically be provided. This pseudolikelihood approach should be used with caution if the sampling weights apply to units at a lower level than the highest level in the multilevel model. The weights are not rescaled; scaling is the responsibility of the user.

Constraints

`constraint(`*clist*`)` specifies the constraint numbers of the linear constraints to be applied. Constraints are defined using the `constraint()` command; see [R] **constraint**. To find out the equation names needed to specify the constraints, run `gllamm` with the `noest` and `trace` options.

Integration method

`adapt` specifies adaptive quadrature; the default is ordinary quadrature.

`ip(m)` specifies spherical quadrature that can be more efficient than the default cartesian product quadrature when there are several random effects.

`nip(`*#*`,...,`*#*`)` specifies the number of integration points or masses to be used for each integral or summation. When quadrature is used, a value may be given for each random effect. When freely estimated masses are used, a value may be given for each level of the model. If only one argument is given, the same number of integration points will be used for each summation. The default value is 8. When used with `ip(m)`, `nip()` specifies the degree d of the approximation (corresponds in accuracy approximately to $(d+1)/2$ points per dimension for cartesian product quadrature). Only certain values are available for spherical quadrature: for two random effects, 5, 7, 9, 11, and 15; for more than two random effects, 5 and 7.

SE/Robust

`robust` specifies that the Huber/White/sandwich estimator of the covariance matrix of the parameter estimates is to be used. If a model has been fitted without the `robust` option, the robust standard errors can be obtained by simply typing `gllamm, robust`.

`cluster(`*varname*`)` specifies that the highest-level units of the GLLAMM model are nested in even higher-level clusters, where *varname* contains the cluster identifier. Robust standard errors will be provided that take this clustering into account. If a model has been fitted without this option, the robust standard errors for clustered data can be obtained using the command `gllamm, cluster(`*varname*`)`.

Reporting

`eform` displays the exponentiated coefficients and corresponding standard errors and confidence intervals. For `family(binomial) link(logit)` (i.e., logistic regression), exponentiated coefficients are odds ratios; for `family(poisson) link(log)` (i.e., Poisson regression), exponentiated coefficients are rate ratios.

`level(#)` specifies the confidence level as a percentage for confidence intervals of the fixed coefficients. The default is 95.

`trace` displays details of the model being fitted, as well as details of the maximum likelihood iterations.

`nolog` suppresses output for maximum likelihood iterations.

`nodisplay` suppresses output of the estimates but still shows the iteration log, unless `nolog` is used.

`allc` causes all estimated parameters to be displayed in a regression table, which may be transformations of the parameters usually reported, in addition to the usual output.

`lf0(##)` gives the number of parameters and the log likelihood for a likelihood-ratio test to compare the model to be fitted with a simpler model. A likelihood-ratio chi-squared test is only performed if the `lf0()` option is used.

`eval` causes the program to evaluate the log likelihood for values passed to `gllamm` using the `from(matrix)` option.

`init` sets random part of model to zero so that only the fixed part is fitted. This is also how initial values or starting values for the fixed part are computed in `gllamm`.

`noest` is used to prevent the program from carrying out the estimation. This may be used with the `trace` option to check that the model is correct and get the information needed to set up a matrix of initial values. Global macros are available that are normally deleted. Particularly useful may be `M_initf` and `M_initr`, matrices for the parameters (fixed part and random part, respectively).

`dots` causes a dot to be printed (if used together with `trace`) each time the likelihood-evaluation program is called by `ml`. This helps the user to assess how long `gllamm` is likely to take to run and reassures the user that it is making some progress when it is very slow.

Starting values

`from(`*matrix*`)` specifies the matrix (one row) to be used as starting values. The column names and equation names must be correct (see `help matrix`), unless the `copy` option is used. The parameter values given may be previous estimates, obtained using `e(b)`. This is useful if new covariates are added or if the number of integration points (or locations in a discrete distribution) is increased. The `skip` option must be used if the model to be fitted contains fewer parameters than are included in *matrix*, for instance if covariates are dropped.

`copy`; see `from(`*matrix*`)`.

`skip`; see `from(`*matrix*`)`.

`long` can be used with the `from(`*matrix*`)` option when parameter constraints are used to indicate that the matrix of initial values corresponds to the unconstrained model; i.e., it has more elements than will be fitted.

`search(#)` causes the program to search for initial values for the random-effects variances at level 2 (in range 0 to 3). The argument specifies the number of random searches. This option may only be used with `ip(`*g*`)` and when `from(`*matrix*`)` is not used.

Max options

`iterate(#)` specifies the maximum number of iterations. With the `adapt` option, using the `iterate(#)` option will cause `gllamm` to skip the Newton–Raphson iterations usually performed at the end without updating the quadrature locations. `iterate(0)` is like `eval`, except that standard errors are computed.

`adoonly` causes `gllamm` to use only ado-code instead of internalized code. `gllamm` will be faster if it uses internalized versions of some of the functions available from Stata 7 (if updated on or after 26 October 2001).

gateaux derivative

`gateaux(###)` can be used with the `ip(f)` or `ip(fn)` options to increase the number of mass points by one from a previous solution with parameter estimates specified using `from(`*matrix*`)`. The number of parameters and log likelihood of the previous solution must be specified using the `lf0(##)` option. The program searches for the location of the new mass point by placing a small mass at the location given by the first argument and moving it to the second argument in the number of steps specified by the third argument. (If there are several random effects, this search is done in each dimension, resulting in a regular grid of search points.) If the maximum increase in likelihood is greater than 0, the location corresponding to this maximum is used as the initial value of the new location; otherwise, the program stops. If the program stops, this suggests that the nonparametric maximum likelihood estimator has been obtained.

C Syntax for gllapred

After you fit a model using `gllamm`, you can use `gllapred` to obtain predictions of various quantities.

Title

gllapred — predict command for gllamm

Syntax

`gllapred` *varname* [*if*] [*in*] [, *statistic options*]

statistic	description
`u`	empirical Bayes predictions of random effects; disturbances if there is a structural model
`corr`	posterior correlations between random effects or disturbances
`fac`	empirical Bayes predictions of random effects
`ustd`	standardized empirical Bayes predictions
`xb`	fixed part of linear predictor
`linpred`	linear predictor with empirical Bayes predictions of random effects plugged in
`mu`	mean of response; by default posterior mean
`pearson`	Pearson residual; by default posterior mean
`deviance`	deviance residual; by default posterior mean
`anscombe`	Anscombe residual; by default posterior mean
`cooksd`	Cook's distance for top level clusters
`p`	posterior probabilities if random effects are discrete
`s`	standard deviations if `s()` option was used
`ll`	log-likelihood contributions from top-level clusters

options	description
marginal	combined with mu, gives marginal or population-averaged mean
us(*varname*)	substitute specific values for random effects in *varname*1, etc.
above(#,...,#)	for ordinal responses with mu option, return probability that y exceeds # (several values if more than one ordinal link)
outcome(#)	for mlogit link with mu option, probability that y=#
nooffset	suppress offset
fsample	predict for full sample, not just estimation sample
from(*matrix*)	use parameters in *matrix* instead of estimated parameters
adapt	use adaptive quadrature even if gllamm did not
adoonly	use ado-version of gllapred

Description

See description for gllapred on page 513.

Options

u returns posterior means (empirical Bayes predictions) and posterior standard deviations of the random effects in *varname*m1, *varname*m2, etc., and *varname*s1, *varname*s2, etc., respectively, where the order of the random effects is the same as in the call to gllamm. In the case of continuous random effects, the integration method (ordinary versus adaptive quadrature) and the number of quadrature points used is the same as in the previous call to gllamm. If the gllamm model includes equations for the random effects (geqs or bmatrix), the posterior means and standard deviations of the disturbances ζ are returned.

corr returns posterior correlations of the random effects in *varname*c21, etc. This option only works together with the u option. If there is a structural model, posterior correlations of the disturbances are returned.

fac returns posterior means (empirical Bayes predictions) and posterior standard deviations of the random effects in *varname*m1, *varname*m2, etc., and *varname*s1, *varname*s2, etc. If there is a structural model, predictions of the random effects on the left-hand side of the structural equations are returned.

xb returns the fixed-effects part of the linear predictor in *varname* including the offset (if there is one), unless the nooffset option is used.

ustd returns standardized posterior means (empirical Bayes predictions) of the disturbances in *varname*m1, *varname*m2, etc. Each posterior mean is divided by the square root of the difference between the prior and posterior variances, which approximates the sampling standard deviation.

`cooksd` returns Cook's distances for the top-level units in *varname*.

`linpred` returns the linear predictor including the fixed- and random-effects part where posterior means (empirical Bayes predictions) are substituted for the random effects.

`mu` returns the expectation of the response, for example, the predicted probability in the case of dichotomous responses. By default, the expectation is with respect to the posterior distribution of the random effects; also see the `marginal` and `us()` options. The offset is included (if there is one in the `gllamm` model), unless the `nooffset` option is specified.

`pearson` returns Pearson residuals. By default, the posterior expectation with respect to the random effects is returned. The `us()` option can be used to obtain the conditional residual when specific values are substituted for the random effects.

`deviance` returns deviance residuals. By default, the posterior expectation with respect to the random effects is returned. The `us()` option can be used to obtain the conditional residual when specific values are substituted for the random effects.

`anscombe` returns Anscombe residuals. By default, the posterior expectation with respect to the random effects is returned. The `us()` option can be used to obtain the conditional residual when specific values are substituted for the random effects.

`p` can only be used for two-level models fitted using the `ip(f)` or `ip(fn)` option. `gllapred` returns the posterior probabilities in *varname*1, *varname*2, etc., giving the probabilities of classes 1, 2, etc. `gllapred` also displays the (prior) probability and location matrices to help interpret the posterior probabilities.

`s` returns the scale. This is useful if the `s()` option was used in `gllamm` to specify level-1 heteroskedasticity.

`ll` returns the log-likelihood contributions of the highest-level (level L) units.

`marginal` together with the `mu` option gives the expectation of the response with respect to the prior distribution of the random effects. This is useful for looking at the marginal or population-averaged effects of covariates.

`us(`*varname*`)` specifies values for the random effects to calculate conditional quantities, such as the conditional mean of the responses (`mu` option), given the values of the random effects. Here *varname* specifies the prefix for the variables, and `gllapred` will look for *varname*1, *varname*2, etc.

`above(`#`,...,`#`)` returns probabilities that *depvar* exceeds # (ordinal responses). If there are several ordinal responses, a different value can be specified for each ordinal response or a single value given for all ordinal responses.

`outcome(`#`)` specifies the outcome for which the predicted probability should be returned (`mu` option) if there is a nominal response. This option is not necessary if the `expanded()` option was used in `gllamm` since in this case predicted probabilities are returned for all outcomes.

`nooffset` excludes the offset from the predictions (with the `xb`, `linpred`, or `mu` options). It will only make a difference if the `offset()` option was used in `gllamm`.

`fsample` causes `gllapred` to return predictions for the full sample (except units or observations excluded due to the `if` and `in` qualifiers), not just the estimation sample. The returned log likelihood may be missing since `gllapred` will not exclude observations with missing values on any of the variables used in the likelihood calculation. It is up to the user to exclude these observations using `if` or `in`.

`from(`*matrix*`)` specifies a row matrix of parameter values for which the predictions should be made. The column and equation names will be ignored. Without this option, the parameter estimates from the last `gllamm` model will be used.

`adapt` specifies that numerical integration should be performed using adaptive quadrature instead of ordinary quadrature. This option is not necessary if estimation in `gllamm` used adaptive quadrature.

`adoonly` causes `gllamm` to use only ado-code. This option is not necessary if `gllamm` was run with the `adoonly` option.

D Syntax for gllasim

After you fit a model using `gllamm`, you can use `gllasim` to simulate responses from the model.

Title

gllasim — simulate command for gllamm

Syntax

`gllasim` *varname* [*if*] [*in*] [, *statistic options*]

statistic	description
`y`	response
`u`	random effects (or disturbances if there is a structural model)
`fac`	random effects
`linpred`	linear predictor
`mu`	mean of response (substituting simulated values for random effects)

options	description
`us(`*varname*`)`	substitute specific values for random effects in *varname*`1`, etc.
`above(#,...,#)`	for ordinal responses with `mu` option, return probability that y exceeds # (several values if more than one ordinal link)
`outcome(#)`	for `mlogit` link with `mu` option, probability that y=#
`nooffset`	suppress offset
`fsample`	simulate for full sample, not just estimation sample
`from(`*matrix*`)`	use parameters in *matrix* instead of estimated parameters
`adoonly`	use ado-version of `gllasim`

Description

gllasim is the simulation command for gllamm. The command is somewhat similar to gllapred, except that random effects are simulated from the prior random-effects distribution instead of predicted as the posterior means.

By default, the response is simulated. If other statistics are requested, the response can be simulated as well if the y option is used. With the u, fac, linpred, and mu options, *varname* is just the prefix of the variable names in which results are stored.

Options

y returns simulated responses in *varname*. This option is only necessary if u, fac, linpred, or mu is also specified.

u returns simulated random effects in *varname*p1, *varname*p2, etc., where the order of the random effects is the same as in the call to gllamm (in the order of the equations in the eqs() option). If the gllamm model includes equations for the random effects (geqs or bmatrix), the simulated disturbances are returned.

fac returns the simulated random effects in *varname*p1, *varname*p2, etc., instead of the disturbances if the gllamm model includes equations for the random effects (geqs() or bmatrix() options in gllamm); that is, the random effects on the left-hand side of the structural model.

linpred returns the linear predictor, including the fixed and simulated random parts in *varname*p. The offset is included (if there is one in the gllamm model), unless the nooffset option is specified.

mu returns the expected value of the response conditional on the simulated values for the random effects; e.g., a probability if the responses are dichotomous.

us(*varname*) specifies that, instead of simulating the random effects, gllasim should use the variables in *varname*1, *varname*2, etc.

above(#,...,#) returns probabilities that *depvar* exceeds # (with the mu option) if there are ordinal responses. A single number can be given for all ordinal responses.

outcome(#) specifies the outcome for which the predicted probability should be returned (mu option) if there is a nominal response and the expanded() option has not been used in gllamm (with the expanded() option, predicted probabilities are returned for all outcomes).

nooffset can be used with the linpred and mu options to exclude the offset from the simulated value. This will only make a difference if the offset() option was used in gllamm.

fsample causes gllasim to simulate values for the full sample (except observations excluded due to the if and in qualifiers), not just the estimation sample.

from(*matrix*) specifies a matrix of parameters for which the simulations should be made. The column and equation names will be ignored. Without this option, the parameter estimates from the last gllamm model will be used.

adoonly causes gllasim to use only ado-code. This option is not necessary if gllamm was run with the adoonly option.

References

Abrevaya, J. 2006. Estimating the effect of smoking on birth outcomes using a matched panel data approach. *Journal of Applied Econometrics* 21: 489–519.

Acitelli, L. K. 1997. Sampling couples to understand them: Mixing the theoretical with the practical. *Journal of Social and Personal Relationships* 14: 243–261.

Adams, M. M., H. G. Wilson, D. L. Casto, C. J. Berg, J. M. McDermott, J. A. Gaudino, and B. J. McCarthy. 1997. Constructing reproductive histories by linking vital records. *American Journal of Epidemiology* 145: 339–348.

Agresti, A. 2002. *Categorical Data Analysis.* 2nd ed. Hoboken, NJ: Wiley.

Agresti, A., J. Booth, J. P. Hobert, and B. Caffo. 2000. Random-effects modeling of categorical response data. In *Sociological Methodology 2000*, ed. R. M. Stolzenberg, 27–80. Oxford: Blackwell.

Agresti, A., and B. Finlay. 2007. *Statistical Methods for the Social Sciences.* 4th ed. Englewood Cliffs, NJ: Prentice Hall.

Agresti, A., and R. Natarajan. 2001. Modeling clustered ordered categorical data: A survey. *International Statistical Review* 69: 345–371.

Aitkin, M. 1978. The analysis of unbalanced cross-classifications. *Journal of the Royal Statistical Society, Series A* 41: 195–223.

Allison, P. D. 1982. Discrete-time methods for the analysis of event histories. In *Sociological Methodology 1982*, ed. S. Leinhardt, 61–98. San Francisco: Jossey-Bass.

———. 1984. *Event History Analysis. Regression for Longitudinal Event Data.* Sage University Paper Series on Quantitative Applications in the Social Sciences, Newbury Park, CA: Sage.

———. 1995. *Survival Analysis Using SAS: A Practical Guide.* Cary, NC: SAS Institute.

———. 2005. *Fixed Effects Regression Methods for Longitudinal Data Using SAS.* Cary, NC: SAS Institute.

Baltagi, B. H. 2005. *Econometrics of Panel Data.* 3rd ed. Chichester, UK: Wiley.

Baltagi, B. H., S. H. Song, and B. C. Jung. 2001. The unbalanced nested error component regression model. *Journal of Econometrics* 101: 357–381.

Balzer, W., N. Boudreau, P. Hutchinson, A. M. Ryan, T. Thorsteinson, J. Sullivan, R. Yonker, and D. Snavely. 1996. Critical modeling principles when testing for gender equity in faculty salary. *Research in Higher Education* 37: 633–658.

Barber, J. S., S. Murphy, W. G. Axinn, and J. Maples. 2000. Discrete-time multilevel hazard analysis. In *Sociological Methodology 2000*, ed. R. M. Stolzenberg, 201–235. Oxford: Blackwell.

Bland, J. M., and D. G. Altman. 1986. Statistical methods for assessing agreement between two methods of clinical measurement. *Lancet* I: 307–310.

Bollen, K. A., and P. J. Curran. 2006. *Latent Curve Models: A Structural Equation Perspective*. New York: Wiley.

Boot, J. C. G., and G. M. DeWit. 1960. Investment demand: An empirical contribution to the aggregation problem. *International Economic Review* 1: 27–28.

Bottai, M., and N. Orsini. 2004. Confidence intervals for the variance component of random-effects linear models. *Stata Journal* 4: 227–233.

Boudreau, N., J. Sullivan, W. Balzer, A. M. Ryan, R. Yonker, T. Thorsteinson, and P. Hutchinson. 1997. Should faculty rank be included as a predictor variable in studies of gender equity in university faculty salaries. *Research in Higher Education* 38: 297–312.

Box-Steffensmeier, J. M., and B. S. Jones. 2004. *Event History Modeling: A Guide for Social Scientists*. Cambridge: Cambridge University Press.

Breslow, N. E., and D. G. Clayton. 1993. Approximate inference in generalized linear mixed models. *Journal of the American Statistical Association* 88: 9–25.

Breslow, N. E., and N. Day. 1987. *Statistical Methods in Cancer Research. Vol II – Design and Analysis of Cohort Studies*. Lyon: IARC.

Bryk, A. S., and S. W. Raudenbush. 1988. Toward a more appropriate conceptualization of research on school effects: A three-level hierarchical linear model. *American Journal of Education* 97: 65–108.

Cameron, A. C., and P. K. Trivedi. 1998. *Regression Analysis of Count Data*. Cambridge: Cambridge University Press.

———. 2005. *Microeconometrics: Methods and Applications*. Cambridge: Cambridge University Press.

Capaldi, D. M., L. Crosby, and M. Stoolmiller. 1996. Predicting the timing of first sexual intercourse for at-risk adolescent males. *Child Development* 67: 344–359.

Caudill, S. B., J. M. Ford, and D. L. Kaserman. 1995. Certificate-of-need regulation and the diffusion of innovations: A random coefficient model. *Journal of Applied Econometrics* 10: 73–78.

Clayton, D. G., and J. Kaldor. 1987. Empirical Bayes estimates of age-standardized relative risks for use in disease mapping. *Biometrics* 43: 671–681.

Collett, D. 2003a. *Modelling Binary Data.* 2nd ed. Boca Raton, FL: Chapman & Hall/CRC.

———. 2003b. *Modelling Survival Data in Medical Research.* 2nd ed. Boca Raton, FL: Chapman & Hall/CRC.

Crowder, M. J., and D. J. Hand. 1990. *Analysis of Repeated Measures.* London: Chapman & Hall.

Curran, P. J., T. Hartford, and B. O. Muthén. 1996. The relation between heavy alcohol use and bar patronage: A latent growth model. *Journal of Studies on Alcohol* 57: 410–418.

Davies, R. B. 1993. Statistical modelling for survey analysis. *Journal of the Market Research Society* 35: 235–247.

Davies, R. B., P. Elias, and R. Penn. 1992. The relationship between a husband's unemployment and his wife's participation in the labour force. *Oxford Bulletin of Economics and Statistics* 54: 145–171.

Davis, C. S. 1991. Semi-parametric and non-parametric methods for the analysis of repeated measurements with applications to clinical trials. *Statistics in Medicine* 10: 1995–1980.

———. 2002. *Statistical Methods for the Analysis of Repeated Measurements.* New York: Springer.

De Backer, M., C. De Vroey, E. Lesaffre, I. Scheys, and P. De Keyser. 1998. Twelve weeks of continuous oral therapy for toenail onychomycosis caused by dermatophytes: A double-blind comparative trial of terbinafine 250 mg/day versus itraconazole 200 mg/day. *Journal of the American Academy of Dermatology* 38: 57–63.

De Boeck, P., and M. Wilson, ed. 2004. *Explanatory Item Response Models: A Generalized Linear and Nonlinear Approach.* New York: Springer.

DeMaris, A. 2004. *Regression with Social Data: Modeling Continuous and Limited Response Variables.* Hoboken, NJ: Wiley.

Dempster, A. P., M. R. Selwyn, C. M. Patel, and A. J. Roth. 1984. Statistical and computational aspects of mixed model analysis. *Journal of the Royal Statistical Society, Series C* 33: 203–214.

Diggle, P. J., P. Heagerty, K.-Y. Liang, and S. L. Zeger. 2002. *Analysis of Longitudinal Data.* Oxford: Oxford University Press.

Dohoo, I. R., W. Martin, and H. Stryhn. 2003. *Veterinary Epidemiologic Research.* Charlottetown, Canada: Atlantic Veterinary College.

Dohoo, I. R., E. Tillard, H. Stryhn, and B. Faye. 2001. The use of multilevel models to evaluate sources of variation in reproductive performance in dairy cattle in Reunion Island. *Preventive Veterinary Medicine* 50: 127–144.

Dunn, G. 1992. Design and analysis of reliability studies. *Statistical Methods in Medical Research* 1: 123–157.

———. 2004. *Statistical Evaluation of Measurement Errors: Design and Analysis of Reliability Studies*. London: Arnold.

Embretson, S. E., and S. P. Reise. 2000. *Item Response Theory for Psychologists*. Mahwah, NJ: Erlbaum.

Everitt, B. S., and A. Pickles. 1999. *Statistical Aspects of the Design and Analysis of Clinical Trials*. London: Imperial College Press.

Fahrmeir, L., and G. Tutz. 2001. *Multivariate Statistical Modelling Based on Generalized Linear Models*. 2nd ed. New York: Springer.

Fitzmaurice, G. M. 1998. Regression models for discrete longitudinal data. In *Statistical Analysis of Medical Data: New Developments*, ed. B. S. Everitt and G. Dunn, 175–201. London: Arnold.

Fitzmaurice, G. M., M. Davidian, G. Molenberghs, and G. Verbeke, ed. 2008. *Longitudinal Data Analysis*. Boca Raton, FL: Chapman & Hall/CRC.

Fitzmaurice, G. M., N. M. Laird, and J. H. Ware. 2004. *Applied Longitudinal Analysis*. New York: Wiley.

Flay, B. R., B. R. Brannon, C. A. Johnson, W. B. Hansen, A. L. Ulene, D. A. Whitney-Saltiel, L. R. Gleason, S. Sussman, M. D. Gavin, K. M. Glowacz, D. F. Sobol, and D. C. Spiegel. 1989. The television, school and family smoking cessation and prevention project: I theoretical basis and program development. *Preventive Medicine* 17: 585–607.

Fox, J. 1997. *Applied Regression Analysis, Linear Models, and Related Methods*. Thousand Oaks, CA: Sage.

Frees, E. W. 2004. *Longitudinal and Panel Data: Analysis and Applications in the Social Sciences*. Cambridge University Press: Cambridge.

Frets, G. P. 1921. Heredity of head form in man. *Genetica* 3: 193–384.

Garner, C. L., and S. W. Raudenbush. 1991. Neighborhood effects on educational attainment: A multilevel analysis of the influence of pupil ability, family, school, and neighborhood. *Sociology of Education* 64: 252–262.

Gelman, A., and J. Hill. 2007. *Data Analysis Using Regression and Multilevel/Hierarchical Models*. Cambridge: Cambridge University Press.

Gibbons, R. D., and D. Hedeker. 1994. Application of random-effects probit regression models. *Journal of Consulting and Clinical Psychology* 62: 285–296.

Gibbons, R. D., D. Hedeker, C. Waterneaux, and J. M. Davis. 1988. Random regression models: A comprehensive approach to the analysis of longitudinal psychiatric data. *Psychopharmacology Bulletin* 24: 438–443.

Goldberg, D. P. 1972. *The Detection of Psychiatric Illness by Questionnaire*. Oxford: Oxford University Press.

Goldstein, H. 1986. Efficient statistical modelling of longitudinal data. *Human Biology* 13: 129–142.

———. 1987. Multilevel covariance component models. *Biometrika* 74: 430–431.

———. 2003. *Multilevel Statistical Models*. 3rd ed. London: Arnold.

Goldstein, H., P. Huiqi, T. Rath, and N. Hill. 2000. The use of value added information in judging school performance. Technical report, Institute of Education, London. Downloadable from http://www.mlwin.com/hgpersonal/value-added-school-performance.html.

Goldstein, H., J. Rasbash, M. Yang, G. Woodhouse, H. Pan, D. Nutall, and S. Thomas. 1993. A multilevel analysis of school examination results. *Oxford Review of Education* 19: 425–433.

Greene, W. H. 1997. Frontier production functions. In *Handbook of Applied Econometrics. Volume II: Microeconometrics*, ed. M. H. Pesaran and P. Schmidt, 81–166. London: Blackwell.

———. 2008. *Econometric Analysis*. 6th ed. Upper Saddle River, NJ: Pearson.

Greenland, S. 1994. Alternative models for ordinal logistic regression. *Statistics in Medicine* 13: 1665–1677.

Greenland, S., J. J. Schlesselman, and M. H. Criqui. 1986. The fallacy of employing standardized regression coefficients and correlations as measures of effect. *American Journal of Epidemiology* 123: 203–208.

Grilli, L. 2005. The random-effects proportional hazards model with grouped survival data: A comparison between the grouped continuous and continuation ratio versions. *Journal of the Royal Statisticial Society, Series A* 168: 83–94.

Grunfeld, Y. 1958. The determinants of corporate investment. Ph.D. thesis, University of Chicago.

Guo, G. 1993. Event-history analysis for left-truncated data. In *Sociological Methodology 1993*, ed. P. V. Marsden, 217–243. Oxford: Blackwell.

Guo, G., and G. Rodriguez. 1992. Estimating a multivariate proportional hazards model for clustered data using the EM algorithm, with an application to child survival in Guatemala. *Journal of the American Statistical Association* 87: 969–976.

Guo, G., and H. Zhao. 2000. Multilevel modeling of binary data. *Annual Review of Sociology* 26: 441–462.

Hall, B. H., Z. Griliches, and J. A. Hausman. 1986. Patents and R&D: Is there a lag? *International Economic Review* 27: 265–283.

Hamerle, A. 1991. On the treatment of interrupted spells and initial conditions in event history analysis. *Sociological Methods & Research* 19: 388–414.

Hancock, G. R., and K. M. Samuelsen. 2007. *Advances in Latent Variable Mixture Models*. Charlotte, NC: Information Age Publishing.

Hand, D. J., F. Daly, A. D. Lunn, K. J. McConway, and E. Ostrowski. 1994. *A Handbook of Small Data Sets*. London: Chapman & Hall.

Hardin, J. W., and J. M. Hilbe. 2003. *Generalized Estimating Equations*. Boca Raton, FL: Chapman & Hall/CRC.

Hausman, J. A. 1978. Specification tests in econometrics. *Econometrica* 46: 1251–1271.

Hausman, J. A., B. H. Hall, and Z. Griliches. 1984. Econometric models for count data with an application to the patents - R&D relationship. *Econometrica* 52: 909–938.

Hausman, J. A., and W. E. Taylor. 1981. Panel data and unobservable individual effects. *Econometrica* 49: 1377–1398.

Hedeker, D. 1999. MIXNO: A computer program for mixed-effects logistic regression. *Journal of Statistical Software* 4: 1–92.

———. 2005. Generalized linear mixed models. In *Encyclopedia of Statistics in Behavioral Science*, ed. B. Everitt and D. Howell, 729–738. London: Wiley.

Hedeker, D., and R. D. Gibbons. 1996. MIXOR: A computer program for mixed-effects ordinal probit and logistic regression analysis. *Computer Methods and Programs in Biomedicine* 49: 157–76.

———. 2006. *Longitudinal Data Analysis*. Hoboken, NJ: Wiley.

Hedeker, D., O. Siddiqui, and F. B. Hu. 2000. Random-effects regression analysis of correlated grouped-time survival data. *Statistical Methods in Medical Research* 9: 161–179.

Hilbe, J. M. 2007. *Negative Binomial Regression*. Cambridge: Cambridge University Press.

Hill, H. C., B. Rowan, and D. L. Ball. 2005. Effect of teachers' mathematical knowledge for teaching on student achievement. *American Educational Research Journal* 42: 371–406.

Hosmer, D. W., Jr., and S. Lemeshow. 1999. *Applied Survival Analysis: Regression Modeling of Time to Event Data*. New York: Wiley.

———. 2000. *Applied Logistic Regression*. 2nd ed. New York: Wiley.

Hox, J. 2002. *Multilevel Analysis: Techniques and Applications*. Mahwah, NJ: Erlbaum.

Hsiao, C. 2002. *Analysis of Panel Data*. 2nd ed. Cambridge: Cambridge University Press.

Huster, W. J., R. Brookmeyer, and S. G. Self. 1989. Modelling paired survival data with covariates. *Biometrics* 45: 145–156.

Jenkins, S. P. 1995. Easy estimation methods for discrete-time duration models. *Oxford Bulletin of Economics and Statistics* 57: 129–137.

Johnson, V. E., and J. H. Albert. 1999. *Ordinal Data Modelling*. New York: Springer.

Karim, M. R., and S. L. Zeger. 1992. Generalized linear models with random effects: Salamander mating revisited. *Biometrics* 48: 631–644.

Kelly, P. J., and L. L. Lim. 2000. Survival analysis for recurrent event data: An application to childhood infectious diseases. *Statistics in Medicine* 19: 13–33.

Kenny, D. A., D. A. Kashy, and W. L. Cook. 2006. *Dyadic Data Analysis*. New York: Guilford.

Klein, J. P., and M. L. Moeschberger. 2003. *Survival Analysis: Techniques for Censored and Truncated Data*. 2nd ed. New York: Springer.

Koch, G. G., G. J. Carr, I. A. Amara, M. E. Stokes, and T. J. Uryniak. 1989. Categorical data analysis. In *Statistical Methodology in the Pharmaceutical Sciences*, ed. D. A. Berry, 389–473. New York: Marcel Dekker.

Kohler, U., and F. Kreuter. 2005. *Data Analysis Using Stata*. College Station, TX: Stata Press.

Kreft, I. G. G., and J. de Leeuw. 1998. *Introducing Multilevel Modeling*. London: Sage.

Lancaster, T. 1990. *Econometric Analysis of Transition Data*. Cambridge: Cambridge University Press.

Langford, I. H., G. Bentham, and A. McDonald. 1998. Multilevel modelling of geographically aggregated health data: a case study on malignant melanoma mortality and UV exposure in the European community. *Statistics in Medicine* 17: 41–58.

Langford, I. H., and T. Lewis. 1998. Outliers in multilevel data. *Journal of the Royal Statistical Society, Series A* 161: 121–160.

Larsen, K., and J. Merlo. 2005. Appropriate assessment of neighborhood effects on individual health: Integrating random and fixed effects in multilevel logistic regression. *American Journal of Epidemiology* 161: 81–88.

Larsen, K., J. H. Petersen, E. Budtz-Jørgensen, and L. Endahl. 2000. Interpreting parameters in the logistic regression model with random effects. *Biometrics* 56: 909–914.

Lawson, A. B., A. Biggeri, D. Böhning, E. Lesaffre, J. Viel, and R. Bertollini, ed. 1999. *Disease Mapping and Risk Assessment for Public Health*. New York: Wiley.

Lawson, A. B., W. J. Browne, and C. L. Vidal-Rodeiro. 2003. *Disease Mapping with WinBUGS and MLwiN*. New York: Wiley.

Lesaffre, E., and B. Spiessens. 2001. On the effect of the number of quadrature points in a logistic random-effects model: An example. *Journal of the Royal Statistical Society, Series C* 50: 325–335.

Leyland, A. H. 2001. Spatial analysis. In *Multilevel Modelling of Health Statistics*, ed. A. H. Leyland and H. Goldstein, 127–140. Chichester, UK: Wiley.

Leyland, A. H., and H. Goldstein, ed. 2001. *Multilevel Modelling of Health Statistics*. Chichester, UK: Wiley.

Littell, R. C., G. A. Milliken, W. W. Stroup, R. D. Wolfinger, and O. Schabenberger. 2006. *SAS for Mixed Models*. Cary, NC: SAS Institute.

Long, J. S. 1997. *Regression Models for Categorical and Limited Dependent Variables*. Thousand Oaks, CA: Sage.

Long, J. S., P. D. Allison, and R. McGinnis. 1993. Rank advancement in academic careers: Sex differences and the effects of productivity. *American Sociological Review* 58: 703–722.

Long, J. S., and J. Freese. 2006. *Regression Models for Categorical and Limited Dependent Variables using Stata*. 2nd ed. College Station, TX: Stata Press.

Lorr, M., and C. J. Klett. 1966. *Inpatient Multidimensional Psychiatric Scale*. Palo Alto, CA: Consulting Psychologists Press.

MacDonald, A. M. 1996. An Epidemiological and Quantitative Genetic Study of Obsessionality. Ph.D. thesis, Institute of Psychiatry, University of London.

Machin, D., T. Farley, B. Busca, M. Campbell, and C. d'Arcangues. 1988. Assessing changes in vaginal bleeding patterns in contracepting women. *Contraception* 38: 165–179.

MacNab, S. J., M. L. Levine, N. Glick, N. Phillips, L. Susak, and M. Elliott. 1994. The Vancouver sedative recovery scale for children: Validation and reliability of scoring based on videotaped instruction. *Canadian Journal of Anaesthesia* 41: 913–918.

Magnus, P., H. K. Gjessing, A. Skrondal, and R. Skjærven. 2001. Paternal contribution to birth weight. *Journal of Epidemiology and Community Health* 55: 873–877.

Mare, R. D. 1994. Discrete-time bivariate hazards with unobserved heterogeneity: A partially observed contingency table approach. In *Sociological Methodology 1994*, ed. P. V. Marsden, 341–383. Oxford: Blackwell.

McCullagh, P., and J. A. Nelder. 1989. *Generalized Linear Models.* 2nd ed. London: Chapman & Hall.

McCulloch, C. E., and S. R. Searle. 2001. *Generalized, Linear and Mixed Models.* New York: Wiley.

McKnight, B., and S. K. van den Eeden. 1993. A conditional analysis for two-treatment multiple-period crossover designs with binomial or Poisson outcomes and subjects who drop out. *Statistics in Medicine* 12: 825–834.

Molenberghs, G., and G. Verbeke. 2005. *Models for Discrete Longitudinal Data.* New York: Springer.

Mundlak, Y. 1978. On the pooling of time series and cross section Data. *Econometrica* 84: 69–85.

Munnell, A. 1990. Why has productivity growth declined? Productivity and public investment. *New England Economic Review* 3–22.

Neuhaus, J. M., and J. D. Kalbfleisch. 1998. Between- and within-cluster covariate effects in the analysis of clustered data. *Biometrics* 54: 638–645.

OECD. 2000. *Manual for the PISA 2000 Database.* Paris: OECD. Downloadable from http://www.pisa.oecd.org/dataoecd/53/18/33688135.pdf.

Pan, W. 2002. A note on the use of marginal likelihood and conditional likelihood in analyzing clustered data. *American Statistician* 56: 171–174.

Papke, L. E. 1994. Tax policy and urban development: Evidence from the Indiana enterprise zone program. *Journal of Public Economics* 54: 37–49.

Paterson, L. 1991. Socio-economic status and educational attainment: A multidimensional and multilevel study. *Evaluation and Research in Education* 5: 97–121.

Pebley, A. R., N. Goldman, and G. Rodriguez. 1996. Prenatal and delivery care and childhood immunization in Guatemala: Do family and community matter? *Demography* 33: 231–247.

Pebley, A. R., and P. W. Stupp. 1987. Reproductive patterns and child mortality in Guatemala. *Demography* 24: 43–60.

Pothoff, R. F., and S. N. Roy. 1964. A generalized multivariate analysis of variance model useful especially for growth-curve problems. *Biometrika* 51: 313–326.

Prosser, R., J. Rasbash, and H. Goldstein. 1991. *ML3 Software for Three-level Analysis, Users' Guide for V.2.* London: Institute of Education, University of London.

Quine, S. 1975. Achievement orientation of aboriginal and white Australian children. Ph.D. thesis, Australian National University.

Rabe-Hesketh, S., and B. S. Everitt. 2007. *Handbook of Statistical Analyses using Stata.* 4th ed. Boca Raton, FL: Chapman & Hall/CRC.

Rabe-Hesketh, S., A. Pickles, and A. Skrondal. 2003. Correcting for covariate measurement error in logistic regression using nonparametric maximum likelihood estimation. *Statistical Modelling* 3: 215–232.

Rabe-Hesketh, S., and A. Skrondal. 2006. Multilevel modeling of complex survey data. *Journal of the Royal Statistical Society, Series A* 169: 805–827.

———. 2008. Generalized linear mixed effects models. In *Longitudinal Data Analysis*, ed. G. M. Fitzmaurice, M. Davidian, G. Molenberghs, and G. Verbeke. Boca Raton, FL: Chapman & Hall/CRC.

Rabe-Hesketh, S., A. Skrondal, and H. K. Gjessing. 2008. Biometrical modeling of twin and family data using standard mixed model software. *Biometrics* 64 (forthcoming).

Rabe-Hesketh, S., A. Skrondal, and A. Pickles. 2002. Reliable estimation of generalized linear mixed models using adaptive quadrature. *Stata Journal* 2: 1–21.

———. 2005. Maximum likelihood estimation of limited and discrete dependent variable models with nested random effects. *Journal of Econometrics* 128: 301–323.

Rabe-Hesketh, S., T. Touloupulou, and R. M. Murray. 2001a. Multilevel modeling of cognitive function in schizophrenics and their first degree relatives. *Multivariate Behavioral Research* 36: 279–298.

Rabe-Hesketh, S., S. Yang, and A. Pickles. 2001b. Multilevel models for censored and latent responses. *Statistical Methods in Medical Research* 10: 409–427.

Rasbash, J. 2005. Cross-classified and multiple membership models. In *Encyclopedia of Statistics in Behavioral Science*, ed. B. Everitt and D. Howell, 441–450. London: Wiley.

Rasbash, J., F. Steele, W. J. Browne, and R. Prosser. 2005. *A User's Guide to MLwiN Version 2.0.* Bristol: University of Bristol. Downloadable from http://www.mlwin.com/download/manuals.html.

Raudenbush, S. W., and C. Bhumirat. 1992. The distribution of resources for primary education and its consequences for educational achievement in Thailand. *International Journal of Educational Research* 17: 143–164.

Raudenbush, S. W., and A. S. Bryk. 2002. *Hierarchical Linear Models*. Thousand Oaks, CA: Sage.

Raudenbush, S. W., A. S. Bryk, Y. F. Cheong, R. Congdon, and M. du Toit. 2004. *HLM 6: Hierarchical Linear and Nonlinear Modeling*. Lincolnwood, IL: Scientific Software International.

Rodriguez, G., and N. Goldman. 2001. Improved estimation procedures for multilevel models with binary response: A case study. *Journal of the Royal Statistical Society, Series A* 164: 339–355.

Ross, E. A., and D. Moore. 1999. Modeling clustered, discrete, or grouped time survival data with covariates. *Biometrics* 55: 813–819.

Sham, P. 1998. *Statistics in Human Genetics*. London: Arnold.

Shavelson, R. J., and N. M. Webb. 1991. *Generalizability Theory: A Primer*. Newbury Park, CA: Sage.

Singer, J. D. 1993. Are special educators' career paths special? Results from a 13-year study. *Exceptional Children* 59: 262–279.

Singer, J. D., and J. B. Willett. 1993. It's about time: Using discrete-time survival analysis to study duration and the timing of events. *Journal of Educational Statistics* 18: 155–195.

———. 2003. *Applied Longitudinal Data Analysis: Modeling Change and Event Occurrance*. Oxford: Oxford University Press.

———. 2005. Growth curve modeling. In *Encyclopedia of Statistics in Behavioral Science*, ed. B. Everitt and D. Howell, 772–779. London: Wiley.

Skrondal, A., and S. Rabe-Hesketh. 2003a. Some applications of generalized linear latent and mixed models in epidemiology: Repeated measures, measurement error, and multilevel modelling. *Norwegian Journal of Epidemiology* 13: 265–278.

———. 2003b. Multilevel logistic regression for polytomous data and rankings. *Psychometrika* 68: 267–287.

———. 2004. *Generalized Latent Variable Modeling: Multilevel, Longitudinal, and Structural Equation Models*. Boca Raton, FL: Chapman & Hall/CRC.

———. 2007. Redundant overdispersion parameters in multilevel models. *Journal of Behavioral and Educational Statistics* 32: 419–430.

———. 2008. Multilevel and related models for longitudinal data. In *Handbook of Multilevel Analysis*, ed. J. de Leeuw and E. Meijer, 275–299. New York, NY: Springer.

Smans, M., C. S. Muir, and P. Boyle. 1992. *Atlas of Cancer Mortality in the European Economic Community*. Lyon, France: IARC Scientific Publications.

Snijders, T. A. B., and R. J. Bosker. 1994. Modeled variance in two-level models. *Sociological Methods & Research* 22: 342–363.

———. 1999. *Multilevel Analysis: An Introduction to Basic and Advanced Multilevel Modeling*. London: Sage.

SOEP Group. 2001. The German Socio-Economic Panel (SOEP) after more than 15 years—Overview. *Proceedings Of The 2000 Fourth International Conference of German Socio-Economic Panel Study Users (GSOEP2000), Vierteljahrshefte zur Wirtschaftsforschung* 70: 7–14.

Spiegelhalter, D. J., A. Thomas, N. G. Best, and W. Gilks. 1996a. *BUGS 0.5 Examples, Volume 1.* Cambridge: MRC-Biostatistics Unit.

———. 1996b. *BUGS 0.5 Examples, Volume 2.* Cambridge: MRC-Biostatistics Unit.

Spielberger, C. D. 1988. *State-Trail Anger Expression Inventory Research Edition. Professional Manual.* Odessa, FL: Psychological Assessment Resources.

StataCorp. 2007. *Stata Longitudinal/Panel-Data Reference Manual, Release 10.* College Station, TX: Stata Press.

Steenbergen, M. R., and B. S. Jones. 2002. Modeling multilevel data structures. *American Journal of Political Science* 46: 218–237.

Streiner, D. L., and G. R. Norman. 2003. *Health Measurement Scales: A Practical Guide to Their Development and Use.* 3rd ed. Oxford: Oxford University Press.

Thall, P. F., and S. C. Vail. 1990. Some covariance models for longitudinal count data with overdispersion. *Biometrics* 46: 657–671.

Vaida, F., and X. L. Meng. 2005. Two slice-EM algorithms for fitting generalized linear mixed models with binary response. *Statistical Modelling* 5: 229–242.

Vansteelandt, K. 2000. Formal models for contextualized personality psychology. Ph.D. thesis, Katholieke Universiteit Leuven, Belgium.

Vella, F., and M. Verbeek. 1998. Whose wages do unions raise? A dynamic model of unionism and wage rate determination for young men. *Journal of Applied Econometrics* 13: 163–183.

Vermunt, J. K. 1997. *Log-Linear Models for Event Histories.* Thousand Oaks, CA: Sage.

Vermunt, J. K., and J. Magidson. 2005. *Technical Guide for Latent GOLD 4.0: Basic and Advanced.* Belmont, MA: Statistical Innovations.

Vittinghoff, E., D. V. Glidden, S. C. Shiboski, and C. E. McCulloch. 2006. *Regression Methods in Biostatistics: Linear, Logistic, Survival, and Repeated Measures Models.* New York: Springer.

von Bortkiewicz, L. 1898. *Das Gesetz der kleinen Zahlen.* Leipzig: Teubner.

Ware, J. H., D. W. Dockery, A. Spiro, F. E. Speizer, and B. G. Ferris. 1984. Passive smoking, gas cooking, and respiratory health in children living in six cities. *American Review of Respiratory Disease* 129: 366–374.

West, B. T., K. B. Welch, and A. T. Galecki. 2007. *Linear Mixed Models: A Practical Guide Using Statistical Software.* Boca Raton, FL: Chapman & Hall/CRC.

Wiggins, R. D., K. Ashworth, C. A. O'Muircheartaigh, and J. I. Galbraith. 1991. Multilevel analysis of attitudes to abortion. *Journal of the Royal Statistical Society, Series D* 40: 225–234.

Willms, J. D. 1986. Social class segregation and its relationship to pupils' examination results in Scotland. *American Sociological Review* 51: 224–241.

Winkelmann, R. 2003. *Econometric Analysis of Count Data.* 4th ed. New York: Springer.

———. 2004. Health care reform and the number of doctor visits: An econometric analysis. *Journal of Applied Econometrics* 19: 455–472.

Wooldridge, J. M. 2002. *Econometric Analysis of Cross Section and Panel Data.* Cambridge, MA: MIT Press.

Yang, M. 2001. Multinomial regression. In *Multilevel Modelling of Health Statistics,* ed. A. H. Leyland and H. Goldstein, 107–123. Chichester, UK: Wiley.

Zheng, X., and S. Rabe-Hesketh. 2007. Estimating parameters of dichotomous and ordinal item response models with gllamm. *Stata Journal* 7: 313–333.

Author index

O

P

Q

R

S

T

U

V

W

Y

Z

Subject index